ITALIAN PHYSICAL SOCIETY

PROCEEDINGS

OF THE

INTERNATIONAL SCHOOL OF PHYSICS
« ENRICO FERMI »

Course LXIII
edited by D. SETTE
Director of the Course

VARENNA ON LAKE COMO
VILLA MONASTERO
5th - 17th AUGUST 1974

New Directions
in Physical Acoustics

1976

NORTH–HOLLAND PUBLISHING COMPANY, AMSTERDAM · NEW YORK · OXFORD

SOCIETA' ITALIANA DI FISICA

RENDICONTI
DELLA
SCUOLA INTERNAZIONALE DI FISICA
«ENRICO FERMI»

LXIII Corso
a cura di D. SETTE
Direttore del Corso

VARENNA SUL LAGO DI COMO
VILLA MONASTERO
5 - 17 AGOSTO 1974

Nuove Tendenze
dell'Acustica Fisica

1976

SOCIETÀ ITALIANA DI FISICA
BOLOGNA - ITALY

PUBLISHER: SOCIETÀ ITALIANA DI FISICA - BOLOGNA

SOLE DISTRIBUTORS: NORTH-HOLLAND PUBLISHING COMPANY -
AMSTERDAM - NEW YORK - OXFORD

ISBN NORTH-HOLLAND 0 7204 0489 4

INDICE

J. Joffrin et A. Levelut – Les échos de phonons.

J. J. McCoy – Wave propagation in random media.

R. R. Goodman – Propagation in fluctuating media.

H. O. Berktay – Finite-amplitude effects in acoustic propagation in fluids.

T. G. WANG – Acoustics in space.

Preface.

D. SETTE

Istituto di Fisica della Facoltà d'Ingegneria dell'Università - Roma

Recent years have witnessed a genuine flourishing of research in acoustics. The close connection of acoustics with other fields of physics has always given acoustics a strong interdisciplinary character, so that specialists in other branches of science have traditionally entered into an involvement into acoustic research.

The summer course of the E. Fermi School of the Italian Physical Society, *New Directions in Physical Acoustics*, has been conceived as a place for presenting and examining the physical aspects of a part of the current interdisciplinary research in acoustics.

The proceedings are openend by an essay on the *Historical Development of Physical Acoustics and Future Perspectives* by the distinguished scholar LINDSAY.

The material of the course can be grouped in three main areas: 1) sound propagation and the structure of matter, 2) propagation in fluctuating media of large dimensions (air, ocean), 3) nonlinear acoustics and surface waves.

In a large number of cases the interpretation of experiments on sound propagation requires the consideration of the co-operative collective nature of the molecular processes involved. Montrose's contribution on *Correlation Functions in Molecular Acoustics* gives the basic notions on time correlation functions, the elements of the linear theory for both cases of the response of an equilibrium system to external forces (Kubo method) and of the relaxation to equilibrium of a perturbed system (Mori theory). The results of this theoretical treatment are applied to sound propagation in dense fluids in the context of linearized hydrodynamics; the various kinds of relaxations are discussed. Some aspects of collective mode dynamics in fluids which are related to the study of sound propagation are pointed out, especially where they are closely connected with molecular dynamics.

Yip's lecture, *High-Frequency Short-Wavelength Fluctuations in Fluids*, is in a sense an application of Montrose's more general treatment to the nature of the sound processes which give rise to dispersion and absorption in the different regions of frequencies and wave numbers, with particular attention to the high-frequency and short-wavelength range. Three different regions are considered: 1) the hydrodynamics region where wavelengths are large compared with the mean free path, and frequencies small compared with collision frequency; 2) Knudsen region, where k and ω are very large, the effects of free-molecule flow predominate and sound propagation in the traditional sense does not occur; 3) kinetic, or transition, region where k and ω are of the orders of magnitude of the mean free path and the collision frequency. The third is naturally the most interesting both from the theoretical point of view and because of the striking experimental findings. In the lectures the kinetic model (simplified transport equations as an approximation to the Boltzmann equations) is used to extract normal-mode solutions for dilute gases and to show the existence for them of continuum modes in addition to the familiar discrete modes. The latter (among which the two sound modes) can disappear at sufficiently large k and ω. The analysis of the experiments of GREENSPAN and of MEYER and SESSEN in monoatomic gases is used to show the roles of the discrete and continuum normal modes.

In dense fluids the existence of significant spatial correlation among molecules give rise to restoring forces which allow collective motion even in the absence of collisions, i.e. even when the frequency of oscillation becomes large compared with the collision frequency. These collective modes which occur in the high-frequency region, called the collisionless regime, constitute the zero sound. The existence of such modes in a neutral classical fluid and its connection with the normal sound modes are discussed. The collective modes in liquids at short wavelengths (k comparable with the inverse intermolecular spacing) are considered by using a generalized viscoelastic description of density and current fluctuations.

The scattered light which emerges from a transparent medium illuminated by a monocromatic beam of light carries in its spectrum information on the dynamics of the scattering centres (molecules) and can be used to extract information on the acoustical modes of the density fluctuations and on the acoustical properties of the medium. The analysis of the experiments and the indications of the cases in which it confirms information also obtainable with ultrasonic experimentation are given by MONTROSE in his contribution on *Light Scattering and Molecular Acoustics*.

RUDNICK offers a thorough and up-to-date presentation of research on *Sound Propagation in Superfluid Helium*. The first part deals with a theoretical description of the various kinds of sounds and of their behaviour, while the second part is a comprehensive review of the relevant experiments carried on up to now.

The subject of *Sound Propagation in Liquid Crystals* is presented by CANDAU and MARTINOTY. The results of absorption measurements for both longitudinal and transversal waves in nematics are compared with the theoretical indications to derive various viscosity coefficients and elastic constants of ne-

matic liquid crystals; the relaxation processes in nematic and isotropic phases
are discussed and some references to recent research in smectic liquid crystals are also included.

Mason's contribution on *Acoustical Properties of Solids* examines the application of sound waves in a wide frequency range to the study of many solid-state
motions such as domain wall motion, point imperfection and dislocation motion.
The author examines in detail the attenuation of sound waves in perfect crystals, the effects of structure in a solid (grain, domain wall, phase transitions) and
the effect of imperfections (point defects, dislocations). Some consideration is also
given to fatigue in metals and to the fast-developing field of acoustic emission.

CAROME has considered the special case of *Superconducting Transducers in
the* 50 *to* 1000 GHz *Range*. He discusses the use of phonon fluorescent elements and tunnel junctions as phonon sources, as well as bolometers and
tunnel junctions as phonon detectors.

A very comprehensive review of the method developed in the last ten years
for the *Production and Detection of Very-High-Frequency Sound Waves* has
been offered by DRANSFELD. The treatment excludes the case of surface wave
(see de Klerk's contribution). The production and detection of coherent phonons can be made in a variety of ways: *a*) piezoelectric methods in traditional
transducers, in depletion layer transducers, in high-polymer transducers;
b) magnetostrictive methods at microwave frequencies; *c*) electromagnetic
generation at microwave ultra-sound; *d*) by scattering of light and X-rays
(spontaneous or stimulated Brillouin scattering, X-ray scattering by phonons,
Raman scattering by phonons, microwave-induced Raman scattering).

JOFFRIN and LEVELUT, in *Phonon Echoes*, have discussed a unique type of
phonon echo experiment which looks promising both for physical application
(measurements of sound absorption, of characteristic relaxation times) and for
engineering applications (signal processing and memory devices). Here the
term echo has to be taken with the same meaning that is applied to spin systems.
A simple experiment consists in launching a pulse of coherent phonons into
a specimen by means of a transducer (frequency ω_1) at time $t = 0$, and in
applying at time τ an electrical field of the same frequency to part of, or to
the entire specimen. The nonlinear interaction between the electric field and
the square of the elastic deformation creates reflected ultra-sonic waves (a
reverse of the wave vector) which travel exactly in reverse of the previous
waves and create an echo at the source position at time 2τ. The echo is independent of the specimen shape and of the source location.

The group of lectures related to sound propagation in fluctuating media
of large dimension opens with the McCoy contribution: *Wave Propagation in
Random Media*. It is concerned first with the analysis of the way in which a
random medium can be statistically described *per se* and, second, with the
propagation of an acoustic field; a two-point coherence function applicable to
ocean studies is given special attention.

An analysis of the physical processes responsible for fluctuations of interest in sound propagation in the sea is the first part of the lectures offered by GOOD-MAN: *Propagation in Fluctuating Media*. The establishment and the characteristics of internal waves receive special attention in the discussion of the dynamics of the medium; turbulence plays an important role also; the sea suface behaviour and the connected capillary and gravity waves are considered in defining the dynamic properties of the sea, in the proximity of the surface. At greater depths, there exists the possibility of trapping the acoustic energy emanating from a source into a channel, allowing transmission over thousands of miles.

The results of some experiments on ocean dynamic fluctuations observed with the acoustical method are discussed. The experiments refer to propagation in the mixed layer where turbulence dominates the phenomena of fluctuations as well as to propagation to much larger depths. The reflection of waves from the sea surface is also discussed.

The third group of contributions includes Berktay's treatment of *Finite-Amplitude Effects in Sound Propagation in Fluids*, with special reference to water. The consequences of the nonlinearity of differential equations describing acoustic disturbances are examined in a lossless medium and successively in a thermo-viscous medium. In the last case and with reference to sinusoidal boundary conditions the development of a « weak shock » wave form, the conservation of the « saw tooth » form and the saturation of the fundamental component of particle velocity at a given range are discussed. The analysis is then applied to the monochromatic radiation from a transducer.

Special attention is given to the parametric acoustic arrays where nonlinear interactions of sound waves are used to produce low-frequency acoustic waves. The parametric transmitters which use two monochromatic primary waves in order to produce a wave at the difference frequency are shown to produce a low-frequency source with beam width of the same order as those of the primary transducers and a level suitable for sonar applications.

Stephens' contribution is a brief review of the last-decade interest on finite wave propagation in solids. After reference to a few positions of the higher-order elasticity theory, the nonlinear effects in wave propagation, resulting either from large wave amplitude or from induced as well as local nonlinearities of the medium, are discussed. Theoretical and experimental results are presented. Harmonic generation in piezoelectric crystals, optical-acoustical interaction in photoconducting piezoelectrics, nonlinearity in surface waves are considered. Research on surface waves in liquids, acoustical streaming, nonlinearities in cochlear hydrodynamics, scattering of sound by sound and on the relation between macrosonics and nonlinear effects in crystalline solids is also reviewed.

The great interest that surface waves have recently provoked for their many valuable applications has led to a frequent presentation of the subject in terms of equivalent circuits of devices which are especially useful for engineers. This method, however, does not give much insight into the physics involved. DE KLERK has therefore developed *A Physical Approach to Elastic Surface Waves*. Both isotropic and anisotropic material are considered. An analytical method for the optimum in crystal orientations and propagation directions is presented. Moreover the effect of piezoelectric properties of the material on the action of interdigital grids on the surface and the basic principles of two surface wave devices are discussed.

A review of the present status of developments of the more interesting *Surface Acoustic-Wave Devices* is given by ATZENI and MASOTTI: interdigital transducers, delay lines, filters, oscillators, multistrip couplers, convolution using parametric interaction, interaction with light and display systems are considered in detail.

The proceedings are closed by a contribution by WANG on *Acoustics in Space*; the acoustical chamber developed for manipulating and controlling a liquid system in a zero-*G* environment is becoming an important tool for space research and technology.

① ② ③ ④ ⑤ ⑥ ⑦ ⑧ ⑨ ⑩ ⑪
⑫ ⑬ ⑭ ⑮ ⑯ ⑰ ⑱ ⑲ ⑳ ㉑ ㉒
㉓ ㉔ ㉕ ㉖ ㉗ ㉘ ㉙ ㉚ ㉛ ㉜ ㉝ ㉞

1. D. Bacci
2. J. Heisermann
3. M. Gross
4. G. Guenin
5. B. Lambert
6. D. Salin
7. S. Garrett

8. G. Natale
9. K. Dransfeld
10. J. Joffrin
11. G. H. Bauer
12. J. Krüger
13. C. Luponio
14. A. Bracco

15. T. G. Wang
16. R. Goodman
17. J. de Klerk
18. S. M. Lee
19. S. Yip
20. C. Montrose
21. R. Piazza

22. Schellino
23. M. Cutroni
24. B. Nilsson
25. S. Gasse
26. B. Lambert
27. J. L. Hunter
28. R. B. Lindsav

29. W. P. Mason
30. D. Sette
31. I. Rudnick
32. H. O. Berktav
33. R. Cook
34. E. Carome

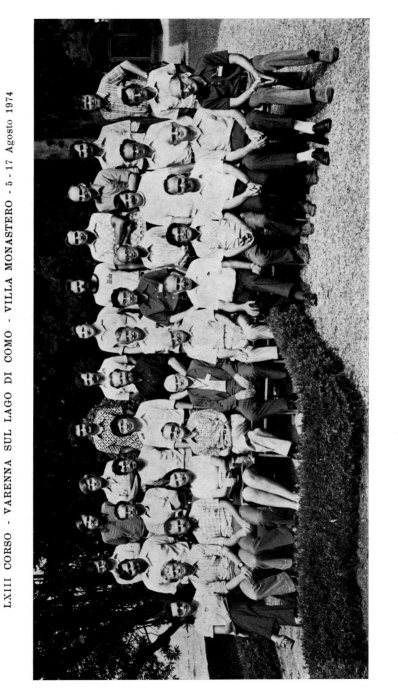

Historical Development of Physical Acoustics and Future Perspectives.

R. B. LINDSAY

Brown University - Providence, R. I.

1. – Introduction.

In organizing the course on « New Directions in Physical Acoustics » in the International School of Physics « Enrico Fermi » in Varenna in August, 1974, the Director, Professor D. SETTE, deemed it desirable to introduce the lectures with a review of the history of the subject and some suggestions for future development. The following paper represents the substance of two lectures delivered by the author on the first day of the course, August 5, 1974.

In approaching any branch of the history of science, one becomes keenly aware of the essential difficulty of the subject. As in the case of history in general, the history of science provides serious problems not merely in the location of relevant documentary material but also in its interpretation in the light of current knowledge. Unless we can effectively relate early developments with the present state of science, their value to the scientist would appear to be minimal. Moreover, there always exists the difficulty of establishing priority of discovery, that is, who really first produced the seminal results on which future developments depended. Pessimists in this area, and there are many among scientists, especially physicists, may at times feel that the history of physics and its various branches is in some danger of deserving the definition which Bertrand RUSSELL once gave for mathematics as the « subject in which we never know what we are talking about nor whether what we are saying is true ». They may not indeed be willing to go so far as Henry FORD, who contented himself with the blunt statement: « History is bunk ».

In this essay we shall take a more optimistic point of view but must warn the reader that in the space allowed only a few of the highlights can be adequately presented.

2. – What is physical acoustics?

The subject can be interpreted in both a broad and a narrow sense. In the broad sense physical acoustics consists of the description of all phenomena relating to the production, propagation and reception of sounds of all kinds, together with appropriate theoretical explanations. To clarify this consider for example the analytic subject index of a journal of acoustics, such as the *Journal of the Acoustical Society of America*. This contains 17 main categories with 215 subcategories, covering all aspects of acoustics from general linear acoustics to acoustic instrumentation and including such diverse fields as physiological and psychological acoustics, speech communication, ultrasonics and the physical effects of sound, noise and its control, underwater sound and architectural acoustics. Basic physics enters into each category, no matter what the detailed applications may be, and this justifies the broad definition of physical acoustics. The subject of acoustics never could have made any progress without physics, and physics still enters into every aspect of it.

In this essay we shall find it convenient to treat physical acoustics in the narrower sense as the experimental and theoretical study of the propagation of mechanical radiation in material media. This puts emphasis on such problems as the velocity and attenuation of sound of all available frequencies as well as reflection, refraction, diffraction, interference and scattering. It also includes the relation between wave and ray acoustics and covers both small- and large-amplitude disturbances, the latter being the subject of so-called nonlinear acoustics. Of very great importance in modern physical acoustics is the study of the interaction between sound waves and the physical properties of the media through which they pass, such as the temperature and pressure, elastic properties and the molecular and atomic constitution. Other related problems are the interaction of sound waves with electric and magnetic fields and light of all wavelengths.

3. – Historical origins.

Evidence indicates that in very early times it was realized that sound is the result of a mechanical disturbance originating at some place and able to affect the ear of a human being at some other place at a later time. But physical acoustics in the sense in which we are using the term did not begin until the attempt was made to understand *how* sound travels from one place to another. Here from early times two theories competed with each other. According to the emission theory sound is emitted from the source in small particles or atoms of sound, something like the atoms of light in the corpuscular theory of light, made famous by NEWTON. The more favored theory, however, was the wave theory, according to which sound is propagated through the air by wave motion in analogy with waves on the surface of water. By stretching matters a bit, one can read this notion into the writings of ARISTOTLE (384-322 B.C.) [1]. The emission theory, though not widely held, did not completely die out until the 17th century. It was supported, for example, by the well-known French philosopher GASSENDI (1592-1655) [2], who did his best to revive Greek atomism in the Renaissance, using atoms to explain practically all physical phenomena. Some might be willing to credit GASSENDI with the introduction of the *phonon*, though this kind of enthusiastic attribution in the history of science should probably be deprecated.

It was early recognized that some sounds are more pleasant than others, and the notion of the pitch of a musical sound is a very old one. In this connection the story of PYTHAGORAS (6th century B.C.) and his association of pitch with the dimensions of struck bars as sources of sound is legendary, though suggestive of the early appreciation of the connection between pitch and frequency. This connection seems to have been understood by the philosopher ARCHYTAS of Tarentum in Southern Italy around 375 B.C. He was one of the followers of PYTHAGORAS. The relation between pitch and frequency was clearly recognized by the Roman philosopher BOETHIUS (480-524) [3] and is of course set forth in detail by GALILEI (1564-1642) at the end of the « First Day » of his famous « Dialogues Concerning Two New Sciences» (1638) [4]. Here the great Italian presents in delightfully attractive form actual experiments to demonstrate the relation between pitch and frequency and also goes into the subject of resonance in clear-cut fashion.

4. – The velocity of sound in air.

From a theoretical standpoint the first important landmark in the history of physical acoustics was connected with the velocity of sound. Though the fact that sound requires a perceptible time to travel was recognized in very early times, actual measurements were not reported until the middle of the 17th century in France and Italy. The first attempt at a theoretical derivation of the velocity of sound was apparently made by NEWTON (1642-1727) [5]. His deduction has been considered by most authorities to be hard to follow. It must be emphasized that he restricted his attention to harmonic sound waves. He took it for granted that the action of a sound wave is to disturb particles of air from their equilibrium positions so as to form compressions and rarefactions, and that, due to the elasticity of the fluid medium, there is a tendency for equilibrium to be restored, just as when a simple pendulum bob is pulled aside from its equilibrium position it tends to move back to this position when released. NEWTON also assumed that the wavelength of the sound wave is the distance between successive compressions or rarefactions. The analogy with the motion of the simple pendulum provided NEWTON with the basis of his theoretical analysis. Effectively he used the fundamental relation between velocity V, wavelength λ and period T for any harmonic wave. He also effectively assumed that the wavelength is proportional to the length l of the simple pendulum.

Analytically this becomes (g being the acceleration of gravity)

$$(1) \qquad V = \frac{\lambda}{T} = \frac{2\pi l}{2\pi \sqrt{l/g}} = \sqrt{gl}\,.$$

The choice of the period of the simple pendulum of length l as the period of the wave was plausible. The choice of 2π as the proportionality constant between λ and l is somewhat mysterious. The final assumption is that l is effectively the thing that supplies the elasticity of the fluid medium. For air this is the height of a uniform atmosphere providing the observed atmospheric pressure at the surface of the Earth. It was then not difficult for NEWTON to show that for the case of air

$$(2) \qquad V = \sqrt{gl} = \sqrt{p/\varrho}\,,$$

where p is the atmospheric pressure and ϱ is the density. It must be stressed that in deriving his famous formula NEWTON made no use of Boyle's gas law, nor did he mention any possible connection between the temperature of the fluid and V. The fact that his result implies isothermal propagation of sound was evidently not in Newton's mind at all. When he substituted the appropriate values for p and ϱ for air, NEWTON found $V = 968$ ft/s as compared with the experimental value known to him at the time of 1142 ft/s. At first the discrepancy does not appear to have worried him greatly. He may well have thought, as theorists have felt through the ages, that the experiments on the velocity of sound in air were a bit uncertain and that more precise measurements would bring the value down to the one his theory predicted. However, when he re-examined the problem at the time of the appearance of the second edition of his « Principia » (1713) he realized that more precise experimental measurements only confirmed the previous value for V. How he tried to solve this difficulty is one of the mysteries of the history of physical acoustics. He evidently felt that actual air as distinct from « ideal » air is full of solid particles which effectively increase the elasticity more than the density decreases it. He also assumed that humidity would have a similar effect. By introducing a « fudge » factor for which no precise reason can be assigned NEWTON managed to get his original value raised to the experimental value.

It is not surprising that the great continental mathematicians and natural philososphers of the eighteenth century were not satisfied with Newton's theory. The first to emit a blast was the youthful EULER (1707-1783) who as a lad of twenty wrote « Dissertatio Physica de Sono », published in Basel in 1727 [6]. EULER criticized Newton's derivation of the expression for the velocity of sound in a fluid as too specialized. He himself then gave a general theory of sound propagation which is qualitatively in agreement with our modern point of view. Unfortunately when he came to providing the analysis for the velocity deduction, he left out some of the algebra and simply quoted the result, which is

(3)
$$V = \frac{4}{\pi}\sqrt{p/\varrho}\,.$$

EULER comments on the fact that this agrees with the observed values better than Newton's formula, though, whereas Newton's expression gives too low a value by about 20 %, Euler's value is too high by about 10 %.

It became obvious to thoughtful people that the problem of sound propagation would be solved only by the application of Newton's second law of motion to an *element* of a continuous fluid medium. NEWTON himself had never done this. The first to do it was apparently the Englishman TAYLOR (1685-1731) who in 1713 wrote $F = ma$ for an element of a stretched vibrating string and obtained the fundamental frequency of vibration [7]. However, he failed to set up the differential equation for wave motion along the string. D'ALEMBERT (1717-1783) succeeded in doing this in 1747, and for the first time the wave equation for a mechanical disturbance in one dimension appeared in print [8]. D'ALEMBERT also provided the general solution of this equation in terms of arbitrary functions of the form $f(x \pm Vt)$.

By the middle of the 18th century LAGRANGE (1736-1813) and EULER had derived the wave equation for the propagation of a mechanical disturbance in a one-dimensional fluid medium. Euler's presentation in his memoir of 1759 is in some respects the clearer [9]. Here he derived the wave equation, more or less as it is done today, in the form

(4)
$$\frac{\mathrm{d}^2y}{\mathrm{d}t^2} = \frac{gh(\mathrm{d}^2y/\mathrm{d}x^2)}{1 + (\mathrm{d}y/\mathrm{d}x)^2}\,.$$

Here y is the fluid particle displacement from equilibrium, x is the spatial coordinate in the direction of propagation, t is the time, g the acceleration of gravity and h the height of the equivalent uniform atmosphere which would produce the actual atmospheric pressure on the ground. Note that the partial-derivative notation had not been invented when EULER wrote this paper. Note also the interesting term in the denominator on the right-hand side, indicating that, in his analysis, EULER made no approximation assumption limiting his analysis to disturbances of very small magnitudes. By writing this equation EULER was taking the first step in the development of nonlinear acoustics. EULER recognized that for small disturbances he could safely neglect the term $(\mathrm{d}y/\mathrm{d}x)^2$ in the denominator, and hence found what has come to be regarded as the one-dimensional wave equation for small disturbances from equilibrium. The wave velocity then turns out to be

(5)
$$V = \sqrt{gh}\,.$$

Since h has the same meaning here as l in Newton's formula (eq. (2)), the velocity is identical with that deduced by NEWTON by a quite different method. This greatly puzzled EULER and he devoted some space in his memoir to wondering about it. He provided no solution to the mystery, though it is interesting to observe that he thought the nonlinear denominator might have something to do with the difficulty.

The problem of the velocity of sound remained unsolved for the next half-century, though various ineffectual attempts were made to solve it. It was the mathematician, physicist and astronomer LAPLACE (1749-1827) who first thought of assuming that heat is produced in the compression involved in the propagation of a sound wave, and that due to the quickness with which the successive compressions and rarefactions follow each other the heat may not have a chance to flow away between compressions. This would suggest that the thermal behavior of a fluid medium traversed by a sound wave is effectively adiabatic rather than isothermal. The full significance of this in modern terms was not apparent at the time of Laplace's work, since the prevailing theory of heat was still the caloric theory, and the theory of heat as a form of energy was still nearly half a century in the future. That LAPLACE was thinking early in the nineteenth century that the problem of the velocity of sound was somehow bound up with the two specific heats of a gas is clear from references to his speculations in Poisson's famous memoir on the theory of sound of 1808 [10]. However, it was not until 1816 that LAPLACE published a short paper setting forth his view that Newton's expression for the velocity of sound in a gas must be corrected by a multiplicative factor equal to the square root of the ratio of the specific heat at constant pressure to the specific heat at constant volume, a quantity which later has been denoted by γ [11]. Thus Laplace's formula for the velocity of sound was

$$(6) \qquad\qquad V = \sqrt{\gamma}\,\sqrt{p/\varrho}\,.$$

Laplace's explanation is certainly no model of clarity; but it is clear that his reasoning, in the light of later developments, was on the right track. Though the experimental value of γ in Laplace's time was very uncertain, the value 1.5 seemed not unreasonable and this gave a value of V in rather good agreement with the measured value. It is scarcely surprising, however, that Laplace's theory and formula were not taken too seriously by his contemporaries. For many years the quantity $\sqrt{\gamma}$ was simply known as Laplace's « factor ». LA-PLACE continued to think about the problem, and when he brought out the final edition of his *magnum opus* « Mécanique Celeste » in 1823, in the final volume he presented a more detailed derivation of his velocity-of-sound formula. This seemed more plausible than his 1816 version, but he was still hampered by his adherence to the caloric theory of heat. The real significance of Laplace's formula in terms of the mechanical theory of heat had to await the investigation of CLAUSIUS (1822-1888) and RANKINE (1820-1872) around 1850. Thereafter it was agreed by theoreticians that for ordinary audible frequencies sound propagation in a gas is an adiabatic process. On this basis confidence in the validity of Laplace's formula became so great that it grew common to use it to obtain γ by measuring the velocity of sound. The question as to the validity of Laplace's formula for very low and very high frequencies produced some confusion in the acoustical literature in the 20th century. It was held by some that at very low frequencies the heat developed in sound propagation would be conducted away fast enough to guarantee isothermal conditions and hence Newton's formula would hold. It was also held that the rate of heat conduction increases with frequency, and hence at very high frequencies adiabaticity would no longer prevail, and Newton's formula would again hold. This, however, neglects the effect of viscosity, which acts to increase the velocity as the frequency increases at a faster rate than the effect of heat conduction decreases it. Hence, the net effect of increasing frequency in a gas is to *increase* the velocity beyond Laplace's value and not to reduce it to the Newtonian value. This matter was well discussed in 1928 by HERZFELD and RICE [12] and should remove further uncertainty and confusion. It should be added that all measurements of V (for not too closely confined gases) as a function of frequency confirm the result here presented. Of course the dispersion of sound in polyatomic gases is a rather complicated affair. This will be discussed later in this essay.

5. – The attenuation of sound.

It was early recognized that the loudness of sound from a given source in the open air decreases with the distance from the source. That this is largely due to the spreading of the sound in all directions was apparently recognized by ARISTOTLE and was more precisely stated by MERSENNE (1588-1648) [13] in France in the 17th century. No precise measurement of this attenuation could be made until an adequate definition of sound intensity was available and appropriate instrumentation had been invented. This did not happen until well after the middle of the 19th century. In this case, however, we are confronted by the surprising fact that a theory of sound attenuation in a gas due to viscosity was developed before there was any experimental evidence for its relevance, and indeed before intensity had been defined in terms of energy flow in a sound wave. This theory was introduced by STOKES (1819-1903). In 1845 in a long paper on the effect of viscosity (called by him fluid friction) on the motion of fluids STOKES applied his theory to the motion of the small disturbances from equilibrium constituting the propagation of sound [14]. He showed that as a result of viscosity the amplitude of the oscillatory displacement of a small element of fluid will decay in time through an exponential term of the form $\exp[-\beta t]$, where β, the temporal attenuation coefficient, is directly proportional to the square of the frequency and to the viscosity of the fluid. In his article STOKES pointed out that there was at that time no way of experimentally testing his formula.

The next development in the theory of sound attenuation resulted from investigations of sound propagation in tubes of small cross-section. In some experiments in 1867 KUNDT (1839-1894) [15] showed that the velocity of sound in such tubes filled with air decreases as the area of cross-section decreases and increases with the frequency. In 1868 HELMHOLTZ (1821-1897) provided a theory to account for these results. Qualitative agreement was found but the quantitative agreement was not satisfactory [16]. The difficulty stimulated KIRCHHOFF (1824-1887) to re-examine the whole problem of sound propagation in a fluid and to consider the influence of heat conduction as well as viscosity [17]. In his famous memoir of 1868 KIRCHHOFF presented a complete theory of the influence of these two important transport processes on sound attenuation and showed that the attenuation due to heat conduction in gases is of the same order of magnitude as that due to viscosity. He also confirmed Stokes' result that the effect of transport processes on attenuation is a first-order effect, whereas that on velocity of propagation is a second-order effect. But when Kirchhoff's general theory was applied to Kundt's experiments on the sound velocity in narrow tubes, once again only qualitative agreement was found.

The work of STOKES and KIRCHHOFF indicated clearly that the numerical values of sound attenuation calculated from transport properties are far less than those due to spherical spreading or to the well-known effects of reflection, refraction and scattering produced in the open air by atmospheric irregularities. Of course, since the estimations of attenuation at that time were very crude, the whole matter was in an unsatisfactory state. The situation by the end of the 19th century is clearly indicated by the experiments of the American physicist DUFF (1864-1951). In 1898 he tried [18] to measure sound attenuation in the open air by the study of the loss of audibility with distance of the sound from whistles of different frequencies placed in the open under extremely quiet atmospheric conditions. After deducting for the effect of spherical spreading, DUFF concluded that viscosity and heat conduction could not possibly account for the equivalent attenuation he observed. For an explanation he suggested *radiation* of heat, a possible attenuation-producing process hitherto neglected. Duff's work was reviewed by Lord RAYLEIGH (1842-1919) in a paper « On the Cooling of Air by Radiation and Conduction and on the Propagation of Sound » [19]. He showed that the magnitude of the radiation effect demanded by Duff's results was far too great to correspond to any reasonable theory of heat radiation in a gas. RAYLEIGH confessed that the problem of the excess attenuation of sound in a gas like air was a puzzle. However, he made the interesting suggestion that the energy of the translational motion of the gas molecules, ultimately responsible for the propagation of sound, may on collision pass in part to internal energy states of the polyatomic molecules involved, *i.e.* states of vibration and rotation. The return of this energy to the translational form would be subject to a certain lag (relaxation time), and this would produce effective attenuation of sound. This was certainly one of the first clear-cut mentions of the possibility of a relaxation time theory of sound attenuation. It is true that in 1881 LORENTZ wrote a long paper on the equations of motion of a gas and the kinetic theory of sound propagation [20] in which he introduced the notion of internal energy states of molecules. But the paper did not introduce any explicit use of the relaxation time idea and evidently did not lead to work along this line. The question arises: why did not RAYLEIGH follow up his own suggestion? Evidently his interest in the attenuation of sound did not extend that far, or perhaps he had too many other more intriguing things on his mind at the time.

Actually Rayleigh's suggestion bore some fruit, or so it might seem. JEANS (1877-1946) devoted Chapter 16 of his book « The Dynamical Theory of Gases » (1904) [21] to the molecular theory of the propagation of sound. Without any reference to RAYLEIGH (or indeed to any other previous investigator) he calculated the effect of the trapping of translational energy in internal-energy states for the case where the molecules are loaded spheres, and found an effect on attenuation only a relatively small fraction of that due to viscosity. He concluded that the same order of magnitude would prevail in the case of diatomic molecules and hence missed the boat with respect to the relaxation time theory of sound attenuation. It may be significant that he omitted this chapter on sound propagation in subsequent editions of his book.

At the turn of the century there were still no precise laboratory measurements of the attenuation of transmitted sound. This seems a bit odd, since the Rayleigh disc had been invented in 1882. But in the history of science as in general history, hindsight is better than foresight. Apparently the first laboratory measurement of sound attenuation in a gas under reasonably well-controlled conditions, with particular reference to the dependence on frequency, was made by the Russian NEKLAPAJEV in 1911 [22]. He used a spark source in air and estimated the frequency by measuring the wavelength by means of an acoustic diffraction grating. The intensity was determined by what appears to have been essentially a Rayleigh disc. Frequencies up to $4 \cdot 10^5$ Hz were studied, and it was found that the attenuation in this range varies as the square of the frequency. The numerical results were considerably in excess of those predicted by viscosity and heat conduction.

It appears that the possible use of the idea of molecular relaxation in the theory of sound propagation was not seriously investigated again until EINSTEIN (1879-1955) decided to employ it in the study of the velocity of chemical reactions. In 1920 he published an article [23] on sound propagation in partially dissociated gases. He showed that the relaxation between dissociated and undissociated molecules in a gas leads to a velocity dispersion formula from which the chemical reaction rates can be determined once the velocity dispersion has been measured.

In the mid nineteen twenties PIERCE (1872-1956) took advantage of the technique newly developed by LANGEVIN (1872-1946) and CADY (1874-1974) of using the piezoelectric effect in quartz and Rochelle salt crystals to produce high-frequency sound radiation to study the velocity of sound in carbon dioxide. His results showed a definite dispersion hitherto undetected at lower frequencies [24]. He carried his measurements up to a frequency of about 1 MHz, and should have detected absorption but did not measure it. These experiments of PIERCE provided the incentive for further experimental studies of the properties of what came to be called supersonic radiation, though the name was later changed to ultrasonics.

The next important contribution to the molecular relaxation theory of sound propagation was that of HERZFELD (1892) and RICE (1890) who in 1928 published a paper [12] in which for the first time the relaxation between the translational energy states of polyatomic molecules and the internal energy states was used to obtain expressions for both the attenuation and velocity of sound as a function of frequency. The relaxation time was introduced as an important factor in both attenuation and velocity though this name was not yet assigned to it. Though there was little experimental evidence with which to compare the theoretical results the authors showed that it was possible to choose a plausible value of the relaxation time in order to account for the excess attenuation found by NEKLAPAJEV. This was definitely ground-breaking research and set the stage for the further development of what has come to be known as molecular acoustics.

6. – The development of molecular acoustics.

The fundamental research of HERZFELD and RICE of 1928, followed immediately thereafter by the somewhat similar work of BOURGIN (1900) [25] was succeeded in the early 1930's by a spate of papers sparked by the pioneer research of KNESER (1901), who in 1931 published a paper on the effect of thermal relaxation on sound propagation in gases [26]. KNESER assigned external and internal specific heats to the external (translational) and internal molecular energy states respectively and set up a « reaction » equation for the transfer of energy back and forth between external and internal states. This equation involved significantly the quantity later to be called the « relaxation time » though KNESER did not use this terminology, preferring the term « lifetime of state ». Though KNESER realized that his analysis led to attenuation of sound as a function of frequency and relaxation time, he did not derive the relation but contented himself with the associated relation for the dispersion of sound.

The next step in the development of molecular acoustics was probably taken by the young English chemistry student HENRY (1906) in Cambridge University. In 1932 he was much interested in specific-heat anomalies at high temperature. He realized that information on these could be obtained from the attenuation and dispersion of sound, and, by introducing the concept of complex specific heat, arrived at equations for attenuation and dispersion of sound in terms of specific heats and the quantity τ, which for the first time he denominated the relaxation time for transition between external and internal energy states of molecules [27].

It is clear that the theoretical development of molecular acoustics would hardly have been a relevant advance had it not been accompanied by the introduction of precise methods of measurement of attenuation in gases. It is an interesting reflection on the diverse character of the science of acoustics that such a precise measurement came out of investigations in architectural acoustics. In early studies on room acoustics following the work of SABINE (1868-1919) [28] it was assumed that the total attenuation of sound in a room, which along with the volume of the room controls the reverberation time, is provided by the material on the surfaces of the room, but KNUDSEN (1893-1974) of the University of California in Los Angeles, who carried out extensive investigations in architectural acoustics from 1925 onwards, showed that this assumption is not correct in the case of large halls, but that the attenuation of sound in the air of the hall enters vitally into the reverberation time. By precise measurements of the latter he was able to obtain accurate measurements of the attenuation of sound in air as a function of frequency [29]. This stimulated KNESER to further theoretical research and KNUDSEN and KNESER carried on important collaborative work on attenuation in air, water vapor as well as other gases during the early 30's [30]. The result was that by the middle 30's molecular acoustics, so far as gases are concerned, was pretty firmly established. The term thermal relaxation began to be applied to the type of relaxation associated with the transitions between external and internal energy states of molecules. Success in this work coupled with increasing interest in the attenuation of sound in sea water led to the extension of the relaxation theory to the explanation of attenuation in liquids. Here different types of relaxation mechanisms were found to be effective. Thus the term structural relaxation was applied to the case in which a liquid possesses two kinds of local structure between which an equilibrium persists under normal conditions: this equilibrium is disturbed by the passage of a sound wave, leading to a relaxation time lag in its restoration, which in turn produces sound attenuation. Chemical dissociation relaxation was also invoked to account for attenuation in electrolytes. A great deal of work of this kind was carried on in the late 30's and the 40's. A review of these developments will be found in a review article on « The Absorption of Sound in Fluids » by MARKHAM, BEYER and LINDSAY [31]. By the middle 1950's the molecular-acoustics relaxation theory for attenuation and dispersion in both gases and liquids was rather firmly established and only specific details remained to be worked out.

During the past twenty years it is fair to say that the chief interest in molecular acoustics has been concentrated on acoustical properties of solids. Further reference to this will be made later in this essay.

7. – Experimental background of modern physical acoustics.

It is obvious that physical acoustics could not have made the progress it has during the past 50 years if it had not been for the development of electroacoustics with the instrumentation it has provided for the production of coherent and pulsed radiation over an extremely wide range of intensity and frequency. For it is only in this way that acoustic radiation of appropriate flexibility can be produced so as to be effective in giving information about the structure and internal behavior of matter.

We have already commented on the fact that there was much excellent acoustical radiation theory developed in the middle and toward the end of the 19th century by people like STOKES, KIRCHHOFF, HELMHOLTZ and RAYLEIGH. But these theoretical developments could not be effectively applied and tested because controlled radiation with respect to intensity and frequency was not available. It was difficult to do much with tuning forks, resonators and bird whistles as sources of sound. What physical acoustics needed was a set of adequately controllable transducers with electric drive. As one looks back over the past hundred years, it seems strange that though magnetostriction was discovered by JOULE in 1847 and piezoelectricity by the CURIES in 1880, neither phenomenon was used acoustically until the end of the second decade of the 20th century. The reason for this long hiatus is doubtless to be sought in the relatively long delay in the development of electrical circuits capable of producing oscillations of controlled frequency over a wide range of frequencies. Though the Edison effect was discovered in 1883, its significance in terms of the emission of electrons was, of course, not recognized at that time. It was not until early in the 20th century that the thermionic valve or vacuum tube was invented and shown to serve as a suitable device for the production of electrical oscillations. Controlled frequency oscillations for scientific and technological uses had to await the work of CADY at the end of the second decade of the century. As a result of these developments the use of piezoelectric quartz oscillations as sources of acoustic radiation in fluids became rather common from the late 1920's onward.

8. – The development of ultrasonics.

Often a new field of physics (or science in general) gains great impetus when someone comes along and shows that there is something really spectacular associated with it. This happened in the case of ultrasonics through the celebrated early work of WOOD (1868-1955) and LOOMIS (1887), who summarized their experiments in a celebrated 1927 paper « The Physical and Biological Effects of High Frequency Sound Waves of Great Intensity » [32]. Here they showed the spectacular effects of ultrasound, including the fountain effect, flocculation and emulsification properties and its destructive action on biological tissue. This emphasized that there is a lot of fascinating physics still left in acoustics, if only one is prepared to push the frequency up to high enough values (WOOD and LOOMIS reached a limit of 300 kHz). The stimulus thus provided for further work was profound. It led to efforts to increase the range of usable frequencies. In the early 1970's the maximum usable ultrasonic frequency is about 10^{11} Hz. Experiments are under way to push this figure even higher. At the same time the technologists were led to realize the possible industrial applications of ultrasonics, including ultrasonic cleaning, drilling, soldering and various metallurgical processes. On the biological side great interest was developed in the possible diagnostic and therapeutic uses of ultrasonics in medicine. This gave a great fillip to the instrumentation field, which in turn produced a wide variety of acoustic transducers for purely scientific work, the kind of fertile interaction between science and technology which has had so much to do with the advancement of science.

9. – Macroscopic physical acoustics.

Molecular acoustics and its use of ultrasonics may be thought of as microscopic physical acoustics. In considering the development of the subject we must not overlook what may be called its macroscopic aspects. First and foremost here is the physics involved in the process of radiation, that is, getting the sound out of a vibrating body into the surrounding medium. The problem is, of course, a very old one. It was first tackled in a fundamental sense in a paper by STOKES in 1868 [33] in which he showed clearly the importance of the relation between the acoustic impedances of the vibrator and medium respectively, though he did not introduce the impedance terminology. This work as extended by RAYLEIGH and others was fundamental for the understanding of the action of all acoustic transducers. Detailed work along this line continues to this day.

Early acoustical theory was confined to waves of very small amplitudes, though there was some interest in the mid and late 19th century in the so-called finite-amplitude acoustic waves [34]. During the last 25 years there has been increasing interest in this branch of the subject, now generally referred to as nonlinear acoustics, from the nonlinear character of the mathematical equations involved.

Among the interesting phenomena associated with nonlinear acoustics, still under study today, may be mentioned acoustic cavitation in liquids, acoustic

streaming, radiation pressure, aerodynamic sound, shock waves and the interaction of sound with sound. Good review articles on these subjects will be found
in the series of books « Physical Acoustics » edited by W. P. MASON (more
recently with the collaboration of R. N. THURSTON [35]). For important sem-
inar papers see also « Physical Acoustics », edited by R. B. LINDSAY [36].
Very recent developments in nonlinear acoustics are covered in « Finite-Amplitude Wave Effects in Fluids », edited by L. BJØRNØ (Proceedings of a 1973
Symposium in Copenhagen) [37].

10. – Recent developments in molecular acoustics.

There has been an increased interest in recent years in the interaction of
light and sound. Here again the problem is an old one, going back to light
scattering in material media and the early work of TYNDALL, RAYLEIGH,
SMOLUCHOWSKI and EINSTEIN. The scattering was attributed to fluctuations
in density of the medium through which the light passes. In a paper published
by BRILLOUIN (1889-1969) [38] in 1914, in which he attempted to improve on
the earlier results of EINSTEIN, BRILLOUIN introduced the quantum-theory
distribution of energy among the various modes of oscillation of the particles
of the scattering medium and treated the latter oscillations as if they were
sound waves of very high frequency. This was the origin of what came to be
called Brillouin scattering, which has stirred up new attention in very recent
times. It also served as an impetus to the search for light scattering by beams
of artificially produced coherent sound in material media. This led to the
discovery of the Debye-Sears effect in 1932 (DEBYE (1884-1966) and SEARS
(1898)) [39], also independently observed by LUCAS and BIQUARD [40]. The
field of acoustical optics is now well established.
    In connection with sound propagation in liquids mention must of course
be made of the fascinating researches on sound transmission through liquid
helium, particularly liquid helium II. Theoretical and experimental research
on the properties of this remarkable liquid have demonstrated the existence of
four different kinds of acoustic propagation (denominated zero, second, third
and fourth sounds) in addition to the sound of ordinary experience. For seminal
papers in this field the book « Physical Acoustics » edited by LINDSAY, previously mentioned [17] may be consulted (see pp. 419-457).
    Probably the most exciting developments in physical acoustics on the
molecular level during the past two decades have taken place in the domain
of solids. This has, of course, been concomitant with the tremendous progress
of solid-state physics in general. In a brief review of this character it is impossible to do justice to this contemporary development. One can mention only
a few special cases. One obvious one is the use of low-temperature ultrasonic
attenuation measurements in the study of superconductivity, and the use of

such data in the plotting of Fermi surfaces in solids. An associated development is the use of ultrasonic radiation to study dislocations and other defects in solids. Surface wave acoustics has been successfully applied to the study of thin films on the surface of solids, leading to new designs of delay lines and acoustical storage devices.

The study of the interaction between acoustic radiation in solids and applied magnetic fields, in which curious oscillations in sound velocity are found, has led to the development of magnetoacoustics as an important branch of solid-state physics. Similar remarks could be made for the acousto-electric effect and acousto-optics. Acoustics has made its way into quantum theory with the recent studies of phonon behavior in crystal lattices, especially phonon-electron and phonon-phonon interactions. Nuclear-spin interactions with ultrasonic radiation have not been overlooked. The best reference to recent work in this whole category of what may be called quantum acoustics is the series of books on physical acoustics, edited by MASON and MASON and THURSTON, already referred to [35].

11. – Research perspectives.

To project the future development of physical acoustics is neither a safe nor an easy thing to do. It was a shrewd American farmer who remarked that you should « never prophesy unless you know ». What we do not know far exceeds what we know in any given field. One is indeed tempted to feel that solid-state acoustics will continue to make great strides, especially if the frequency of coherent acoustic radiation can be pushed above the THz limit. However, the present author has a strong feeling that physical acoustics could make a great contribution to biophysics and yield much valuable information about the behavior of living things, particularly on the neurological level. Here is a domain that offers a tremendous challenge to physicists and biologists alike.

REFERENCES

[1]   ARISTOTLE: *Selections from* De Anima *and* De Audibilibus (M. R. COHEN and I. E. DRABKIN: *A Source Book of Greek Science* (New York, N. Y., 1948), p. 288). Reprinted in *Acoustics, Historical and Philosophical Development*, edited by R. B. LINDSAY, *Benchmark Papers in Acoustics* (Stroudsburg, Pa., 1973), p. 21. This last-named work is referenced hereinafter as Lindsay I.
[2]   R. B. LINDSAY: *Amer. Journ. Phys.*, **13**, 235 (1945).
[3]   BOETHIUS: *Concerning the Principles of Music*. Translated and reprinted in Lindsay I, p. 35-39.

[4] G. GALILEI: *Mathematical Discourses Concerning Two New Sciences*, translated by T. WESTON (London, 1730), p. 138-157. Reprinted in Lindsay I, p. 42-61.

[5] I. NEWTON: *Mathematical Principles of Natural Philosophy*, translated by A. MOTTE (New York, N. Y., 1848), p. 356-367. Reprinted in Lindsay I, p. 75-86.

[6] L. EULER: *Dissertation on Sound* (Basel, 1727). Translated and reprinted in Lindsay I, p. 105-117.

[7] B. TAYLOR: *Phil. Trans. Roy. Soc.*, **28**, 26 (1713). Translated and reprinted in Lindsay I, p. 96-102.

[8] D'ALEMBERT: *Hist. Acad. Sci., Berlin*, **3**, 214 (1747). Translated and reprinted in Lindsay I, p. 119-123.

[9] L. EULER: *On the propagation of sound*, in *Memoirs of the Academy of Sciences* (Berlin, 1766). Translated and reprinted in Lindsay I, p. 136-154.

[10] S. D. POISSON: *Journ. Ecole Polytech.*, **7**, 319 (1808). Introduction translated and reprinted in Lindsay I, p. 173-179.

[11] P. S. LAPLACE: *Velocity of sound in air and water*, in *Annales de Chimie et de Physique*, III (1816). Translated and reprinted in Lindsay I, p. 181-182.

[12] K. F. HERZFELD and F. O. RICE: *Phys. Rev.*, **31**, 691 (1928). Reprinted in Lindsay II, p. 298-302. See under ref. [17].

[13] M. MERSENNE: *On the velocity of sound in air*, in *Cogitata Physico Mathematica* (1644). Translated and reprinted in Lindsay I, p. 64-66.

[14] G. G. STOKES: *Trans. Cambridge Phil. Soc.*, **8**, 287 (1845). Reprinted in part in Lindsay I, p. 262-289.

[15] A. KUNDT: *Monatsberichte, Berliner Akademie, Dec. 19, 1867.*

[16] H. HELMHOLTZ: *Verhandlungen des natur-historisch-medizinischen Vereins zu Heidelberg vom Jahre 1868*, Bd. III, p. 8, 16.

[17] G. KIRCHHOFF: *Ann. der Phys. Chem.*, **134**, 177 (1868). Translated and reprinted in *Physical Acoustics, Benchmark Papers in Acoustics*, edited by R. B. LINDSAY (Stroudsburg, Pa., 1974), p. 7-21. This last-named work is referenced hereinafter as Lindsay II.

[18] A. W. DUFF: *Phys. Rev.*, **6**, 129 (1898).

[19] Lord RAYLEIGH: *Phil. Mag.*, **47**, 308 (1899). Reprinted in Lindsay I, p. 411-416.

[20] H. A. LORENTZ: *Archiv. Neerl.*, **16**, 1 (1881). Reprinted in *Collected Papers of H. A. Lorentz*, Vol. **6** (The Hague, 1938), p. 1.

[21] J. H. JEANS: *The Dynamical Theory of Gases* (Cambridge, 1904), p. 302.

[22] N. NEKLEPAJEW: *Ann. der Phys.*, **35**, 175 (1911).

[23] A. EINSTEIN: *Sitz. Ber. Preuss. Akad. Wiss. Berlin*, **24**, 380 (1920). Translated and reprinted in Lindsay II, p. 268-272.

[24] G. W. PIERCE: *Proc. Amer. Acad. Arts and Sci.*, **60**, 271 (1925). Reprinted in part in Lindsay II, p. 277-297.

[25] D. G. BOURGIN: *Nature*, **122**, 133 (1928).

[26] H. O. KNESER: *Ann. der Phys.*, **11**, 761 (1931). Translated and reprinted in Lindsay II, p. 303-315.

[27] P. S. H. HENRY: *Proc. Cambridge Phil. Soc.*, **28**, 249 (1932). Reprinted in Lindsay II, p. 316-322.

[28] W. C. SABINE: *Reverberation (Collected Papers in Acoustics of W. C. Sabine* (New York, N. Y., 1964), p. 3). Reprinted in part in Lindsay I, p. 417-457.

[29] V. O. KNUDSEN: *Journ. Acoust. Soc. Amer.*, **3**, 126 (1931). Reprinted in Lindsay II, p. 323-335.

[30] V. O. KNUDSEN and H. O. KNESER: *Journ. Acoust. Soc. Amer.*, **5**, 4 (1933).

[31] J. J. MARKHAM, R. T. BEYER and R. B. LINDSAY: *Rev. Mod. Phys.*, **23**, 353 (1951). Reprinted in Lindsay II, p. 346-404.

[32] R. W. Wood and A. L. Loomis: *Phil. Mag.*, **4**, 417 (1927). Reprinted in Lindsay II, p. 240-266.

[33] G. G. Stokes: *Phil. Trans. Roy. Soc.*, **158**, 447 (1868). Reprinted in part in Lindsay II, p. 22-31.

[34] Lord Rayleigh: *Proc. Roy. Soc.*, A **84**, 247 (1910). Reprinted in Lindsay II, p. 136-173.

[35] W. P. Mason, Editor: *Physical Acoustics, Principles and Methods*, Vol. **1-5** (New York, 1964-1969). Edited by W. P. Mason and R. N. Thurston with same title, Vol. **6-10** (1970-1973).

[36] Lindsay II, p. 135-219.

[37] L. Bjørmø, Editor: *Finite-Amplitude Wave Effects in Fluids, Proceedings of the 1973 Symposium in Copenhagen* (London, 1974).

[38] L. Brillouin: *Compt. Rend.*, **158**, 1331 (1914). Translated and reprinted in Lindsay II, p. 416-418.

[39] P. Debye and F. W. Sears: *Proc. Nat. Acad. Sci.*, **18**, 409 (1932). Reprinted in Lindsay II, p. 410-415.

[40] R. Lucas and P. Biquard: *Compt. Rend.*, **194**, 2132 (1932).

Correlation Functions in Molecular Acoustics.

C. J. MONTROSE

Department of Physics, Catholic University of America - Washington, D. C.

1. – Introduction.

The propagation of high-frequency acoustic waves in fluids has proven to be a valuable probe for studying a rather wide variety of molecular relaxation processes. In a typical experiment the absorption and dispersion of ultrasonic waves are measured as functions of the ultrasonic frequency and the results are interpreted in terms of complex frequency-dependent bulk and/or shear viscosities. The problem is to identify the microscopic mechanism responsible for the relaxation of the viscosities, and, from the specific nature of the frequency dependence, to infer its dynamical behavior. In some cases, notably vibrational and rotational isomeric relaxation, the microscopic processes are rather well understood and this program can be carried out straightforwardly. In other instances, for example structural or shear relaxation, the situation is much less clear, although a substantial body of data is extant. The principal difficulties in these cases stem from the highly co-operative collective nature of the molecular processes involved.

In attempting to understand and eventually to propose a microscopic picture of these processes it is reasonable to expect that the theoretical framework provided by linear response theory will play a central role. This approach has been exploited rather successfully in the interpretation of relaxation spectra in a number of situations; in particular, the description of frequency-dependent transport coefficients in terms of time correlation functions has led to an increased understanding of the molecular motions involved in various relaxation phenomena.

The purpose of this paper is didactic; completeness and rigor have been given secondary status to what I hope is a clear pedagogical development. This tutorial format is aimed at encouraging further inquiry with the hope that molecular acousticians will view their results, through the vehicle of time correlation functions, in conjunction with those of complementary spectroscopic studies in an effort to synthesize a coherent picture of molecular dynamics in fluids.

In Sect. **2** the basic notions of time correlation functions are set out and a few of their properties are considered. Section **3** outlines the major elements of linear response theory. Both the response of an equilibrium system to an external perturbing force (Kubo theory) and the relaxation of a system in an initially nonequilibrium state (Mori theory) will be treated. These ideas are applied in Sect. **4** to the consideration of sound propagation in dense fluids. This will be done within the context of linearized hydrodynamics generalized to allow for relaxation processes. (For the most part the discussion will be restricted to the case of single-component isotropic systems.) Relaxation associated with both internal molecular co-ordinates and external « structural » degrees of freedom is discussed. Finally, in Sect. **5,** a brief discussion of several aspects of collective-mode dynamics in fluids, in which the close relationship of general molecular dynamics to acoustical phenomena is emphasized, is presented.

Notably absent from this precis is any mention of acoustic relaxation and dispersion phenomena in the neighborhood of critical points. This omission is justified only on the basis that this particular area has grown so rapidly in recent years that its inclusion would require discussion at a depth far beyond the bounds that I have set for myself in this presentation.

2. – Time correlation functions.

Several excellent review articles on time correlation functions and their connection through linear response theory with dynamic, *i.e.* frequency-dependent, transport coefficients exist in the literature. Of particular interest from a tutorial standpoint are those of ZWANZIG [1], BERNE and FORSTER [2] and BERNE [3]. It is not my purpose to reproduce these authors' work here; rather, only certain key ideas and formulae will be given. For more extensive (and intensive) study, the above-mentioned articles and the references cited therein are recommended.

2˙1. *General description.* – Examine some property A of a physical system that varies with time t, for example the total kinetic energy of the molecules in a monoatomic gas. We take the average value of A, denoted by $\langle A \rangle$, to be zero. If $\langle A \rangle$ does not vanish, the property A is replaced by $A - \langle A \rangle$ for which the average is zero. In the example of a system of N monoatomic gas molecules with total kinetic energy E, the appropriate property would be $E - \frac{3}{2} N k_{\mathrm{B}} T$, where k_{B} is Boltzmann's constant and T is the temperature. An experiment is performed in which the quantity A is measured as a function of time; a large number N_0 of trials are performed and the results are displayed as shown in Fig. 1. Suppose that on the j-th trial the result for time t_1 is $A = A_j(t_1)$, *i.e.* some numerical value, and at time t_2 $(> t_1)$ the result is $A = A_j(t_2)$. Then the time autocorrelation function of A is defined as

$$(2.1) \qquad C_A(t_1, t_2) = \frac{1}{N_0} \sum_{j=1}^{N_0} A_j(t_1) A_j(t_2) = \langle A(t_1) A(t_2) \rangle \,.$$

The average $\langle ... \rangle$ is over the « ensemble of repetitions » [4] and is equivalent to an ensemble average in statistical mechanics as $N_0 \to \infty$. The time corre-

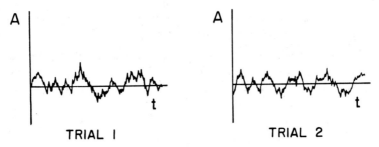

Fig. 1. – Typical results obtained in several repetitions of an experiment in which some property of a system A is measured as a function of time.

lation functions of interest in this article will be descriptive of *stationary* processes and consequently depend only upon the time difference $t = |t_2 - t_1|$. Thus we may write the correlation function in (2.1) as

$$(2.2) \qquad C_A(t) = \langle A(t_1) A(t_1 + t) \rangle = \langle A(0) A(t) \rangle \,,$$

the last equality following because $C_A(t)$ is invariant with respect to a translation of the time origin.

In a similar fashion we can examine time correlation functions for the two properties, say A and B. Proceeding along lines essentially the same as those outlined above, we picture the record for the j-th trial as being composed of two displays such as are sketched in Fig. 2. The time correlation

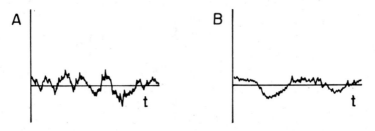

Fig. 2. – Typical results obtained for a given trial of an experiment in which two properties of a system, A and B, are measured as functions of time.

function of A and B is given by

(2.3)
$$C_{AB}(t) = \frac{1}{N_0} \sum_{j=1}^{N_0} A_j(t_1) B_j(t_1 + t) = \langle A(0) B(t) \rangle .$$

Note that

(2.4)
$$\langle A(0) B(t) \rangle = \pm \langle B(0) A(t) \rangle ,$$

the choice of sign depending upon whether the product AB changes sign under the time-reversal operation $t \to -t$. For the most part we shall deal with time autocorrelation functions of the form $\langle A(0) A(t) \rangle$ (or perhaps $\langle A(0)^* A(t) \rangle$ if A is a complex quantity); for these functions we have obviously

(2.5)
$$\langle A(0) A(t) \rangle = \langle A(0) A(-t) \rangle .$$

The positive sign in (2.1) is to be employed.

We shall also be interested in the derivatives of correlation functions. A few simple results can be derived and are useful in certain manipulations. These include

(2.6)
$$\frac{\mathrm{d}}{\mathrm{d}t} \langle A(0) A(t) \rangle = \langle A(0) \dot{A}(t) \rangle = - \langle \dot{A}(0) A(t) \rangle ,$$

(2.7)
$$\frac{\mathrm{d}^2}{\mathrm{d}t^2} \langle A(0) A(t) \rangle = \langle A(0) \ddot{A}(t) \rangle = \langle \ddot{A}(0) A(t) \rangle = - \langle \dot{A}(0) \dot{A}(t) \rangle ,$$

and so on. Notice that if we put $t = 0$ in (2.6), we have $\langle A(0) \dot{A}(0) \rangle = -\langle \dot{A}(0) A(0) \rangle$ from which it follows that

(2.8)
$$\left[\frac{\mathrm{d}}{\mathrm{d}t} \langle A(0) A(t) \rangle \right]_{t=0} = \langle A(0) \dot{A}(0) \rangle = 0 .$$

From an operational standpoint, a time correlation function is constructed by taking the product of a quantity $A(t)$ with a second quantity $B(t')$ evaluated at some other time and averaging the result. We may now legitimately ask what all this means. Or looked at another way, if we know a correlation function $C(t) = \langle A(0) A(t) \rangle$, what do we know about the physical system? To get a feeling for this, we consider a subset of the ensemble of trials for which, at $t = 0$, the quantity A is constrained to some specific value, say A_0. For $t > 0$, the behavior of A for any given trial within this subensemble will be random, but the average over the subensemble will exhibit certain definite trends. For one thing, the average $\langle A(t) \rangle_c$ (the subscript c on $\langle ... \rangle$ denotes an average over the constrained subset of the total ensemble) must vanish as

$t \to \infty$. The behavior might be similar to that shown in Fig. 3. With this we can form the correlation function $C_A(t)$. We multiply $\langle A(t) \rangle_c$ by the initial value $A(0) = A_0$ and average over all initial states in the full ensemble of trials, *i.e.*

$$(2.9) \qquad \langle A_0 \langle A(t) \rangle_c \rangle = \langle A(0) A(t) \rangle = C_A(t) .$$

If the displacements from equilibrium, that is, the set of A_0, are not too large, then the dynamical parameters of the motion (relaxation times, etc.) are am-

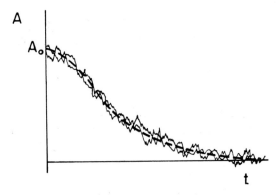

Fig. 3. – Two trials of an experiment to measure $A(t)$ where the initial value is constrained to A_0. The dashed curve shows the average behavior of $A(t)$ subject to this constraint.

plitude independent, and the time variation of both $\langle A(t) \rangle_c$ and $C_A(t)$ should be identical. Since $C_A(0) = \langle A^2 \rangle$, $C_A(t)$ gives the average behavior of $A(t)$ for an initial displacement $\langle A^2 \rangle^{\frac{1}{2}}$, *i.e.* the r.m.s. fluctuation in A multiplied by $\langle A^2 \rangle^{\frac{1}{2}}$. Thus if we separate the correlation function into the product of an « intensity factor » $\langle A^2 \rangle$ and a dynamical factor $c_A(t)$, *i.e.*

$$(2.10) \qquad C_A(t) = \langle A^2 \rangle c_A(t) ,$$

the dynamical factor—or normalized correlation function—$c_A(t)$ describes the time evolution of $\langle A(t) \rangle_c = A_0 c_A(t)$ for small displacements A_0.

2'2. *Correlation functions in statistical mechanics.* – Our view of time correlation functions has so far been a strictly operational one. In this part we translate this viewpoint into the language of classical statistical mechanics. The problems that concern us are usually adequately described within the confines of classical physics and we shall ignore those corrections and complications occasioned by quantum mechanics.

The dynamical variables of interest are expressible as *phase functions*; that is, they depend upon the instantaneous state of the system. This state is described by giving all the phase-space co-ordinates q_1, q_2, ..., q_M and momenta p_1, p_2, ..., p_M for a system with M degrees of freedom. We shall use the symbol X to denote this complete set of phase-space co-ordinates, and integration over all the co-ordinates will be written as

$$(2.11) \qquad \int dX \leftrightarrow \int dq_1 \int dq_2 \ldots \int dq_M \int dp_1 \int dp_2 \ldots \int dp_M .$$

Now suppose that at some (initial) instant of time the state of the system is given by X_0; this is represented by a point in the $2M$-dimensional phase space

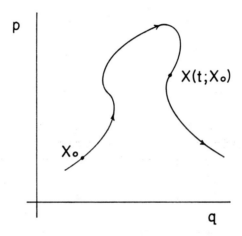

Fig. 4. – A schematic representation showing the time evolution of the state of a system in phase space.

of the system. As time evolves the state of the system changes; the representative point moves tracing out a path in phase space. This is shown schematically in Fig. 4. At a time t the state of the system can be written as

$$(2.12) \qquad X_t = X(t; X_0);$$

the new state depends upon the initial state as well as the time interval. The time variation of a dynamical variable $a(t)$ or phase function $A(X_t)$—$a(t)$ is the numerical value of A in the state X_t—is expressed by

$$(2.13) \qquad a(t) = A(X_t) = A[X(t; X_0)] = A(t, X_0) .$$

The average value of A is calculated in some convenient ensemble, characterized by a distribution function $f(t, X)$ which, in general, depends upon time as well as on the co-ordinates X. Mainly we will be concerned with averages with respect to an *equilibrium* ensemble; in this case the distribution function is not an explicit function of t. For example, in the canonical ensemble

$$(2.14a) \qquad\qquad f(X) = Q^{-1} \exp\left[-\beta H(X)\right]$$

with

$$(2.14b) \qquad\qquad Q = \int dX \exp\left[-\beta H(X)\right],$$

where $\beta = 1/k_{\mathrm{B}}T$ and $H(X)$ is the Hamiltonian of the system. Thus, the (equilibrium) ensemble average of A is

$$(2.15) \qquad\qquad \langle A \rangle = \int dX f(X) A(X),$$

and the time correlation function in the equilibrium ensemble is

$$(2.16) \qquad C_A(t) = \langle a(0) a(t) \rangle = \int dX_0 f(X_0) A(X_0) A(t, X_0).$$

In (2.16) the subscript $_0$ could have been suppressed since the choice of the initial instant is clearly arbitrary. The averaging described by this equation is the statistical mechanical equivalent of the operational definition given in (2.1).

Formal manipulations involving the type of many-body dynamics implied by considering the time evolution of a state X are greatly facilitated by the introduction of operator techniques. These arise naturally and provide a convenient shorthand; consider, for example the Liouville equation of motion for the distribution function $f(t, X)$. Essentially this is a statement of probability conservation:

$$(2.17) \qquad \frac{df}{dt} = \frac{\partial f}{\partial t} + \sum_\alpha \left(\frac{\partial f}{\partial q_\alpha} \dot{q}_\alpha + \frac{\partial f}{\partial p_\alpha} \dot{p}_\alpha \right) = 0.$$

Using Hamilton's equations of motion gives

$$(2.18) \qquad \frac{\partial f}{\partial t} = -\sum_\alpha \left(\frac{\partial H}{\partial p_\alpha} \frac{\partial}{\partial q_\alpha} - \frac{\partial H}{\partial q_\alpha} \frac{\partial}{\partial p_\alpha} \right) f \equiv -iLf,$$

where this defines the *Liouville operator* L. (The imaginary unit i is introduced as this makes L a Hermitian operator, *i.e.*

$$\int dX\, B(X)^* LA(X) = \int dX [A(X)^* LB(X)]^*,$$

where A and B are any phase functions.) The Liouville equation (2.18) has the formal solution

(2.19)
$$f(t, X) = \exp[-itL]f(0, X);$$

this is easily understood when we recognize that $\exp[-itL]$ generates a Taylor series for $f(t, X)$:

(2.20)
$$f(t, X) = [f]_{t=0} + t\left[\frac{\partial f}{\partial t}\right]_{t=0} + \frac{1}{2!}t^2\left[\frac{\partial^2 f}{\partial t^2}\right]_{t=0} + \cdots$$

with

$$[\partial^n f/\partial t^n]_{t=0} = (iL)^n f(0, X).$$

The same notation is useful for describing the time variation of a phase function $A(X)$. Since there is no explicit time dependence, the equation of motion is

(2.21)
$$dA/dt = iLA,$$

with the formal solution

(2.22)
$$A(t, X) = \exp[itL]A(0, X).$$

Note the sign difference in the exponential compared with (2.19). Suppose we are interested in the average value of A at time t, i.e. $\langle a(t) \rangle$. In terms of the initial distribution $f(0, X) = f(X)$ we can generate $f(t, X)$ using (2.19) and thus

(2.23)
$$\langle a(t) \rangle = \int dX \, A(X) f(t, X) = \int dX \, A(X) \exp[-itL]f(X).$$

Using the Hermitian property of L allows rewriting this in the form

(2.24)
$$\langle a(t) \rangle = \int dX \, f(X) \exp[itL]A(X) = \int dX \, f(X) A(t, X).$$

Equations (2.23) and (2.24) represent two viewpoints for examining dynamical behavior. In the form (2.23) we picture the system as evolving in time (as characterized by a time-varying f) and the average is calculated with respect to the instantaneously appropriate $f(t, X)$. In the form (2.24) we envision calculating $A(t, X)$—the value of A at time t given some initial state X—and then averaging over the initial states, i.e. with respect to the distribution function for $t = 0$.

3. – Linear response theory; transport coefficients.

The purpose of this Section is to describe a mathematical structure within the framework of classical statistical mechanics in which we can analyze two types of experimental situations: a) the response of a system in equilibrium to an external mechanical driving force and b) the relaxation of a system initially displaced from equilibrium. We shall assume that we are dealing with « small » forces or deviations from equilibrium and shall try to present the essence of the theoretical approaches to analyzing these situations. Our aim is to convey the flavor of what kind of results can be obtained and how they can be useful in interpreting experimental results.

The motivation for this is that we obtain expressions for complex frequency-dependent transport coefficients in terms of molecular properties—specifically in the form of time correlation functions. Since ultrasonic absorption and dispersion data (the basic product of molecular-acoustics experiments) yield just such transport coefficients, namely viscosities, linear response theory provides a structure within which one can (speaking optimistically) infer molecular properties and dynamics from these data.

3`1. *Response to external forces: the Kubo method.* – The basic references for this Section (in addition to the reviews mentioned above) include the classic articles by KUBO [4-6], the elegant discussion of thermal transport coefficients by LUTTINGER [7] as well as the work by CALLEN and his colleagues [8, 9] on the fluctuation-dissipation theorem.

The theory is constructed to describe an experiment in which an external force is applied and the time-dependent response of the system is measured. For times $t < 0$ the system is in equilibrium and is characterized by a distribution function $f_0(X)$ (*e.g.* given, in the canonical ensemble, by (2.14)). At time $t = 0$ the system is in a state X (which is characterized only by certain average properties such as temperature) and a perturbing force field, say $E(t)$, is switched on. This produces a response in the system which is characterized by a flux $J(t)$. The average for many repetitions of the experiment, that is $\langle J(t) \rangle$, is obtained. We presume that $\langle J(t) \rangle = 0$ in the absence of the field, that is in equilibrium, and that the field is sufficiently weak so that $\langle J(t) \rangle$ can be taken as linear in E. With the additional requirement of causality, the most general form relating $\langle J(t) \rangle$ and $E(t)$ is

$$(3.1) \qquad \langle J(t) \rangle_E = \int_0^t \mathrm{d}t' \, \varphi(t - t') \, E(t') ,$$

where $\varphi(t - t')$ is the « response » or « after-effect » function giving the response at time t to an impulsive force of unit strength at t'. The subscript E is appended to the average on the left-hand side of (3.1) to indicate averaging with respect to an ensemble in which the effect of the forcing field E is taken into account. We expect that, for $t \gg t'$, $\varphi(t-t') \to 0$ since memory is presumed to exist only for a finite duration. Note that, if there is no memory, that is if $\varphi(t-t') = \sigma_0 \delta(t-t')$, then $\langle J(t) \rangle_E = \sigma_0 E(t)$ and σ_0 is an ordinary transport coefficient.

Generally, however, this is not the case, and the description of (3.1) is usefully considered in terms of the Fourier frequency components of $\langle J(t) \rangle_E$ and $E(t)$:

$$(3.2a) \qquad \langle \tilde{J}_\omega \rangle_E = \int_{-\infty}^{\infty} \mathrm{d}t \, \exp\left[i\omega t\right] \langle J(t) \rangle_E \,,$$

$$(3.2b) \qquad \tilde{E}_\omega = \int_{-\infty}^{\infty} \mathrm{d}t \, \exp\left[i\omega t\right] E(t) \,.$$

In terms of these, the Fourier transform of (3.1) is

$$(3.3) \qquad \langle \tilde{J}_\omega \rangle_E = \sigma(\omega) \tilde{E}_\omega \,,$$

where

$$(3.4) \qquad \sigma(\omega) = \int_{0}^{\infty} \mathrm{d}t \, \exp\left[i\omega t\right] \varphi(t)$$

appears as a complex frequency-dependent transport coefficient. Generally speaking, this is the property determined by the experimental measurements.

The question now revolves around what this tells us about the physical system; or phrased slightly differently: upon what aspect of the molecular dynamics does $\varphi(t)$ depend? Kubo's method provides the answer to this; the principal results are sketched here. For $t < 0$, the system is described by a Hamiltonian $H_0(X)$; for $t \geqslant 0$ the perturbed system has a Hamiltonian

$$(3.5) \qquad H(X) = H_0(X) - M(X) E(t) \,,$$

where $M(X)$ is the coupling between the system and the external field. (More complicated forms for the interaction have been treated; the paper by LUT-TINGER [7] considers a number of these. For our purposes the simple form given in (3.5) is adequate to illustrate the principles.) The ensemble distribution function is $f_0(X)$ for $t \leqslant 0$ and is $f(t, X)$ for $t > 0$, and thus the average flux is

$$(3.6a) \qquad \langle J(t) \rangle_E = \int \mathrm{d}X \, J(X) f_0(X) = \langle J(t) \rangle = 0$$

for $t \leqslant 0$, and

$$(3.6b) \qquad \langle J(t) \rangle_E = \int \mathrm{d}X \, J(X) f(t, X)$$

for $t > 0$. In these $J(X)$ is the appropriate phase function for the current or flux under consideration. For example, the phase function corresponding to an electrical current due to ions in a solution would be $\sum_j e_j(\mathbf{p}_j/m_j)\,\delta(\mathbf{r} - \mathbf{r}_j)$, where m_j, $\mathbf{p}_j$ and $\mathbf{r}_j$ are respectively the mass, momentum and position of the j-th ion and the sum runs over all the ions in the solution.

From (3.6) it is clear that the problem is to discover how the distribution function is pushed around by the external field E. The approach is to solve the Liouville equation (2.18) by a perturbation scheme. The Liouville operator can be written as

$$(3.7) \qquad\qquad L = L_0 + L_1 ,$$

where L_0 involves only the unperturbed part of the Hamiltonian and L_1 contains the term $- M(X)\,E(t)$, and the distribution function is expanded with only terms linear in E being retained:

$$(3.8) \qquad\qquad f = f_0 + f_1 .$$

Here f_0 is of order zero in E (and is thus the equilibrium distribution) and f_1 is linear in E. The Liouville equation can now be separated into a zeroth- and a first-order equation

$$(3.9a) \qquad\qquad \partial f_0/\partial t = - iL_0 f_0 ,$$

$$(3.9b) \qquad\qquad \partial f_1/\partial t = - iL_0 f_1 - iL_1 f_0 .$$

Our initial condition is that $f(t = 0, X) = f_0(X)$, i.e. that $f_1(t = 0) = 0$, so that, when we solve (3.9b), an inhomogeneous linear differential equation of first order gives

$$(3.10) \qquad f_1(t, X) = \beta \int_0^t dt'\, \exp\left[- i(t - t')\,L_0\right] iL_0\, M(X) f_0(X)\, E(t') .$$

Since the average current in equilibrium is zero,

$$(3.11) \qquad\qquad \langle J(t)\rangle_E = \int d(X)\, J(X) f_1(t, X) ,$$

and thus

$$(3.12) \quad \langle J(t)\rangle_E = \int_0^t dt' \left\{ \beta \int dX \left[\exp\left[itL_0\right] J(X)\right]\left[\exp\left[it'L_0\right] iL_0\, M(X)\right] f_0(X) \right\} E(t') .$$

Here we have made use of the Hermitian character of L_0 and of the fact that $\partial f_0/\partial t = 0$. The quantity in $\{...\}$ in (3.12) can be identified by a comparison of this result with (3.1); it is just $\varphi(t-t')$. Noting that

$$\exp[it L_0] J(X) = J(t, X)$$

and

$$\exp[it' L_0] i L_0 M(X) = \exp[it' L_0] \dot{M}(X) = \dot{M}(t', X)$$

gives

(3.13) $$\varphi(t-t') = \beta \langle \dot{M}(t') J(t) \rangle .$$

The frequency-dependent transport coefficient $\sigma(\omega)$—see (3.4)—is

(3.14) $$\sigma(\omega) = \beta \int_0^\infty dt \, \exp[i\omega t] \langle \dot{M}(0) J(t) \rangle ,$$

a one-sided Fourier transform of the correlation function $\beta \langle M(0) J(t) \rangle$. Observe that the average is an *equilibrium* average, but that dynamical behavior, *i.e.* $J(t)$, is needed.

Probably the most familiar example of the use of the Kubo method in general and of (3.14) in particular is the calculation of the mobility of an ion in solution [1]. For a single ion (charge e) in solution with an electric field E aligned parallel to the z-axis, the flux is just the ion's velocity:

$$J = v = dz/dt .$$

The perturbing Hamiltonian is $-ezE$ so that $M = -ez$. Thus $\dot{M} = e\dot{z} = ev$, so the mobility σ is

(3.15) $$\sigma(\omega) = e\beta \int_0^\infty dt \, \exp[i\omega t] \langle v(0) v(t) \rangle .$$

The relevant dynamics are contained in the velocity autocorrelation function. The ordinary, *i.e.* static, mobility σ_0 is just the $\omega \to 0$ limit of this expression:

(3.16) $$\sigma_0 = e\beta \int_0^\infty dt \, \langle v(0) v(t) \rangle .$$

The mobility can be expressed in terms of the average relaxation time τ for the velocity correlation function. This is defined as

(3.17) $$\tau \equiv \int_0^\infty dt \, \langle v(0) v(t) \rangle / \langle v(0)^2 \rangle ,$$

so that

(3.18) $$\sigma_0 = e\beta \langle v(0)^2 \rangle \, \tau \, .$$

This example illustrates several important results of much greater generality.

1) The transport coefficient is given by the one-sided Fourier transform of a time *autocorrelation* function, *i.e.* $M \propto J$. All of the transport coefficients of interest here are so determined.

2) The ordinary or $\omega = 0$ transport coefficient is proportional to the integral of the time autocorrelation function and hence to the relaxation time.

3) The transport coefficient is complex

$$\sigma(\omega) = \sigma'(\omega) + i\sigma''(\omega) \, ,$$

with the real and imaginary parts being related by Kramers-Kronig relations.

4) For $\omega \to \infty$, both σ' and σ'' vanish, ultimately going to zero at least exponentially in the frequency. (This follows from the vanishing of the first derivative of the correlation function at $t = 0$, but is often not a useful criterion since the onset of the exponential region occurs at frequencies that are inaccessible for many techniques ($\omega \sim 10^{13}\,\mathrm{s}^{-1}$). It is usually well below this that both σ' and σ'' have decayed to less than a few percent of σ_0.) At $\omega = 0$ σ' goes to its low frequency value σ_0 and σ'' vanishes, varying linearly with ω. (In certain formulations there occurs the possibility of a pole at $\omega = 0$; this situation is treated by LANDAU and LIFSHITZ [10]. It does not concern us here.)

3‘2. *Relaxation to equilibrium; generalized Langevin theory.* – The formulation in this Subsection is designed to describe situations where, at $t = 0$, the system is in a state of constrained equilibrium, the constraint being that some property of interest A has some nonzero average value, *i.e.* $\langle A(0) \rangle_c = a_0$. Its equilibrium average $\langle A \rangle$ is taken to be zero. For times $t > 0$ the constraint is released and the system relaxes to equilibrium; this is accompanied by the relaxation of $\langle A(t) \rangle_c$ to its equilibrium zero value. The description of this type of process has been expounded most thoroughly by MORI [11] and is reviewed rather well in ref. [2, 3]. Many ideas common to ordinary Langevin- or Brownian-motion theory are involved in the development and we shall begin by reviewing these.

3‘2.1. Elementary Langevin theory. Examine a particle (mass m, velocity v) acted upon by an external force $G(t)$ as well as by frictional forces. Newton's second law takes the form

(3.19) $$m \frac{dv}{dt} = -\zeta v + G(t) \, ,$$

where ζ is a friction coefficient. This equation works quite well as long as the particle's energy is large in comparison with $k_{\mathrm{B}} T$. When the energy is comparable to $k_{\mathrm{B}} T$, there is also a random component in the motion; this is described by including an additional stochastic force $F(t)$ on the right-hand side of (3.19):

$$(3.20) \qquad m \frac{\mathrm{d}v}{\mathrm{d}t} = -\zeta v + G(t) + F(t) .$$

From here on we shall take $G(t) = 0$. Certain statistical properties of $F(t)$ are generally specified. These may be enumerated as

1) $\langle F(t) \rangle = 0$. With this the average of (3.20) implies that $\langle v(t) \rangle = \exp[-\zeta t/m] \langle v(0) \rangle$. Thus $F(t)$ is clearly the origin of the fluctuating component of $v(t)$.

2) $\langle v(0) F(t) \rangle = 0$. The stochastic force and the velocity are statistically independent, *i.e.* uncorrelated.

3) $\langle F(0) F(t) \rangle = 2\chi \delta(t)$, where χ is a constant. The force spectrum is taken to be « white noise ». Physically this corresponds to assuming that the force autocorrelation function relaxes very rapidly compared with $\langle v(t) \rangle_c$ as sketched in Fig. 5b).

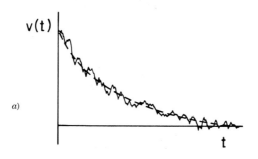

a)

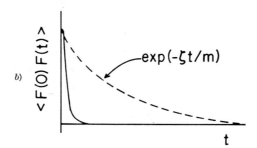

b)

Fig. 5. – *a)* The velocity of a Brownian particle as a function of time. The dashed curve shows the average behavior. *b)* The autocorrelation function of the random force, normalized to $\langle |F(0)|^2 \rangle = 1$.

We can now derive several useful results. If we multiply (3.20) by $v(0)$ and average, letting $C(t) = \langle v(0)\,v(t)\rangle$, the result is

$$\text{(3.21)} \qquad \frac{\mathrm{d}C}{\mathrm{d}t} = -\,(\zeta/m)\,C(t)\,.$$

$C(t)$ satisfies the same differential equation as $\langle v(t)\rangle_c$; the time variation of the correlation function and the constrained average are identical (see Subsect. 2'1). An explicit form for the friction coefficient ζ can be found. The most direct procedure is to solve (3.20) for $v(t)$, square the result, average and examine the limit $t \to \infty$. The result is a *fluctuation-dissipation* theorem

$$\text{(3.22)} \qquad \zeta = \beta\chi = \beta\int\limits_0^\infty \mathrm{d}t\,\langle F(0)\,F(t)\rangle\,.$$

The form on the right in this equation is particularly suggestive and will be seen to have a straightforward generalization.

There is a serious difficulty with the treatment just given. For short times the velocity correlation function has the form

$$\langle v(0)\,v(t)\rangle = (k_\mathrm{B}\,T/m)(1 - \zeta t/m + \ldots)\,;$$

clearly $\langle v(0)\,\dot v(0)\rangle \neq 0$ which is contrary to our result of Sect. **2**. To remedy this some modification of the simple Langevin equation is needed.

A simple generalization of the Langevin equation that eliminates this inconsistency is the replacement of $-\zeta v(t)$ on the right-hand side of (3.20) by a term of the form

$$\text{(3.23)} \qquad -\int\limits_0^t \mathrm{d}t'\,K(t-t')\,v(t')\,,$$

where $K(t-t')$ is termed the memory kernel or relaxation function. As long as $K(0)$ is finite, $\langle v(0)\,\dot v(0)\rangle = 0$ and our problem with respect to its not vanishing is removed. Additional changes are required, however. While we may still put $\langle F(t)\rangle_c = 0$ and $\langle v(0)\,F(t)\rangle = 0$, we may no longer take $\langle F(0)\,F(t)\rangle \propto \delta(t)$. In fact, we find that the force correlation function and the memory kernel are related by a generalized fluctuation-dissipation theorem

$$\text{(3.24)} \qquad K(t) = \beta\langle F(0)\,F(t)\rangle\,.$$

This is stated without proof and without further elaboration. We regard it as an interesting result but we postpone its discussion for a few paragraphs when it will reappear in a more general context.

3'2.2. Generalized Langevin theory. The context we have in mind is that of the more general problem that we initially addressed in this Section, namely the relaxation of a variable $a(t) = A(X_t)$ following an initial displacement from equilibrium. The equation of motion governing the relaxation is the Liouville equation (2.21). Generally, the solution to this is not possible, nor indeed is the full time variation $a(t)$ desired. What we seek are more modest results—specifically the constrained average $\langle a(t) \rangle_c$ and the correlation function $\langle a(0) a(t) \rangle$. We have already seen that, if the phase function $A(X)$ is just the momentum (or velocity) of a Brownian particle, a great simplification—the Langevin equation modified as in (3.23)—results. What we will now show is that *for any phase function* a similar simplification can be effected, subject to certain limitations. Specifically we shall obtain the following results which we now summarize, in advance.

1) For any dynamical variable $a(t) = A(X_t) = A(t, X)$ initially constrained to a nonzero average a_0, the relaxation to equilibrium for $t > 0$ (which is governed by the Liouville equation) can be cast into an equation of motion of the form of a *generalized Langevin equation*:

$$(3.25) \qquad \frac{\mathrm{d}}{\mathrm{d}t} a(t) = i\Omega a(t) - \int_0^t \mathrm{d}t'\, K(t-t') a(t') + F(t),$$

where Ω is a « resonant » frequency, $K(t-t')$ is a memory kernel and $F(t)$ is the stochastic force conjugate to A.

2) The random force is uncorrelated with A, *i.e.*

$$(3.26) \qquad \langle a(0)^* F(t) \rangle = 0.$$

(Note the complex conjugate included in the correlation function.) Thus the correlation function $C(t) = \langle a(0)^* a(t) \rangle$ satisfies

$$(3.27) \qquad \frac{\mathrm{d}C}{\mathrm{d}t} = i\Omega C(t) - \int_0^t \mathrm{d}t'\, K(t-t') C(t').$$

3) The frequency is given by

$$(3.28) \qquad i\Omega = \langle a(0)^* \dot{a}(0) \rangle / \langle a(0)^* a(0) \rangle.$$

Observe that for a single variable this implies $i\Omega = 0$. However, there is a direct generalization of (3.25) to the case of several dynamical variables $a(t)$, $b(t), \dots$ In this case $a(t)$ in (3.25) is replaced by a vector with elements $a(t)$,

$b(t)$, ...; similarly $F(t)$ is replaced by a vector with elements $F_a(t)$, $F_b(t)$, ..., the stochastic forces conjugate to $a(t)$, $b(t)$, ... The frequency $i\Omega$ and memory kernel $K(t-t')$ are matrices with elements

$$i\Omega_{ab} = \langle a(0)^* b(t)\rangle/\langle|a(0)|^2\rangle^{\frac{1}{2}}\langle|b(0)|^2\rangle^{\frac{1}{2}},$$

for example. In such a case the off-diagonal elements of the frequency matrix are nonvanishing. (The diagonal elements are still zero.) As a reminder of this type of situation we shall carry the frequency term in subsequent manipulations which for the sake of brevity will be performed for the case of a single variable.

4) The memory kernel is given by

$$(3.29) \qquad\qquad K(t) = \langle F(0)^* F(t)\rangle/\langle|a(0)|^2\rangle.$$

This is the generalized form of the result presented in (3.24). The integral $\int_0^\infty dt\, K(t)$ will be seen to take the part of a transport cofficient—as expected in the form of the integral of a time correlation function; the one-sided Fourier transform or Laplace transform of the memory function, *i.e.* of $K(t)$, is the complex frequency-dependent transport coefficient

$$(3.30) \qquad\qquad \hat{K}(-i\omega) = \int_0^\infty dt\, \exp[i\omega t]\, K(t).$$

The caret $\char"5E$ is used to designate the Laplace transform of a function; this is defined for a function $f(t)$ by

$$(3.31) \qquad\qquad \hat{f}(s) = \int_0^\infty dt\, \exp[-st]\, f(t).$$

In (3.30) we have put $s = -i\omega$.

5) The equilibrium ensemble average of the stochastic force is zero; the constrained ensemble average $\langle F(t)\rangle_c$ vanishes up to terms of second order in the initial displacement from equilibrium. If this deviation is small we may put $\langle F(t)\rangle_c \approx 0$, which leads to a linear equation of the generalized Langevin form

$$(3.32) \qquad\qquad \frac{d}{dt}\langle a(t)\rangle_c = i\Omega\langle a(t)\rangle_c - \int_0^t dt'\, K(t-t')\langle a(t')\rangle_c$$

for the relaxation of the dynamical variable to equilibrium.

6) An explicit form for $F(t)$ and thus for $K(t)$ and $\hat{K}(-i\omega)$ can be written down. Since this is done most conveniently in terms of Zwanzig-Mori projection operators [11, 12], this result will be deferred until these are treated below. A useful form of $\hat{K}(-i\omega)$ has been obtained by BERNE, BOON and RICE [13] and this can be given without recourse to projection techniques. Their result is

(3.33) $$\hat{K}(-i\omega) = \hat{\Phi}(-i\omega)[1 + \hat{\Phi}(-i\omega)/i\omega]^{-1},$$

where $\hat{\Phi}(s)$ is the Laplace transform of the correlation function

$$\Phi(t) = \langle \dot{a}(0)^* \dot{a}(t)\rangle / \langle |a(0)|^2\rangle.$$

7) From (3.27) it follows immediately that the Laplace transform of the correlation function $C(t)$ can be written as

(3.34) $$\hat{C}(s)/\langle |a(0)|^2\rangle = [s + i\Omega + \hat{K}(s)]^{-1}.$$

MORI [11] has demonstrated that this can be expressed as an infinite continued fraction:

(3.35) $$\frac{\hat{C}(s)}{\langle |a(0)|^2\rangle} = \cfrac{1}{s + i\Omega_1 + \cfrac{\Delta_1^2}{s + i\Omega_2 + \cfrac{\Delta_2^2}{s + i\Omega_3 + \dots}}},$$

where the Ω_i's and the Δ_i^2's are given by well-defined equilibrium (equal-time) averages.

3'2.3. Mori-Zwanzig derivation of the generalized Langevin equation. The Liouville operator L operates upon phase functions (dynamical variables) which we can regard as elements in a vector space—« Liouville space ». We denote the (ket) vector corresponding to $A[X(t)]$ by $|A(t)\rangle$. The inner product in the space is defined as the *equilibrium average*

(3.36) $$\langle B|A\rangle \equiv \langle B^* A\rangle = \int dX\, f(X)\, B^*(X)\, A(X).$$

Here B is a member of the corresponding dual space. For simplicity we normalize all the vectors:

$$\langle A|A\rangle = 1, \quad \textit{i.e. we let} \quad A \rightarrow A/\langle A^* A\rangle^{\frac{1}{2}}.$$

In this view the time evolution of A is

(3.37) $$|A(t)\rangle = \exp[itL]|A(0)\rangle = \exp[itL]|A\rangle,$$

and since the time propagation operator $\exp[itL]$ is unitary, this can be thought of as a *rotation* in Liouville space; $|A(t)\rangle$ has components parallel and orthogonal to $|A\rangle$.

This kind of thinking suggests defining a projection operator to separate these components. We define such an operator

$$P = |A\rangle\langle A|$$

and its complement $1 - P$, which respectively project onto the vector $|A(0)\rangle = |A\rangle$ and its orthogonal subspace:

(3.38) $$P|B(t)\rangle = |A\rangle\langle A|B(t)\rangle \to A(0)\langle A(0)^* B(t)\rangle.$$

Next we regard the dynamical variable $A(t)$ as a sum, $A(t) = A_1(t) + A_2(t)$, where $A_1(t) = P|A(t)\rangle$ and $A_2(t) = (1-P)|A(t)\rangle$. Then we separate the Liouville equation (which A obeys exactly) into two parts, first by operating on it with P:

(3.39) $$\frac{\mathrm{d}A_1}{\mathrm{d}t} = PiL[A_1(t) + A_2(t)],$$

and then by operating on it with $1 - P$:

(3.40) $$\frac{\mathrm{d}A_2}{\mathrm{d}t} = (1 - P)\,iL[A_1(t) + A_2(t)].$$

The second of these is then solved (probably most conveniently by means of the Laplace transform), and noting that $A_2(0) = 0$ gives

(3.41) $$A_2(t) = \int_0^t \mathrm{d}t'\, \exp[(t - t')(1 - P)\,iL](1 - P)\,iL\,A_1(t').$$

Next we introduce the correlation function $c(t)$ by observing that

$$A_1(t) = |A\rangle\langle A|A(t)\rangle = |A\rangle c(t).$$

Using this along with (3.39) and (3.41) yields

(3.42) $$\frac{\mathrm{d}c}{\mathrm{d}t} = \langle A|iL|A\rangle c(t) + \int_0^t \mathrm{d}t'\,\langle A|iL\exp[(t - t')(1 - P)iL](1 - P)\,iL|A\rangle c(t').$$

This has the form of the generalized Langevin equation (3.27) with

(3.43) $$i\Omega = \langle A|iL|A\rangle = \langle A|iLA\rangle = \langle A(0)^* \dot{A}(0)\rangle$$

and

(3.44) $$K(t) = -\langle A | iL \exp[t(1-P)iL](1-P)iL | A \rangle .$$

If we note that

$$\exp[t(1-P)iL](1-P)iL | A \rangle = (1-P)\exp[t(1-P)iL](1-P)iL | A \rangle ,$$

$K(t)$ can be written as

(3.45) $$K(t) = \langle (1-P)\dot{A} | \exp[t(1-P)iL] | (1-P)\dot{A} \rangle .$$

Now if we presume that $A(t)$ is governed exactly by a generalized Langevin equation we can work backwards to get $F(t)$. So, if we begin with an equation of the form of (3.25) and operate with P upon it, comparison with (3.42) then yields $P|F(t)\rangle = 0$. This implies $\langle A|F(t)\rangle = \langle A(0)^* F(t)\rangle = 0$ as required. It also implies that $(1-P)|F(t)\rangle = |F(t)\rangle$; thus operating with $1-P$ on (3.25) gives a solution for $A_2(t)$ in terms of $F(t)$; comparing this with the result in (3.41) allows one to identify

(3.46) $$|F(t)\rangle = \exp[t(1-P)iL](1-P)iL | A \rangle .$$

Then using (3.44) we have the fluctuation-dissipation theorem

(3.47) $$K(t) = \langle F(0)|F(t)\rangle = \langle F(0)^* F(t)\rangle$$

as anticipated.

4. – Sound propagation.

We examine the propagation of sound in fluids for situations in which the acoustic wavelength is large in comparison with the « scale of graininess » of the fluid. Roughly speaking this scale can be taken on the order of the molecular mean free path in gases, or, in liquids, as the range of the pair correlation function (a few times the average interparticle separation). The treatment of short-wavelength phenomena is contained in the article by YIP [14] in this volume. The linearized equations of hydrodynamics describing the conservation of mass, momentum and energy in the fluid can be taken as governing the propagation of low-amplitude acoustic disturbances. For a single-component fluid, if we ignore relaxation phenomena, these include the equation of continuity

(4.1) $$\partial \varrho / \partial t + \nabla \cdot \boldsymbol{J} = 0 ,$$

the Navier-Stokes equation

$$(4.2) \qquad \partial \boldsymbol{J}/\partial t = -(1/\varrho_0 K_T)\nabla\varrho - (\beta/K_T)\nabla T + (\eta/\varrho_0)\nabla(\nabla\cdot\boldsymbol{J})$$

and the energy transport law

$$(4.3) \qquad \partial T/\partial t = -[(r-1)/\varrho_0\beta]\nabla\cdot\boldsymbol{J} + (\lambda/\varrho_0 c_v)\nabla^2 T.$$

In these equations ϱ is the density, the average value of which is denoted by ϱ_0, $\boldsymbol{J}$ is the momentum density, *i.e.* $\boldsymbol{J} = \varrho_0\boldsymbol{v}$, where $\boldsymbol{v}$ is the particle velocity, K_T is the isothermal compressibility, β is the coefficient of thermal expansion, η is the longitudinal viscosity, *i.e.*

$$(4.4) \qquad \eta = \eta_v + \tfrac{4}{3}\eta_s,$$

where η_v and η_s are the bulk and shear viscosities, respectively, γ is the ratio of the specific heat at constant pressure c_p to that at constant volume c_v and λ is the thermal conductivity. It is convenient to rewrite these in terms of the spatial Fourier components of the density, momentum density (or current) and temperature ϱ_k, $\boldsymbol{J}_k$ and T_k respectively. These are defined by

$$(4.5) \qquad \varrho_k(t) = \int d\boldsymbol{r}\,\exp\,[+i\boldsymbol{k}\cdot\boldsymbol{r}]\varrho(\boldsymbol{r}, t)$$

with similar formulae for $\boldsymbol{J}_k$ and T_k. Equations (4.1) to (4.3) lead to

$$(4.6) \qquad \partial\varrho_k/\partial t = ikJ_k^L,$$

$$(4.7) \qquad \partial J_k^L/\partial t = ik(1/\varrho_0 K_T)\varrho_k + ik(\beta/K_T)T_k - k^2(\eta/\varrho_0)J_k^L,$$

$$(4.8) \qquad \partial T_k/\partial t = ik[(\gamma-1)/\varrho_0\beta]J_k^L - k^2(\lambda/p_0 c_v)T_k.$$

Here J_k^L is the longitudinal component of $\boldsymbol{J}$, *i.e.* $J_k^L = \boldsymbol{k}\cdot\boldsymbol{J}_k/k$. We can also write an equation of motion for the transverse component of $\boldsymbol{J}_k$, denoted as J_k^T:

$$(4.9) \qquad \partial J_k^T/\partial t = -k^2(\eta_s/\varrho_0)J_k^T.$$

It must be emphasized that these equations apply only in the long-wavelength, that is $k \to 0$, limit. Practically speaking, this means wave vectors $k \leqslant 10^7$ cm^{-1} or wavelengths longer than, say, 100 Å. Generalizations of these hydrodynamic equations, applicable for larger k values, have been developed [15-17], but these will not concern us here.

In view of what we have done in the preceding Section, several comments are in order. Equations (4.6)-(4.9) are essentially in the form of generalized Langevin equations (3.32). The only change needed is to replace the terms involving transport coefficients in (4.7)-(4.9) by convolution integrals. For example (4.9) would become

$$(4.10) \qquad \partial J_k^T / \partial t = - k^2 \int_0^t \mathrm{d}t' \, \nu_s(t - t') \, J_k^T(t') \,.$$

Generalizations of this type are used to include the effects of relaxation and are treated below. For now we shall simply restrict our attention to cases where the memory kernels decay rapidly so that the simple equations (4.6)-(4.9) are adequate.

We should also note that, strictly speaking, the variables ϱ_k, J_k and T_k are not quite suitable for the type of analysis outlined in Sect. 3. This derives from the fact that the variable T_k is *not* the average of a phase function as is required by the derivation leading up to (3.32). The variables ϱ_k and J_k are satisfactory in this respect. Specifically, if a microscopic density is defined by

$$(4.11) \qquad \varrho^\mu(\boldsymbol{r}, t) = \sum_{j=1}^{N} \delta[\boldsymbol{r} - \boldsymbol{r}_j(t)] \,,$$

where the sum runs over the particles in the system, the variable $\varrho_k(t)$ is just the (constrained ensemble) average of its Fourier transform, *i.e.* of

$$(4.12) \qquad \varrho_k^\mu = \sum_{j=1}^{N} \exp [i\boldsymbol{k} \cdot \boldsymbol{r}_j(t)] \,.$$

In a similar fashion forms can be obtained for the longitudinal- and transverse-momentum currents. Suppose the wave vector $\boldsymbol{k}$ is taken to lie along the z-axis, then the phase functions that upon averaging give $J_k^L(t)$ and $J_k^T(t)$ are respectively

$$(4.13) \qquad j_k^L(t) = \sum_j p_j^z(t) \exp [ikz_j(t)]$$

and

$$(4.14) \qquad j_k^T(t) = \sum_j p_j^x(t) \exp [ikz_j(t)] \,,$$

where p_j^x and p_j^z are the x and z components of momentum of the j-th particle.

The solution of the hydrodynamic equations is straightforward and the results are familiar ones. The question of whether to consider the spatial variation for a given harmonic time dependence or the time evolution of a particular spatial Fourier component is largely a matter of taste. The physics of the two situations is understood and the fluid properties entering the analysis (viscosity, compressibility, etc.) are, of course, not dependent upon the method of analysis chosen. We adopt the second procedure here and evaluate the density autocorrelation function (or intermediate scattering function as it is sometimes called)

$$(4.15) \qquad F(k, t) = \langle \varrho_k(0)^* \, \varrho_k(t) \rangle \, .$$

We delay treating the transverse current correlation function

$$(4.16) \qquad C_T(k, t) = \langle J_k^T(0)^* \, J_k^T(t) \rangle$$

until a later Section and we shall not deal directly with either the temperature or the longitudinal current correlation functions at this time as these provide no new information about acoustic processes that is not contained in $F(k, t)$. For instance, from the equation of continuity it follows that

$$(4.17) \qquad k^2 C_L(k, t) = k^2 \langle J_k^L(0)^* J_k^L(t) \rangle = \langle \dot{\varrho}_k(0)^* \, \dot{\varrho}_k(t) \rangle = - \frac{\mathrm{d}^2}{\mathrm{d}t^2} F(k, t) \, ,$$

where the last step follows with the help of (2.7). That we choose to discuss the acoustic phenomena in terms of correlation functions is, at this point, a matter of convenience. However it does emphasize the point that identical information is obtained by studying the dynamics of the fluctuations, in the density (as is done here), and the propagation of an externally driven density disturbance in the fluid.

The calculation of $F(k, t)$ is straightforward [18, 19]; in the case of greatest physical interest where the transport coefficients are not too large and where the thermal conductivity is small ($\lambda/\varrho_0 c_v \ll \eta/\varrho_0$) the result is

$$(4.18) \qquad F(k, t) \approx \langle |\varrho_k(0)|^2 \rangle \left[\left(1 - \frac{1}{\gamma} \right) \exp \left[- \lambda k^2 t/\varrho_0 c_p \right] + \frac{1}{\gamma} \exp \left[- \Gamma_B t \right] \cos \left(c_0 k t \right) \right] ,$$

where

$$(4.19) \qquad c_0 = (\gamma/\varrho_0 \, K_T)^{\frac{1}{2}}$$

is the sound speed,

$$(4.20) \qquad \Gamma_B = \tfrac{1}{2} \left[(\eta/\varrho_0) + (\gamma - 1) \, \lambda/\varrho_0 \, c_p \right] k^2 \approx \eta k^2/2\varrho_0$$

is the attenuation coefficient and we have neglected terms of order $\Gamma_B/c_0 k$. The first term in (4.18) represents the degradation of a density fluctuation by thermal diffusion; the second term corresponds to the propagating sound mode of frequency $\omega_B = c_0 k$. The attenuation coefficient Γ_B is related to the perhaps more common absorption coefficient α by $\alpha \approx \Gamma_B/c_0$ to the same degree of approximation as (4.18), and thus

$$(4.21) \qquad \alpha = \eta \omega_B^2/2\varrho_0 \, c_0^3 \, ,$$

which is just the usual formula.

These results are a reminder that $\eta = \eta_v + 4\eta_s/3$ is the quantity directly accessible in acoustic experiments. From an observational viewpoint, the experimental result is that α generally does not vary as the square of the frequency but exhibits an «unexpected» and thus more interesting behavior [20-22]. Moreover, it is found that the sound speed c is not a constant given by $(\gamma/\varrho_0 K_T)^{\frac{1}{2}}$ but is frequency dependent increasing from this value at low frequencies to a high-frequency limiting value c_∞. These observations are interpreted in terms of a complex frequency-dependent longitudinal viscosity which we designate by $\eta(\omega) + i\zeta(\omega)$. In terms of these, the absorption and dispersion formulae are

(4.22)
$$\alpha \approx \omega^2 \eta(\omega)/2\varrho_0 c(\omega)^3$$

and

(4.23)
$$c(\omega)^2 \approx c_0^2 + \omega\zeta(\omega)/\varrho_0 ,$$

where we have neglected terms of order $(\alpha c/\omega)^2$, i.e. $O(\Gamma_B/c_0^2 k)$, in comparison with unity. The general type of behavior of $\eta(\omega)$ and $\zeta(\omega)$ is shown in Fig. 6.

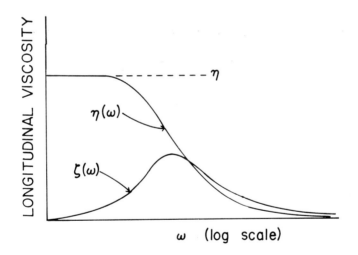

Fig. 6. – General behavior of the real and imaginary parts of the longitudinal viscosity, $\eta(\omega)$ and $\zeta(\omega)$ repectively, as a function of the frequency.

By interpreting experimental results in this form the acoustic spectroscopist attempts to get an understanding of molecular motions and processes in fluids. It is therefore essential to investigate what aspects of molecular behavior are reflected in the shear and bulk viscosities.

The dynamical shear viscosity is defined by (4.10); denoting the real and imaginary parts by $\eta_s(\omega)$ and $\zeta_s(\omega)$ respectively, we have

(4.24) $$\eta_s(\omega) + i\zeta_s(\omega) = \varrho_0 \hat{v}_s(-i\omega) = \varrho_0 \int_0^{\infty} dt \, \exp\left[i\omega t\right] v_s(t) .$$

The key to the dynamics is contained within the memory function $v_s(t)$. From the preceding Section we know that $v_s(t)$ can be written as the time correlation function of an appropriate random force $F(t)$; our task is thus to identify the physical meaning and microscopic form of that force.

The phase function corresponding to the transverse-momentum current has been given in (4.14); it satisfies a generalized Langevin equation of the form of (3.25) and thus the memory kernel is given by

(4.25) $$v_s(t)\, k^2 = \langle (1-P)\, iLj_k^T | \exp\left[t(1-P)\, iL\right] | (1-P)\, iLj_k^T \rangle / \langle |j_k^T|^2 \rangle$$

as indicated by (3.45)-(3.47). If we make use of the definition of P, (3.37), it is clear that

$$|(1-P)\, iLj_k^T\rangle = |iLj_k^T\rangle - |j_k^T\rangle \langle j_k^T | iLj_k^T\rangle = |iLj_k^T\rangle .$$

Using (4.14) explicitly and taking the long-wavelength $(k \to 0)$ limit gives

(4.26) $$|(1-P)\, iLj_k^T\rangle = \sum_j ik\,[p_j^x \dot{z}_j + \dot{p}_j^x z_j] + O(k^2) ,$$

where we have used the fact that there is no net force on the system to set $\sum_j \dot{p}_j^x = 0$. Then we observe that to lowest order in k

(4.27) $$\exp\left[t(1-P)\, iL\right] \to \exp\left[tiL\right]$$

in (4.25), so that

(4.28) $$v_s(t) = \frac{1}{Nk_{\mathrm{B}}T} \langle \sigma^{xz}(0)^* \sigma^{xz}(t)\rangle ,$$

where we have substituted $\langle |j_k^T|^2\rangle = Nk_{\mathrm{B}}T$ (for $k \to 0$) and have set

(4.29) $$\sigma^{xz} = \sum \frac{d}{dt}\,(p_j^x z_j) .$$

The dynamic shear viscosity (4.24) can now be written as

(4.30) $$\eta_s(\omega) + i\zeta_s(\omega) = (Vk_{\mathrm{B}}T)^{-1}\int_0^{\infty} dt \, \exp\left[-i\omega t\right] \langle \sigma^{xz}(0)^* \sigma^{xz}(t)\rangle ,$$

where V is the volume of the system. The physical interpretation of σ^{xz} can be seen by writing the Navier-Stokes equation in the form $\mathrm{d}\boldsymbol{J}/\mathrm{d}t = -\nabla\cdot\boldsymbol{\sigma}$, where $\boldsymbol{\sigma}$ is the stress tensor. For the transverse current J^x_k this gives, via eqs. (4.26) and (4.29), in the limit $k \to 0$

(4.31) $$\frac{\mathrm{d}J^x_k}{\mathrm{d}t} = ik\sigma^{xz} ;$$

σ^{xz} is an element of the stress tensor. The frequency dependence of the shear viscosity thus arises from the finite relaxation time of shear stresses. Note that the ordinary, i.e. $\omega \to 0$, shear viscosity is

(4.32) $$\eta_s = (Vk_\mathrm{B}T)^{-1}\int_0^\infty \mathrm{d}t\,\langle\sigma^{xz}(0)^*\,\sigma^{xz}(t)\rangle ,$$

which agrees with the familiar visco-elastic formula [20]

(4.33) $$\eta_s = G_\infty \tau_s ,$$

in which τ_s is the average shear relaxation time:

(4.34) $$\tau_s = \int_0^\infty \mathrm{d}t\,\langle\sigma^{xz}(0)^*\,\sigma^{xz}(t)\rangle / \langle|\sigma^{xz}(0)|^2\rangle ,$$

and

(4.35) $$G_\infty = \langle|\sigma^{xz}(0)|^2\rangle / Vk_\mathrm{B}T$$

is the high-frequency ($k \to 0$, $\omega \to \infty$) shear rigidity.

A similar sort of argument can be employed to obtain the bulk viscosity (and hence the longitudinal viscosity). Equations (4.2) and (4.7) are generalized by replacing the viscous term by a convolution integral in which the memory kernel is of the form $\nu_v(t) + 4\nu_s(t)/3$; by a procedure analogous to that leading to (4.30), an expression for $\nu_v(t)$ and its transform $\eta_v(\omega) + i\zeta_v(\omega)$ can be obtained. The result is

(4.36) $$\eta_v(\omega) + i\zeta_v(\omega) = (Vk_\mathrm{B}T)^{-1}\int_0^\infty \mathrm{d}t\,\exp\left[i\omega t\right]\langle\Delta\sigma(0)^*\Delta\sigma(t)\rangle ,$$

where $\Delta\sigma$ is the *fluctuation* in the mean normal stress. That is

(4.37) $$\Delta\sigma = \tfrac{1}{3}\mathrm{Tr}\,\boldsymbol{\sigma} - \tfrac{1}{3}\langle\mathrm{Tr}\,\boldsymbol{\sigma}\rangle ,$$

and is given by

(4.38)
$$\Delta\sigma = \frac{\mathrm{d}}{\mathrm{d}t}\left(\sum_j \boldsymbol{p}\cdot r_j/3\right) - PV - \frac{\partial PV}{\partial\langle E\rangle}(E - \langle E\rangle).$$

In this last equation P is the pressure of the N-particle system with volume V and average energy $\langle E\rangle$, *i.e.* temperature T. Note that this expression is appropriate for canonical-ensemble calculations [1]. In the microcanonical ensemble the last term would be absent, while in the grand canonical ensemble a term $-(\partial PV/\partial\langle N\rangle)(N - \langle N\rangle)$ would be included. It is clear that

$$\langle(\mathrm{d}/\mathrm{d}t)\sum_j(\boldsymbol{p}_j\cdot\boldsymbol{r}_j/3)\rangle = PV,$$

so that $\Delta\sigma$ is essentially the fluctuation in the pressure.

The shear and bulk viscosities are seen to be determined by time correlation functions of *collective* variables; that is, the correlation functions are of the form

(4.39)
$$\sum_{j,l}\langle\boldsymbol{O}_l(0)\cdot\boldsymbol{O}_j(t)\rangle,$$

where $\boldsymbol{O}_j(t)$ is determined by the momentum and co-ordinates of the j-th molecule at time t. The terms in (4.39) describing co-operative effects, *i.e.* the $l \neq j$ terms, are not zero, and, in fact, chiefly govern the behavior. This complicates the situation considerably in comparison with a number of other relaxation processes, *e.g.* nuclear quadrupole relaxation, in which only the single-particle terms are of importance. Largely as a result of this, the calculation of the dynamic viscosities in a fundamental way, that is starting from an intermolecular potential, has proven to be intractable. In some instances one can model the relevant dynamics in terms of some ordering parameter which obeys some assumed equation of motion and is coupled dynamically to the quantities σ^{xz} and/or $\Delta\sigma$. Generally speaking, however, probably the major value of expressing viscosities in terms of time correlation functions is that one understands (in principle, at least) what a measurement of, *e.g.*, $\eta_s(\omega)$ is saying about the fluid and the data can therefore be interpreted in this light. Moreover, it enables one to make reasonable comparisons with other relaxational data that probe a different aspect of the molecular motions in the fluid.

That ultrasonic absorption data reflect frequency-dependent viscosity coefficients is not surprising—the *ad hoc* introduction of a relaxing bulk viscosity has long been used to rationalize this kind of experimental result [23]. The point here is that definite physical notions are involved; the observed frequency dependence is caused by specific dynamical behavior in the fluid. This point has been made rather eloquently by ZWANZIG [24] in an article in which he examines the relaxation of the bulk viscosity caused by weak coupling between internal and external degrees of freedom.

Without going into too many details, it is instructive to reproduce a few of the steps in that derivation as they indicate the kind of approximations that are involved and also at what point a model of the kinetics must be inserted. By allowing only weak coupling between internal and external co-ordinates, the pressure fluctuation $\Delta\sigma$ in (4.38) can be separated into parts which depend separately on these co-ordinates. The external co-ordinates are assumed to equilibrate rapidly and thus contribute a constant « background » to the viscosity in the frequency range of importance. The internal co-ordinates appear only in the last term on the right of (4.38) and lead to a relaxational viscosity given by

$$(4.40) \qquad (Vk_{\mathrm{B}}T)^{-1}\left(\frac{\partial PV}{\partial\langle E\rangle}\right)^{2}\int_{0}^{\infty}dt\,\exp\left[i\omega t\right]\langle\Delta E_{i}(0)\,\Delta E_{i}(t)\rangle\,,$$

where ΔE_i is the fluctuation in the energy of the internal degrees of freedom. By using some standard thermodynamic formulae and writing

$$(4.41) \qquad \langle\Delta E_{i}(0)\,\Delta E_{i}(t)\rangle = \langle\Delta E_{i}(0)^{2}\rangle\,\psi(t) = k_{\mathrm{B}}T^{2}\,c_{i}\,\psi(t)\,,$$

where c_i is the internal specific heat, one gets for the expression in (4.40)

$$(4.42) \qquad \frac{1}{K_{T}}\frac{c_{p}-c_{v}}{c_{v}-c_{i}}\frac{c_{i}}{c_{v}}\int_{0}^{\infty}dt\,\exp\left[i\omega t\right]\psi(t)\,.$$

If we introduce an exponential ansatz, $\psi(t) = \exp\left[-t/\tau\right]$, the familiar single relaxation time frequency variation $\tau/(1 - i\omega\tau)$ is recovered [25, 26]. Rather than introducing an *ad hoc* form for $\psi(t)$, one can model the dynamics for specific instances. For vibrational relaxation, for example, a harmonic-oscillator system with the dynamical behavior treated in first-order perturbation theory leads to an exponential form for $\psi(t)$ [20]. Similar procedures can be used to describe the relaxation associated with fluctuations in the relative populations of two (or more) energetically differing structural forms of certain molecules [21]. The point is that, at some point in the calculation, the Hamiltonian of the specific system must be specified and the resulting dynamical problem solved. The relaxation of an internal degree of freedom presents a relatively simple calculational problem because the co-operative terms, *i.e.* the $l \neq j$ terms in (4.39), do not contribute.

This simplification cannot be extended to the case of structural relaxation in liquids where it is the dynamics of the external co-ordinates that govern the situation. If we ignore the contribution from the energy fluctuation term in (4.38), *i.e.* the relaxing « structural specific heat », the problem is to understand the type of molecular motions described by the correlation function $\langle \Delta\sigma(0)^* \Delta\sigma(t) \rangle$ with $\Delta\sigma = \frac{1}{3} \sum_j (d/dt)(\boldsymbol{p}_j \cdot \boldsymbol{r}_j) - PV$.

From an experimental standpoint there are a number of clues available. These come largely from measurements made on liquids of relatively high viscosity undercooled below their normal freezing points and may not be generally characteristic of « normal » liquids [27]. The order of magnitude of the relaxation time

$$(4.43) \qquad \tau_v = \int_0^\infty dt \, \langle \Delta\sigma(0)^* \Delta\sigma(t) \rangle / \langle |\Delta\sigma|^2 \rangle$$

is comparable with that associated with the relaxation of the single-particle velocity autocorrelation function [22], so that we may reasonably suppose that relatively short-range motions involving relatively small groups of molecules are important. Moreover, since generally

$$(4.44) \qquad \tau_v \approx \tau_s$$

with τ_s as defined in (4.34), it follows that similar processes are involved in the relaxation of shear stresses. The short-range character of the relaxation mechanism follows also from the experimentally observed k-independence of the bulk and shear viscosities as well as the relaxation times. Only in molecular-dynamics calculations which examine wave vectors on the order of 10^{-7} cm^{-1} do these quantities become functions of k [17]. An additional fact is that the normalized correlation function $\langle \Delta\sigma(0)^* \Delta\sigma(t) \rangle / \langle |\Delta\sigma|^2 \rangle$ depends on temperature almost solely through the variation of τ_v, the shape remaining unchanged. That is, if one were to write the correlation function in terms of a reduced time t/τ_v, it would be essentially independent of the temperature. Moreover, this shape varies remarkably little from one liquid to another. It consists of a rapid drop-off for short times $t < \tau_v$ and goes over to an essentially exponentially decay for longer times. This is illustrated in Fig. 7.

A commonly employed practice is to represent this kind of nonexponential behavior in terms of a « distribution of relaxation times » $f(\tau)$:

$$(4.45) \qquad c_\sigma(t) = \int_0^\infty d\tau \, f(\tau) \exp[-t/\tau].$$

A simple two-parameter function that is often used and which fits the data rather well is the Davidson-Cole [29] function

$$(4.46) \qquad f(\tau) = \begin{cases} \dfrac{\sin b\pi}{\pi\tau} \left(\dfrac{\tau/\tau_0}{1 - \tau/\tau_0} \right)^b & (0 \leqslant \tau < \tau_0), \\ 0 & (\tau \geqslant \tau_0). \end{cases}$$

The parameter τ_0 is closely related to the relaxation time τ_v ($\tau_v = b\tau_0$) and b is a width parameter which varies from zero to unity and describes the deviation of the correlation function for a simple exponential decay. (As $b \to 1$, $f(\tau) \to \delta(\tau - \tau_0)$ and thus $c_\sigma(t) \to \exp[-t/\tau_0]$.) Typically a value of $b \approx \frac{1}{2}$ is used to describe the data.

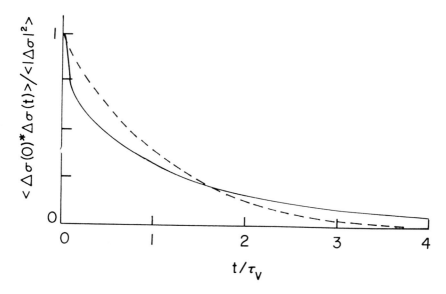

Fig. 7. – The general shape of the correlation function $\langle \Delta\sigma(0)^* \Delta\sigma(t) \rangle$ (normalized to unity at $t = 0$) obtained in acoustic experiments, plotted vs. t/τ_v. The dashed curve is an exponential decay $\exp[-t/\tau_v]$.

While this kind of scheme is useful in terms of classifying and parametrizing the experimental results, it contributes little to our understanding of the molecular nature of structural relaxation. Various attempts to model the dynamics of structural relaxation have been published in the last few years, but, while these do a reasonably good job of fitting the data, they do not contribute much to our understanding of the molecular motions involved. The type of approach that is generally used is to relate the structural relaxation to the dynamics of an ordering parameter (which is usually not well defined). The ordering parameter is assumed to obey a relatively simple equation of motion (often a diffusion equation) subject to certain assumed boundary and/or initial conditions. One next writes the coupling to the stress as some nonlinear relationship and turns the crank. If things have been chosen « properly », the resulting form for the correlation function will agree rather well with the experimental results. The utility of these models is that they phrase the problem of structural relaxation in terms of a single parameter « characteristic of the local order in the liquid »—an intuitively satisfying notion. One anticipates that this parameter will explicitly focus on relatively small groups of molecules, for which approximate dynamical modes may be formulated. The problem is, of course, to arrive at some precise specification of the ordering parameter and to justify the manner in which it couples to the stress.

5. – Collective-mode dynamics.

The dynamical behavior of molecules in fluids are specified, as we have seen, in terms of a number of modes or phase functions. Generally speaking, we are concerned with the spontaneous fluctuations in quantities whose equilibrium average values vanish. The dynamics of these fluctuations are described in terms of a time correlation function; if A, the phase function corresponding to the quantity of interest, is written in the form

$$(5.1) \qquad A = \sum_{i=1}^{N} a_i(t),$$

where the sum runs over the molecules in the system, then the autocorrelation function is of the form $\langle A(0)^* A(t) \rangle = \sum_{i,j} \langle a_i(0)^* a_j(t) \rangle$. The sum is separable into two parts—the self- and distinct terms—and is thus

$$(5.2) \qquad \langle A(0)^* A(t) \rangle = \sum_{i=1}^{N} \langle a_i(0)^* a_i(t) \rangle + N \sum_{j=2}^{N} \langle a_1(0)^* a_j(t) \rangle.$$

For those quantities for which the distinct term, *i.e.* the second term in (5.2), vanishes, the variable A is said to describe a *single-particle* mode of the system. If the distinct term is important, the mode is termed *collective*. In this Section we shall examine two such collective modes—the transverse- and longitudinal-momentum currents, J_k^T and J_k^L, defined in (4.13) and (4.14). The correlation functions corresponding to these are

$$C_T(k, t) = \langle J_k^T(0)^* J_k^T(t) \rangle \qquad \text{and} \qquad C_L(k, t) = \langle J_k^L(0)^* J_k^L(t) \rangle,$$

respectively.

Let us examine the transverse currents first as the situation here is a bit simpler. In the hydrodynamic ($k \to 0$) regime, the correlation function obeys an equation of motion

$$(5.3) \qquad \frac{d}{dt} C_T(k, t) = - k^2 \int_0^t dt' \, \nu_s(t - t') \, C_T(k, t'),$$

which is easily obtained from (4.10) and serves as a definition of the dynamic shear viscosity. The power spectrum of the transverse currents is just $C_T(k, -i\omega)$ and this is easily expressed in terms of the complex frequency-dependent shear viscosity $\hat{v}_s(-i\omega) = \varrho_0[\eta_s(\omega) + i\zeta_s(\omega)]$:

$$(5.4) \qquad C_T(k, -i\omega) = C_T(k, 0)[-i\omega + k^2\hat{v}_s(-i\omega)]^{-1} .$$

This kind of expression contains the usual visco-elastic behavior as can be seen by taking an exponential ansatz for the memory function, *i.e.*

$$v_s(t) \propto \exp[-t/\tau_s] ,$$

which gives

$$(5.5) \qquad \eta_s(\omega) + i\zeta_s(\omega) = \eta_s/(1 - i\omega\tau) .$$

If τ_s is very short compared to the time scale of interest (*i.e.* $\omega\tau_s \ll 1$), then $\eta_s(\omega) \approx \eta_s$ and

$$(5.6) \qquad C_T(k, t) \simeq C_T(k, 0) \exp[-\eta_s k^2 t/\varrho_0] .$$

The shear currents simply decay as an « evanescent wave » as expected. (The spatial character of the « wave » is contained in the factor k^2 appearing in the exponential.) In the opposite limit of a long relaxation time, say $\eta_s k^2 \tau_s/\varrho_0 \gg 1$, the result is

$$(5.7) \qquad C_T(k, t) \simeq C_T(k, 0) \exp[-t/2\tau_s] \cos\left(\sqrt{\eta_s/\varrho_0\tau_s} \, kt\right) .$$

This implies « solidlike » elastic-wave propagation at a speed $c_{s\infty} = \sqrt{\eta_s/\varrho_0\tau_s}$. In terms of usual visco-elastic parameters

$$(5.8) \qquad \eta_s/\tau_s = G_\infty = \varrho_0 c_{s\infty}^2 ,$$

where G_∞ is the high-frequency (but $k \to 0$) limiting shear rigidity. The region intermediate between these two limits is that of visco-elastic relaxation. Of course the exponential form of $v_s(t)$ is not generally appropriate; still the two limiting cases described above, *i.e.* of $\eta_s k^2 \tau_s/\varrho_0 \ll 1$ and $\eta_s k^2 \tau_s/\varrho_0 \gg 1$, are correct. ($\tau_s$ should now be interpreted as the average relaxation time, as in (4.34).) The frequency variation of $\eta_s(\omega) + i\zeta_s(\omega)$ in the visco-elastic region will, of course, be more complicated. We shall return to this point.

Next we look at the longitudinal currents; these are somewhat more complicated as they are coupled through the hydrodynamic equations to variations in density and temperature. If we ignore the term in the energy transport equation involving the thermal conductivity (we have seen that this simply gives an exponential decay term corresponding to thermal diffusion), then we can obtain a rather straightforward interpretation of $C_L(k, t)$. Laplace transforming as before (and noting that $\langle J_k^L(0)^* \varrho_k(0) \rangle = 0$, etc.) gives

$$(5.9) \qquad \hat{C}_L(k, -i\omega) = -i\omega C_L(k, 0)/[-\omega^2 + i\omega k^2 \hat{v}(-i\omega) + c_0^2 k^2],$$

where $\hat{v}(-i\omega) = \eta(\omega) + i\zeta(\omega)$ is the complex longitudinal viscosity and is the Laplace transform of the memory function involved in generalizing the viscous term in (4.7) to the form of (4.10). We can invert this to obtain $C_L(k, t)$. Observe that the quadratic nature of the denominator ensures the existence of propagating waves even in the limit of « fast » relaxation. In this limit we get the « usual » result

$$(5.10) \qquad C_L(k, t) = C_L(k, 0) \exp[-\eta k^2 t/2\varrho_0] \cos(c_0 kt).$$

When $v(t)$ relaxes slowly, things become considerably more complicated. But if we examine the limiting case $\tau \equiv \int_0^\infty dt v(t)/v(0) \gg 1/c_0 k$, we can obtain a useful approximation. We let c_∞ be the high-frequency limiting longitudinal wave speed and make use of the visco-elastic formula

$$(5.11) \qquad \eta/\varrho_0 = (c_\infty^2 - c_0^2)\,\tau\,;$$

then we get for the correlation function a sum of two types of terms:

1) Wave propagation terms which are of the form

$$\exp[-\eta t/\varrho_0 c_\infty^2 \,\tau^2] \cos(c_\infty kt).$$

Note that in this limit the attenuation factor is independent of wave number.

2) A purely decaying term which is determined almost exclusively by the form of $v(t)$. For example, if $v(t) \propto \exp[-t/\tau]$, then this term is [35]

$$\exp[-c_0^2 t/c_\infty^2 \,\tau].$$

It is clear that the dynamical behavior of these currents is closely related to the acoustical properties of the medium. Consequently it follows that acoustics provides a valuable probe for eliciting these dynamics. The equivalent information, *i.e.* the form of $v_s(t)$ and $v(t)$, are available in a number of other experiments. In particular optical scattering [18, 35] and thermal-neutron scattering [36] provide means of studying the longitudinal currents. Whereas acoustical studies are essentially limited to the range $\omega \leqslant 10^{10}\,\mathrm{s}^{-1}$, $k \leqslant 10^5\,\mathrm{cm}^{-1}$, these other techniques extend this ω, k range to $\approx 10^{14}\,\mathrm{s}^{-1}$ and $10^9\,\mathrm{cm}^{-1}$ respectively. Therefore essentially the entire range of frequencies (time) and wave vectors (distances) important in the study of the molecular motions associated with longitudinal currents is accessible.

In the case of transverse currents the experimental picture is not nearly so bright. Acoustical techniques still provide a means of probing the hydrodynamic regime up to frequencies on the order of, say, $2 \cdot 10^9$ Hz ($\omega \sim 10^{10}$ s^{-1}); thus the form of $\nu_s(t)$ can be inferred for time scales as short as approximately 10^{-10} s. Unfortunately, this means that the decay of $\nu_s(t)$ can only be observed in rather viscous liquids, typically $\eta_s \geqslant 2$ to 10 poise. At present, however, acoustics provides the only experimental method of investigating transverse currents in fluids. (Several authors [37-38] have suggested that depolarized light scattering is also sensitive to these currents, but the experimental confirmation of this is still somewhat uncertain.) The only nonacoustic information available is that derived from the so-called molecular-dynamics (MD) computer « experiments ». It is interesting to compare the results of these experiments with what one would expect on the basis of acoustical data.

The details of molecular-dynamics experiments are rather well delineated in the literature [39, 40]. The basic idea is that one uses a high-speed computer to solve the classical equations of motion for N interacting particles in volume V at a temperature T (determined in terms of the mean kinetic energy per particle) subject to periodic boundary conditions. The number of particles is usually taken $\sim 10^3$, the volume is adjusted to simulate liquid densities, and initial conditions are chosen to give the temperature. The interaction between molecules is generally taken to be given in terms of a pair potential; for example, the Lennard-Jones potential

$$\varphi(R) = -4\varepsilon[(\sigma_0/R)^6 - (\sigma_0/R)^{12}],$$

where ε and σ_0 are constants and R is the interparticle separation, is often used to simulate the noble gases. The output of such a computation is a set of numbers giving the positions and momenta of each of the N particles at various instants of time usually separated by intervals $\sim 10^{-15}$ s. Of themselves, these numbers are not of much interest; what is of interest are the correlation functions that can be constructed from them.

RAHMAN [41, 42] has determined the current correlations in several MD experiments; AILAWADI, RAHMAN and ZWANZIG [17] have carried through a particularly thorough analysis of current correlations obtained using MD on a system chosen to simulate slightly undercooled liquid argon ($\varepsilon/k_B = 120$ °K, $\sigma_0 = 3.4$ Å, mass per particle = atomic mass of argon) at a density 1.407 g/cm^3 and a temperature 76 °K. Because of the periodic boundary conditions, reliable data can be obtained only for $k \geqslant 2/V^{\frac{1}{3}}$. In this case the smallest k values reported were $k = 5 \cdot 10^7$ cm^{-1}, clearly out of the hydrodynamic regime. Nevertheless, the comparison with acoustical data is interesting. Typical results for the current correlation functions are given in Fig. 8. The most pronounced difference between the longitudinal and transverse current correlations is the obvious oscillatory character of the former which is barely apparent in the latter. From these it is possible to infer sound speeds and attenuation coefficients.

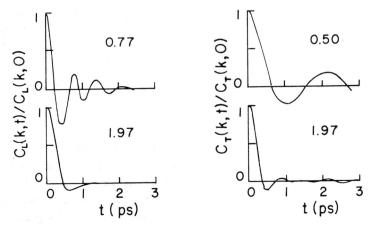

Fig. 8. – Normalized longitudinal and transverse current correlation functions (adapted from ref. [17]) plotted for several values of wave vector k. The values of k are given on the plots in units of 10^8 cm^{-1}.

For example for the transverse currents one can obtain $c_{s\infty}(k) \approx 6 \cdot 10^4$ cm/s and thus $G_\infty(k) \approx 5 \cdot 10^9$ dyn/cm^2. (Note that explicit dependences of $c_{s\infty}$ and G_∞ on k are indicated since this is what one should expect for such non-hydrodynamic, *i.e.* large, wave vectors.) These results are rather reasonable in view of the calculation of ZWANZIG and MOUNTAIN [43] who obtain values of this same order of magnitude in the $k \to 0$ limit for a high-density «argon-like» Lennard-Jones fluid. AILAWADI *et al.* also use these data to estimate k-dependent zero-frequency shear and longitudinal viscosities. Extrapolating their data to $k = 0$ gives $\eta_s \approx 3 \cdot 10^{-3}$ P and $\eta \approx 5.6 \cdot 10^{-3}$ P; these compare favorably with the experimental results of NAUGLE, LUNSFORD and SINGER [44]. From these one can use (4.33) to estimate a shear relaxation time of $\tau_s \approx 6 \cdot 10^{-13}$ s. This result is quite reasonable in view of what would be expected on the basis of (5.7).

What is perhaps most interesting, however, is the observation of AILAWADI *et al.* that their data are most satisfactorily represented if the memory functions $\nu(t)$ and $\nu_s(t)$ are taken to be of a Gaussian form, *e.g.*

$$(5.12) \qquad\qquad \nu_s(t) \propto \exp\left[-\alpha(k)\,t^2\right],$$

where $\alpha(k)$ is a fitting parameter. There is apparently no hint of the long-time exponential «tail» (see Fig. 7) which so dominates the acoustically obtained data in complex molecular systems. The origin of this strikingly different behavior in the two cases is not clear. One may hazard the guess that the exponential tail in the molecular systems may be associated with a coupling between molecular reorientation and shear flow, which would of course not appear for the spherically symmetric argon molecules. A great deal more experimental work is needed, however, before any such explanation can be put forth with confidence.

* * *

The support of the National Science Foundation (USA) is gratefully acknowledged by the author.

REFERENCES

[1] R. Zwanzig: *Ann. Rev. Phys. Chem.*, **16**, 67 (1965).
[2] B. J. Berne and D. Forster: *Ann. Rev. Phys. Chem.*, **22**, 563 (1971).
[3] B. J. Berne: *Physical Chemistry*, edited by D. Henderson, Vol. **8** B, Chap. 9 (New York, N. Y., 1971).
[4] R. Kubo: *Journ. Phys. Soc. Japan*, **12**, 570 (1957).
[5] R. Kubo: *Lectures in Theoretical Physics*, Vol. **1** (New York, N. Y., 1961), p. 120.
[6] R. Kubo: *Rep. Prog. Phys.*, **29**, 255 (1965).
[7] J. M. Luttinger: *Phys. Rev.*, **135**, A 1505 (1964).
[8] H. B. Callen and T. A. Welton: *Phys. Rev.*, **83**, 34 (1951).
[9] W. Bernard and H. B. Callen: *Rev. Mod. Phys.*, **31**, 1017 (1959).
[10] L. D. Landau and E. M. Lifshitz: *Statistical Physics*, Chap. 12 (Reading, Mass., 1958).
[11] H. Mori: *Prog. Theor. Phys.*, **33**, 423 (1965); **34**, 399 (1965).
[12] R. Zwanzig: *Lectures in Theoretical Physics*, Vol. **3** (New York, N. Y., 1961), p. 135.
[13] B. J. Berne, J. P. Boon and S. A. Rice: *Journ. Chem. Phys.*, **45**, 1086 (1966).
[14] S. Yip: this volume, p. 55.
[15] C. H. Chung and S. Yip: *Phys. Rev.*, **182**, 323 (1969).
[16] A. Z. Akcasu and E. Daniels: *Phys. Rev. A*, **2**, 962 (1970).
[17] N. K. Ailawadi, A. Rahman and R. Zwanzig: *Phys. Rev. A*, **4**, 1616 (1971).
[18] R. D. Mountain: *Rev. Mod. Phys.*, **38**, 205 (1966).
[19] C. J. Montrose, V. A. Solovyev and T. A. Litovitz: *Journ. Acoust. Soc. Amer.*, **43**, 117 (1968).
[20] K. F. Herzfeld and T. A. Litovitz: *Absorption and Dispersion of Ultrasonic Waves* (New York, N. Y., 1959).
[21] *Physical Acoustics*, edited by W. P. Mason, Vol. **2** A (New York, N. Y., 1965).
[22] R. T. Beyer and S. V. Letcher: *Physical Ultrasonics* (New York, N. Y., 1969).
[23] See, for example, R. Meister, C. Marhoeffer, R. Sciamanda, L. Cotter and T. A. Litovitz: *Journ. Appl. Phys.*, **31**, 854 (1960).
[24] R. Zwanzig: *Journ. Chem. Phys.*, **43**, 714 (1965).
[25] K. F. Herzfeld and F. O. Rice: *Phys. Rev.*, **31**, 691 (1928).
[26] H. O. Kneser: *Ann. der Phys.*, **16**, 360 (1933).

[27] See, for example, J. LAMB: in *Molecular Motions in Liquids*, edited by J. LASCOMBE (Boston, Mass., 1974).

[28] C. J. MONTROSE and T. A. LITOVITZ: in *Neutron Inelastic Scattering*, Vol. 1 (Vienna, 1968), p. 623.

[29] D. W. DAVIDSON and R. H. COLE: *Journ. Chem. Phys.*, **19**, 1484 (1951).

[30] A. J. BARLOW, A. ERGINSAV and J. LAMB: *Proc. Roy. Soc.*, A **298**, 481 (1967).

[31] M. C. PHILLIPS, A. J. BARLOW and J. LAMB: *Proc. Roy. Soc.*, A **329**, 193 (1972).

[32] M. A. ISAKOVICH and I. A. CHABAN: *Sov. Phys. Dokl.*, **10**, 1055 (1966); *Sov. Phys. JETP*, **23**, 893 (1966).

[33] V. P. ROMANOV and V. A. SOLOVYEV: *Sov. Phys. Acoust.*, **11**, 68 (1965).

[34] C. J. MONTROSE and T. A. LITOVITZ: *Journ. Acoust. Soc. Amer.*, **47**, 1250 (1970).

[35] R. D. MOUNTAIN: *Journ. Res. Nat. Bur. Stand.*, A **70**, 207 (1966).

[36] *Thermal Neutron Scattering*, edited by P. A. EGELSTAFF (New York, N. Y., 1965).

[37] T. KEYES and D. KIVELSON: *Journ. Chem. Phys.*, **54**, 1786 (1971).

[38] H. C. ANDERSEN and R. PECORA: *Journ. Chem. Phys.*, **54**, 2584 (1971).

[39] B. J. ALDER and T. E. WAIWRIGHT: *Journ. Chem. Phys.*, **31**, 459 (1959).

[40] A. RAHMAN: *Phys. Rev.*, **136**, A 405 (1964).

[41] A. RAHMAN: *Phys. Rev. Lett.*, **19**, 420 (1967).

[42] A. RAHMAN: in *Neutron Inelastic Scattering*, Vol. 1 (Vienna, 1968), p. 561.

[43] R. ZWANZIG and R. D. MOUNTAIN: *Journ. Chem. Phys.*, **43**, 4464 (1965).

[44] D. G. NAUGLE, J. H. LUNSFORD and J. R. SINGER: *Journ. Chem. Phys.*, **45**, 4669 (1966).

High-Frequency Short-Wavelength Fluctuations in Fluids.

S. Yip

Department of Nuclear Engineering
Massachusetts Institute of Technology - Cambridge, Mass.

List of symbols.

c	sound speed,
$\boldsymbol{c}$	dimensionless velocity,
c_0	adiabatic sound speed (low-frequency limit),
$c_0(k)$	wavelength-dependent adiabatic sound speed,
$c_\infty(k)$	high-frequency sound speed,
$C(k)$	direct correlation function,
$c_B(k),\ c_l(k),\ c_a(k)$	propagation speeds (see Sect. **4**),
D_T	thermal diffusivity,
D_l	longitudinal viscosity,
$D'(k,\omega),\ D''(k,\omega)$	real and imaginary parts of generalized susceptibility,
f	Eucken number,
$f(\boldsymbol{r},\boldsymbol{v},t)$	distribution function,
$f_0(c)$	Maxwellian velocity distribution function,
$J[f]$	collision integral,
$J(k,\omega)$	longitudinal-current correlation function,
k	wave number,
l_c	collision mean free path,
n	equilibrium number density,
P	pressure,
P_{zz}	zz-component of pressure field,
r	dimensionless frequency parameter,
$\boldsymbol{r}$	position vector,
S	entropy,
$S(k)$	structure factor,
$S(k,\omega)$	density correlation function (dynamic structure factor),
T	temperature,
v_0	thermal speed,
x	dimensionless frequency parameter,
y	wavelength to collision mean free path ratio,
Z	$x + iy$,
α	sound attenuation constant,
β	inverse temperature in units of Boltzmann's constant,
γ	specific-heat ratio,

Γ	classical sound attenuation coefficient,
δ	dimensionless wave number,
η	dimensionless sound attenuation constant,
η_s	shear viscosity,
η_v	bulk viscosity,
$\varkappa$	thermal conductivity,
λ	sound wavelength,
ν	collision frequency,
ξ	dimensionless frequency,
$\tau(k)$	shear relaxation time,
ω	frequency.

1. – Introduction.

The concept of sound propagation in fluids can be simply described as coherent motions involving groups of molecules. It is not difficult to visualize that the random molecular motions can become organized as a result of inter-molecular collisions. As long as the wavelength of the sound wave is much longer than the collision mean free path in the fluid, the system will be able to sustain the disturbance and one can speak of propagation. However, when the phenomenon is examined more closely one finds subtleties that depend on the structural and dynamical properties of the fluid. For this reason the study of sound dispersion and attenuation in simple fluids is a subject of considerable interest in nonequilibrium statistical mechanics.

The processes giving rise to sound dispersion and attenuation are clearly dynamical in nature. The damping of sound waves depends on the dissipative processes in the system. At low frequencies and in simple fluids these are known to be viscous flow and thermal conduction. Sound dispersion, on the other hand, refers to deviation in the sound speed from the low-frequency value. This occurs when the frequency of the disturbance is such that the fluctuations in the fluid can no longer be calculated from thermodynamics. Although one generally speaks of frequency dependence in sound propagation, one can equally well consider wave number dependence as in the time decay of spontaneous density fluctuations of different wavelengths. At sufficiently large wave numbers one can anticipate significant differences between the propagation characteristics of a dense fluid and those of a dilute gas since the effects of the local structure in a dense fluid now become important.

The purpose of these lectures is to examine those processes which give rise to sound dispersion and attenuation in different frequency and wave number regions. In particular we are concerned with the understanding of the propagation of density and current fluctuations at high frequencies and short wavelengths. It is clear that we need to deal with the theory of thermal fluctuations at the molecular level if we are to obtain results valid at arbitrary frequencies and wave numbers. Moreover, this theory must take into account the existence of local order in a dense fluid if we are to explain the distinction between propagation in a dense fluid and a dilute gas. Since our intention is only to illustrate the essential ideas, we will restrict our consideration to simple gases and liquids, and use, whenever possible, simplified models.

The study of the behavior of sound dispersion and attenuation at high frequencies and short wavelengths leads naturally to the question of the nature of sound under these conditions. It is helpful to consider different frequency (ω) and wave number (k) regions in such a discussion. The region of small (k, ω) is the hydrodynamic region, where wavelengths are long compared to the mean free path and frequencies small compared to collision frequency. Here sound propagation is well understood in terms of dispersion relations derived from the hydrodynamic equations. The opposite region of very large (k, ω) is the Knudsen region where the effects of free-molecule flow predominate over the effects of collision. In this region the hydrodynamic equations completely break down and it is questionable that sound, in the conventional sense, exists. The intermediate region is the kinetic or transition region where the wavelengths and frequencies are comparable to the collision mean free paths and collision frequencies. Now neither the effects of free flow nor those of collisions can be neglected, and this implies that a transport equation has to be used in analyzing the dispersion relations.

From both standpoints of theory and experiment the kinetic region offers the greatest challenge. A successful calculation in this region must preserve the correct free-particle and the hydrodynamic behavior, so in effect it has to be valid at arbitrary wavelengths and frequencies. One can distinguish two kinds of problems, the dilute gases and the dense fluids. For dilute gases a reasonably general description is provided by the linearized Boltzmann equation. The calculation is therefore well posed and the problem is to analyze the normal modes and interpret the data properly. For dense fluids it is much more difficult to obtain a tractable transport equation which incorporates structural correlations and collisional dynamics in a fashion that does not restrict the description to low frequencies and long wavelengths. But even here some progress has been made in recent years in attempts to understand the interesting results of neutron inelastic scattering and computer molecular-dynamics calculations.

The plan of these lectures is as follows. In Sect. **2** we review briefly the analysis of dispersion relations in power series of the frequency or the wave number. For dilute gases the Boltzmann equation can be used to generate successively higher-order approximations to hydrodynamics; however, the results tell us little about the general behavior at arbitrary k or ω. In Sect. **3** we employ kinetic models to demonstrate the existence of continuum normal modes in addition to the familiar discrete modes, and to show that at sufficiently large k or ω the discrete modes can disappear. The importance of the

two kinds of modes in sound propagation is illustrated in Subsect. **3˙**2 where
we describe the results of analyzing the sound propagation experiment as a
boundary-value problem. One sees that the pressure field in the gas continues
to behave smoothly even when the discrete modes no longer exist and one
cannot speak of a dispersion equation. In the last part of Sect. **3** we examine
the phenomenon of zero sound in dense fluids. This occurs because the average
potential field generated by the neighboring molecules gives rise to a restoring force which can sustain collective oscillations even though ω is larger than
the collision frequency. In Sect. **4** we consider the collective modes in liquids
at short wavelengths. Using a generalized visco-elastic description of density
and current fluctuations we show how one can interpolate between the region
of neutron scattering and computer molecular dynamics and the small-(k, ω)
region of hydrodynamics. Lastly in Sect. **5** we offer a number of concluding
remarks which are an attempt to put into perspective the relation of these
lectures to current developments.

2. – Sound propagation at low frequencies.

2˙1. *Hydrodynamic dispersion relations.* – At low frequencies one can calculate sound dispersion and attenuation from the linearized equations of hydrodynamics. The most convenient system of local variables for this calculation are the entropy S, the pressure P and the velocity $\boldsymbol{U}$. For normal or
plane-wave solutions of the form $S = S_{k\omega} \exp[i(\boldsymbol{k}\cdot\boldsymbol{r} - \omega t)]$, etc., the Navier-
Stokes equations are of the form [1]

$$(2.1) \quad \begin{pmatrix} \left(\dfrac{\partial \varrho}{\partial S}\right)_P \omega & \dfrac{\omega}{c_0^2} & -\varrho k \\ 0 & -k & \varrho\omega + i\varrho D_l k^2 \\ \omega + iD_T k^2 & iD_T \dfrac{C_p}{T}\left(\dfrac{\partial T}{\partial P}\right)_S k^2 & 0 \end{pmatrix} \begin{pmatrix} S_{k\omega} \\ P_{k\omega} \\ U_{k\omega} \end{pmatrix} = 0,$$

where

$$(2.2) \quad \begin{cases} c_0^2 = \left(\dfrac{\partial P}{\partial \varrho}\right)_S, \\[2mm] D_T = \dfrac{\varkappa}{\varrho C_p}, \\[2mm] D_l = \left(\dfrac{4}{3}\eta_s + \eta_v\right)\Big/\varrho. \end{cases}$$

Here ϱ, c_0, C_p, $\varkappa$, η_s and η_v are respectively the mass density, the (zero-frequency)
adiabatic sound speed, the constant-pressure specific heat, the thermal conduc-

tivity, the shear and bulk viscosity coefficients. Only the longitudinal (along the direction of k) component of the velocity, $U_{k\omega}$, is coupled to the entropy and pressure.

The dispersion equation is obtained by setting the determinant of the coefficient matrix equal to zero, the solvability condition for (2.1). Using the thermodynamic identity

$$(2.3) \qquad \left(\frac{\partial T}{\partial P}\right)_s \left(\frac{\partial \varrho}{\partial S}\right)_P = -\frac{T}{c_0^2 C_p}(\gamma - 1),$$

we can put the dispersion equation into the form

$$(2.4) \qquad (\omega + iD_T k^2)(\omega^2 - c_0^2 k^2 + i\omega \Gamma k^2) - \omega D_T(\gamma - 1)(\Gamma - \gamma D_T) k^4 = 0,$$

where

$$(2.5) \qquad \Gamma = D_l + D_T(\gamma - 1),$$

and γ is the specific-heat ratio C_p/C_v. Notice that this is a cubic equation in the frequency but a quartic equation in the wave number. As is well known, (2.4) can be solved with either real frequency ω or real wave number k. The case of real frequency and complex wave number corresponds to the propagation of excitations which are generated by a steady-state source at a fixed frequency. This is a boundary-value problem where one measures the spatial dependence of the amplitude and phase of the sound waves. It is conventional to write

$$(2.6) \qquad k = \frac{2\pi}{\lambda} + i\alpha,$$

where λ is the sound wavelength which is related to the phase velocity by

$$(2.7) \qquad c = \frac{\lambda}{2\pi}\omega = \left[\frac{1}{\omega}\operatorname{Re}(k)\right]^{-1},$$

and α is the attenuation constant, the reciprocal of absorption length.

When the wave number is real the dispersion relations then describe the evolution of an excitation at a fixed wavelength as in an initial-value problem. In this case one writes

$$(2.8) \qquad \omega = \omega' - i\tau^{-1},$$

and the phase velocity is now given by

$$(2.9) \qquad c = \frac{1}{k}\operatorname{Re}(\omega).$$

In general (2.7) is not the same as (2.9); however, if the damping is small, they are then equal and also $\alpha^{-1} = c\tau$.

The most familiar solutions of (2.4) are those obtained from a power series expansion in k or ω. For k real the dispersion relations are

$$(2.10) \qquad \omega = -iD_T k^2 (1 - \lambda_1 k^2 + ...) ,$$

$$(2.11) \qquad \omega = \pm c_0 k (1 - \lambda_2 k^2 + ...) - i(\Gamma/2) k^2 (1 - \lambda_3 k^2 + ...) .$$

Equation (2.10) describes the nonpropagating heat mode, whereas (2.11) gives the two propagating sound modes. At long wavelengths the damping of the heat mode is governed by $D_T k^2$, where D_T is the thermal diffusivity; the sound modes propagate at the adiabatic sound speed c_0 and they are damped according to Γk^2, where Γ is the classical sound attenuation coefficient.

The dispersion relations (2.10) and (2.11) are expansions in powers of the wave number, or more precisely the ratio of the wavelength to mean free path. The terms with coefficients λ_i represent dispersion effects. These correction terms are not calculated correctly by (2.4) because the Navier-Stokes equations are known to give results correctly only to order k^2. Higher-order terms in the series can be obtained if we use dispersion equations which are better approximations to hydrodynamics. We can simplify the calculations by considering a monoatomic gas at sufficiently low density that its thermodynamic properties are those of an ideal gas, $C_p = 5k_B/2M$, $\gamma = 5/3$, $c_0^2 = 5k_B T/3M$, where k_B is the Boltzmann's constant and M is the particle mass. Equation (2.4) then becomes

$$(2.12) \qquad \left(\xi + i\frac{3}{5}f\delta^2\right)\left[\xi^2 - \delta^2 + i\left(\frac{4}{3} + \frac{2}{5}f\right)\xi\delta^2\right] + \frac{2}{5}f\left(\frac{3}{5}f - \frac{4}{3}\right)\xi\delta^4 = 0 ,$$

where $\eta_v = 0$, $f = \varkappa/\eta_s C_v$ is the Eucken number, and $l_c = \eta_s/\varrho c_0$ is the collision mean free path, and

$$(2.13) \qquad \delta = kl_c ,$$

$$(2.14) \qquad \xi = \frac{\omega l_c}{c_0}$$

are respectively the mean free path to wavelength ratio and the frequency to collision frequency ratio. Equation (2.12) is the dimensionless Navier-Stokes dispersion equation for a dilute gas. The dispersion equation correct to order k^3 can be obtained from the so-called Burnett equations [2]. The result is the same as (2.12) except the last term is replaced by

$$\left[\frac{2}{5}f\left(\frac{3}{5}f - \frac{4}{3}\right) + \frac{16}{225}f^2 C_1 + \frac{16}{45}fC_2 - \frac{10}{9}C_3\right]\xi\delta^4 ,$$

where C_i are dimensionless constants which can be evaluated once the inter-atomic potential function is specified. It is clear that to order k^2 the additional terms introduced by the Burnett equations do not affect the dispersion relations.

As a further specialization we consider the system of Maxwell molecules where the interaction potential is purely repulsive and varies as $1/r^4$. In this case it is known [2] that $f = 5/2$, $C_1 = -45/8$, $C_2 = 3$ and $C_3 = 2$. Using these values one can find the leading correction terms to the adiabatic sound speed and the classical attenuation constant [1]:

(2.15)
$$\frac{c_0}{c} = \frac{1}{\xi} \operatorname{Re}(\delta) = \begin{cases} 1 - \dfrac{141}{72}\, \xi^2\,, \\[2mm] 1 - \dfrac{215}{72}\, \xi^2\,, \end{cases}$$

(2.16)
$$\frac{\alpha c_0}{\omega} = \frac{1}{\xi} \operatorname{Im}(\delta) = \begin{cases} \dfrac{7}{6}\,\xi - \dfrac{1559}{432}\, \xi^3\,, \\[2mm] \dfrac{7}{6}\,\xi - \dfrac{3483}{432}\, \xi^3\,, \end{cases}$$

where upper and lower lines correspond to Navier-Stokes and Burnett results respectively. This shows that the dispersion equation (2.4) or (2.12) cannot be used to discuss dispersion effects. We will see from kinetic-theory calculations that the Burnett dispersion equation does not give correctly the correction term in the damping (2.16).

2˙2. Kinetic-theory dispersion relations. – Hydrodynamic equations are valid only when the wavelengths of fluctuation are long compared to the collision mean free path in the fluid. For this reason the calculations just described cannot be expected to describe the dispersion behavior at high frequencies and short wavelengths. In the case of dilute gases molecular processes such as collisions and free flow are treated by the Boltzmann transport equation [3, 4]. Since this equation is believed to be valid even when the wavelengths are comparable to the mean free path, it should be capable of giving reliable results beyond the hydrodynamic regime.

For the analysis of sound propagation it is appropriate to work only with the linearized Boltzmann equation [1, 3]. The equation describing the normal solution

(2.17)
$$f(\boldsymbol{r}, \boldsymbol{c}, t) = f_{k\omega}(\boldsymbol{c}) \exp\left[i(\boldsymbol{k}\cdot\boldsymbol{r} - \omega t)\right]$$

is of the form

(2.18)
$$\left(-i\frac{\omega}{\sqrt{2}\, v_0} + i\boldsymbol{k}\cdot\boldsymbol{c}\right) f_{k\omega}(\boldsymbol{c}) = J[f_{k\omega}]\,,$$

where $J[f]$ is the collision integral

(2.19) $J[f_{k\omega}] = n \int \mathrm{d}^3 c_1 f_0(c_1) \int \mathrm{d}\Omega |c - c_1| I(|c - c_1|, \theta) \cdot$

$\cdot [f_{k\omega}(c') + f_{k\omega}(c_1') - f_{k\omega}(c) - f_{k\omega}(c_1)] ,$

and n is the number density. Throughout these lectures we will use dimensionless velocity, $c = v/\sqrt{2} v_0$, where $v_0 = (k_B T/M)^{\frac{1}{2}}$ is the thermal speed. In (2.19) I is the binary collision cross-section and $f_0(c)$ is the Maxwellian function $(\pi)^{-\frac{3}{2}} \cdot$ $\cdot \exp[-c^2]$. To extract the dispersion equation from (2.18) one can expand the solution in a set of axially symmetric functions [2]

(2.20) $f_{k\omega}(c) = \sum_{r,l=0}^{\infty} a_{rl}(k, \omega) \psi_{rl}(c, c_z) .$

The functions ψ_i, $i \equiv (rl)$, are orthonormal

(2.21) $(\psi_i, \psi_j) \equiv \pi^{-\frac{3}{2}} \int \mathrm{d}^3 c \exp[-c^2] \psi_i(c) \psi_j(c) = \delta_{ij} .$

Equation (2.21) defines the inner-product notation. Inserting the expansion into (2.18) one obtains

(2.22) $\sum_{r',l'} \left[\omega \delta_{rr'} \delta_{ll'} - \sqrt{2} k v_0 (\psi_{rl}, c_z \psi_{r'l'}) - i\sqrt{2} v_0 (\psi_{rl}, J[\psi_{r'l'}]) \right] a_{r'l'}(k, \omega) = 0 .$

Equation (2.22) may be compared with (2.1). As before the dispersion equation is obtained by setting the determinant of the coefficient matrix equal to zero. Notice, however, that (2.22) is an infinite system of coupled equations. A portion of the coefficient matrix is shown in Table I where we have taken

TABLE I. – *Elements of the coefficient matrix in eq. (2.22).*

	(00)	(01)	(10)	(02)	(11)
(00)	ω	$-kv_0$	0	0	0
(01)	$-kv_0$	ω	$(2/3)^{\frac{1}{2}} kv_0$	$-(4/3)^{\frac{1}{2}} kv_0$	0
(10)	0	$(2/3)^{\frac{1}{2}} kv_0$	ω	0	$-(10/6)^{\frac{1}{2}} kv_0$
(02)	0	$-(4/3)^{\frac{1}{2}} kv_0$	0	$\omega - i\sqrt{2} v_0 J_{02}$	$(8/15)^{\frac{1}{2}} kv_0$
(11)	0	0	$-(10/6)^{\frac{1}{2}} kv_0$	$(8/15)^{\frac{1}{2}} kv_0$	$\omega - i\sqrt{2} v_0 J_{11}$

the interaction potential to be $1/r^4$. For this case (Maxwell molecules) the collision operator J is known to be diagonal, $(\psi_{rl}, J[\psi_{r'l'}]) = J_{rl} \delta_{rr'} \delta_{ll'}$.

In Table I the states are arranged in the order of increasing polynomial order, $2r + l$. The collision operator has zero eigenvalues in the first three states as a consequence of conservation of particle number, momentum and energy. All other eigenvalues are negative. One can truncate the coefficient matrix at any order and examine the resulting dispersion equation [1]. If we consider only the first three states, we would find

$$(2.23) \qquad \begin{cases} \omega = 0 \, , \\ \omega = \pm \left(\tfrac{5}{3}\right)^{\frac{1}{2}} k v_0 \, , \end{cases}$$

the results corresponding to the Euler approximation to hydrodynamics. Keeping the first five states gives

$$(2.24) \qquad \begin{cases} \omega = i \dfrac{v_0}{\sqrt{2} J_{11}} \, k^2 + \ldots \, , \\ \omega = \pm \left(\dfrac{5}{3}\right)^{\frac{1}{2}} k v_0 + i \sqrt{2} \, v_0 \left(\dfrac{1}{6 J_{11}} + \dfrac{1}{3 J_{02}}\right) k^2 + \ldots \, . \end{cases}$$

In addition there are two highly damped modes which are of no interest to us. Equation (2.24) agrees with the Navier-Stokes dispersion relations to order k^2 if the shear viscosity and thermal conductivity are identified as

$$(2.25) \qquad \eta_s = -\frac{n M v_0}{\sqrt{2}} \frac{1}{J_{02}} \, , \qquad \varkappa = -\frac{5 \sqrt{2} n v_0 k_B}{4} \frac{1}{J_{11}} \, .$$

These relations are consistent with transport coefficient expressions derived using the Chapman-Enskog procedure [4]. If one includes the first eight states in Table I, then the results are found to correspond with the Burnett approximation to hydrodynamics. We see therefore that successive truncations of the coefficient matrix in (2.22) yield successively higher-order approximations to hydrodynamics. The leading terms of the series are

$$(2.26) \qquad \frac{c_0}{c} = 1 - \frac{215}{72} \xi^2 + \frac{4\,115\,101}{72\,576} \xi^4 - \ldots \, ,$$

$$(2.27) \qquad \frac{\alpha c_0}{\omega} = \frac{7}{6} \xi - \frac{5\,155}{432} \xi^3 + \ldots \, .$$

Comparing these with (2.15) and (2.16) we see that the second term in (2.16) is not calculated correctly by the Burnett approximation.

The results of this Section show that the Boltzmann equation can be used to generate in a systematic fashion series expansions for the sound dispersion and attenuation. Extensive numerical calculations have been carried out using Maxwell molecule and hard-sphere interaction potentials [5]. Comparison with experimental data on rarefied gases indicates that in the hydrodynamic region these results are extremely accurate, but in the kinetic region

the theoretical attenuation rates diverge very badly. The implication is that
series solutions like (2.26) and (2.27) may actually converge, but the radius
of convergence is so small that the utility of this method of analysis is rather
limited.

3. – Sound propagation at high frequencies.

3˙1. *Dispersion relations of kinetic models.* – In the preceding Section we
have discussed sound dispersion and attenuation in terms of power series expansions of the dispersion relations. These expansions are useful for understanding how the sound speed c and the attenuation α vary as the frequency
increases or the wavelength decreases. On the other hand, they do not give any
information about the behavior of c and α at arbitrary wavelengths and frequencies. We now consider methods for extracting general solutions to the
dispersion equation.

In the Navier-Stokes approximation the dispersion equation (2.12) is sufficiently simple that one can readily obtain the dispersion relations numerically [1].
Figure 1 shows the variation of the sound mode with wave number. When the

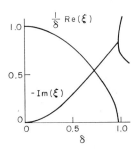

Fig. 1. – Sound dispersion and attenuation in a dilute gas of Maxwell molecules,
Navier-Stokes approximation to hydrodynamics (ref. [1]). Dimensionless frequency ξ
and wave number δ are defined in (2.13) and (2.14).

wavelength decreases the sound speed decreases until the phase velocity (2.9)
disappears at $\delta = \delta_0 = 0.979 \dots$. At this point the damping bifurcates with one
mode increasing and the other decreasing with k. Asymptotically the former
takes on a k^2-dependence while the latter approaches a finite value [6].

The behavior of the dispersion equation obtained from the linearized Boltzmann equation at finite k and ω is different from that shown in Fig. 1. A general
discussion of this problem would entail considerations of the eigenvalue spectrum of the collision operator which would be beyond the scope of these lectures [7]. To indicate the qualitative behavior of the kinetic-theory dispersion
relations we will carry out the analysis on simplified transport equations which
we will call kinetic models. These models in general possess essentially all

the important features of the Boltzmann equation, and for most purposes they are often regarded as adequate approximations to the Boltzmann equation [8].

Since we will be using kinetic models in much of our analysis, it seems worth-while to review here the basic approximations involved in deriving these transport equations from the linearized Boltzmann equation. Since the collision integral is a function in c, we can expand it in the same way as (2.20):

$$(3.1) \qquad J[f_{k\omega}] = \sum_i J_i(k, \omega)\, \psi_i(c)\ .$$

The expansion coefficients follow from the orthonormality of ψ_i and the use of (2.20):

$$(3.2) \qquad J_i(k, \omega) = (\psi_i, J[f_{k\omega}]) = \sum_j a_j(k, \omega)(\psi_i, J[\psi_j]) = \sum_j J_{ij}\, a_j(k, \omega)\ .$$

Now we insert (3.2) into (3.1) and introduce two approximations [9, 10]:

$$(3.3) \qquad J[f_{k\omega}] \simeq \sum_{i,j=1}^{M} J_{ij}\, a_i(k, \omega)\, \psi_j(c) + \sum_{i=M+1} J_{ii}\, a_i(k, \omega)\, \psi_i(c) \simeq$$

$$\simeq \sum_{i,j=1}^{M} J_{ij}\, a_i(k, \omega)\, \psi_j(c) - \nu \sum_{i=M+1} a_i(k, \omega)\, \psi_i(c) = \sum_{i,j=1}^{M} (J_{ij} + \nu\delta_{ij})\, a_i(k, \omega)\, \psi_j(c) - \nu f_{k\omega}(c)\ .$$

The first approximation consists of assuming that J_{ij} is diagonal for $i, j \geqslant \geqslant M+1$, and the second approximation is the replacement of all diagonal elements J_{ii}, $i \geqslant M+1$, by a constant $-\nu$, where ν is positive. We can regard ν as the collision frequency of the model, and it is proportional to the density.

Equation (3.3) is the collision integral for a kinetic model of order M. Its utility is that the corresponding transport equation is much more amenable to analytical analysis than the linearized Boltzmann equation (2.18). One can discuss the convergence properties of these models as successively larger values of M are taken [11]. We will not dwell on this issue since our interest is in the qualitative behavior of the sound mode at arbitrary k and ω.

In the case of Maxwell-molecule interaction J_{ij} is diagonal in the states $\psi_i(c)$, so one only needs the second approximation to arrive at

$$(3.4) \qquad J[f_{k\omega}] \simeq \sum_{i=1}^{M} \mu_i\, a_i(k, \omega)\, \psi_i(c) - \nu f_{k\omega}(c)$$

with

$$(3.5) \qquad \mu_i = J_{ii} + \nu\ .$$

To make the collision integral more explicit we note that the expansion coefficients $a_i(k, \omega)$ are

$$(3.6) \qquad a_{rl}(k, \omega) = \pi^{-\frac{3}{2}} \int \mathrm{d}^3 c\, \exp\left[-c^2\right] \psi_{rl}(c)\, f_{k\omega}(c)\ .$$

The first three Burnett functions are [2]

$$
(3.7) \qquad
\begin{cases}
\psi_{00}(\boldsymbol{c}) = 1 \,, \\[4pt]
\psi_{01}(\boldsymbol{c}) = \sqrt{2}\, c_z \,, \\[4pt]
\psi_{10}(\boldsymbol{c}) = \sqrt{\tfrac{2}{3}}\,(\tfrac{3}{2} - c^2) \,.
\end{cases}
$$

The corresponding a_i are therefore the local number density $n(k, \omega)$, the current density along the $\hat{\boldsymbol{k}}$-direction $J(k, \omega)$, and the temperature $T(k, \omega)$:

$$
(3.8) \qquad
\begin{cases}
a_{00}(k, \omega) = n(k, \omega) \,, \\[4pt]
a_{01}(k, \omega) = \sqrt{2}\, J(k, \omega) \,, \\[4pt]
a_{10}(k, \omega) = -\sqrt{\tfrac{2}{3}}\, T(k, \omega) \,.
\end{cases}
$$

Using (3.8) we can write out the kinetic model for $M = 3$. Recall that since the collisions conserve particle number, momentum and energy, one has $J_{00} = J_{01} = J_{10} = 0$. Equation (3.4) now becomes

$$
(3.9) \qquad J[f_{k\omega}] \simeq \nu[n(k, \omega) + 2c_z J(k, \omega) + \tfrac{2}{3}(c^2 - \tfrac{3}{2})\, T(k, \omega)] - \nu f_{k\omega}(\boldsymbol{c}) \,.
$$

This is the well-known model of BHATANAGAR, GROSS and KROOK [12]. It gives hydrodynamic equations which have the correct structure, but, because the model has only one parameter, the collision frequency ν, one cannot fit ν to both the shear viscosity and the thermal conductivity. As a result (3.9) gives a Eucken number of $f = \tfrac{5}{3}$ instead of $\tfrac{5}{2}$, the correct value for Maxwell molecules. This defect is of no importance to us.

To minimize the algebra we will introduce another simplification. Since we are interested only in the behavior of the sound mode, we will neglect temperature fluctuations in the gas by dropping the term containing $T(k, \omega)$ in (3.9). The effect is that the heat mode is eliminated from the dispersion equation and the zero-frequency sound speed is changed from the adiabatic value c_0 to the isothermal value $\sqrt{\tfrac{3}{5}}\, c_0$. The attenuation and dispersion of course will be slightly affected when the temperature fluctuations are ignored.

The isothermal kinetic model which describes the normal solution (2.17) is

$$
(3.10) \quad (-i\omega + i\sqrt{2}\, v_0 \boldsymbol{k} \cdot \boldsymbol{c})\, f_{k\omega}(\boldsymbol{c}) = \nu \int d^3 c'\, f_0(c')(1 + 2\boldsymbol{c} \cdot \boldsymbol{c}')\, f_{k\omega}(\boldsymbol{c}') - \nu f_{k\omega}(\boldsymbol{c}) \,,
$$

or

$$
(3.11) \qquad (Z - c_z)\, f_{xy}(\boldsymbol{c}) = iy[n(x, y) + 2\boldsymbol{c} \cdot \boldsymbol{J}(x, y)] \,,
$$

where

(3.12)
$$
\begin{cases}
Z = x + iy \,, \\[2mm]
x = \dfrac{\omega}{\sqrt{2}\,kv_0} \,, \\[2mm]
y = \dfrac{\nu}{\sqrt{2}\,kv_0} \,,
\end{cases}
$$

and

(3.13)
$$
n(x, y) = \int \mathrm{d}^3 c\, f_0(c)\, f_{xy}(\boldsymbol{c}) \,,
$$

(3.14)
$$
\boldsymbol{J}(x, y) = \int \mathrm{d}^3 c\, f_0(c)\, \boldsymbol{c} f_{xy}(\boldsymbol{c}) \,.
$$

Notice that we have introduced two dimensionless variables, x and y. The use of x and y variables is convenient for initial-value problems where k is real and ω complex, for then y is real and x is complex. For boundary-value problems where ω is real and k complex it would be simpler to work in terms of the variables δ and ξ as we did in Sect. 2. We will therefore think of x as a complex frequency scaled by the inverse time required for a particle to move a distance equal to a wavelength $1/k$, and regard y as a measure of the ratio of wavelength to collision mean free path. We will refer to y as the collision parameter. When y is large we expect the fluctuations to exhibit hydrodynamic behavior; that is, the hydrodynamic regime corresponds to the region of $y \gg 1$, and the free-flow regime corresponds to $y \ll 1$.

To obtain the dispersion equation from (3.11) we consider separately the case of real Z from the case of complex Z. The distinction arises because if Z is complex, then the factor $Z - c_z$ cannot vanish for any c_z and we can divide through (3.12) by this factor; otherwise the division is not well defined. As we will see, these two cases give very different results. Notice that a real Z implies that the imaginary part of x is just $-y$, or, in other words, the relaxation time is $1/\nu$.

We examine first the case of real Z. The solution to (3.11) can be written as [13]

(3.15)
$$
f_{xy}(\boldsymbol{c}) = A(x, y)\,\delta(Z - c_z) + iy[n(x, y) + 2\boldsymbol{c}\cdot\boldsymbol{J}(x, y)]\, P\,\frac{1}{Z - c_z}\,,
$$

where $\delta(x)$ is the Dirac delta-function, P denotes the principal value and $A(x, y)$ is an arbitrary function. Observe that (3.15) is a generalized function in the sense of a distribution. We next integrate (3.15) to obtain $n(x, y)$ as in (3.13). This gives

(3.16)
$$
n(x, y) =
$$
$$
= A(x, y)\pi^{-\frac{1}{2}}\exp[-Z^2] + iy P \int_{-\infty}^{\infty} \mathrm{d}t\, \pi^{-\frac{1}{2}}\exp[-t^2]\frac{n(x, y) + 2t J_z(x, y)}{Z - t} \,.
$$

We can elimintate $J_z(x, y)$ by using the continuity equation

(3.17) $$J_z(x, y) = xn(x, y) \, .$$

Equation (3.16) is then an inhomogeneous equation in $n(x, y)$, and a solution exists for any y and $x = \text{Re}\,(x) - iy$. The absence of any condition relating x and y means that there is no dispersion equation. For this case the eigenvalue spectrum is a continuum (see Fig. 4). The corresponding modes are known as single-particle modes [3], or Knudsen modes since they are closely related to the « normal modes » in a Knudsen gas [1]. It is clear that these modes are different from the normal modes discussed in Sect. **2**; in fact they do not occur in the hydrodynamic descriptions of thermal fluctuation.

When Z is complex the solution to (3.11) is

(3.18) $$f_{xy}(\boldsymbol{c}) = \frac{iy}{Z - c_z}\left[n(x, y) + 2\boldsymbol{c}\cdot\boldsymbol{J}(x, y)\right].$$

Applying (3.13) and (3.17) we obtain a homogeneous equation in $n(x, y)$ and the condition for nontrivial solution is

(3.19) $$F(x, y) = 1 - iy\pi^{-\frac{1}{2}}\int_{-\infty}^{\infty}\mathrm{d}t\,\frac{\exp\left[-t^2\right]}{Z - t}\,(1 + 2xt) = 0 \, .$$

This is the dispersion equation for the isothermal kinetic model. Comparing (3.19) with (2.12) we see that the frequency and wave number dependence is no longer explicit, so the calculation of dispersion relations is now more complicated. Fortunately, the analysis is still manageable because we are dealing with kinetic models and not with the full Boltzmann equation.

The significance of (3.19) is that for a given wave number the complex frequency is determined from the zeros of F. The corresponding normal solutions (3.18) are well-behaved functions in contrast to the continuum-mode solutions (3.15). One can make use of the *winding theorem* [14] to find the number of zeros admitted by (3.19). We first observe that $F(x, y)$ is an analytic function except when $x + iy$ is on the real axis in the complex Z-plane. If we restrict our attention to only the upper half Z-plane, then F is analytic everywhere. As one moves around the boundary of the upper half Z-plane the complex function F traces out a trajectory in the F-plane. According to the winding theorem, the number of times this trajectory encircles the origin in the F-plane is the number of zeros. It is to be expected that different y values will give rise to different trajectories. It is found that there is a critical value of y_c $(y_c = 1.952\,59\,...)$ above which the trajectory circles the origin twice and below which the trajectory does not circle the origin at all [1]. This means that as Z is carried around the upper half-plane there will be two discrete modes if the wave number is below a critical values k_c, but these modes disappear when k exceeds k_c. If we repeat the analysis for Z in the lower half-plane where F is also analytic, we find that there are no zeros.

The winding theorem affords a qualitative understanding of the behavior of the dispersion relation. The precise variation of frequency with wave number can only be obtained by studying (3.19) numerically [1]. The results in terms of dimensionless variables x and y are shown in Fig. 2. At the same time one can obtain a power series expansion

$$(3.20) \qquad \frac{\omega}{kv_0} = \pm \left(1 + \frac{1}{4y^2} - \frac{33}{32y^4} + ...\right) - i\frac{1}{\sqrt{2}y}\left(1 - \frac{1}{y^2} + \frac{1}{y^4} - ...\right).$$

The isothermal kinetic model gives two discrete modes when the wavelengths are longer than the critical length $\propto 1/k_c$. These are clearly the sound modes.

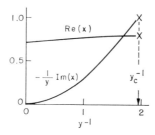

Fig. 2. – Sound dispersion and attenuation according to the isothermal kinetic-model dispersion equation (3.19) (ref. [1]). Dimensionless variables x and y are defined in (3.12). The mode ceases to exist at a wavelength corresponding to $y_c^{-1} = 1.952\,59\,....$

We do not expect a heat mode because of our neglect of temperature fluctuations. As we anticipated, the long-wavelength sound speed has the isothermal value $c_0/\sqrt{\gamma} = v_0$ and the damping constant does not contain thermal conduction effects.

We summarize the findings of this calculation. By looking for normal-mode solutions (2.17) to the isothermal kinetic model we find the normal modes consist of two kinds. For a fixed wave number or y there is a continuous spectrum of complex frequencies which corresponds to a fixed relaxation time $1/\tau = \nu$. For these modes one does not have a dispersion equation. For a fixed $y < y_c$ there are also two discrete modes which are the sound modes of the system. One can analyze the dispersion behavior of these modes by power series expansion or numerically. The variation of the continuous spectrum and of the discrete modes is summarized in Fig. 3. In the limit of long wavelengths $(y \gg 1)$ the sound modes are located on the real frequency axis and the continuum line is essentially at minus infinity. As the wave number increases (y decreases) the sound modes move into the lower half frequency plane while the continuum line moves upward toward the real axis. When y approaches the value y_c, the discrete modes approach the continuum line, and, as we saw in Fig. 2, at $y = y_c$ the discrete modes disappear. As y becomes less than y_c we have only the continuum spectrum which continues to move toward the real axis.

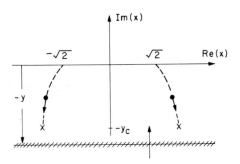

Fig. 3. – Schematic behavior of the discrete modes (circles) and the continuum mode (slashed line) in the isothermal kinetic-model approximation. Arrows indicate the direction of movement as the wave number increases.

The normal-mode behaviors of higher-order kinetic models [15] are qualitatively similar to those noted here. When more terms in the M-sum in (3.4) are included more discrete modes appear in the dispersion relations. For example, the kinetic model (3.9) leads to three discrete modes, the additional one being the heat mode. Each discrete mode has its own critical wave number cut-off beyond which the mode does not exist, and this cut-off is in general model dependent. In the next Subsection we will investigate the significance of the discrete and continuum modes in the interpretation of ultrasonic experiments.

3'2. *Analysis of ultrasonic experiments.* – The most precise and extensive measurements of sound dispersion and attenuation in monoatomic gases are those reported by GREENSPAN [16] and by MEYER and SESSLER [17]. These data have long been regarded as the standard experimental results for testing kinetic-theory calculations. It is not our aim to review the many interesting analyses that have appeared in the literature [18]. What we wish to describe here is a boundary-value calculation in which kinetic models of the type just discussed were employed to analyze the sound propagation data [19]. This calculation is an excellent illustration of the role of the discrete and continuum normal modes in describing the observable properties of the gas.

The experimental data are conventionally presented in terms of the sound speed ratio c_0/c, where c is given by (2.7), and the attenuation coefficient

$\eta = \alpha c_0/\omega$, with α defined by (2.6). The independent variable is taken to be the dimensionless parameter

$$(3.21) \qquad r = \frac{P_0}{\mu\omega},$$

where ω is the oscillator frequency, P_0 is the equilibrium pressure and μ the viscosity. Notice that r is related to the inverse Knudsen number. For hard spheres the Knudsen number is

$$(3.22) \qquad K_n = \frac{8}{5\pi} \sqrt{\frac{3}{10\pi} \frac{1}{r}}.$$

We can relate the collision frequency to the shear viscosity in terms of our collision parameter and obtain

$$(3.23) \qquad y = 1.52 \sqrt{\frac{10}{27}} r \sim 0.925 r.$$

Notice that one can vary r by changing either the gas pressure or the oscillator frequency. The fact sound propagation is characterized by r is a consequence of the scaling properties of gas dynamics, as is evident from an examination of the Boltzmann equation.

Since the sound propagation experiment is a boundary-value problem, it is more appropriate to look for solutions to the kinetic-model equations in the form

$$(3.24) \qquad f(z, \mathbf{c}, t) = f_{k\omega}(\mathbf{c}) \exp[i\omega t - kz],$$

where the direction of propagation is along the z-axis, ω is real and k is complex. The kinetic-model equation (cf. (3.4)) now becomes

$$(3.25) \qquad (\theta - c_z) f_{k\omega}(\mathbf{c}) = \frac{1}{\sqrt{2}\, kv_0} \sum_{i=1}^{M} \mu_i a_i(k, \omega) \psi_i(\mathbf{c}),$$

where we define

$$(3.26) \qquad \theta = \frac{i\omega + \nu}{\sqrt{2}\, kv_0}.$$

As before we distinguish two regions in the complex θ-plane. When θ is not real we have solutions corresponding to the discrete modes

$$(3.27) \qquad f_{k\omega}^D(\mathbf{c}) = \frac{1}{\sqrt{2}\, kv_0} \sum_{i=1}^{M} \mu_i a_i(k, \omega) \frac{\psi_i(\mathbf{c})}{\theta - c_z};$$

this leads to the dispersion equation

(3.28) $\det \left(\delta_{ij} - C_{ij}(k, \omega) \right) = 0$,

where

(3.29) $C_{ij}(k, \omega) = \dfrac{\mu_i}{\sqrt{2}\, k v_0} \displaystyle\int d^3c\, f_0(c)\, \dfrac{\psi_i(\boldsymbol{c})\, \psi_j(\boldsymbol{c})}{\theta - c_z}$,

and the indices $i, j = 1, ..., M$. Figure 4 shows the heat and sound modes
for kinetic models with $M = 3$ and $M = 5$ [19]. Notice that the critical fre-

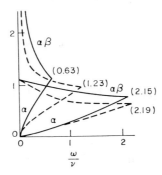

Fig. 4. – Dispersion relations of sound and heat modes derived from kinetic models
with $M = 3$ (solid curves) and $M = 5$ (dashed curves) (ref. [19]). The numbers in
parentheses are the values of ω/ν where the modes vanish. The sound-mode branches
are distinguished by their finite $\alpha\beta$ value at $\omega/\nu = 0$. Parameters $\alpha\beta$ and α are
defined in (3.31).

quencies for the sound modes show much less variation with M than the heat
mode. The particular appearance of these dispersion relations is the result
of our notation:

(3.30) $\begin{cases} \alpha = \dfrac{\sqrt{2}\, v_0}{\nu} \operatorname{Re}(k), \\[2ex] \alpha\beta = \dfrac{\sqrt{2}\, v_0}{\omega} \operatorname{Im}(k). \end{cases}$

The relations to sound speed and attenuation are

(3.31) $\begin{cases} \alpha\beta = \dfrac{\sqrt{2}\, v_0}{c}, \\[2ex] \alpha = \dfrac{\sqrt{2}\, v_0 \omega}{c_0 \nu}\, \eta. \end{cases}$

At the critical frequency θ is on the real axis and therefore $\operatorname{Re}(k) = \nu \operatorname{Im}(k)/\omega$.
It can be seen from (3.30) that β becomes unity at this point.

The solutions to (3.25) when θ is real are in the form of (3.16)

$$(3.32) \qquad f_{k\omega}^c(\boldsymbol{c}) = \frac{1}{\sqrt{2}\,kv_0} \sum_{i=1}^{M} \mu_i P \left[\frac{a_i(k,\,\omega)\,\psi_i(\boldsymbol{c})}{\theta - c_z} \right] + A(k,\,\omega)\,\delta(\theta - c_z) ,$$

and we know that in this case no dispersion equation exists. One can show that solutions (3.27) and (3.32) form a complete set of functions in velocity space for any real ω. They can therefore be used to expand the solutions to an appropriate boundary-value problem. Although the measurement of sound propagation is actually a half-space problem, it can be analyzed as an infinite-medium problem provided a suitable source distribution is introduced at $z = 0$ for all c_z such that the infinite-medium solution is identical to the half-space problem for $0 \leqslant z < \infty$ and is zero everywhere for $-\infty < z \leqslant 0$ [19]. For a source distribution the approximation

$$(3.33) \qquad S(\boldsymbol{c},\,t) = Ac_z \exp[i\omega t] ,$$

where A is a parameter related to the amplitude of the oscillator velocity, has been used. A source distribution corresponding to diffuse reflection also has been studied. It turns out that the essential results are not very sensitive to the precise details of the source.

The experimentally measured quantity is the pressure at the detector. Neglecting the effects of the detector we can write the pressure field as

$$(3.34) \qquad P_{zz}(z,\,t) = 2\int \mathrm{d}^3 c\, f_0(c)\, c_z^2\, f(z,\,\boldsymbol{c},\,t) .$$

The evaluation of P_{zz} is quite involved [19], and, since the details are of no interest in the present discussion, we summarize here only the final results.

It is of interest to ask if any discontinuity in the pressure field appears at the critical frequency where the sound modes disappear. Figure 5 shows

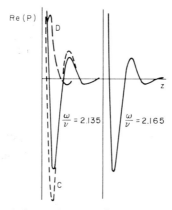

Fig. 5. – Spatial behavior of the real part of the pressure field $P_{zz}(t)$ at ω/ν values just below and just above the critical value $\omega/\nu = 2.15$ (ref. [19]). The continuum- and discrete-mode contributions below the cut-off are labeled C and D respectively.

the real part of P_{zz} at frequencies just below and above the cut-off computed using a kinetic model with $M = 3$. There is no discontinuity that can be observed. Detailed analysis of the transition behavior reveals a cancellation between the discrete mode and a portion of the continuum-mode contribution as ω approaches ω_c. At ω_c the canceling terms both disappear leaving the remaining portion of the continuum-mode contribution which behaves smoothly as a function of frequency.

The procedure used to extract the sound speed and attenuation is as follows. The complex pressure P_{zz} is evaluated numerically and its spatial variation is analyzed according to (3.24). This gives the real and imaginary parts of k, which are then used in (3.31) to give c_0/c and η. The results obtained from kinetic models with $M = 3$ and $M = 5$ are compared with the Greenspan-Meyer-Sessler data in Fig. 6 [19]. Also shown in Fig. 6 are the results derived

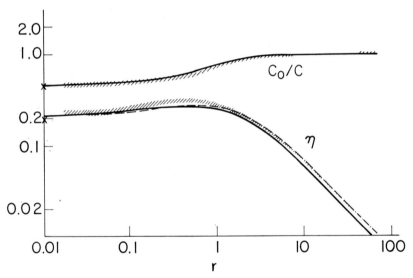

Fig. 6. – Comparison of kinetic-model results of sound speed c_0/c and attenuation η with the experimental data of GREENSPAN and MEYER and SESSLER (ref. [19]), experiment (slashed lines), kinetic model $M = 3$ (solid curves), model $M = 5$ (dashed curves). The crosses denote values when only free flow is considered (ref. [20]). The sound speed result for model $M = 5$ is indistinguishable from the $M = 3$ model and is therefore not shown. For $r \geqslant 1$ the $M = 3$ model does not give the attenuation correctly but the $M = 5$ model does.

from free-flow calculations [20]. We have noted earlier that the $M = 3$ model cannot give the correct shear viscosity and thermal conductivity simultaneously because it has only one collision frequency parameter. This means that the attenuation constant Γ will not be calculated correctly, and therefore it is understandable that the η values calculated from the $M = 3$ model do not agree with experiment in the hydrodynamic region. Aside from this discrepancy the two models are seen to give essentially indistinguishable results which are in remarkably good agreement with the observations. It is noteworthy that this agreement extends over practically all three regions of the dimensionless frequency, hydrodynamic (continuum), kinetic (transition) and free particle (Knudsen).

The results of this Section greatly help to clarify the relation between the experimental data in the large–Knudsen-number region and the phenomenon of sound propagation. Unlike the previous analyses, where the dispersion relations are implicitly assumed to be directly measurable, the present calculations deal directly with the pressure field which is the quantity actually measured. The pressure field is evaluated in terms of contributions from all the normal modes. Because the normal modes now comprise a continuum portion and a discrete spectrum, the existence of pressure disturbance clearly does not require the existence of the discrete spectrum.

At small Knudsen numbers the calculations show that the discrete-mode contributions dominate and there is no difficulty in interpreting the experimental results as sound propagation. In this region the kinetic-model results also agree with hydrodynamic calculations as one would expect. When one leaves the hydrodynamic region and enters the transition region, the continuum-mode effects become more and more important as the damping of the discrete modes continues to increase. The important conclusion, brought out only by the particular calculation [19] described in this Section, is that the pressure field behaves smoothly even when the discrete modes disappear. There appears to be no discontinuity of any type in the propagation of pressure disturbance as both calculations and measurements give phase velocities and attenuation rates that extend well into the Knudsen region.

Above the cut-off frequencies the pressure field is completely determined by the continuum-mode solutions (3.32). Even though it is still possible to extract a phase velocity and attenuation, the « wavelike » solutions are not truly plane waves. Since the gas is very rarefied in this region, collisional damping effects become less important than phase mixing. Moreover, experimental effects such as boundary effects at the source and detector can have appreciable influence on the results. Under these conditions the question of the existence of sound becomes moot since the answer depends on how one wishes to define sound. On the other hand, it is now clear that the discrete and continuum modes both play an important role in our understanding of propagation phenomena in gases (*).

(*) Very recently measurements on rarefied monoatomic gases have been carried out at the University of Stuttgart. Special attention was given to the effects of the source-detector separation, and a detailed analysis of the behavior of the observed sound speed and attenuation in the kinetic and free-flow regimes was mode. The calculations of ref. [19] were found to be in excellent agreement with the data. See the contribution of H. G. BAUER in the present Proceedings and an article by R. SCHÖTTNER: *Phys. Fluids*, **20**, 40 (1974).

Finally it should be mentioned that the existence of a cut-off frequency for each discrete mode is a general property of kinetic equations. The cut-off frequencies are known to occur whenever the collision frequency or the cross-section is finite. This applies to kinetic-model solutions of the type we have discussed as well as to more general solutions of the Boltzmann equation [21]. On physical grounds the requirement of finite collision frequency appears to be quite reasonable. Even though one does encounter infinite cross-sections in classical dynamics, such infinities do not appear when quantum mechanics is used.

3˙3. *Propagation in the collisionless regime—zero sound.* – Thus far we have analyzed sound propagation in dilute gases which are structureless fluids as far as their equilibrium properties are concerned. We now turn our attention to dense fluids and consider a phenomenon which depends on the existence of appreciable local order in the system. We have seen that, when the frequency of oscillation is large compared to collision frequencies, the collisions have little effect on the way in which the fluctuations can be propagated. In a dilute gas this means that free-flow effects are dominant and it would be difficult to maintain collective modes of motion. The situation is different in a dense fluid because there are significant spatial correlations among neighboring molecules. These correlations give rise to a restoring force which enables the system to sustain collective motions even in the absence of collisions. Such collective modes are called zero sound and they occur in the region of high frequencies called the collisionless regime.

Zero sound is an important collective mode in the dynamical theories of liquid helium. Its properties have long been studied theoretically [22] and its experimental observation seems well established [23]. The counterpart of zero sound in charged-particle fluids is called plasma oscillations or plasmons [24]. The phenomenon also occurs in solids where it is known that elastic constants can exhibit different temperature dependence if they are measured at frequencies high compared to the reciprocal phonon lifetimes [25]. Our interest in zero sound is to point out its existence in a neutral classical fluid and to briefly examine its connection with the sound mode that we have been discussing up to now. To distinguish the two kinds of sound modes it is conventional to refer to the latter as first sound.

It is clear that the linearized Boltzmann equation does not take into account any equilibrium spatial correlation effects between molecules. One can use the Enskog equation, which is an empirical modification of the Boltzmann equation for dense hard-sphere fluids [4]. The Enskog equation leads to the proper equation of state for hard spheres which implies that it gives a much more realistic low-frequency sound speed for dense fluids than the Boltzmann equation. However, we will not work with the Enskog equation mainly because this equation is too complicated for the analysis we have in mind. Also, there are defects in the Enskog equation which can cause unnecessary confusion in a discussion which is intended to be brief and qualitative.

We will instead adopt the simple approach of modifying the isothermal model of Subsect. 3'1 by adding a mean-field term [26, 27]. This term represents the effects of interaction between a molecule and an average potential field generated by its neighbors. Self-consistency arguments then require the potential to reflect the spatial correlations in the fluid. The mean-field term occurs frequently in transport theory problems, particularly when the random-phase approximation is used. It should be emphasized that, despite our apparently empirical treatment of this term, its appearance in the kinetic equation is well established by first-principles considerations.

We will again be concerned with the normal solutions of the form (2.17). The modified isothermal kinetic-model equation is

$$(3.35) \qquad (-i\omega + i\sqrt{2}kv_0 c_z + \nu)f_{k\omega}(\mathbf{c}) = \nu \int \mathrm{d}^3 c' f_0(c')(1 + 2\mathbf{c} \cdot \mathbf{c}')f_{k\omega}(\mathbf{c}') +$$
$$+ i\sqrt{2}kv_0 nC(k)c_z \int \mathrm{d}^3 c' f_0(c')f_{k\omega}(\mathbf{c}') ,$$

where $C(k)$ is the direct correlation function in wave number space. This function characterizes the extent to which molecules are spatially correlated in the fluid. In the molecular theory of equilibrium properties of fluids the basic quantity is the radial-distribution function $g(r)$. Physically $4\pi ng(r)r^2\,\mathrm{d}r$ is the average number of molecules in a spherical shell of thickness $\mathrm{d}r$ and radius r centred about a molecule located at $r = 0$. In X-ray and neutron diffraction experiments on fluids one can measure the Fourier transform of $g(r)$ [28]:

$$(3.36) \qquad S(k) = 1 + n\int \mathrm{d}^3 r \exp[i\mathbf{k} \cdot \mathbf{r}]g(r) ,$$

which is called the static-structure factor. The direct correlation function is related to $S(k)$ according to

$$(3.37) \qquad nC(k) = \frac{S(k) - 1}{S(k)} .$$

In the long-wavelength limit one knows [28]

$$(3.38) \qquad S(0) = \frac{n}{\beta}\chi_T ,$$

where χ_T is the isothermal compressibility of the fluid. It then follows that in this limit $nC(k)$ is always negative; its magnitude is large at high densities and it is proportional to the density in a dilute gas. When we analyze (3.35) for the case of complex wave number, we will take the argument of $C(k)$ to be real.

This procedure introduces no serious error since the wave numbers under consideration in this Section are sufficiently small that $C(k)$ can be replaced by its long-wavelength limit $C(0)$.

Equation (3.35) closely resembles the kinetic equations used to discuss collective oscillations in ionized gases [12]. One can regard the factor $nC(k)$ as an effective potential of the fluid system. In the case of charged particles it becomes the self-consistent Coulomb potential and (3.35) is the Vlasov equation with collisions treated by the isothermal kinetic model of Subsect. **3'**1. In analyzing (3.35) we have in effect two parameters at our disposal, the collision frequency ν and the density n. We expect that the conditions most favorable for zero-sound propagation are high density and $\omega \gg \nu$. These are the same requirements for plasma oscillation in a charged fluid.

It is convenient to express (3.35) in terms of the dimensionless variables given in (3.12):

$$(3.39) \qquad (Z - c_z) f_{k\omega}(\boldsymbol{c}) = \int \mathrm{d}^3 c' \, f_0(\boldsymbol{c}') f_{k\omega}(\boldsymbol{c}') \left[iy(1 + 2\boldsymbol{c} \cdot \boldsymbol{c}') - nC(k) c_z \right].$$

As before we recognize that there are two types of modes, the discrete modes and the continuum modes. We will consider only the behavior of the discrete modes. Therefore y is real and x complex, then we exclude the line $\mathrm{Im}\,(x) = -y$ in the complex x-plane. The dispersion equation that follows from (3.39) is

$$(3.40) \qquad D(x, y) = 1 - \pi^{-\frac{1}{2}} \int\limits_{-\infty}^{\infty} \mathrm{d}t \, \frac{\exp\left[-t^2/2\right]}{Z - t/\sqrt{2}} \left[\frac{iy}{\sqrt{2}} \left(1 + \sqrt{2}xt\right) - \frac{nC(k)}{2} t \right].$$

The integrals in (3.40) can all be expressed in terms of a tabulated function $F(Z)$ [29]:

$$(3.41) \qquad \frac{1}{\sqrt{2\pi}} \int\limits_{-\infty}^{\infty} \mathrm{d}t \, \frac{\exp\left[-t^2/2\right]}{Z - t/\sqrt{2}} = -2iF(-iZ),$$

where F is related to the error function of complex argument by

$$(3.42) \qquad F(Z) = \left(\frac{\sqrt{\pi}}{2} - \int\limits_{0}^{z} \mathrm{d}t \exp\left[-t^2\right] \right) \exp\left[Z^2\right].$$

We now seek an expression for the dispersion relation under the condition of long wavelengths, or more precisely $|Z| \gg 1$. There are two ways to satisfy this condition. One is to have the wavelengths long compared to the collision mean free path ($y \gg 1$). This is the familiar condition for hydrodynamic behavior. The second way is for x to be much larger than unity. As we will see in the following, the real part of x is proportional to $|nC|^{\frac{1}{2}}$ when $|nC| \gg 1$, and this occurs when we are dealing with dense fluids. Therefore, only in the case of dense fluids can we have $|Z| \gg 1$ and still obtain results valid for arbitrary y. There are several large-argument expansions of F. For our purposes the approximation [12]

$$(3.43) \qquad F(Z) \simeq \frac{Z^2 + 1}{Z(2Z^2 + 3)}$$

is adequate. Applying (3.43) to (3.40) we find that the dispersion equation becomes

$$(3.44) \qquad 2ix\left(x^2 - \frac{3}{2} + \frac{nC}{2}\right) = y(2x^2 - 1 + nC) + \mathscr{L},$$

where $\mathscr{L}$ is a small term which corresponds to the Landau damping term in plasma oscillations [24]. Since $\mathscr{L}$ is generally quite small we will not include it in our analysis.

The dispersion relations derived from (3.44) are quite simple. For real ω and complex k we find

$$(3.45) \qquad [\mathrm{Re}\,(k)]^2 =$$

$$= \left(\frac{\omega}{v_0}\right)^2 \frac{\frac{1}{2}[v^2(1 - nC) + \omega^2(3 - nC)] + \{\frac{1}{4}[v^2(1 - nC) + \omega^2(3 - nC)]^2 + \omega^2 v^2\}^{\frac{1}{2}}}{v^2(1 - nC)^2 + \omega^2(3 - nC)^2},$$

$$(3.46) \qquad [\mathrm{Im}\,(k)]^2 = \left(\frac{\omega}{v_0}\right)^2 \omega^2 v^2 [v^2(1 - nC)^2 + \omega^2(3 - nC)^2]^{-1} \cdot$$

$$\cdot \left[\tfrac{1}{2}\{v^2(1 - nC) + \omega^2(3 - nC)\} + \{\tfrac{1}{4}[v^2(1 - nC) + \omega^2(3 - nC)]^2 + \omega^2 v^2\}^{\frac{1}{2}}\right]^{-1}.$$

It is interesting to observe two limits. First by considering $|nC| \ll 1$ we can see the effects of the mean-field term on the Boltzmann-equation results. In this case (3.45) and (3.46) are valid only if $v \gg \omega$:

$$(3.47) \qquad \mathrm{Re}\,(k) \simeq \pm \frac{\omega}{v_0}(1 - nC)^{-\frac{1}{2}},$$

$$(3.48) \qquad \mathrm{Im}\,(k) \simeq \frac{\omega}{v v_0}(1 - nC)^{-\frac{3}{2}}.$$

The resulting sound speed and attenuation are

$$(3.49) \qquad c = v_0(1 - nC)^{\frac{1}{2}},$$

$$(3.50) \qquad \eta = \frac{\omega c_0}{v v_0}(1 - nC)^{-\frac{3}{2}}.$$

We see that the mean-field term causes the sound speed to increase and the damping to decrease. The other limit is $|nC| \gg 1$ where (3.45) and (3.46) are valid for arbitrary $\Omega = \omega/\nu$:

$$(3.51) \qquad \frac{v_0}{\omega} \, \mathrm{Re}\,(k) \simeq \pm \left[\frac{3\Omega^2 + 1 - nC(\Omega^2 + 1)}{(1 - nC)^2 + \Omega^2(3 - nC)^2} \right]^{\frac{1}{2}},$$

$$(3.52) \qquad \frac{v_0}{\omega} \, \mathrm{Im}\,(k) \simeq \Omega \left[\frac{1}{(1 - nC)^2 + \Omega^2(3 - nC)^2} \right]^{\frac{1}{2}} \left[\frac{1}{3\Omega^2 + 1 - nC(\Omega^2 + 1)} \right]^{\frac{1}{2}}.$$

The corresponding sound speed ratio and attenuation coefficient are shown in Fig. 7. For definiteness we have used the value of 19.16 for $|nC(0)|$, which corresponds to liquid argon at 85 °K (cf. Fig. 8). The regions of small and large ω/ν are respectively the first-sound and zero-sound regimes. One can observe that at $\omega = 4\nu$ the sound speed has already attained the zero-sound

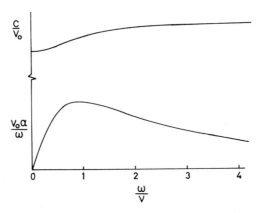

Fig. 7. – Variation of sound speed and attenuation from the first-sound regime (small ω/ν) to the zero-sound regime (large ω/ν). Results are based on the isothermal kinetic-model description of a dense fluid, $|nC(0)| = 19.16$.

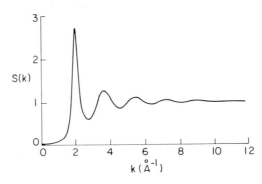

Fig. 8. – Structure factor of liquid argon at 85 °K as measured by a neutron diffraction experiment (ref. [30]).

value which in this case is an increase of 4.66% over the speed of first sound. The damping $v_0 \alpha / \omega$ behaves like $\omega / \nu (1 - nC)^{\frac{3}{2}}$ in the collision-dominated region and reaches a maximum when $\omega \sim \nu$. In the collisionless region it falls off like $\nu / \omega (3 - nC)^{\frac{3}{2}}$ which implies that the damping coefficient is independent of frequency in the propagation of zero sound.

From (3.44) we can also readily obtain the dispersion relations for the case of real k. In the first-sound regime, $y \gg x$, then

$$(3.53) \qquad \omega = \pm \, kv_0 \sqrt{1 - nC} - i \, \frac{(kv_0)^2}{\nu} \, .$$

The damping is seen to be independent of the mean field. In the zero-sound regime, $y \ll x$, we find

$$(3.54) \qquad \omega = \pm \, kv_0 \sqrt{3 - nC} - i \, \frac{\nu}{3 - nC} \, .$$

Notice that the damping is independent of the wave number.

In summary we have shown that the presence of a mean-field term in the kinetic-equation description leads to collective modes in the frequency region where collision effects are not important. The phase velocity of zero sound is only slightly greater than the first-sound velocity but the damping has a distinctly different behavior. In dense fluids $|nC(0)|$ is sufficiently large that to a good approximation both sound velocities are given by $|nC(0)|^{\frac{1}{2}} v_0$, or $v_0/(S(0))^{\frac{1}{2}}$. In view of (3.38) we see that this is just the isothermal sound speed in a fluid.

From our discussions in Subsect. 3'1, we know that the low-frequency sound speed is the isothermal value because we are using a kinetic model which does not take into account the energy or temperature fluctuations. In a dilute gas one can obtain the adiabatic sound speed simply by using a kinetic model with $M = 3$ or higher. If we used (3.9) to treat the collisional effects, then we would find that in the first term in (3.53) $1 - nC$ would be replaced by $\frac{5}{3} - nC$ [27]. Notice, however, that this does not give the correct sound speed value in a dense fluid. The difficulty lies in the proper description of static and dynamical properties of dense fluids. We will comment on this further in Sect. **5**.

4. – Propagation of density and current fluctuations at short wavelengths.

We have seen that spatial correlations among molecules in a dense fluid result in a restoring force which significantly affects the long-wavelength collective oscillations in the system. We now investigate the effects of dynamical and structural correlations on the propagation of collective modes at short wavelengths. By short wavelengths we mean values of wave number k which are comparable to inverse intermolecular spacing. In this region information on density and current fluctuations is available from neutron inelastic-scattering experiments and computer molecular-dynamics calculations. Our aim is to set up an appropriate description of these fluctuations for discussing basic propagation characteristics such as phase velocity and damping.

To appreciate the behavior of thermal fluctuations at short wavelengths one should keep in mind those properties of the fluid which show significant variations with k. The most important quantity in this respect is the structure factor $S(k)$ which appeared in Subsect. **3·3**. Fig. 8 shows the experimental structure factor of liquid argon at 85 °K determined by neutron diffraction [30]. We have already noted that $S(0)$ is related to the isothermal compressibility, or $v_0/\sqrt{S(0)}$ is the isothermal sound speed. In this case $S(0) \simeq 0.0522$, so that $|nC(0)| \simeq 19.16$. The oscillatory behavior of $S(k)$ is a direct reflection of spatial correlation between neighboring atoms; the Fourier transform $g(r)$ will show similar oscillations in position space. The position of the first maximum in $S(k)$ occurs at $k = 1.995$ Å^{-1}. We can expect that structural correlation effects will manifest themselves most strongly at this value of k. Detailed examination of the data shows that, although $S(k)$ is not independent of k for $k < 1$ Å^{-1}, the rapid oscillations in $S(k)$ do not set in until k is greater than 1 Å^{-1}. We therefore expect the results of Subsect. **3·3** to be valid in a qualitative sense for $k < 1$ Å^{-1}.

The quantity which is measured in neutron scattering experiments is the spectral distribution of density fluctuations $S(k, \omega)$. In the experimental context $\hbar k$ and $\hbar \omega$ appear as the momentum and energy losses of the neutron, but for our discussion we can regard k and ω as the wave number and the frequency of fluctuations. The molecular definiton of $S(k, \omega)$ is as follows. For a fluid system where the j-th molecule at time r is located at $\mathbf{R}_j(t)$ we define the time correlation function

$$(4.1) \qquad G(\mathbf{r} - \mathbf{r}', t) = \frac{1}{n} \left\langle \left[\sum_j \delta(\mathbf{r} - \mathbf{R}_j(t)) - n \right] \left[\sum_{j'} \delta(\mathbf{r}' - \mathbf{R}_{j'}(0)) - n \right] \right\rangle,$$

where $\langle \ \rangle$ denotes an average over a thermodynamic ensemble. Physically $G(\mathbf{r} - \mathbf{r}', t)$ is the density of molecules at $\mathbf{r}$ at time t given that a molecule was located at $\mathbf{r}'$ at time zero [31]. One can also regard $G(r, t)$ as the time-dependent generalization of the radial-distribution function $g(r)$ in the sense that at $t = 0$ one has from (4.1) $G(r, 0) = \delta(r) + ng(r)$. The double Fourier transform of G is S:

$$(4.2) \qquad S(k, \omega) = \frac{1}{2\pi} \int\limits_{-\infty}^{\infty} \mathrm{d}t \int \mathrm{d}^3 r \exp\left[-i(\mathbf{k} \cdot \mathbf{r} - \omega t) \right] G(r, t).$$

A function which is closely related to $S(k, \omega)$ and is more useful in a discussion of propagation at large k is the spectral distribution of the longitudinal-current fluctuations. The current density is defined by

$$(4.3) \qquad j(r, t) = \sum_j V_j(t)\, \delta(r - R_j(t)) \,,$$

where $V_j(t)$ is the velocity of molecule j at time t. The longitudinal component of the current density is that component along the direction of k. Thus the longitudinal-current correlation function is

$$(4.4) \qquad J(r - r', t) = \frac{1}{k^2} \langle [k \cdot j(r, t)][k \cdot j(r', 0)] \rangle \,,$$

and its Fourier transform will be denoted by $J(k, \omega)$. Since the number density and the current density are related by the continuity equation we have the rigorous and useful expression (cf. (3.17))

$$(4.5) \qquad J(k, \omega) = \frac{\omega^2}{k^2} S(k, \omega) \,.$$

It should be noted that we are dealing with classical correlation functions. Therefore S and J are both real functions and even in ω.

The study of density and current fluctuations in dense fluids is a problem of considerable current interest. In the case of dilute gases one can extract S and J from the initial-value solution to the linearized Boltzmann equation. For example,

$$(4.6) \qquad S(k, \omega) = \frac{2}{\pi} \operatorname{Re} \left[\int d^3c\, f_{k\omega}(c) \right] \,,$$

where $f_{k\omega}(c)$ is the solution to (2.18) subject to the initial condition $f_k(c, t = 0) = f_0(c)$. This approach has been studied quite extensively, particularly in connection with light scattering [32]. For dense fluids and the large-(k, ω) region there also exist kinetic-theory calculations [33]. Since these are quite involved and require considerable discussion to develop properly, we will follow a more physical approach which gives qualitatively the same results. The approach is essentially a modification of the Navier-Stokes equations to take into account two effects, nonlocal compressibility and (k, ω)-dependent transport processes [34].

Before examining the results of the generalized hydrodynamic calculations it is important to point out that the frequency moments of the density and current correlation functions have been studied [35]. For our purposes the relevant moments are

(4.7)
$$\int_{-\infty}^{\infty} d\omega\, J(k, \omega) = v_0^2 \,,$$

(4.8)
$$\int_{-\infty}^{\infty} d\omega\, \omega^2 J(k, \omega) = \frac{k^2}{\beta M n} \left[\frac{4}{3} G_\infty(k) + K_\infty(k) \right] \equiv \omega_l^2(k)/\beta M \,,$$

(4.9)
$$\omega_l^2(k) = \frac{3k^2}{\beta M} + \frac{n}{M} \int d^3 r\, g(r)(1 - \cos \boldsymbol{k}\cdot\boldsymbol{r}) \frac{(\boldsymbol{k}\cdot\boldsymbol{\nabla})^2}{k^2} \varphi(r) \equiv c_\infty^2(k) k^2 \,,$$

where $\varphi(r)$ is the pair interaction potential. Equations (4.7) and (4.8) correspond to the second and fourth frequency moments of $S(k, \omega)$. Equation (4.8) shows that the second moment of $J(k, \omega)$ is related to the wavelength-dependent generalization of the high-frequency shear and bulk moduli, G_∞ and K_∞ [36]. Equation (4.9) is useful for computing the moduli at any wave number when the potential and the pair distribution function of the fluid are known. The two terms in (4.9) clearly represent the kinetic and potential contributions. In the long-wavelength limit $Mn\omega_l^2/k^2$ becomes the elastic constant C_{11}.

In eq. (4.9) we have defined a high-frequency sound speed by $c_\infty(k)$ which is explicitly k-dependent. Its variation is shown in Fig. 9 in the case of

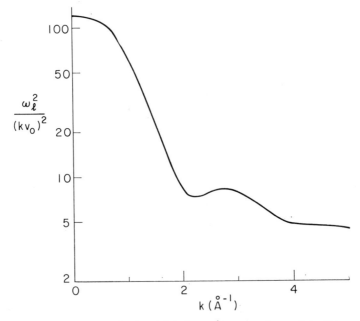

Fig. 9. – The square of dimensionless high-frequency sound speed $c_\infty(k)/v_0$ as computed from eq. (4.9) for liquid argon at 85 °K.

liquid argon at 85 °K [37]. The calculation was carried out using the Lennard-Jones potential and the experimental pair distribution derived from the results of Fig. 8. The essential behavior of $c_\infty^2(k)$ is a strong fall-off with increasing k, and in the vicinity of the first diffraction maximum $(k \sim 2 \text{ Å}^{-1})$ the effects of structural correlations can be seen. Both $S(k)$ and $c_\infty^2(k)$ will appear in the dispersion relations we now obtain from the results of generalized hydrodynamics.

As mentioned above, an empirical method of extending the hydrodynamic description of fluctuations to the large-(k, ω) region is to introduce nonlocal compressibility and (k, ω)-dependent transport coefficients. The compressibility has to be modified because at short wavelengths the local structure in the fluid cannot be ignored. Since the isothermal compressibility is directly proportional to $S(k)$ in the limit $k \to 0$, we can use (3.38) to define a k-dependent isothermal compressibility in terms of $S(k)$. Modification of the transport coefficients is carried out by replacing the longitudinal viscosity coefficient in the Navier-Stokes equation by an integral operator in space and time. The operator is characterized by a displacement-type kernel in both space and time so that after the equation is transformed to k and ω variables a complex damping function $D(k, \omega)$ appears. The net result of these modifications is an expression for the longitudinal-current correlation function [38]

$$(4.10) \qquad J(k, \omega) = 2v_0^2 \frac{\omega^2 k^2 D'(k, \omega)}{[\omega^2 - v_0^2 k^2/S(k) + \omega k^2 D''(k, \omega)]^2 + [\omega k^2 D'(k, \omega)]^2},$$

where D'' and D' are the imaginary and real parts of the damping function or generalized susceptibility which satisfy the Kramers-Kronig relations; for example,

$$(4.11) \qquad D''(k, \omega) = P \int_{-\infty}^{\infty} \frac{d\omega'}{\pi} \frac{D'(k, \omega')}{\omega' - \omega}.$$

Until we specify the damping function (4.10) is only a formal expression defining one unknown function, $J(k, \omega)$, in terms of another, $D'(k, \omega)$. As such no approximation has been made. The utility of the expression stems from the general belief that the damping function is a less varied quantity than the correlation function; therefore it should be more amenable to approximation. The form of eq. (4.10) suggests the interpretation of a generalized dispersion formula. It is tempting to think of $kv_0/\sqrt{S(k)}$ as the resonant frequency associated with the average potential field effects discussed in Subsect. **3˙3**. Then the D' and D'' terms in the denominator represent the damping and frequency shift arising from collisional dynamics.

The form of (4.10) is also quite convenient for analyzing the dispersion relations of current fluctuations in dense fluids. Because of (4.5) the same dispersion relations also apply to density fluctuations. We consider the damping function which is motivated by the hydrodynamic behavior of $J(k, \omega)$ and $S(k, \omega)$ [38, 39]:

$$(4.12) \quad D'(k, \omega) = [c_\infty^2(k) - c_0^2(k)] \frac{\tau(k)}{1 + [\omega\tau(k)]^2} + \frac{\gamma - 1}{\gamma} c_0^2(k) \frac{1/\gamma D_T k^2}{1 + (\omega/\gamma D_T k^2)^2},$$

where $c_0^2(k) = \gamma v_0^2/S(k)$. For this choice of D' the imaginary part of the damping function is

$$(4.13) \quad D''(k, \omega) = -[c_\infty^2(k) - c_0^2(k)] \frac{\omega\tau^2(k)}{1 + [\omega\tau(k)]^2} - \frac{\gamma - 1}{\gamma} c_0^2(k) \frac{\omega/(\gamma D_T k^2)^2}{1 + (\omega/\gamma D_T k^2)^2}.$$

In (4.12) the two terms represent the damping effects of viscous flow and the thermal conduction respectively. It is assumed that dissipation effects decay in time in a simple exponential manner. The relaxation time of thermal conduction processes $(\gamma D_T k^2)^{-1}$ follows directly from the hydrodynamic structure of the correlation functions, whereas the shear relaxation time $\tau(k)$ is a phenomenological quantity that has yet to be specified.

We first examine the behavior of this approximate damping function in the hydrodynamic region. Inserting (4.12) and (4.13) into (4.10) one finds that the denominator gives two resonant frequencies which corresond to the heat mode and sound modes. Since the current correlation function contains a factor of ω^2 in the numerator, the heat mode is suppressed and the spectral distribution peaks in the vicinity of the sound frequency. However, the density correlation function $S(k, \omega)$, by virtue of (4.5), will show a peak centred about $\omega = 0$ and a peak at $\omega = c_0(k)k$. These two peaks are well known in light scattering in fluids as the Rayleigh and Brillouin components respectively.

It is instructive to analyze the effects of D' and D'' on the sound frequency and damping. The sound frequency in this case is determined by the condition

$$(4.14) \quad \omega^2 - \frac{c_0^2(k)k^2}{\gamma} + \omega k^2 D''(k, \omega) = 0 .$$

In the vicinity of the sound peak ω is of order k, therefore at small k the thermal conduction term in (4.13) contributes a term $(\gamma - 1)c_0^2(k)k^2/\gamma$ in (4.14). Equation (4.14) then becomes

$$(4.15) \quad \omega^2 = c_0^2(k)k^2 + [c_\infty^2(k) - c_0^2(k)]k^2 \frac{\omega^2\tau^2(k)}{1 + \omega^2\tau^2(k)}.$$

Although we have written c_0, c and τ as k-dependent quantities, (4.15) is only valid when $D_T k^2 \ll \omega$. This implies that k should be small because (4.15) shows that ω is indeed proportional to k.

We now observe that at frequencies such that $\omega \ll 1/\tau(k)$ (4.15) gives the sound frequency at

(4.16)
$$\omega = \pm c_0(k)\, k \;,$$

which is the adiabatic sound frequency if we ignore the k-dependence in $c_0(k)$. Since $c_0(k)$ is $v_0[\gamma/S(k)]^{\frac{1}{2}}$ for dense fluids and the typical behavior of $S(k)$ is shown in Fig. 8, we see that unless k is of order 10^8 cm^{-1} this neglect is justified. At high frequencies, $\omega \gg 1/\tau(k)$, (4.15) gives

(4.17)
$$\omega = \pm c_\infty(k)\, k \;,$$

where $c_\infty(k)$, as defined in (4.9), is related to the high-frequency shear and bulk moduli. The present situation is completely analogous to the discussion of Subsect. 3`3 with the shear relaxation time replacing $1/\nu$. Therefore, according to our earlier terminology, $c_0(k)$ is the extension of the speed of first sound to finite wavelength and $c_\infty(k)$ is the corresponding speed of zero sound. In this context (4.15) is an interpolation formula valid for both low- and high-frequency fluctuations.

We can also show that the damping term D' is a frequency- and wavelength-dependent generalization of the classical sound attenuation Γ defined in (2.5). At long wavelengths and in the vicinity of the sound peak, D' behaves like

(4.18)
$$D' - [c_\infty^2(0) - c_0^2(0)]\, \tau(0) + (\gamma - 1)\, D_T \;.$$

It is reasonable to identify $\tau(0)$ as

$$c_\infty^2(0)\, \tau_M/[c_\infty^2(0) - c_0^2(0)] \;,$$

where τ_M is the Maxwell relaxation time for the longitudinal viscosity [40]. Then

(4.19)
$$D' = \frac{1}{Mn}\left(\frac{4}{3}\eta_s + \eta_v\right) + (\gamma - 1)\, D_T \;,$$

which is identical to Γ.

We have established that (4.12) and (4.13) describe correctly the behavior of density and current fluctuations in the region of small wave numbers. This particular damping function also enables us to discuss collective-mode fluctuations at wave numbers of order 10^8 cm^{-1} where experimental information from neutron scattering and computer molecular dynamics is available. The extension of our calculations to the short-wavelength region is achieved by treating the full k-dependence of $c_\infty(k)$ and $c_0(k)$. In addition we will allow the shear relaxation time to vary with k. The value of $\tau(0)$ is already fixed by the shear and bulk viscosities and one expects that at finite k the relaxation time will be somewhat shorter. It has been found that $\tau(k)$ decreases with increasing k much less rapidly than $c_\infty(k)$ [34].

In discussing collective oscillations at short wavelengths it has become conventional to consider the current correlation function, and in particular to focus attention on the wavelength variation of $\omega_{max}(k)$, the frequency where $J(k, \omega)$ has its maximum value. We therefore define a sound speed

$$(4.20) \qquad\qquad c_l(k) = \frac{\omega_{max}(k)}{k} \, ,$$

which holds for all k. In the small-k limit $\omega_{max}(k)$ is also the frequency of the sound peak in the density correlation function $S(k, \omega)$. We introduce another sound speed

$$(4.21) \qquad\qquad c_B(k) = \frac{\omega_B(k)}{k} \, ,$$

where $\omega_B(k)$ is the position of the Brillouin component in $S(k, \omega)$. Notice that, unlike $c_l(k)$, $c_B(k)$ is well defined only if the sound peak in $S(k, \omega)$ is reasonably well resolved. When one is no longer in the hydrodynamic region the fluctuations will be sufficiently damped that $S(k, \omega)$ shows only a broad peak centred at $\omega = 0$. We therefore expect that $c_l(k)$ and $c_B(k)$ will be the same at small k and, when they deviate from each other, $c_l(k)$ will be greater than $c_B(k)$ because of the ω^2 factor in the numerator of (4.10). At sufficiently large k, $c_B(k)$ will cease to exist while $c_l(k)$ is well defined at any k value.

The various propagation velocities we have introduced are useful in discussing the collective behavior of thermal fluctuations at short wavelengths. The low- and high-frequency sound speeds $c_0(k)$ and $c_\infty(k)$ are both well-defined quantities at any wavelength. Since $c_0^2(k)$ varies inversely as the structure factor and $c_\infty^2(k)$ is directly proportional to the generalized elastic moduli, they appear to represent somewhat different structural properties of the fluid. (Recall also the long-wavelength limits of $c_0(k)$ and $c_\infty(k)$.) The velocity $c_l(k)$ is also well defined for all k. This is a more complicated quantity since it is determined by (4.10), which describes both structural and dynamical correlations in the current density correlation function. Therefore one would not expect $c_l(k)$ to have the same behavior as either $c_0(k)$ or $c_\infty(k)$. Lastly, when $c_B(k)$ can be defined, we expect its behavior to be similar to that of $c(k)$.

We have examined numerically all of these velocities over a wide range of wave numbers in the case of liquid argon at 76 °K. For $k < 10^6$ cm^{-1} none of the velocities shows any dispersion, so we need to consider only the region of larger k values. Figure 10 shows the variation with wave number of the velocities up to $k = 4 \cdot 10^8$ cm^{-1} [41]. In addition to the above velocities we have also included the velocity obtained from the approximate dispersion equation (4.15); this result is denoted as $c_a(k)$.

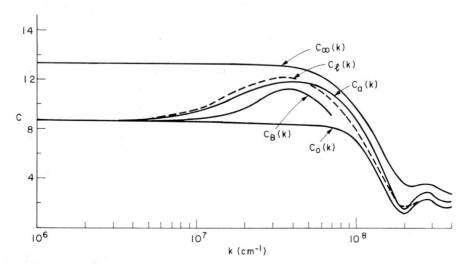

Fig. 10. – Calculated propagation velocities (in units of 10^4 cm/s) in liquid argon at 76 °K, with $c_0(k) = v_0(\gamma/S(k))^{\frac{1}{2}}$, $c_\infty(k)$, $c_l(k)$, and $c_B(k)$ are defined respectively by (4.9), (4.20) and (4.21), and $c_a(k)$ is calculated from (4.15) by setting $\omega(k) = kc_a(k)$ (ref. [41]).

In the calculations shown the relaxation time was estimated using the expression [42]

$$(4.22) \qquad \tau^{-2}(k) = \frac{8}{3}[c_\infty^2(k) - c_0^2(k) - v_0^2]k^2 + \frac{\tau^{-2}(0) + \frac{8}{3}[c_\infty^2(k) - c_0^2(k) - 2v_0^2]k^2}{1 + (k/k_0)^2},$$

where

$$(4.23) \qquad \tau^{-1}(0) = \frac{c_\infty^2(0) - c_0^2(0)}{\frac{4}{3}\eta_s + \eta_v}$$

and $k_0 = 1.5$ Å^{-1}. This expression was motivated by considering the behavior of $J(k, \omega)$ at large k, while still retaining the property of $\tau(0)$ as a Maxwell relaxation time. Other expressions are available, but (4.22) has been found to give quite reasonable results. There are presently no satisfactory theories for calculating $\tau(k)$.

In Fig. 9 one can discern basically two types of behaviors: a positive dispersion in the region $4 \cdot 10^6 \leqslant k \leqslant 4 \cdot 10^7$, and a strongly negative dispersion with subsequent oscillations occurring for $k > 4 \cdot 10^7$. It is significant that $c_l(k)$ and $c_a(k)$ display both types of behaviors, whereas $c_0(k)$ and $c_\infty(k)$ show only the negative dispersion. The behavior of $c_B(k)$ is consistent with our expectations; it is perhaps a little surprising that collective oscillations in $S(k, \omega)$ can be discerned up to wave numbers as large as $0.7 \cdot 10^8$ cm^{-1}. Notice also that $c_a(k)$ is a reasonably good approximation up to about $k = 2 \cdot 10^8$.

From the previous discussions it is clear that the process giving rise to the positive dispersion effect is shear relaxation. This is the only dynamical effect treated in both (4.10) and (4.15). Such an effect is not present in (4.9) or in the structure factor, so it is not surprising that $c_\infty(k)$ and $c_0(k)$ do not show any positive dispersion. It should be clear also that the rapid decrease in all the velocities is a result of the strong spatial correlations among near neighbors in the liquid. Although the velocities seem to depend on the local structure in the liquid in different ways, the fact that the results all behave similarly implies that the sharp decrease and subsequent oscillations are a general feature characteristic of short-wavelength collective modes in a dense fluid.

There has been considerable interest in the results on $\omega_{\max}(k)$ as observed by coherent neutron inelastic scattering and calculated by computer molecular-dynamics experiments [43]. The variation of $\omega_{\max}$ with k in the region $k > 0.4 \cdot 10^8$ is often interpreted as evidence for a dispersion relation similar to the acoustic-phonon branch in solids [44]. To compare our model calculations with neutron [45] and computer [46] data we have redrawn the large-k region of Fig. 10 on a linear scale. This comparison is shown in Fig. 11 [41]. We see that the data

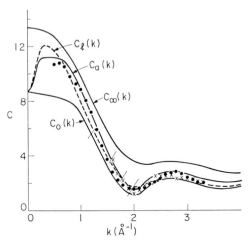

Fig. 11. – Comparison of the calculated propagation velocities of Fig. 10 with experimental results, neutron scattering measurements ($\times$ ref. [45], / ref. [46]) and computer molecular-dynamics results ($\bullet$ ref. [47]) (ref. [41]). No corrections for the small differences in liquid temperature and density have been made.

points all lie between $c_\infty(k)$ and $c_0(k)$. The fact that $c_i(k)$ and also $c_a(k)$ are in better agreement with the data than $c_\infty(k)$ or $c_0(k)$ indicates that dynamical or lifetime effects are important in a quantitative description of fluctuations in the short-wavelength region.

We may conclude that (4.10) constitutes a simple but semi-quantitatively correct description of density and current fluctuations in a dense fluid. While it is not a molecular theory in the sense of the Boltzmann equation, it does contain molecular quantities which depend explicitly on the interatomic potential and the pair distribution function. Lest the reader become too complacent about the apparently simple state of affairs, a number of comments on recent developments are in order. We offer these as concluding remarks in the next Section.

5. – Concluding remarks.

In these lectures our main concern has been to elucidate the various processes which govern sound propagation at different frequencies in classical fluids ranging from dilute gases to liquids. In the interest of pedagogy we have attempted to stress only the essential ideas, and in so doing we have ignored a number of recent developments which, although relevant, would have made the discussions more complicated. It seems appropriate that we should conclude the lectures with some comments on these developments.

In Sect. 2 we noted the limited utility of power series expansions of the dispersion relations, but we did not question the existence of such series. Recent theoretical studies in kinetic theory have shown that the hydrodynamic dispersion relations are in fact nonanalytic, *i.e.* they do not possess a power series expansion [48]. The source of this difficulty lies in the effects of certain sequences of binary collisions in which particle pairs recollide after a number of other collisions in the medium. These events, called the ring collisions, are not treated in the conventional Boltzmann equation, but they become important when one tries to extend the Boltzmann description to higher densities. Because of these terms one can show that the transport coefficients which appear in the Navier-Stokes equations do not have a power series expansion in the density [49]. Another consequence of the ring collisions is that time correlation functions can decay at long times like $t^{\frac{3}{2}}$ instead of exponentially [50]. This long-time tail behavior, which was first oberved in computer molecular-dynamics studies of hard-sphere fluids [51], implies that time integrals involving the correlation function multiplied by powers of t in the integrand will not exist; in particular, the transport coefficients in the Burnett and higher-order approximations to hydrodynamics have been found to be divergent [52].

In the light of these results it is perhaps not surprising that the ring collisions are also responsible for the nonanalyticity of the hydrodynamic dispersion relations. To demonstrate this property one can analyze the Boltzmann equation with the addition of the collision operator describing the ring collisions. The presence of the ring terms leads to an expansion of the sound modes in the form [48]

$$\omega(k) = \pm c_0 k - i\frac{\Gamma}{2} k^2 - \Delta k^{\frac{5}{2}} + \ldots ,$$

where Δ is complex. In addition to fractional powers, terms like $k^r \ln k$, $r \geqslant 2$, also appear. The magnitude of Δ was estimated by evaluating the first nonzero term in its density expansion in the case of a hard-sphere gas [48]. It was found that the ratio $2|\Delta|/\Gamma$ in dimensionless form is of order $(n\sigma^3)\varkappa^2$, where n is the gas density, σ the hard-sphere diameter and $\varkappa$ the ratio of mean free path to wavelength ($\varkappa \propto 1/y$). For a gas at STP and assuming $\varkappa \sim 0.1$, one sees that the $k^{\frac{5}{2}}$ term is smaller than the k^2 term by a factor of 10^7. One may conclude that, while terms like $k^{\frac{5}{2}}$ are present in an expansion of the dispersion relations, their effects are negligible so far as possible experimental detection is concerned. The overall implication of the nonanalyticity property is that one must now question the theoretical basis of hydrodynamic approximations (beginning with the Burnett equations) which are derived by considering higher powers of wavelength to mean free path ratio. However, the theoretical results do not preclude the fact that Burnett and higher-order approximations may still be useful in practice.

We have considered two different levels of analyzing propagation phenomena in fluids. The kinetic-equation approach enabled us to treat the effects of molecular flow and collisions on an equal footing, and this has led to a detailed understanding of the sound propagation data on dilute gases. The generalized hydrodynamics method is a more physical approach to the study of collective modes in dense fluids. Being at the more macroscopic level, one cannot calculate the frequency dependence of the damping function explicitly in terms of collisional dynamics. On the other hand, the formalism is very convenient for introducing models to describe the features that have been observed in neutron scattering and computer experiments. From a practical standpoint generalized hydrodynamics calculations are much more accessible than kinetic theory except in the case of dilute gases.

The use of kinetic theory is appropriate when studying low-density fluids where neither the collisional effects nor free molecular flow should be ignored. The kinetic models of Sect. **3** provide a tractable means of analyzing propagation phenomena for frequencies up to collision frequencies. However, when the wavelengths are as short as the molecular dimension and the frequencies are comparable to inverse duration of collision, the Boltzmann collision operator needs to be modified. The properly generalized collision operator has been derived recently [53] and analyzed in the case of hard spheres [54]. It is interesting that, within the framework of this generalization, the conventional Boltzmann equation appears as the limit when both $n\sigma^3$ and $k\sigma$ approach zero but their ratio is finite and arbitrary.

The kinetic-theory approach becomes much more difficult when applied to dense fluids. Here the mean-field effects are large, but they present no difficulty. In fact one knows from general theoretical considerations that the mean-field term of Subsect. **3'3** correctly describes the static correlations in the system. However, the kinetic-model calculations of Subsect. **3'3** are much too simplified to give quantitative results even in the low-frequency limit. The difficulty lies in the proper treatment of collisions. Recently, significant progress has been made in deriving kinetic equations which consider collisions in a more general way than the Boltzmann equation [55]. Analysis of the low-frequency region has shown that the adiabatic sound speed is the result of effects involving both the mean-field term and collisions. This kind of coupling effects is not present if the collision operator is not k and ω dependent. In this connection one should notice that collisions do not conserve energy unless $k, \omega \to 0$ and they do not conserve momentum unless $k \to 0$. The high-frequency region has yet to be studied in detail, although model calculations have given good results in the analysis of computer and neutron data [56].

The generalized hydrodynamics method is particularly well suited to the study of high-frequency wave propagation in dense fluids [57]. In (4.10) the term $k^2 v_0^2 / S(k)$ plays the same role as the mean-field term in a kinetic equation. However, the presence of the complex damping function significantly affects the propagation properties at all frequencies. At low frequencies the sound speed is not the isothermal value as would be the case if the damping function were k and ω independent. The approximation (4.13) is seen to shift the sound speed to the correct adiabatic value. This is an example of mode-coupling effects [38, 39] and corresponds to the interaction between the mean-field term and the collision operator in the kinetic-theory analysis.

Our results in Sect. **4** show that there exists an intermediate region of wavelengths where shear relaxation effects give rise to a positive dispersion and the structural effects related to $S(k)$ have not yet set in. This region is characterized by $k v_0 \tau(k) \sim 1$ provided k is still appreciably smaller than the position of the first diffraction maximum in $S(k)$. In the case of liquid argon this means that k is of the order 10^7 cm^{-1}. These k values are too large for laser light scattering and are somewhat too small for most neutron scattering experimental facilities [58]. Experimental study of this region will be most interesting. The behavior of correlation functions here is likely to be more complicated than that shown in Fig. 10 because of the presence of another characteristic relaxation time representing the dynamics of small clusters of molecules. Computer studies have indicated that such a relaxation time is needed if one is to extrapolate from the large-(k, ω) region to obtain correctly hydrodynamics [59]. Recent neutron data also provide evidence for this second relaxation time [60]. It is certainly tempting to interpret this collective effect as having the same origin as the long-time tails of correlation functions and other related phenomena.

In conclusion we note that computer molecular-dynamics and neutron scattering studies of the high-frequency and short-wavelength region will continue to stimulate the development of a molecular theory of fluid dynamics. A basic requirement of such a theory is that it must be valid for fluids of different densities and is applicable to various regions of frequency and wavelength. The understanding of how collective modes behave under various conditions is clearly an important part of the investigation. On this basis we can expect new ideas and interesting results to emerge for quite some time to come.

* * *

This work was supported by a grant from the National Science Foundation.

REFERENCES

[1]   J. D. FOCH and G. W. FORD: in *Studies in Statistical Mechanics*, edited by J. DEBOER and G. E. UHLENBECK (Amsterdam, 1970).

[2]   C. S. WANG CHANG and G. E. UHLENBECK: in *Studies in Statistical Mechanics*, edited by J. DEBOER and G. E. UHLENBECK, Chap. IV (Amsterdam, 1970).

[3]   G. E. UHLENBECK and G. W. FORD: *Lectures in Statistical Mechanics* (Providence, R. I., 1963).

[4]   S. CHAPMAN and T. G. COWLING: *The Mathematical Theory of Nonuniform Gases*, third edition (Cambridge, 1970).

[5]   C. L. PEKERIS, Z. ALTERMAN, L. FINKELSTEIN and K. FRANKOWSKI: *Phys. Fluids*, **5**, 1608 (1962).

[6]   L. SIROVICH: *Phys. Fluids*, **6**, 10 (1963).

[7]   H. GRAD: in *Rarefied Gas Dynamics*, edited by J. A. LAURMAN, Vol. **1** (New York, N. Y., 1963); J. H. FERZIGER: *Phys. Fluids*, **8**, 426 (1965).

[8]   See, for example, C. CERCIGNANI: *Mathematical Methods in Kinetic Theory* (New York, N. Y., 1969).

[9]   E. P. GROSS and E. A. JACKSON: *Phys. Fluids*, **2**, 432 (1959).

[10] L. SIROVICH: *Phys. Fluids*, **5**, 908 (1962).

[11] See the discussions in Chap. V of ref. [1]. For numerical results see ref. [15] and A. SUGAWARA, S. YIP and L. SIROVICH: *Phys. Fluids*, **11**, 925 (1968).

[12] P. F. BHATNAGAR, E. P. GROSS and M. KROOK: *Phys. Rev.*, **94**, 511 (1954).

[13] K. M. CASE and P. F. ZWEIFEL: *Linear Transport Theory* (Reading, Mass., 1967).

[14] E. T. WHITTAKER and G. N. WATSON: *Modern Analysis*, fourth edition (Cambridge, 1958), p. 119.

[15] L. SIROVICH and J. K. THURBER: *Journ. Acoust. Soc. Amer.*, **37**, 329 (1965).

[16] M. GREENSPAN: *Journ. Acoust. Soc. Amer.*, **22**, 568 (1950); **28**, 644 (1955).

[17] E. MEYER and G. SESSLER: *Zeits. Phys.*, **149**, 15 (1957).

[18] The interested reader should see, besides ref. [15, 19, 20], the discussions in G. MAIDANIK and H. L. FOX: *Journ. Acoust. Soc. Amer.*, **38**, 477 (1965); L. SIROVICH and J. K. THURBER: *Journ. Acoust. Soc. Amer.*, **38**, 478 (1965). See also the papers on sound propagation in the *Proceedings of the International Symposium on Rarefied Gas Dynamics*.

[19] J. K. BUCKNER and J. H. FERZIGER: *Phys. Fluids*, **9**, 2315 (1966).

[20] D. KAHN and D. MINTZER: *Phys. Fluids*, **8**, 1090 (1965).

[21] N. CORNGOLD: *Nucl. Sci. Eng.*, **19**, 80 (1964).

[22] D. PINES: in *Quantum Fluids*, edited by D. F. BREWER (Amsterdam, 1966); D. PINES and P. NOZIÈRES: *Theory of Quantum Liquids* (New York, N. Y., 1966).

[23] R. T. BEYER and S. V. LETCHER: *Physical Ultrasonics* (New York, N. Y·, 1969).

[24] D. C. MONTGOMERY and D. A. TIDMAN: *Plasma Kinetic Theory*, Chap. 5 (New York, N. Y., 1964); see also ref. [39], Sect. G.2.

[25] R. A. COWLEY: *Proc. Phys. Soc.*, **90**, 1127 (1967); E. C. SVENSSON and W. J. L. BUYERS: *Phys. Rev.*, **165**, 1063 (1968).

[26] J. K. PERCUS: in *The Equilibrium Theory of Classical Fluids*, edited by H. L. FRISCH and J. L. LEBOWITZ, Appendix A (New York, N. Y., 1964); R. ZWANZIG: *Phys. Rev.*, **144**, 170 (1966).

[27] M. NELKIN and S. RANGANATHAN: *Phys. Rev.*, **164**, 222 (1967); see also J. CHIHARA: *Prog. Theor. Phys.*, **41**, 285 (1969).

[28] P. A. EGELSTAFF: *An Introduction to the Liquid State* (London, 1967).

[29] V. N. FADDEYEVA and N. M. TERENT'EV: *Tables of Probability Integral for Complex Arguments* (London, 1961); B. D. FRIED and S. D. CONTE: *The Plasma Dispersion Function* (New York, N. Y., 1961).

[30] J. L. YARNELL, M. J. KATZ, R. G. WENZEL and S. H. KOENIG: *Phys. Rev.*, **7**, 2130 (1973).

[31] L. VAN HOVE: *Phys. Rev.*, **95**, 249 (1954).

[32] S. YIP: *Journ. Acoust. Soc. Amer.*, **49**, 941 (1971) and references given therein.

[33] J. J. DUDERSTADT and A. Z. AKCASU: *Phys. Rev. A*, **1**, 905 (1970); D. FORSTER and M. S. JHON: *Phys. Rev.*, *A*, to be published (1975).

[34] C. H. CHUNG and S. YIP: *Phys. Rev.*, **182**, 323 (1969); A. Z. AKCASU and E. DANIELS: *Phys. Rev. A*, **2**, 962 (1970); N. K. AILAWADI, A. RAHMAN and R. ZWANZIG: *Phys. Rev. A*, **4**, 1616 (1971).

[35] See, for example, D. FORSTER, P. C. MARTIN and S. YIP: *Phys. Rev.*, **170**, 155 (1968). Additional calculations have been reported recently by R. BANSL and K. N. PATHAK: *Phys. Rev. A*, **9**, 2773 (1974).

[36] R. ZWANZIG and R. D. MOUNTAIN: *Journ. Chem. Phys.*, **43**, 4464 (1965).

[37] P. FURTADO: private communication.

[38] L. P. KADANOFF and P. C. MARTIN: *Ann. of Phys.*, **24**, 419 (1963); P. C. MARTIN: in *Statistical Mechanics of Equilibrium and Nonequilibrium*, edited by J. MEIXNER (Amsterdam, 1965), p. 100.

[39] P. C. MARTIN: *Measurements and Correlation Functions* (New York, N. Y., 1968).

[40] J. FRENKEL: *Kinetic Theory of Liquids*, Chap. IV (New York, N. Y., 1955).

[41] C. H. CHUNG: Ph. D. Thesis, Massachusetts Institute of Technology (1969); C. H. CHUNG and S. YIP: *Phys. Lett.*, **50** A, 175 (1974).

[42] A. Z. AKCASU and E. DANIELS: *Phys. Rev. A*, **2**, 962 (1970).

[43] For a recent discussion see J. COPLEY and S. W. LOVESEY: *Rep. Prog. Phys.*, to appear, and references given therein.

[44] See, for example, P. D. RANDOLPH and K. S. SINGWI: *Phys. Rev.*, **152**, 99 (1966).

[45] K. SKOLD and K. E. LARSSON: *Phys. Rev.*, **161**, 102 (1967). Similar results have been obtained in a more recent measurement of liquid argon 36 at 85.2 °K, K. SKOLD, J. M. ROWE, G. OSTROWSKI and P. D. RANDOLPH: *Phys. Rev. A*, **6**, 1107 (1972).

[46] S. H. CHEN, O. J. EDER, P. A. EGELSTAFF, B. C. G. HAYWOOD and F. J. WEBB: *Phys. Lett.*, **19**, 269 (1965).

[47] A. RAHMAN: in *Inelastic Scattering of Neutrons*, Vol. **1** (Vienna, 1968), p. 561.

[48] M. H. ERNST and J. R. DORFMAN: *Physica*, **61**, 157 (1972); *Phys. Lett.*, **38** A, 269 (1972).

[49] K. KAWASAKI and I. OPPENHEIM: *Phys. Rev.*, **139**, A 1763 (1965).

[50] See J. R. DORFMAN and E. G. D. COHEN: *Phys. Rev. A*, **6**, 776 (1972) and references therein.

[51] B. J. ALDER and T. E. WAINWRIGHT: *Phys. Rev. A*, **1**, 18 (1970).

[52] J. R. DORFMAN and E. G. D. COHEN: *Phys. Rev. Lett.*, **25**, 1257 (1970).

[53] G. F. MAZENKO: *Phys. Rev. A*, **3**, 2121 (1971); **6**, 2545 (1972).

[54] G. F. MAZENKO, T. Y. C. WEI and S. YIP: *Phys. Rev. A*, **6**, 1981 (1972).

[55] See G. F. MAZENKO: *Phys. Rev. A*, **9**, 360 (1974) and references therein. For formal properties see also D. FORSTER: *Phys. Rev. A*, **9**, 943 (1974).

[56] J. J. DUDERSTADT and A. Z. AKCASU: *Phys. Rev. A*, **1**, 905 (1970); also D. FORSTER and M. JHON: *Phys. Rev.*, to appear.

[57] N. S. GILLIS and R. D. PUFF: *Phys. Rev. Lett.*, **16**, 606 (1966).

[58] Recent measurements have reached a k value as small as $6 \cdot 10^6$ cm^{-1}. See H. BELL, A. KOLLMAR, B. ALEFELD and T. SPRINGER: *Phys. Lett.*, **45** A, 479 (1973).

[59] D. LEVESQUE, L. VERLET and J. KURKIJARVI: *Phys. Rev. A*, **7**, 1690 (1973).

[60] J. COPLEY and J. M. ROWE: *Phys. Rev. A*, **9**, 1656 (1974); see also A. RAHMAN: *Phys. Rev. A*, **9**, 1667 (1974).

Light Scattering and Molecular Acoustics.

C. J. MONTROSE

Department of Physics, Catholic University of America - Washington, D.C.

1. – Introduction.

When a transparent medium is illuminated with a monochromatic beam of light, a small but detectable fraction of the light is scattered away from the beam. The frequency spectrum of the scattered light differs considerably from the monochromatic distribution of the incident beam. Roughly speaking the scattered-light spectrum arises from a modulation of the light field by the motion of the molecules responsible for the scattering. Consequently the study of this spectrum can be used to infer various aspects of molecular dynamics. Certain of these dynamics can be described in terms of the acoustical modes of the system, and as a result light scattering experiments can be used to study the acoustical properties of the scattering medium. The description of how this comes about is the subject of this contribution.

The type of experiment we have in mind involves the scattering through an angle θ of a beam from a continuous-wave laser by a dense fluid. For definiteness, we are thinking of something like an argon ion laser producing about 0.5 watt of optical power at a wavelength 514.5 nm. The frequency spread of

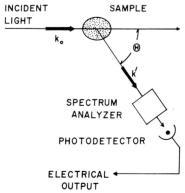

Fig. 1. – Schematic representation of a typical light scattering experiment designed to measure the spectrum of the scattered light.

the laser is on the order of 10^7 Hz, *i.e.* it is monochromatic to about 2 parts in 10^8. We shall be concerned with situations in which the incident beam is polarized vertically, that is perpendicular to the scattering plane defined by the wave vectors of the incident and scattered beams, $\boldsymbol{k}_0$ and $\boldsymbol{k}'$ respectively. The scattered light can be resolved into vertical and horizontally polarized components; we shall deal only with that component that is vertically polarized, and we designate this as vv, *i.e.* vertical-vertical, or polarized scattering.

For the most part we shall be discussing experiments in which the frequency distribution of the scattered light is measured by an optical spectrum analyzer (usually a Fabry-Perot interferometer or a grating spectrometer) before photo-electronic detection. The general features of the scattering arrangement are shown in Fig. 1. We shall briefly consider an alternate procedure in which the scattered light is allowed to fall directly on a photoelectronic detector, *e.g.* a photomultiplier tube, and the characteristics of the photocurrent are measured. The two approaches will be seen to be complementary rather than competitive.

2. – Light scattering.

We consider an incident light beam characterized by an intensity I_0, frequency ω_0, wave vector $\boldsymbol{k}_0$ ($\omega_0 = ck_0/n$, where c is the speed of light and n is the refractive index of the medium) and unit polarization vector $\hat{e}_0$. The light scattered by the sample through an angle θ is described by an intensity I, wave vector $\boldsymbol{k}'$, polarization vector $\hat{e}'$ and a spectral distribution $\sigma(\omega)$, where ω is the *shift* in frequency of the scattered light. The question we wish to address is: what can we learn about the scattering medium (especially about its acoustical properties) by measuring the intensity I and spectral distribution $\sigma(\omega)$ of the scattered light?

The underlying physics of the scattering of light by fluctuations in a fluid can be expressed quite simply [1]. The electric field of a beam of light incident upon the fluid induces oscillating dipole moments in the molecules of the fluid. These serve as source terms for the scattered radiation. The details of the scattered light depend upon the dynamical fluctuations, *i.e.* the molecular motions, in the fluid and hence these details can be used to infer fluid properties once the theory is worked out.

We assume the liquid to be composed of rigid molecules whose electrical properties are characterized by a polarizability tensor $\boldsymbol{\alpha}$. The incident electric field is given as

$$(2.1) \qquad \boldsymbol{E}_{\text{inc}}(\boldsymbol{r}, t) = \hat{e}_0 E_0 \exp\left[i(\boldsymbol{k}_0 \cdot \boldsymbol{r} - \omega_0 t)\right],$$

and the moment induced on the m-th molecule is

$$(2.2) \qquad \boldsymbol{\mu}_m(t) = \boldsymbol{\alpha}_m \cdot \boldsymbol{E}(\boldsymbol{r}_m, t) = (\boldsymbol{\alpha}_m \cdot \hat{e}_0) E_0 \exp\left[i(\boldsymbol{k}_0 \cdot \boldsymbol{r}_m(t) - \omega_0 t)\right],$$

where $\boldsymbol{r}_m$ denotes the position of the molecule. The usual calculation of electro-magnetic-radiation theory leads to the expression

$$(2.3) \qquad \boldsymbol{E}_m(R, t) = -\frac{\exp[ik'R]}{R} \exp[-i\omega_0 t]k_0^2 \sin\chi\, E_0(\boldsymbol{\alpha}_m \cdot \hat{e}_0) \exp[i\boldsymbol{k} \cdot \boldsymbol{r}_m(t)],$$

where R is the distance from the scattering molecule to the detector (we assume $R \gg |\boldsymbol{r}_m|$), χ is the angle between $\hat{e}_0$ and $\boldsymbol{k}'$, and $\boldsymbol{k}$, the scattering vector, is defined by

$$(2.4) \qquad\qquad\qquad \boldsymbol{k} = \boldsymbol{k}_0 - \boldsymbol{k}'.$$

The total scattered field is obtained by summing (2.3) over all the molecules in the scattering volume:

$$(2.5) \qquad\qquad \boldsymbol{E}(\theta, t) \propto \exp[-i\omega_0 t] \sum_m (\boldsymbol{\alpha}_m \cdot \hat{e}_0) \exp[i\boldsymbol{k} \cdot \boldsymbol{r}_m(t)],$$

where we have shown explicitly the dependence of the scattered field on θ. This dependence arises geometrically from the factor $\sin\chi$ (which we have not exhibited in (2.5)) and in terms of the liquid structure through the factor $\exp[i\boldsymbol{k} \cdot \boldsymbol{r}_m(t)]$, since

$$(2.6) \qquad\qquad\qquad k = 2k_0 \sin(\theta/2).$$

Observe that this factor, *i.e.* $\exp[i\boldsymbol{k} \cdot \boldsymbol{r}_m(t)]$, takes into account the phase differences of the radiation scattered by different molecules of the fluid.

In a conventional scattering experiment one measures, as a function of the scattering angle θ, the spectral intensity distribution of scattered light of a selected polarization $\hat{e}'$. Denoting this as $\sigma(\theta, \omega)$ we have (in arbitrary units)

$$(2.7) \qquad\qquad\qquad \sigma(\theta, \omega) = \langle |\hat{e}' \cdot \boldsymbol{E}(\theta, \omega)|^2 \rangle,$$

where ω is the frequency *shift* of the scattered light from ω_0, $\langle ... \rangle$ denotes an ensemble average and $\widetilde{\boldsymbol{E}}(\theta, \omega)$ is the Fourier time transform of $E(\theta, t)$:

$$(2.8) \qquad\qquad \widetilde{\boldsymbol{E}}(\theta, \omega) = \int_{-\infty}^{\infty} dt\, \exp[i(\omega_0 + \omega)t]\, \boldsymbol{E}(\theta, t).$$

Observe that from (2.5) $E(\theta, t)$ contains explicitly the factor $\exp[-i\omega_0 t]$ so that the transform (2.8) depends only on ω. A more useful expression for $\sigma(\theta, \omega)$ is obtained by making use of the Wiener-Khinchin theorem:

$$(2.9) \qquad\qquad \sigma(\theta, \omega) = \int_{-\infty}^{\infty} dt\, \exp[i\omega t] \langle E(0)^* \cdot E(t) \rangle;$$

the spectrum is given as the transform of the time correlation function for the scattered field. In (2.9) we have let $E(t) = \hat{e}' \cdot \boldsymbol{E}(\theta, t)$ and have suppressed the explicit dependence on θ. Combining (2.5) and (2.9) gives

$$(2.10) \quad \sigma(\theta, \omega) = \int_{-\infty}^{\infty} dt \, \exp\left[i\omega t\right] \left\langle \sum_{m,n} (\hat{e}' \cdot \boldsymbol{\alpha}_m \cdot \hat{e}_0)(\hat{e}' \cdot \boldsymbol{\alpha}_n \cdot \hat{e}_0) \, \exp\left[-i\boldsymbol{k} \cdot [\boldsymbol{r}_m(0) - \boldsymbol{r}_n(t)]\right]\right\rangle.$$

Note that the expressions of the type $(\hat{e}' \cdot \boldsymbol{\alpha} \cdot \hat{e}_0)$ are, in general, time dependent since $\boldsymbol{\alpha}$ describes the molecular (electrical) anisotropy and this can change relative to a laboratory-fixed co-ordinate system as the molecules of the fluid reorient.

The total intensity of the light scattered through θ is just

$$(2.11) \qquad\qquad I(\theta) = \int_{-\infty}^{\infty} d\omega \, \sigma(\theta, \omega) = 2\pi \langle |E|^2 \rangle \, .$$

The intensity $I(\theta)$ depends only upon equilibrium properties of the scattering medium; the spectral characteristics that reflect dynamical fluid behavior are eliminated in the integration over frequencies.

We may now focus our attention on the case of polarized or vv scattering. To sort out the spectra corresponding to the different polarization combinations it is convenient to write $\boldsymbol{\alpha}$ as

$$(2.12) \qquad\qquad \boldsymbol{\alpha} = \alpha \mathbf{U} + \boldsymbol{\beta} \, ,$$

where $\mathbf{U}$ is the unit tensor, $\alpha = \frac{1}{3} \, \mathrm{Tr} \, (\boldsymbol{\alpha})$ and $\boldsymbol{\beta}$ is the anisotropy of the polarizability. Then for vv scattering

$$(2.13) \qquad\qquad (\hat{e}' \cdot \boldsymbol{\alpha} \cdot \hat{e}_0) = \alpha + \beta_{vv}(t)$$

with $\beta_{vv}(t) = \hat{e}_v \cdot \boldsymbol{\beta} \cdot \hat{e}_v$. ($\hat{e}_v$ is a unit vector normal to the scattering plane.) In most cases of interest, $\beta_{vv} \ll \alpha$ and therefore the anisotropy makes a negligible contribution to the polarized scattering. (Even when this is not the case, the contribution from the anisotropy can be evaluated from the depolarized spectra and therefore the isotropic scattering, *i.e.* the scattering related to α, can be esamined separately.)

Therefore we consider the spectral distribution function

$$(2.14) \qquad \sigma_{vv}(\theta, \omega) = \alpha^2 \int_{-\infty}^{\infty} dt \, \exp\left[i\omega t\right] \sum_{m,n} \left\langle \exp\left[-i\boldsymbol{k} \cdot [\boldsymbol{r}_m(0) - \boldsymbol{r}_n(t)]\right]\right\rangle$$

as describing the polarized spectrum. The correlation function is easily recognized as $\langle \varrho_{-k}(0)\,\varrho_k(t) \rangle$, where $\varrho_k(t)$ is the spatial Fourier transform of the space and time dependent density of the fluid $\varrho(\boldsymbol{r}, t)$. That is

(2.15)
$$\varrho_k(t) = \int d\boldsymbol{r} \, \exp\left[i\boldsymbol{k}\cdot\boldsymbol{r}\right]\varrho(\boldsymbol{r}, t)$$

with

(2.16)
$$\varrho(\boldsymbol{r}, t) = \sum_{n=1}^{N} \delta[\boldsymbol{r} - \boldsymbol{r}_n(t)] \, .$$

The polarized spectrum therefore reflects the dynamics of a particular Fourier component of the local density fluctuations in the medium. The intensity of the polarized scattering is seen to be, from (2.11),

(2.17)
$$I_{vv} = \int_{-\infty}^{\infty} d\omega \, \sigma_{vv}(\theta, \omega) = \alpha^2 \langle \varrho_{-k}(0)\,\varrho_k(0) \rangle \, .$$

Since k is small, *i.e.* $2\pi/k$ is large compared to the intermolecular spacings, we can evaluate this using equilibrium statistical mechanics in the $k = 0$ limit. The result is just

(2.18)
$$I_{vv} = \alpha^2 \varrho_0^2 \frac{k_{\mathrm{B}} T}{V} \, \varkappa_T \, ,$$

where ϱ_0 is the average density, k_{B} is Boltzmann's constant, T is the temperature, V is the scattering volume and $\varkappa_T$ is the equilibrium isothermal compressibility.

We must now explore further the way in which information concerning fluid properties (especially acoustical properties) can be inferred from the spectral distribution of the polarized scattering.

3. – The scattered-light spectrum and molecular acoustics.

The analysis of the polarized spectrum is carried out most conveniently in terms of (2.14). As with any discussion of molecular light scattering the analysis consists of two parts. First it is necessary to identify the fluctuating molecular quantities which are involved in the scattering. For polarized scattering we have seen that the required variable is the k-th Fourier component of the density. The other part of the theory involves the dynamics of the fluctuations.

Since the wave vectors involved in optical scattering are generally no larger than about $5 \cdot 10^5$ cm^{-1}, it is reasonable to use linearized hydrodynamics to compute the dynamics of the density fluctuations and thus the spectral distribution of the scattered light. Since much of the analytical work can be found elsewhere [3-5], we shall concentrate this discussion on the interpretation of the final results. One writes the equations of mass, momentum and energy conservation (generalized to allow for relaxational effects), solves these for the density and constructs the correlation function $\langle \varrho_{-k}(0)\,\varrho_k(t)\rangle$. The spectral distribution and intensity are then obtained respectively as the Fourier transform and $t = 0$ value of this (see eqs. (2.14) and (2.17)).

The general character of the spectrum $\sigma_{vv}(\theta, \omega)$ is governed by the combined effects of thermal diffusion, longitudinal stress relaxation and sound wave propagation. The first two of these are manifest in the Rayleigh line centered at $\omega = 0$, i.e. at the frequency of the incident light, while the effect of the sound waves of wave vector k are seen in the Brillouin lines, a doublet placed symmetrically about $\omega = 0$. The important parameters in describing the spectrum are the frequency shift of the Brillouin lines ω_B, the longitudinal stress relaxation time τ and the thermal diffusivity $\lambda/\varrho_0\,C_p$. (λ is the thermal conductivity and C_p is the specific heat at constant pressure.) In terms of these we can distinguish three important limiting cases.

Case I: $1/\tau \gg \omega_B \gg \lambda k^2/\varrho_0\,C_p$. Here the Rayleigh line arises entirely from the isobaric portion of the density fluctuations associated with energy transport via thermal diffusion; this line is a Lorentzian of half-width $\lambda k^2/\varrho_0\,C_p$. Because the relaxation time τ is short compared with $1/\omega_B$, the adiabatic portion of the density fluctuations associated with internal or structural relaxation processes remains in equilibrium with the phonon modes and thus contributes only to the Brillouin lines in the spectrum. The ratio of Rayleigh to Brillouin intensity is just $\gamma - 1$, where γ is the ratio of specific heats.

The shape of the spectrum can be constructed as described above. The result is just

$$(3.1) \qquad \sigma_{vv}(\theta, \omega) = \left(1 - \frac{1}{\gamma}\right)\frac{\lambda k^2/\varrho_0 C_p}{\omega^2 + (\lambda k^2/\varrho_0 C_p)^2} + \frac{1}{\gamma}\frac{(\eta k^2/\varrho_0)(C_0^2 k^2)}{\omega^2(\eta k^2/\varrho_0)^2 + (\omega^2 - C_0^2 k^2)^2},$$

where C_0 is the low-frequency sound speed and η is the longitudinal viscosity, *i.e.*

$$(3.2) \qquad \eta = \eta_v + \tfrac{4}{3}\eta_s,$$

where η_v and η_s are the bulk and shear viscosities of the liquid. The general form of this spectrum is sketched in Fig. 2a).

In terms of acoustic parameters the Brillouin shift ω_B is simply determined by the low-frequency sound speed, *i.e.*

$$(3.3) \qquad \omega_B = C_0\,k,$$

and the Brillouin line width Γ_B is related to the sound absorption coefficient α_s by

$$(3.4) \qquad \Gamma_B = \alpha_s\,C_0,$$

that is, Γ_{B} is the temporal absorption coefficient. In this limit the scattering spectrum provides essentially the same information as a high-frequency ultrasonic propagation experiment.

One point of conceptual difference between Brillouin scattering and ultrasonic propagation experiments that might be mentioned centers about the fact that in the former experiment one selects for study a (thermally generated)

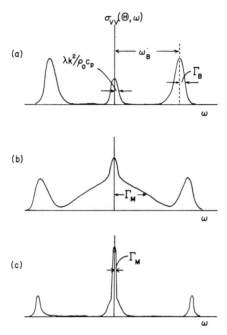

Fig. 2. – Typical polarized scattering spectra for the three cases discussed in the text: a) very short relaxation time, *i.e.* $1/\tau \gg \omega_{\mathrm{B}}$;   b) intermediate relaxation time, *i.e.* $\omega_{\mathrm{B}} \gg 1/\tau \gg \lambda k^2/\varrho_0 C_p$ and c) very long relaxation time, *i.e.* $\lambda k^2/\varrho_0 C_p \gg 1/\tau$.

sound wave of given wave vector k (by choosing a scattering angle) and measures its temporal characteristics, *i.e.* its frequency ω_{B} and temporal decay coefficient Γ_{B}. In the ultrasonic propagation experiment the frequency of the sound wave is selected, and the spatial properties, the wavelength λ_s and the absorption coefficient α_s are measured. However, if the losses are relatively small so that one can take

$$\Gamma_{\mathrm{B}}/\omega_{\mathrm{B}} \approx \alpha_s \lambda_s/2\pi \ll 1 \,,$$

then in this case the two experiments are, in a practical sense, equivalent.

The values of ω_{B} and Γ_{B} can be estimated quite straightforwardly. Taking typical values of $k \sim 2 \cdot 10^5 \, \mathrm{cm^{-1}}$ and $C_0 \sim 10^5 \, \mathrm{cm/s}$ gives a Brillouin shift of

$\omega_{\mathrm{B}} \sim 2 \cdot 10^{10} \mathrm{\,s}^{-1}$. The Brillouin half-width Γ_{B} is given by

$$(3.5) \qquad \Gamma_{\mathrm{B}} \approx (\tfrac{4}{3}\eta_s + \eta_v)\, k^2/2\varrho_0\,,$$

and so taking typical values of $\eta_s \approx \eta_v \sim 10^{-2}$ poise and $\varrho_0 \approx 1 \mathrm{\,g/cm^3}$ gives $\Gamma_{\mathrm{B}} \sim 5 \cdot 10^8 \mathrm{\,s}^{-1}$. (The parameters chosen here imply $\tau \sim 10^{-12} \mathrm{\,s}$, so that the situation corresponds to the case $1/\tau \gg \omega_{\mathrm{B}}$ under consideration.) Thus $\Gamma_{\mathrm{B}}/\omega_{\mathrm{B}} \sim 0.025 \ll 1$, as required.

From an experimental standpoint, the resolution and spectral-range demands implied by these kinds of numbers are rather nicely satisfied if one uses a flat-plate Fabry-Perot interferometer as the spectrum analyzer [6]. For a plate separation of 1.0 cm and plates of 98% reflectivity one can rather straight-forwardly achieve a resolution of 200 MHz with a spectral range of 15 GHz, quite satisfactory for Brillouin analysis of the type just described.

Case II: $\omega_{\mathrm{B}} \gg 1/\tau \gg \lambda k^2/\varrho_0 C_p$. Because of the slow rate of thermal diffusion, the adiabatic and isobaric fluctuations are uncoupled (as in Case I). Consequently the Rayleigh line contains the Lorentzian term

$$(1-1/\gamma)\, \frac{\lambda k^2/\varrho_0 C_p}{\omega^2 + (\lambda k^2/\varrho_0 C_p)^2}\,,$$

just as above. However in this case the internal or relaxational degrees of freedom of the density fluctuations equilibriate slowly and thus cannot contribute to the phonon modes. Rather they constitute the so-called Mountain component of the Rayleigh line.

To describe this situation the usual viscous term appearing in the (spatially transformed) Navier-Stokes equation [7], *i.e.* $-k^2(\eta/\varrho_0)J_K^L$, where J_K^L is the longitudinal-momentum current, is replaced by

$$-k^2 \int_0^t \mathrm{d}t'\, \nu(t-t')\, J_K^L(t')\,.$$

The form of the spectrum $\sigma_{\mathrm{vv}}(\theta, \omega)$ is now

$$(3.6) \qquad \sigma_{\mathrm{vv}}(\theta, \omega) = \left(1 - \frac{1}{\gamma}\right) \frac{\lambda k^2/\varrho_0 C_p}{\omega^2 + (\lambda k^2/\varrho_0 C_p)^2} +$$

$$+ \frac{1}{\gamma} \frac{[\eta(\omega)k^2/\varrho_0] C_0^2 k^2}{[\omega\eta(\omega)k^2/\varrho_0]^2 + [\omega^2 - \omega\zeta(\omega)k^2/\varrho_0 - C_0^2 k^2]^2}\,,$$

where $\eta(\omega)$ and $\zeta(\omega)$ are defined by

$$(3.7) \qquad \eta(\omega) + i\zeta(\omega) = \varrho_0 \int \mathrm{d}t \, \exp\,[i\omega t]\, \nu(t)\,,$$

and represent respectively the real and imaginary parts of the complex longitudinal viscosity. These are exactly the dynamical quantities probed by ultrasonic absorption and velocity experiments. Their appearance in (3.6) means
that, since they govern the shape of the polarized scattering spectrum, it should
be possible to infer their behavior from it and indeed this is the case. The general
form of the spectrum is given in Fig. 2b).

The Brillouin shift frequency ω_B is determined (approximately) by the vanishing of the term

$$(3.8) \qquad\qquad [\omega^2 - \omega\zeta(\omega)\,k^2/\varrho_0 - C_0^2\,k^2]_{\omega=\omega_B} = 0$$

in the denominator of the second term in (3.6). The Brillouin half-width is
given by

$$(3.9) \qquad\qquad \Gamma_B \approx \eta(\omega_B)\,k^2/\varrho_0\,.$$

The general form of the Mountain component can be understood by examining
(3.6) for frequencies $\omega \ll \omega_B$. In this spectral region we may neglect the ω^2
in the second term in the denominator and obtain

$$(3.10) \qquad \sigma_{vv}(\theta,\omega) \approx \left(1 - \frac{1}{\gamma}\right)\frac{\lambda k^2/\varrho_0\,C_p}{\omega^2 + (\lambda k^2/\varrho_0\,C_p)^2} + \frac{\varrho C_0^2}{\gamma}\,\frac{\eta(\omega)}{[\omega\eta(\omega)]^2 + [\varrho C_0^2 + \omega\zeta(\omega)]^2}\,.$$

Apart from the factor $\varrho C_0^2/\gamma$ the second term—the Mountain component—is
the frequency-dependent longitudinal fluidity $\Phi(\omega)$ of the material. Once the
form of $\Phi(\omega)$ is determined one can find the corresponding functions $\eta(\omega)$
and $\zeta(\omega)$.

The function $\Phi(\omega)$ can be represented as the cosine transform of a time correlation function $\varphi(t)$, $i.e.$

$$(3.11) \qquad\qquad \Phi(\omega) = \int_0^\infty dt\,\cos\omega t\varphi(t)\,,$$

and the average relaxation time—the relaxation time for constant adiabatic
longitudinal stress—is

$$(3.12) \qquad\qquad \tau_{\sigma s} = \int_0^\infty dt\,\varphi(t)/\varphi(0)\,.$$

From an order-of-magnitude standpoint, the Mountain half-width

$$(3.13) \qquad\qquad \Gamma_M \approx \frac{1}{\tau_{\sigma s}}\,;$$

of course, the exact value of Γ_M depends upon the precise shape of $\Phi(\omega)$. For relaxation times $\tau_{\sigma s} \lesssim 10^{-8}$ s, Fabry-Perot spectroscopy such as was described above is quite nicely suited to satisfying the necessary resolution and spectral-range requirements.

The important point that should be stressed here is that measurements of $\Phi(\omega)$, and indeed of the total spectrum $\sigma_{vv}(\theta, \omega)$, provide exactly the same

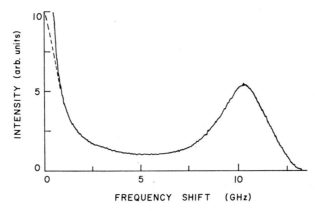

Fig. 3. – The experimental scattered-light spectrum measured in liquid glycerol at 35 °C. The contribution from heat transport has been subtracted from this spectrum so that only the Mountain and Brillouin components are shown.

information concerning the stress (or strain) relaxation dynamics as is obtained from ultrasonic propagation experiments. The effective frequency range over which $\eta(\omega)$, for instance, can be studied is roughly from $5 \cdot 10^{7}$ Hz to 10^{11} Hz as compared with the usual ultrasonic range of roughly 10^{6} Hz to 10^{9} Hz. The two approaches are complementary rather than competitive.

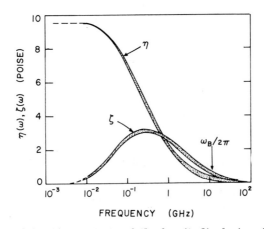

Fig. 4. – The real and imaginary parts of the longitudinal viscosity, $\eta(\omega)$ and $\zeta(\omega)$, derived from the spectrum shown in Fig. 3 for liquid glycerol at 35 °C.

To illustrate the utility of the technique of using the Mountain contribution to the spectrum, data obtained in liquid glycerol [8] are presented. A typical spectrum is shown in Fig. 3; in this spectrum the contribution from isobaric fluctuations has been subtracted so that only the Mountain component of the Rayleigh line is shown. The plots of $\eta(\omega)$ and $\zeta(\omega)$ obtained from this spectrum are presented in Fig. 4. Results similar to this were obtained at temperatures ranging from $+10\,°C$ to $+100\,°C$. At the lower temperatures the spectral *shapes* of $\eta(\omega)$ and $\zeta(\omega)$ exactly correspond with those obtained ultrasonically in the temperature range $-40\,°C$ to $+10\,°C$ [9, 10]. At the higher temperatures a narrowing of the dispersion region is observed; in particular, one notices that the high-frequency tail of the relaxation curves is somewhat suppressed. This is illustrated in Fig. 5 where the reduced quantity $\zeta(\omega)/\zeta_0 (\equiv \zeta^*)$ is plotted as a function of $\omega/\omega_0 (\equiv \omega^*)$, where ω_0 is the frequency at which $\zeta(\omega)$ reaches its maximum value ζ_0.

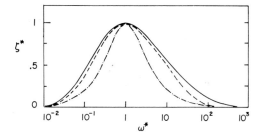

Fig. 5. – The reduced quantity $\zeta^*(\omega)$ plotted *vs.* reduced frequency ω^* for glycerol at $35\,°C$ (——) and $96\,°C$ (— — —). The single-relaxation form $2\omega^*/(1+\omega^{*2})$ is shown in the dash-dotted curve.

Case III: $\omega_B \gg \lambda k^2/\varrho_0 C_p \gg 1/\tau$. The Rayleigh line again consists of two components, but now it is the Mountain component that forms the narrow line, while the normal Lorentzian component associated with heat transport appears as a relatively broader background. In this case because the viscoelastic relaxation process occurs quite slowly compared with the thermal-diffusion rate, the relaxational behavior responsible for the Mountain line shape is *isothermal* rather than adiabatic as in Case II above. In addition, the thermal diffusion must proceed without changes in the internal or structural degrees of freedom and consequently it must be the « unrelaxed » values of such parameters as the specific heat, *i.e.* C_p^∞, that govern this rate. That is, the half-width of the « normal » Rayleigh line is expected to be just $\lambda k^2/\varrho_0 C_p^\infty$. On the other hand, by an argument entirely analogous to that used in developing the result (3.10) for Case II above, the Mountain contribution to the spectrum is just the frequency dependence of the *isothermal* longitudinal fluidity which we shall designate as $\Psi(\omega)$. This is expressible as

$$(3.14) \qquad \Psi(\omega) = \int_0^\infty dt \cos \omega t \, \psi(t) \,,$$

where $\psi(t)$ is the isothermal strain correlation function. The average relaxation time is given by

$$(3.15) \qquad \tau_{\sigma T} = \int_0^\infty dt \, \psi(t)/\psi(0) \,.$$

This differs from the relaxation time $\tau_{\sigma s}$ (see (3.12)) by a factor which is just the ratio of the equilibrium to the unrelaxed specific heat, *i.e.*

$$(3.16) \qquad \tau_{\sigma T} = \tau_{\sigma s}(C_p/C_p^\infty) \,.$$

The measurement of the scattered-light spectrum and thus of $\Psi(\omega)$ is very difficult by normal spectroscopic techniques. The half-width of the Mountain component

$$\Gamma_{\mathrm{M}} \sim \tau_{\sigma T}^{-1} \ll \lambda k^2/\varrho C_p^\infty \sim 10^6 \; \mathrm{s}^{-1} \,,$$

so that an optical resolution better than 10^9 is required.

To avoid the need for such a resolution, one can measure the characteristics of the scattered light directly in the time domain rather than in the frequency domain. Essentially what is done is to allow the scattered light to fall directly upon the photomultiplier, and then to electronically measure the correlation in time of the photoelectron pulses output by the tube [11-13]. In the usual « single-clipped » experimental situation the measured correlation function is

$$(3.17) \qquad G(t) = \langle n_K(0) \, n(t) \rangle \,,$$

where $n(t)$ is the number of photopulses in the interval t to $t+\theta$, where θ is the bin size; $n_K(0) = 0$ if $n(0) \leqslant K$, and $n_K(0) = 1$ if $n(0) > K$. The quantity K is the preset « clipping level » and is usually chosen $\approx \langle n(t) \rangle$. If the measurements are restricted to times t such that $\lambda k^2 t/\varrho_0 C_p \gg 1$, then $G(t)$ is related to $\Psi(t)$ through the relation

$$(3.18) \qquad G(t) = a + b\Psi(t)^2 \,,$$

where a and b are constants which can be computed for a given set of experimental conditions.

To complete the picture it is worth-while to compare some experimental results obtained in this fashion with ones obtained from both ultrasonic and

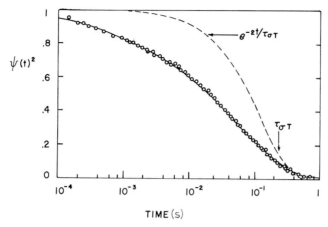

Fig. 6. – The measured correlation function $\Psi(t)^2$ in glycerol at -73.1 °C.

more conventional optical-scattering experiments. Measurements in glycerol have been obtained over the temperature range -80 °C to -45 °C [14]. A typical experimental correlogram—$G(t)$—is shown in Fig. 6. The shape of the correlation function is essentially identical with that obtained ultrasonically

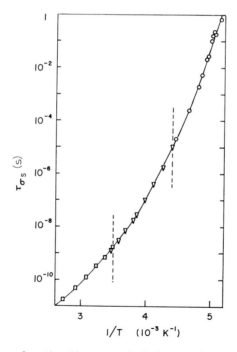

Fig. 7. – The average relaxation times $\tau_{\sigma s}$ plotted vs. reciprocal temperature. Circles indicate data from correlation spectroscopy, triangles indicate ultrasonic data, and squares indicate spectral Mountain-component measurements.

in the temperature range -40 °C to $+10$ °C; that is $\Psi(\omega)$ obtained using (3.14) is of the same form as $\Phi(\omega)$ determined ultrasonically. There is apparently little difference in the form of the adiabatic and isothermal relaxation functions. One can also compare the relaxation times, say τ_{σ_s}, measured

a) using optical spectroscopic analysis of the Mountain component above 10 °C,

b) using ultrasonic propagation techniques in the temperature range -40 °C to $+10$ °C and

c) using time correlation spectroscopy of light scattered by isothermal density fluctuations, below -40 °C. The comparison is shown in Fig. 7. The agreement is quite remarkable; the relaxation times measured by the three methods above join smoothly to form a continuous curve over more than ten orders of magnitude.

The use of light scattering measurements greatly extends the accessible range of time scales of relaxation phenomena that have been traditionally investigated acoustically. When used in concert with one another, they provide the molecular acoustician with a capability for measurement that spans nearly the entire range of time scales that are of interest in relaxation processes.

* * *

The author wishes to acknowledge several stimulating discussions with T. A. LITOVITZ and R. D. MOUNTAIN during the preparation of this manuscript.

The support of the National Science Foundation (USA) is also gratefully acknowledged by the author.

REFERENCES

[1]   See, for example, I. L. FABELINSKII: *Molecular Scattering of Light*, translated from the Russian by R. T. BEYER (New York, N. Y., 1968).
[2]   See, for example, W. B. DAVENPORT and W. L. ROOT: *An Introduction to the Theory of Random Signals and Noise* (New York, N. Y., 1958).
[3]   R. D. MOUNTAIN: *Rev. Mod. Phys.*, **38**, 205 (1966).
[4]   R. D. MOUNTAIN: *Journ. Res. Nat. Bur. Stand.*, **70** A, 207 (1966).
[5]   C. J. MONTROSE, V. A. SOLOVYEV and T. A. LITOVITZ: *Journ. Acoust. Soc. Amer.*, **43**, 117 (1968).
[6]   M. BORN and E. WOLF: *Principles of Optics*, fourth edition, Chap. 7 (Oxford, 1970).
[7]   C. J. MONTROSE: this volume, p. 18.
[8]   B. KATZ: Ph. D. Thesis, Catholic University of America (1973).

[9]   R. PICCIRELLI and T. A. LITOVITZ: *Journ. Acoust. Soc. Amer.*, **29**, 1009 (1957).

[10] W. M. SLIE, A. R. DONFOR and T. A. LITOVITZ: *Journ. Chem. Phys.*, **44**, 3712 (1966).

[11] E. JAKEMAN and E. R. PIKE: *J. Phys. A*, **2**, 411 (1969).

[12] R. J. GLAUBER: *Phys. Rev.*, **131**, 2766 (1963).

[13] An excellent review of optical correlation spectroscopy can be found in *Photon Correlation and Light Beating Spectroscopy*, edited by H. Z. CUMMINS and E. R. PIKE (New York, N. Y., and London, 1974).

[14] C. DEMOULIN, C. J. MONTROSE and N. OSTROWSKY: *Phys. Rev. A*, **9**, 1740 (1974).

Physical Acoustics at UCLA in the Study of Superfluid Helium.

I. Rudnick

University of California - Los Angeles, Cal.

1. – Sound propagation in superfluid helium—theory.

1'1. *Introduction* [1]. – Below the lambda-transition (2.17 °K at the vapor pressure) ^{4}He becomes a superfluid known as He II (the liquid is called He I above T_λ, the temperature of the lambda-transition). It behaves as though it is composed of two fluids, a superfluid component and a normal-fluid component. Its density $\varrho = \varrho_s + \varrho_n$, where ϱ_s and ϱ_n are respectively the densities of the superfluid and normal fluid. Its mass flux $\varrho v = \varrho_s v_s + \varrho_n v_n$, where v, v_s and v_n are the respective velocities of He II, the superfluid and normal-fluid components. The superfluid component has no viscosity but it is more than an ideal Euler liquid since it has the most unusual property of having zero entropy. All the entropy is carried by the normal component.

Two-fluid hydrodynamics has as a basic postulate that eight independent variables are needed to describe the motion. These may be taken to be v_n, v_s, ϱ and s, where s is the entropy. When dissipation can be ignored the eight equations needed to uniquely determine these eight variables are

$$
(1) \qquad \frac{\partial \varrho}{\partial t} + \nabla \cdot (\varrho_s v_s + \varrho_n v_n) = 0 \, ,
$$

$$
(2) \qquad \frac{\partial \varrho s}{\partial t} + \nabla \cdot (\varrho s\, v_n) = 0 \, ,
$$

$$
(3) \qquad \frac{\mathrm{D}_s v_s}{\mathrm{D} t} + \nabla \mu = 0 \, ,
$$

where $\mathrm{D}_s/\mathrm{D}t = \partial/\partial t + v_s \cdot \nabla$,

$$
(4) \qquad \frac{\partial}{\partial t}(\varrho_n v_{ni} + \varrho_s v_{si}) + \frac{\partial \pi_{ij}}{\partial x_j} = 0 \, ,
$$

where $\pi_{ij} = \varrho_n v_{ni} v_{nj} + \varrho_s v_{si} v_{sj} + \delta_{ij} p$.

Equation (1) is the mass conservation condition. Equation (2) is the entropy conservation condition and contains the fact that the entropy is transported only by the normal component. In eq. (3) μ is the chemical potential including all external potentials (such as gravity). The thermodynamic part of $\nabla\mu = (1/\varrho)\nabla p - s\nabla T$, and it is thus clear that the superfluid component can be reversibly accelerated by either a pressure gradient or temperature gradient and in this respect is unique. In eq. (4) π_{ij} is called the stress tensor and we see that eq. (4) is a natural extension of the momentum conservation condition for ordinary liquids.

Equations (1)-(4) are Landau's superfluid dynamic equations which he always supplemented with the additional condition

(5)
$$\nabla \times \boldsymbol{v}_s = 0 \, .$$

With this condition eq. (3) becomes

(6)
$$\frac{\partial \boldsymbol{v}_s}{\partial t} + \nabla\left(\mu + \frac{v_s^2}{2}\right) = 0 \, .$$

1'2. First and second sound. – If we write

$$\varrho = \varrho_0 + \delta\varrho \, , \qquad s = s_0 + \delta s \, , \qquad p = p_0 + \delta p \, , \qquad T = T_0 + \delta T \, ,$$

where the subscript zero refers to the equilibrium value of the parameter, then the acoustic requirement is that $\delta\varrho/\varrho$, $\delta s/s$, $\delta p/p$, $\delta T/T \ll 1$. If $\boldsymbol{v}_n$ and $\boldsymbol{v}_s$ are the acoustic particle velocities and u is the wave velocity, then v_n/u, $v_s/u \ll 1$. To the first order of small quantities, eqs. (1)-(4) respectively become

(7)
$$\frac{\partial \delta\varrho}{\partial t} + \varrho_s \nabla \cdot \boldsymbol{v}_s + \varrho_n \nabla \cdot \boldsymbol{v}_n = 0 \, ,$$

(8)
$$\varrho \frac{\partial \delta s}{\partial t} + s \frac{\partial \delta p}{\partial t} + s\varrho \nabla \cdot \boldsymbol{v}_n = 0 \, ,$$

(9)
$$\frac{\partial \boldsymbol{v}_s}{\partial t} + \frac{1}{\varrho}\nabla\, \delta p - s\nabla\, \delta T = 0 \, ,$$

(10)
$$\frac{\partial \boldsymbol{v}_n}{\partial t} + \frac{1}{\varrho}\nabla\, \delta p + \frac{\varrho_s}{\varrho_n} s\nabla\, \delta T = 0 \, .$$

Equation (10) comes from a combination of eqs. (4) and (3) in the acoustic approximation. From eqs. (7), (9) and (10) we get

(11)
$$\frac{\partial^2 \delta\varrho}{\partial t^2} = \nabla^2\, \delta p \, .$$

Using eqs. (7)-(10) we can obtain

(12)
$$\frac{\partial^2 \delta s}{\partial t^2} = \frac{\varrho_s}{\varrho_n} s^2 \nabla^2 \delta T .$$

If we write

(13)
$$\begin{cases} \delta\varrho = \left(\frac{\partial\varrho}{\partial p}\right)_T \delta p + \left(\frac{\partial\varrho}{\partial T}\right)_p \delta T , \\ \delta s = \left(\frac{\partial s}{\partial T}\right)_p \delta T + \left(\frac{\partial s}{\partial p}\right)_T \delta p , \end{cases}$$

then we note that when the isobaric expansion coefficient $\beta = -(1/\varrho)(\partial\varrho/\partial T)_p$ vanishes, the second terms on the right vanish, $\varrho = \varrho(p)$, $s = s(T)$ and eqs. (11) and (12) become

(14)
$$\begin{cases} \dfrac{\partial^2 \delta\varrho}{\partial t^2} = u_{\mathrm{I}}^2 \nabla^2 \delta\varrho , \\ \text{where} \\ u_{\mathrm{I}}^2 = \left(\dfrac{\partial p}{\partial\varrho}\right)_s , \end{cases}$$

(when the expansion coefficient β vanishes $(\partial p/\partial\varrho)_T = (\partial p/\partial\varrho)_s$),

(15)
$$\begin{cases} \dfrac{\partial^2 \delta s}{\partial t^2} = u_{\mathrm{II}}^2 \nabla^2 \delta s , \\ \text{where} \\ u_{\mathrm{II}}^2 = \dfrac{\varrho_s}{\varrho_n} \dfrac{T s^2}{C_v} , \end{cases}$$

where C_v is the specific heat. (When the expansion coefficient β vanishes $C_p = C_v$.)

We see that (14) is an ordinary compressional wave with density and pressure variation and no temperature or entropy variations. LANDAU called it first sound. Equation (15) is a wave unique to superfluid. LANDAU called it second sound. There are entropy and temperature variations and no pressure or density variations. It is a purely thermal wave which propagates. In Navier-Stokes liquids thermal waves diffuse. From eqs. (9) and (10) it is easy to see that for first sound $v_n = v_s$ and for second sound $\varrho_n v_n + \varrho_s v_s = 0$.

If the expansion coefficient is not zero (it vanishes as $T \to 0$ and between 1.1 and 1.2 °K) then first sound has associated with it some small temperature and entropy changes and second sound has small density and pressure changes. Their wave velocities are accordingly altered from the values given in eqs. (14) and (15). To find these new values we use (11), (12) and (13) and solve for one of the four independent variables. It is not difficult to show then (see LONDON, ref. [1], p. 84) that the new phase velocities are given by the solution to the quadratic equation

(16)
$$\left(\frac{u^2}{u_{\mathrm{I}}^2}-1\right)\left(\frac{u^2}{u_{\mathrm{II}}^2}-1\right)=\frac{\gamma-1}{\gamma},$$

where $\gamma = C_p/C_v$. Since $\gamma - 1 \ll 1$ at all accessible temperatures (γ is weakly singular at T_λ) we can get the following results for the first- and second-sound velocities u_1 and u_2:

(17)
$$u_1^2 = u_{\mathrm{I}}^2 + \frac{\gamma-1}{\gamma}\frac{u_{\mathrm{I}}^2 u_{\mathrm{II}}^2}{u_{\mathrm{I}}^2 - u_{\mathrm{II}}^2},$$

(18)
$$u_2^2 = u_{\mathrm{II}}^2 - \frac{\gamma-1}{\gamma}\frac{u_{\mathrm{I}}^2 u_{\mathrm{II}}^2}{u_{\mathrm{I}}^2 - u_{\mathrm{II}}^2}.$$

Between $1\,°\mathrm{K}$ and T_λ, $u_{\mathrm{I}}^2 \gg u_{\mathrm{II}}^2$ and the following expressions hold with con-

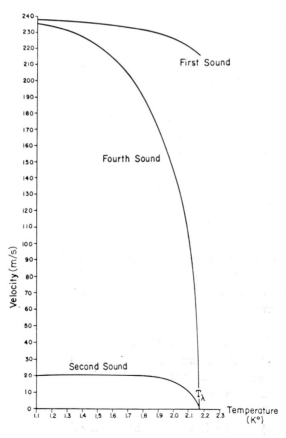

Fig. 1. – The velocity of first, second and fourth sound.

siderable accuracy:

$$(19) \qquad u_1^2 = u_I^2 + \frac{\gamma - 1}{\gamma} u_{II}^2 ,$$

$$(20) \qquad u_2^2 = \frac{u_{II}^2}{\gamma} = \frac{\varrho_s}{\varrho_n} \frac{T s^2}{C_p} .$$

Consider a plane wave. If we write

$$(21) \qquad v_n = a v_s , \qquad \delta p = b v_s , \qquad \delta T = c v_s ,$$

then for first sound

$$(22) \qquad a_1 = 1 + \frac{\beta \varrho}{\varrho_s s} \frac{u_I^2 u_{II}^2}{u_I^2 - u_{II}^2} , \qquad b_1 = \varrho u_I , \qquad c_1 = \frac{\beta T u_I^3}{C_v (u_I^2 - u_{II}^2)} ,$$

and for second sound

$$(23) \qquad a_2 = -\frac{\varrho_s}{\varrho_n} + \frac{\beta \varrho}{\varrho_n s} \frac{u_I^2 u_{II}^2}{u_I^2 - u_{II}^2} , \qquad b_2 = \beta \varrho \frac{u_I^2 u_{II}^3}{s(u_I^2 - u_{II}^2)} , \qquad c_2 = -\frac{u_{II}}{s} .$$

It is sometimes said that when $\beta \neq 0$ first and second sound are coupled. This statement is misleading because clearly first and second sound remain independently propagating modes. They are the two solutions of eq. (16). It is correct to say that when $\beta \neq 0$ pressure (density) variations are coupled to temperature (entropy) variations.

Figure 1 shows the temperature dependence of first and second sound at vapor pressure above 1 °K.

1`3. *Fourth sound.* – When the normal fluid component is locked by viscous forces, sound waves can still propagate through motion of the superfluid component. ATKINS discussed two such waves [2] and named them third and fourth sound. In the context of this paper it is logical to discuss fourth sound before third sound.

When He II is contained in a porous solid matrix, such as a packed fine powder, whose pores are so small that the viscous wavelength for the normal component is larger than the pore diameter, v_n will effectively be zero. The sound wave which propagates in such a system, called a superleak, is fourth sound. We can obtain its wave equation by setting $v_n = 0$ in eqs. (7)-(9). We cannot use eq. (10) because it does not include the dominant viscous force which reduces v_n to zero, but we do not need it because the number of variables has been reduced to five and we have five equations. We get

(24)
$$\frac{\partial \delta \varrho}{\partial t} + \varrho_s \nabla \cdot \boldsymbol{v}_s = 0 \, ,$$

(25)
$$\frac{\partial \varrho s}{\partial t} = 0 \, , \quad \varrho s = \text{const} \, ,$$

(26)
$$\frac{\partial \boldsymbol{v}_s}{\partial t} + \nabla \, \delta \mu = 0 \, .$$

Let ϱ and ϱs be the independent thermodynamic variables. Then (26) can be written in the form

(26')
$$\frac{\partial \boldsymbol{v}_s}{\partial t} + \left(\frac{\partial \mu}{\partial \varrho} \right)_{\varrho s} \nabla \, \delta \varrho = 0 \, .$$

Using (24) and (26') we get

(27)
$$\begin{cases} \dfrac{\partial^2 \delta \varrho}{\partial t^2} - u_4^2 \nabla^2 \, \delta \varrho = 0 \, , \\[2mm] \text{where} \\[2mm] u_4^2 = \varrho_s \left(\dfrac{\partial \mu}{\partial \varrho} \right)_{\varrho s} . \end{cases}$$

Evaluating $(\partial \mu / \partial \varrho)_{\varrho s}$ we find [3]

(28)
$$u_4^2 = \frac{\varrho_s}{\varrho} u_\mathrm{I}^2 + \frac{\varrho_n}{\varrho} u_\mathrm{II}^2 \left(1 - \frac{2\beta u_\mathrm{I}^2}{\gamma s} \right) .$$

The dominant term is the first term on the right. If we write $u_4^2 = u_\mathrm{I}^2 [\varrho_s / \varrho + \varepsilon(T)]$, then $\varepsilon(T) < 0.02$ for all temperatures—it is a maximum at about 2 °K. It is easy to see that if the fourth-sound wave is isothermal, then we get identically

(29)
$$u_4^2 = \frac{\varrho_s}{\varrho} u_\mathrm{I}^2 \, .$$

It is not difficult to show in a plane progressive wave

(30)
$$\begin{cases} v_s = \dfrac{u_4}{\varrho_s} \delta \varrho \, , \\[3mm] \delta s = -\dfrac{s}{\varrho} \delta \varrho \, , \\[3mm] \delta p = u_\mathrm{I}^2 \left(1 - \dfrac{T\beta s}{C_p} \right) \delta \varrho \, , \\[3mm] \delta T = \left[\dfrac{\beta u_\mathrm{I}^2}{\gamma} - s \right] \dfrac{T}{C_v \varrho} \delta \varrho \, . \end{cases}$$

In bulk He II we had two propagating waves—first and second sound. In clamped or locked helium we have only one—fourth sound. This arises from the fact that the number of variables has been reduced to five (since $\boldsymbol{v}_n = 0$) the same number as an ordinary fluid. We may ask which of these two modes, first and second sound, disappears when the helium is clamped. To answer this we allow some motion of $\boldsymbol{v}_n$ and find that both first and second sound become attenuated. As the clamping is tightened the attenuation for both modes increases until a certain point when the attenuation of first sound starts decreasing. In the limit of complete clamping first sound becomes fourth sound and second sound becomes a diffusion wave with vanishingly small diffusive velocity.

1'4. *Third sound.* – A shallow water, or long-wavelength surface wave, travels with a velocity given by $(gy)^{\frac{1}{2}}$, where g is the gravitational acceleration and y is the depth of the water. The quantity g is present because gravity levels the surface. In helium films surface waves, called third sound, also propagate and their velocity is given by

$$(31) \qquad u_3 = \left(\frac{\varrho_s}{\varrho} fy\right)^{\frac{1}{2}}.$$

The quantity f is the van der Waals force of attraction between the substrate and the films and occurs because it is the body force which levels the surface of the film. It is many orders of magnitude greater than the gravitational body force for helium films. The quantity y is the thickness of the film. Just as ϱ_s/ϱ entered the expression for the velocity of fourth sound, it enters here because the normal component is locked by its viscosity and only the superfluid component moves. Before deriving this result, we will briefly discuss the thickness of helium films.

1'5. *The thickness of a helium film.* – Consider an empty container at low temperature into which a small amount of helium gas has been introduced. Some of this gas will be adsorbed on the walls of the container in the form of a film. The thickness of the film is determined by the phase equilibrium requirement that the chemical potential of the surface of the film is equal to that in the vapor. This leads to the equality

$$(32) \qquad \frac{kT}{m} \ln \frac{p_0}{p} = \mu_{\text{v.d.w.}},$$

where k is Boltzmann's constant, m is the atomic mass of helium, p_0 is the saturated vapor pressure, p is the pressure of the vapor and $\mu_{\text{v.d.w.}}$ is the van der Waals potential at the film surface. For thin films

$$(33) \qquad \mu_{\text{v.d.w.}} = \alpha/y^3.$$

The exponent of y approaches 4 for thick films [4, 5]. Except for a slight approximation, eq. (32) says that the difference in the chemical potential of the unsaturated and saturated vapor is equal to the reduction in potential of the helium atoms at the surface of the film due to the attraction of the substrate. In CGS units $\alpha \simeq 10^{-14}$. Known values of α have a range of about a factor of 10 for various substrates. When $p \ll p_0$, y becomes much less than the thickness of an atomic layer usually taken to be 3.6 Å. If $p_0 - p \ll p_0$, y can become large. Practically the range of thickness is from 10^2 atomic layers to submonoatomic layer thickness. If the container, referred to earlier, has enough vapor introduced liquid helium will start collecting on its bottom and there will be a film on the walls. The vapor inside will obey the law of atmospheres

$$p = p_0 \exp \left[- mgh/kT \right],$$

where p_0 is the pressure at the top surface of the bulk helium, and eqs. (32) and (33) can still be used to determine the film thickness as a function of height and we see that this leads to

(34) $$gh = \alpha/y^3 .$$

Such films are said to be saturated. If $h = 10^{-1}$ cm, y is seen to be approximately 100 atomic layers. The body force f is the negative gradient of the potential, *i.e.*

(35) $$f = 3\alpha/y^4 ,$$

and it is easy to verify an earlier statement that this is much greater than the gravitational body force.

1'6. *Simple approximate derivation of the velocity of third sound.* – Consider a film displaced from equilibrium as shown in Fig. 2. We assume 1) $v_n = 0$, 2) $v_{sx} = v_{sy} = 0$, 3) v_{sz} is independent of y and 4) isothermal processes, 5) $\delta y/y \ll 1$, 6) incompressible flow, 7) p is constant at the film vapor surfaces. Then the equation for the acceleration of the superfluid is

(36) $$\frac{\partial v_s}{\partial t} = - \frac{\partial \mu}{\partial z} = - \frac{d\mu}{dy} \frac{\partial y}{\partial z} = f \frac{\partial \delta y}{\partial z} .$$

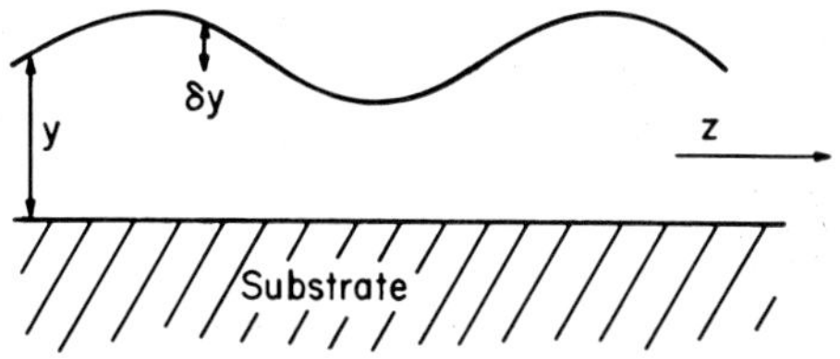

Fig. 2. – Diagram of a third-sound wave.

The equation of continuity is

(37)
$$\varrho_s y \frac{\partial v_s}{\partial z} = \varrho \frac{\partial \delta y}{\partial t}.$$

From (36) and (37) we get

(38)
$$\frac{\partial^2 \delta y}{\partial t^2} = u_3^2 \frac{\partial^2 \delta y}{\partial z^2}$$

with u_3 given by eq. (31).

The isothermal assumption needs discussion. The film and the vapor are in equilibrium in the absence of the third-sound wave. When the superfluid moves out of the region it leaves a trough and creates a mound in the region it enters. The trough is hotter and the mound colder than the vapor and condensation occurs in the latter and evaporation in the former. If the evaporation-condensation process were instantaneous the film would be isothermal but even so the velocity is altered since the defining equations are altered. If L is the latent heat and S the entropy of the liquid, then u_3 is given by

(39)
$$u_3^2 = \frac{\varrho_s}{\varrho} fy \left(1 + \frac{TS}{L}\right).$$

$TS/L = 0.01$ at 1 °K, 0.08 at 2 °K and 0.15 at T_λ. It increases monotonically with temperature.

Besides affecting the velocity the evaporation-condensation process is important because it is not instantaneous and thus is irreversible. Thus there is attenuation. What is very important practically is that there is a residual temperature swing in the wave which we shall see has important experimental applications since it leads to a way to detect the wave.

1˙7. *Doppler shift of first, second, third and fourth sound.* – Suppose we have a convective flow of the superfluid in the direction of sound propagation. (If the two directions are perpendicular, the change in velocity is second order in v/u.) In general v_n and v_s can have different values and this complicates the Doppler shift. Of course in the case of third and fourth sound v_n is necessarily zero. If u_{n0} is the velocity of n-th sound in the stationary medium, the resulting velocity of the four sounds is then given by

(40)
$$u_1 = u_{10} + \left(\frac{\varrho_s}{\varrho} v_s + \frac{\varrho_n}{\varrho} v_n\right),$$

(41)
$$u_2 = u_{20} + \left(\frac{\varrho_s}{\varrho} v_s + \frac{\varrho_n}{\varrho} v_n\right) + (v_n - v_s)\left(\frac{2\varrho_s}{\varrho} - \frac{s}{\varrho_n}\left(\frac{\partial \varrho_n}{\partial T}\right)_p \left(\frac{\partial T}{\partial s}\right)_p\right),$$

(42)
$$u_3 = u_{30} + \left\langle \frac{\varrho_s}{\varrho} \right\rangle v_s,$$

(43)
$$u_4 = u_{40} + \left(\frac{\varrho_s}{\varrho} + \frac{s[(\partial/\partial T)(\varrho_n/\varrho)]_p}{(\partial s/\partial T)_p} - \frac{\varrho[(\partial/\partial p)(\varrho_n/\varrho)]_T}{(\partial \varrho/\partial p)_T}\right) v_s.$$

The derivation of the first two values is by KHALATNIKOV [6]. Refer also to ref. [1] PUTTERMAN. For (42) see ref. [1] PUTTERMAN and ref. [7]. For (43) see ref. [1] PUTTERMAN and ref. [8].

There are several things worth noting. The added velocity of first sound is just the velocity of He II, an eminently reasonable result.

Consider the last term on the right of eq. (41). It can be rewritten as

$$(44) \qquad \left(\frac{\partial \log \varrho_n}{\partial \log s}\right)_p .$$

Since the entropy is carried by the normal component, the following equation is sometimes written:

$$(45) \qquad \varrho s = \varrho_n s_n .$$

The « Tisza approximation » is $s_n = $ constant. Between 1.1 and T_λ, ϱ_n changes 100-fold but s_n changes only by about 20%. If we use this approximation the quantity (44) is unity and eq. (41) becomes

$$(46) \qquad u_2 = u_{20} + \left(\frac{\varrho_s}{\varrho} v_s + \frac{\varrho_n}{\varrho} v_n\right) + (v_n - v_s)\left(\frac{\varrho_s}{\varrho} - \frac{\varrho_n}{\varrho}\right).$$

We note that when $v_n = v_s$ the Doppler shift given by (41) and (46) is just that due to the motion of He II. But even if there is no net flow of He II, i.e. $(\varrho_s/\varrho)v_s + (\varrho_n/\varrho)v_n = 0$, which occurs when there is internal convection due to a heat source, there is a Doppler shift given by the third term on the right. Note that it changes sign when $\varrho_s = \varrho_n$, which is at about 1.95 °K.

In thin films of helium the value of ϱ_s/ϱ is depressed. One explanation which has been advanced is that ϱ_s/ϱ must obey a boundary condition vanishing at solid surfaces and perhaps also at the liquid-vapor surface. Thus where ϱ_s/ϱ is measured by, say, measuring the velocity of third sound it is the value averaged over the thickness which is determined and it is this value which appears in eq. (42). PUTTERMAN [1] proposes that $\frac{1}{2}(\varrho_s/\varrho + \langle\varrho_s/\varrho\rangle)$ should be used instead. In thick films $\varrho_s/\varrho \simeq \langle\varrho_s/\varrho\rangle$ and we see that the sound is Doppler-shifted by the velocity of He II.

In the case of fourth sound it is Doppler-shifted by the velocity of He II only at low temperatures, where the partial derivatives contribute little to the bracketed term. Their value is imperfectly known at higher temperatures.

2. – Sound propagation in superfluid helium—experiments.

2'1. First sound.

2'1.1. First-sound transducers. First sound is ordinary sound and accordingly the transducers are ordinary transducers. Thus more or less conventional electrostatic or electrodynamic units which have the appropriate design compatible with the required low-temperature operation are suitable. Piezoelectric and ferroelectric devices may also be used. The characteristic acoustic impedance of He II is about $3.3 \cdot 10^3$ CGS units (approximately the geometrical mean of that of air and water at room temperature and pressure). As a result the radiation damping of a quartz crystal is small and a common difficulty in their use with pulses is the ringing which occurs. This, of course, is absent in diaphragm-type electrostatic transducers and is one of the reasons why they are receiving increased use.

2˙1.2. Cavitation in liquid helium. The measured cavitation threshold in ordinary liquids is always very much smaller than that required to break molecular bonds. This discrepancy is usually attributed to the presence of microscopic (considerably greater than molecular dimensions) nuclei; for example, gas or vapor cavities on or away from solid surfaces, solid particulate matter suspended in the liquid, or events associated with the passage of cosmic rays. Because of the low temperature and viscosity of the He II, all but the last of these are absent in He II. It is accordingly surprising to find that the cavitation threshold is abnormally low in liquid helium—approximately 10^{-3} atmospheres. Figure 3 shows an arrangement of PZT transducers (resonant

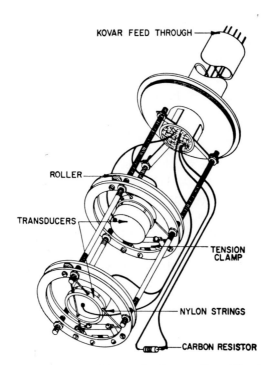

Fig. 3. – Diagram of apparatus used in cavitation studies. The transducers are made of PZT. The carbon resistor is used to measure temperature. Taken from ref. [9].

frequency $= 91$ kHz) that was used in such a study by FINCH *et al.* [9]. The cavitation threshold was taken to be the r.m.s. acoustic pressure at which cavitation noise appeared or disappeared as determined by listening with one of the transducers. Figure 4 shows this threshold as a function of temperature. An interesting feature is the peak at the λ temperature. This has the appearance

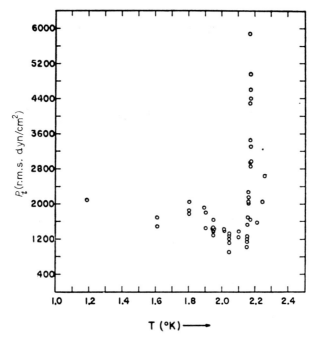

Fig. 4. – Cavitation threshold (in dyn/cm²) of ⁴He as a function of temperature. Taken from ref. [10].

of a singularity and in fact was found to be more strongly temperature dependent than the specific-heat anomaly. By varying the depth of the transducers in the helium, the ambient pressure could be changed by as much as $6 \cdot 10^{-3}$ atmospheres and surprisingly the cavitation threshold was unaffected. Moreover, above T_λ the vapor pressure could be increased by 1 atmosphere and the cavitation threshold rose from $3 \cdot 10^{-3}$ to only $5 \cdot 10^{-3}$ atmospheres. This behavior is unlike that of any other liquid. FINCH and co-workers have investigated the possibility that the nucleation of cavitation is associated with the presence of quantized vortices and have collected information supporting this possibility [11]. Quantized vortices do not occur in He I and so one cannot seek in this process an explanation of why there is an abnormally low threshold and an insensitivity to ambient pressure in both He I and He II. There is much more work to be done on this problem.

2˙1.3. Velocity of first sound at the lambda-transition [12]. At a lambda-transition the Pippard-Buckingham-Fairbank relations (hereafter PBFR) govern the singular behavior of the specific heat C_p, the isobaric expansion coefficient β and the isothermal compressibility K_T. They play the same role as the Ehrenfest relations at a second-order phase transition and in fact can be considered as a strong form of those relations. The PBFR are

(47)
$$C_p = \alpha V_\lambda T_\lambda \beta + C_0\,, \qquad \beta = \alpha K_T + \beta_0\,,$$

where

$$C_0 = T_\lambda \left(\frac{\mathrm{d}S}{\mathrm{d}T}\right)_\lambda, \qquad \beta_0 = \frac{1}{V_\lambda}\left(\frac{\mathrm{d}V}{\mathrm{d}T}\right)_\lambda, \qquad \alpha = \left(\frac{\mathrm{d}P}{\mathrm{d}T}\right)_\lambda.$$

C_0, β_0 and α are evaluated along the λ-line. The statement one makes about PBFR is that there is a region close to the lambda-transition where they hold and they are exact in the limit as one approaches the transition. It is of interest to know how close to T_λ one must be to use PBFR since in this region a measurement of, say, C_p suffices to determine β and K_T. It is clear from the relations that if C_p is singular, so are β and K_T. It is not difficult to show that the ther-

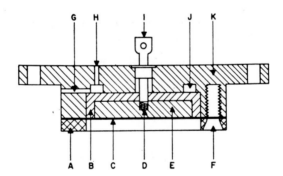

A. COPPER RING
B. NYLON INSULATOR
C. ALUMINUM FILMED MYLAR
D. SPRING FOR ELECTRICAL CONTACT.
E. COPPER BACK PLATE
F. ONE OF SIX EQUALLY SPACED 2-56 SCREWS
G. ONE OF THREE FILLING PORTS TO RESONATOR
H. MAIN FILLING PORT
I. INCONEL – KOVAR ELECTRICAL FEEDTHROUGH
J. CIRCULAR RESERVOIR VOLUME
K. COPPER HOUSING

Fig. 5. – Transducer used in first-sound measurements near T_λ. The aluminized mylar is the vibrating element of a capacitative transducer. Taken from ref. [12].

modynamic value of the sound velocity u is given by

$$(48) \qquad \frac{V^2}{u_1^2} = -\left(\frac{\partial V}{\partial p}\right)_s = -\left(\frac{\partial^2 \mu}{\partial p^2}\right)_T + \left(\frac{\partial^2 \mu}{\partial p \, \partial T}\right)^2 \left(\frac{\partial^2 \mu}{\partial T^2}\right)_p^{-1},$$

where μ is Gibbs' potential. It is thus clear that in the Ehrenfest scheme of ordering phase transitions the velocity of sound is discontinuous at first- and second-order transitions and is continuous at higher-order transitions. Moreover, as we shall now show, it is also continuous at lambda-transitions. By use of PBFR and the relations

$$C_p = -T\left(\frac{\partial^2 \mu}{\partial T^2}\right)_p, \qquad \beta = \frac{1}{V}\frac{\partial^2 \mu}{\partial p \, \partial T}, \qquad K_T = -\frac{1}{V}\left(\frac{\partial^2 \mu}{\partial p^2}\right)_T,$$

we find

$$(49) \qquad \frac{2(u_1 - u_{1\lambda})}{u_{1\lambda}^3} \simeq \frac{1}{u_{1\lambda}^2} - \frac{1}{u_1^2} = \frac{C_0^2}{\alpha^2 V^2 T}\frac{1}{C_p},$$

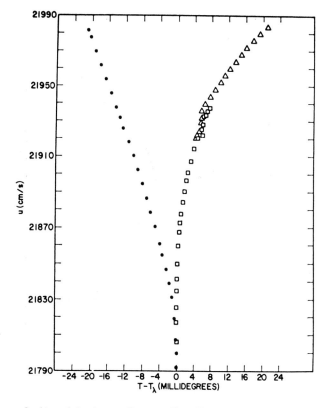

Fig. 6. – The velocity of first sound near T_λ. Taken from ref. [12].

where $u_{1\lambda}$ is the sound velocity at T_λ and is given by

$$(50) \qquad u_{1\lambda} = \frac{\alpha V T^{\frac{1}{2}}}{(C_0 - \alpha\beta_0 V T)^{\frac{1}{2}}} .$$

We see that, since the right-hand side of (49) is necessarily positive, u_1 is a minimum at T_λ. Moreover, u_1 is a linear function of $1/C_p$ in the regime near T_λ where PBFR hold. Thus a necessary, but not sufficient, condition that PBFR hold is that the velocity of sound is a linear function of $(C_p)^{-1}$. At all critical points there is attenuation and dispersion which is more and more apparent as the frequency is increased. The relations we have derived are obtained from reversible thermodynamics and can be expected to hold only for low frequencies.

Figure 5 shows the transducer used in this investigation. It is an electrostatic transducer whose movable element is aluminized mylar $6 \cdot 10^{-4}$ cm thick. A copper cylinder about 4.4 cm long is terminated at both ends by identical transducers, one serving as a source and the other as a receiver. In its most useful mode of operation the output of the receiver is amplified and used to drive the source so that the system self-oscillates at a plane-wave resonance of

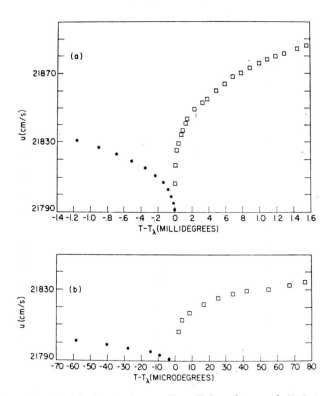

Fig. 7. – The velocity of first sound near T_λ. Taken from ref. [12].

the cylinder. Suitable filtering eliminated all but the desired resonant mode. Figure 6 shows the results in the range $T_\lambda \pm 20$ m °K.

Figure 7 shows the results for the range $T_\lambda \pm 1.4$ m °K and $T_\lambda \pm 70$ μ °K. The graph at the bottom of Fig. 7 shows that the minimum in sound velocity at this frequency (22 kHz) occurs at $T_\lambda - 3$ μ °K. The shift in position of three microdegrees is due to dispersion. Figure 8 is a plot of $u_1 - u_{1(min)}$ against $(C_p)^{-1}$.

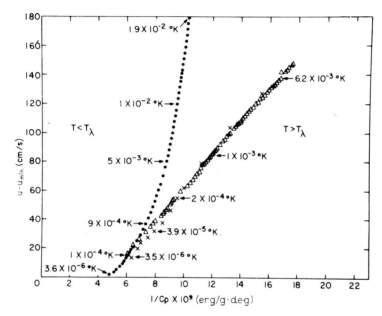

Fig. 8. – Measurement of the velocity difference $u_1 - u_{1min}$ against $1/C_p$. u_{1min} is the lowest value of u_1 measured. o and ∆: steady drift up in temperature; ×: measurements under equilibrium situation.

It is clear that in He I there is a linear relationship between u_1 and $(C_p)^{-1}$ over a temperature range of millidegrees. In the range in which PBFR hold, the data for He II should fall on this same straight line and it is clear that this only happens for temperatures somewhat closer to T_λ than 1 m °K. The value of $u_{1\lambda}$ can be obtained by extrapolating the line in Fig. 8 to $(C_p)^{-1} = 0$ and is found to be 217.3 m/s. Other pertinent thermodynamic properties of the lambda-transition can be obtained from the experimental data (see ref. [12] for these results).

The dispersion which resulted in shifting the minimum in sound velocity to 3 μ °K below T_λ also resulted in a large increase in attenuation near T_λ. A proper study of this requires higher frequencies and this is possible using the same transducers (Fig. 5). Figure 9 shows the results of such measurements from 0.60 to 3.17 MHz [13]. At these high frequencies the Q's of the resonators are much reduced and, moreover, one cannot reliably obtain a pure plane-wave

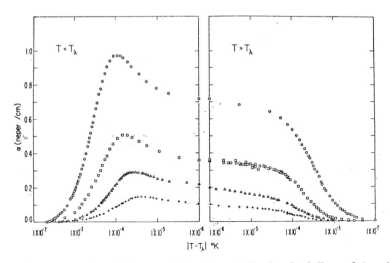

Fig. 9. – The attenuation of first sound in the neighborhood of T_λ as determined by pulse-echo measurements. Beginning with the lowest curves the frequencies are 0.60, 1.00, 1.75 and 3.17 MHz. Taken from ref. [13].

mode. The appropriate method is pulse-echo and, because the transducers do not ring, the pulses are quite sharp. It is seen that the attenuation is asymmetric about T_λ and has a maximum below T_λ. Before these results became available it was generally believed that the peak in attenuation occurred at T_λ; with a resolution of millidegrees only, that is the way the results appear. It is clear from Fig. 9 that with microdegree resolution, additional physical phenomena become apparent. Critical attenuation due to order parameter fluctuations is expected to be symmetrical about T_λ. If we assume that there are two causes of attenuation here—critical attenuation and an attenuation which occurs only below T_λ—then we can get the latter by subtracting $\alpha(\varepsilon)$ from $\alpha(-\varepsilon)$ where $\varepsilon = T - T_\lambda$. Figure 10 gives the results of this subtraction. LANDAU and KHALATNIKOV [14] had proposed that there ought to be a relaxation process associated with the order parameter which in turn is related to ϱ_s. They concluded that the relaxation time τ should be proportional to $(T_\lambda - T)^{-1}$. This, however, was based on an incorrect temperature dependence for ϱ_s. This has been emphasized by POKROVKII and KHALATNIKOV [15], who ascribe the relaxation process to a coupling of first and second sound and find a relaxation time τ and attenuation α given by

$$(51) \qquad\qquad \tau = \xi/u_2 \,,$$

$$(52) \qquad\qquad \alpha = \frac{u_{1\infty} - u_{10}}{u_{1\infty} u_{10}} \frac{\omega^2 \tau}{1 + \omega^2 \tau^2}\,,$$

where ξ is the coherence length, u_{10} and $u_{1\infty}$ are the velocities in the low- and high-frequency limits. Since $\xi \sim \varrho_s^{-1}$, $\tau \sim \varrho_s^{-\frac{3}{2}}$. But $\varrho_s \sim (T_\lambda - T)^{\frac{2}{3}}$. There-

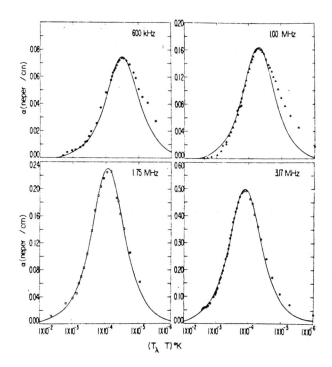

Fig. 10. – The points were obtained by subtracting the attenuation in the He I from that in He II at the same value of $\varepsilon = |T - T_\lambda|$ using the data of Fig. 9. The curves are plots of eq. (52). Taken from ref. [13].

fore τ is expected to vary as $(T_\lambda - T)^{-1}$, just as for the Landau-Khalatnikov result. The curves shown in Fig. 10 are plots of eq. (52) with $\xi = 1.36 \cdot 10^{-8}(T_\lambda - T)^{\frac{2}{3}}$ and $u_{1\infty} - u_{10}$ (in cm/s) $= 18.5$, 24.6, 20.6 and 23.2, in order of ascending frequency.

To summarize: it appears that the attenuation near the lambda point has two components: 1) a critical component associated with order parameter fluctuations which is symmetrical about T_λ and 2) a component associated with a relaxation process involving first sound and the critical wave mode (which in this case is second sound). It is apparent from Fig. 9 that the critical attenuation is temperature independent at T_λ and that it has a weak frequency dependence. Measurements between 16 kHz and 3.17 MHz give $\alpha \sim \omega^{1.15}$. Note also that it is temperature independent at T_λ and not singular as earlier theories indicated.

There is reason to believe that exactly similar phenomena occur at all lambda-transitions, and it is clear that their observation requires high frequencies, extreme temperature resolution and very good sample uniformity. A non-uniform sample will have its transition smeared over a temperature interval and perhaps this is the reason the existence of the two attenuation mechanisms was not discovered prior to this work. Subsequently similar phenomena were seen at magnetic transitions [16].

Attenuation measurements have been made at 1 GHz [17]. The measurement uses identical cadmium-sulphide films on quartz substrates and determines the pulse amplitude as the separation between the transducers is changed. The central experimental problem stems from the fact that the acoustic wavelength is only $2 \cdot 10^3$ Å and accuracy of alignment usually associated with optical rather than acoustical interferometry must be maintained throughout the experiment. The second article cited in ref. [17] deals with the experimental technique which achieved this result. Figure 11 shows the overall temperature dependence of

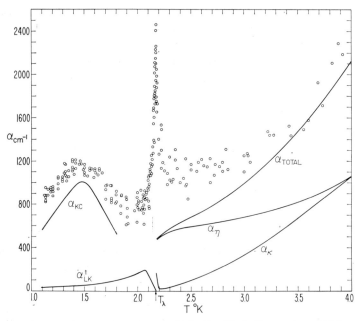

Fig. 11. – The attenuation of first sound in liquid helium at 1.0 GHz. α_η is the attenuation due to shear viscosity. α_K is that due to thermal conductivity. $\alpha_{tot} = \alpha_\eta + \alpha_K$ (the singular behaviour of the thermal conductivity in the neighborhood of T_λ has been excluded in this sum). α_{LK} is that due to the order parameter relaxation (see Fig. 10), α_{KC} is that due to KHALATNIKOV and CHERNIKOVA [19]. Note that the measured attenuation near T_λ is unaccounted for. Taken from the first citation in ref. [17].

the attenuation. From 4.2 to 2.3 °K the attenuation is controlled by the shear viscosity and thermal conductivity. In the critical region from 2.3 to 2.0 °K there is a sharp peak near T_λ. In this measurement, and particularly subsequent measurements [18], the peak is found to occur about $3.3 \cdot 10^{-3}$ °K below T_λ. Like the lower-frequency measurements just discussed, α is independent of $|T_\lambda - T|$ as $T \rightarrow T_\lambda$. It scales as $\omega^{1.3}$ with these measurements. The attenuation due to the Landau-Khalatnikov process is seen to contribute very little to the critical peak. In the superfluid region the attenuation is described in a theory by KHALATNIKOV and CHERNIKOVA [19]. The elementary excitations in superfluid helium are phonons and rotons. At higher temperatures the relaxation times are short and at lower temperatures long. The peak in attenuation at 1.5 °K separates these two regions. The results are seen to be in quite good agreement with theory.

2'2. Second sound.

2'2.1. Second-sound transducers. The first attempts to detect second sound by PESHKOV were unsuccessful—he used quartz crystal transducers. Only after LIFSHITZ pointed out the essentially thermal nature of second sound and PESHKOV used a heater as a source and a thermometer as a receiver was he successful. Until recently these were the only transducers used. A defect of the heater is that it produces a d.c. and a.c. component, since it can only heat—it cannot cool. A defect of the usual thermometer is that it is generally a resistive element which requires a current so that the change in resistance, current, or voltage can be detected—it also produces heating.

There is a novel transducer which is purely mechanical. If one drives the normal and superfluid components, then when $v_n/v_s = 1$ first sound is produced, and when $v_n/v_s = \varrho_s/\varrho_n$ second sound is produced. If the ratio v_n/v_s differs from these two values, then *both* first and second sound are produced. If a porous membrane, whose pore diameter is smaller than the normal fluid viscous wavelength, is substituted for the mylar diaphragm of the first-sound transducer shown in Fig. 5, then it will be opaque to the normal component, but the superfluid component can flow through the pores. Accordingly when used as a source the normal fluid will have the velocity of the membrane but the superfluid component will have a considerably reduced velocity and as a result a pair of such transducers will generate and detect both first and second sound. Because the two sounds are independent wave modes which travel with much different velocities there is no difficulty in distinguishing between them.

WILLIAMS recently pointed out that a diaphragm which has unclamped edges also produces both sounds because the superfluid component does not have to overcome viscous forces in order to flow in and out of the space between the diaphragm and the electrode. Again the ratio of the superfluid-component flow and that of the normal component is neither 1 nor ϱ_s/ϱ_n and both sounds are produced.

2'2.2. Velocity of second sound at the lambda-transition. Scaling theory makes predictions about the singular behavior of physical quantities near critical points [20]. A result which is pertinent to the lambda-line of helium is that if we write $C_p \sim (T_\lambda - T)^{-\alpha}$ and $\varrho_s \sim (T_\lambda - T)^\sigma$ then $\sigma = \frac{1}{3}(2 - \alpha)$.

Now it is observed that $C_p = \text{const} - \text{const} \cdot \log(T_\lambda - T)$. If we approximate this by the power law, then in the limit of vanishing $T_\lambda - T$ we see that $\alpha = 0$. Thus the scaling law we have written results in $\varrho_s = (T - T)^{\frac{2}{3}}$. The first precise measurement of ϱ_s near T_λ was by REPPY *et al.* [21]. They used the angular momentum of persistent currents and verified this result. One of the principal uses of second sound has been to determine ϱ_s. It came as a surprise then that when it was so used the values of ϱ_s determined from eq. (20) departed more and more from the $\frac{2}{3}$ power law as T_λ was approached, and the discrepancy at $T_\lambda - T \sim 50\ \mu\ {}^\circ\text{K}$ was large. If these results were correct then a possible conclusion was that two-fluid hydrodynamics was suffering a break-down near T_λ and (20) was no longer correct (see ref. [22] for bibliography of these measurements). The porous-diaphragm transducer is uniquely suited to such measurements and Fig. 12 shows that there is a substantial agreement between the

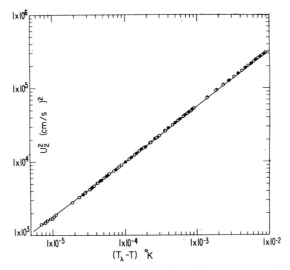

Fig. 12. – The velocity of second sound near T_λ. Taken from ref. [22].

velocity of second sound given by (20) $\left(\text{using } \varrho_s \sim (T_\lambda - T)^{\frac{2}{3}}\right)$ and the measurements for $T_\lambda - T$ as small as $7\ \mu\ {}^\circ\text{K}$. Recently GREYWALL and AHLERS [23] have investigated the temperature dependence of ϱ_s near T_λ over the full pressure range using this technique. There is an interesting feature of this type of transducer. The healing length (which is discussed in the following Subsection on third sound) varies as ϱ_s^{-1} and consequently gets large as $T \to T_\lambda$. When it becomes larger than the pore radius, the superfluid can no longer flow with zero viscosity through the pores and the transducer ceases to operate as a second-sound transducer. Thus the minimum $T_\lambda - T$ attained with diaphragms having a pore diameter 8 μm (open circles) is less than that where the pore diameter was 3 μm (solid circles). With pore diameters of 0.05 μm useful measurements were obtainable only when $T_\lambda - T > 2\ \mu\ ^\circ K$. Clearly one can use this property to obtain a measure of the healing length near T_λ [23, 24].

2˙3. *Third sound.*

2˙3.1. Third-sound transducers. To fix our ideas let us consider a helium film on a flame-smoothed glass surface. If heat is locally applied the superfluid will accelerate toward it and there will be a gradient of force pushing the normal component away which will be overwhelmingly opposed by viscous forces. As a result the helium mounds at the heat source. If the heat is applied periodically the flow will be periodic and third sound will be produced. In the first experiments the heat source was a light beam [25]. In later experiments it was a thin metallic film deposited on the glass [26].

As pointed out earlier the thickest practical films are about 100 atomic layers (3.6 Å = atomic layer) thick. The earliest experiments detected the third-sound waves by the change in polarization of a light beam reflected from the top surface and the substrate. This was responsive to the height modulation of the film. We shall discuss experiments in films as thin as two atomic layers and clearly such a method loses its sensitivity as the film thickness decreases. As pointed out earlier the temperature changes which occur because of the motion of the superfluid are largely, but not wholly, wiped out by the evaporation and condensation of vapor. The residual temperature oscillations can be detected by the use of thin aluminum films. Bulk aluminum is a super-

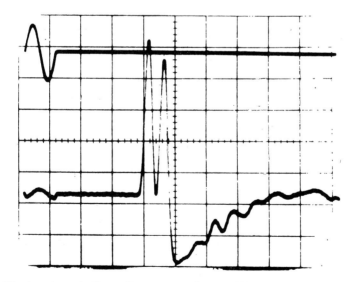

Fig. 13. – The top trace is the voltage across the metallic source film; the cycle period is 0.2 ms. The bottom trace is the time-delayed received pulse and is the voltage across the superconductor receiver film. Temperature increase is up. Source-receiver distance is 1.1 cm. u_3 is 1445 cm/s. Taken from ref. [26].

conductor having a transition temperature of 1.2 °K. Under appropriate circumstances thin films have higher transition temperatures—higher than T_λ. Using such a film one can adjust an external magnetic field and the current flowing in the films so that its transition temperature has a desired value. Its resistance becomes a very strong function of temperature and this is the type of transducer which is currently used to detect third sound. Figure 13 shows oscilloscope traces of the signal applied to a metallic source strip (aluminum operating at a supercritical current)—top trace—and the received signal is the bottom trace. Since the heat developed is proportional to the square of the current, this produced a double-peaked third-sound signal. The velocity is obtained from the time of flight.

2'3.2. The velocity of third sound. In Fig. 14 the velocity of third sound is plotted against $P_0 - P$, the difference between the saturated-vapor pressure and the pressure of the gas in equilibrium with the adsorbed film. As the film gets thinner, u_3 increases but eventually levels off. It is remarkable that the signals improve in quality until the film is quite thin and then they deteriorate. We will understand this when we discuss the attenuation of third sound. When the film is quite thin ($P_0 - P$ large) the poor quality of the signal precludes accurate measurements and this is indicated by the

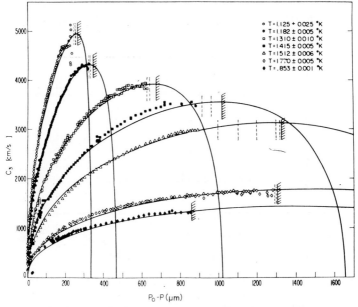

Fig. 14. – The third-sound velocity *vs.* $P_0 - P$. The points are experimental. The hatched vertical lines indicate where all signs of third sound disappear. This is taken to be the onset point for superflow. Between the hatching and the last data points the dashed lines signals are seen but not measured because of their poor quality. The curves are discussed in the text. Taken from ref. [27].

vertical dashed lines. Eventually in the hatched region they disappear. The thickness at this point is called the superfluid onset thickness. The curves through the points are related to a theoretical discussion which will be given later. Figure 15 is a direct plot of u_3 against d (in atomic layers) at two different temperatures. Figure 16 shows the experimentally observed onset thickness as a function of temperature. It is noteworthy that at low temperatures

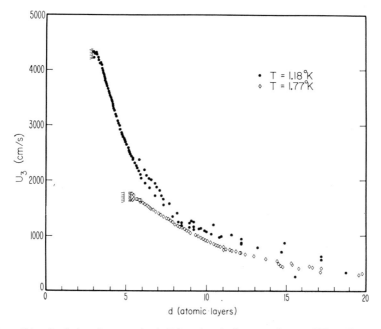

Fig. 15. – Direct plots of u_3 against d in atomic layers at 1.18 °K and 1.77 °K.

films as thin as 2.1 atomic layers support third sound. This is a clear demonstration that superfluidity exists in two dimensions, a result which has been the center of controversy in recent years.

What is not apparent in Fig. 14 is that, especially for the thinnest films, the velocity values are considerably below those one calculates using eq. (31) with the values for ϱ_s/ϱ obtained for bulk helium. One can show that, if ϱ_s/ϱ varies through the width of the film and if $\langle \varrho_s/\varrho \rangle$ is its value averaged over the width, then

$$(53) \qquad u_3 = \left(\frac{\langle \varrho_s \rangle}{\varrho} fy \right)^{\frac{1}{2}},$$

and we conclude that the low values are due to a depletion of ϱ_s in the film. This was expected by many but not always on the same grounds. As mentioned before, one reason advanced for expecting this is that one can argue that ϱ_s

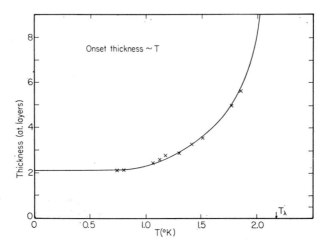

Fig. 16. – Onset thickness in atomic layers *vs.* temperature as determined by third sound.

ought to obey a boundary condition and vanish at the solid substrate. This is formalized in the Ginzburg-Pitaevskii theory where a fundamental characteristic length—the healing length—characterizes the distance over which ϱ_s goes from its value 0 at the boundary to the bulk value in the interior of the film. The curves in Fig. 14 are calculated from this theory using the healing length as a free parameter and it is seen that good agreement is obtainable. One of the most important uses of third sound has been the determination of $\langle \varrho_s \rangle$ through eq. (53). In Fig. 17 the value of $\langle \varrho_s \rangle d/\varrho_s$ is plotted *vs.* d. The line through the points has a slope of 1. If $\langle \varrho_s \rangle = \varrho_s$ then a line of unit slope

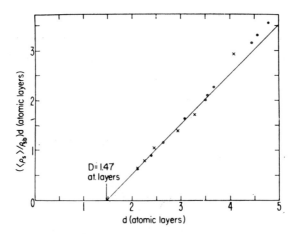

Fig. 17. – Plot of $(\langle \varrho_s \rangle/\varrho) d$ *vs.* d for helium films on glass (circles) and CaF_2 (crosses) at 0.5 °K. The intercept of the line of slope 1 with the abscissa gives D which is plotted in Fig. 18. Taken from ref. [28].

through the origin would describe the results and the depletion of ϱ_s is readily apparent. A heuristic way to describe the results is to say that the film acts as though it has a zero value for a thickness D equal to the horizontal intercept in Fig. 17 and its bulk value everywhere else. The equation of the line is

$$(54) \qquad \langle \varrho_s \rangle = \varrho_s \left(1 - \frac{D}{y} \right).$$

In the Ginzburg-Pitaevskii theory [29] eq. (54) is a good approximation in the region in which third-sound measurements were made and departures (towards a larger intercept) only occur for thickness less than the observed onset thickness. Thus the good fit in Fig. 14 only reflects the fact that the data fall along the unit-slope line in Fig. 17. In the Ginzburg-Pitaevskii theory, if ϱ_s vanishes at the free surface as well as the solid substrate,

$$D = d_s + 2\sqrt{2}\, l,$$

where l is the healing length and d_s is the thickness of helium at the substrate which is solid by virtue of the large van der Waals force. A major difficulty in pushing these ideas further is that there are as yet no experiments which unambiguously determine the thickness of the solid layer on glass with the required accuracy. Figure 18 is a plot of D vs. T. It is temperature independent and equal to 1.47 up to 0.6 °K. Above 1 °K it can be fitted with the empirical curve $D = 0.5 + 2.45(T/T_\lambda)(\varrho/\varrho_s)$.

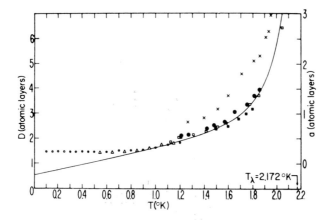

Fig. 18. – A plot of D (see Fig. 17) as a function of temperature. The circles, squares and triangles are from data on third sound. The crosses are persistent-current results by REPPY and co-workers. The circled crosses are the results of a quartz microbalance experiment by CHESTER and YANG. The solid curve is from the empirical formula $D = A + B(T/T_\lambda)\varrho/\varrho_s$. See ref. [28] for further details.

2`3.3. The attenuation of third sound. Third sound attenuates
as it propagates. BERGMAN has calculated [30] the attenuation that arises as
a result of heat conduction, viscosity and evaporation and found that the dominant source of attenuation for thick films is the evaporation and condensation
process occurring between the film and the vapor in equilibrium with it. Because there is a finite time required for this, the process is irreversible. This
gives rise to an attenuation which varies as $\omega^2 y^{11/2}$. Thus it decreases strongly
as the thickness decreases. For thin films the dominant dissipative processes
are heat conduction in the substrate and in the vapor above the film. This
leads to an attenuation which varies as $\omega^{\frac{1}{2}} y^{-\frac{5}{2}}$ and thus increases as the thickness decreases. One consequently expects a minimum in attenuation at an
intermediate thickness. At 1.4 °K this occurs at about 40 atomic layers according to theory.

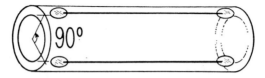

Fig. 19. – Third-sound pulses are launched by one of the two metallic strips 90° apart,
travel in both directions around the cylinder, and are picked up by the other strip.
The helium film adsorbed on the cylinder is an endless medium for the waves generated
in this way. Taken from ref. [31].

Figure 19 is a schematic diagram of the apparatus used to measure the attenuation of third sound. Two axially directed aluminum film third-sound
transducers, of the type already discussed, are evaporated on a glass cylinder
90° apart. A signal pulse applied at the source transducer results in 2 equal
third-sound pulses which propagate around the cylinder in opposite directions.

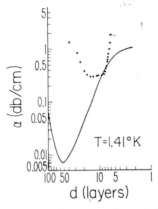

Fig. 20. – The points are the measured attenuation of third sound and the curve is
calculated from Bergman's theory [30]. Taken from ref. [31].

The cylindrical helium film is an endless medium for the third sound, and the exponentially decaying signal at the receiver transducer gives the attenuation. Figure 20 shows the observed attenuation at 1.41 °K (plotted points) as a function of helium film thickness in comparison with Bergman's calculated attenuation. The attenuation is a minimum at an intermediate film thickness but this thickness is much smaller than that at which the attenuation is a minimum in Bergman's theory. Only at a thickness of about 5 atomic layers do the observed and calculated values agree. At all other thicknesses the observed attenuation exceeds Bergman's value. How are we to understand this?

Attenuation is associated with nonlinear processes in the wave propagation and so we inquire about the wave distortion which occurs. Every point on the wave surface travels with a velocity which is the sum of the third-sound velocity for the film thickness at that point and the convective velocity v_s at that point. Thus we can write

$$(55) \qquad u_3 = u_3^0 + \frac{\partial u_3}{\partial y} \delta y + \frac{\partial u_3}{\partial v_s} v_s ,$$

where u_3^0 is the infinitesimal amplitude velocity of third sound. From the continuity equation (37) we obtain

$$(56) \qquad v_s = \frac{u_3}{\langle \varrho_s \rangle / \varrho} \frac{\delta y}{y} .$$

Using (53) and (54) we get

$$(57) \qquad \frac{\partial u_3}{\partial y} = \frac{(\varrho_s / \varrho) u_3}{2 \langle \varrho_s \rangle / \varrho y} \left(\frac{4D}{y} - 3 \right) .$$

From eq. (42) for the Doppler shift of third sound we find

$$(58) \qquad \frac{\partial u_3}{\partial v_s} = \frac{\langle \varrho_s \rangle}{\varrho} .$$

As pointed out earlier (see comments following eqs. (41) and (44)) PUTTERMAN argues that

$$(59) \qquad \frac{\partial u_3}{\partial v_s} = \frac{1}{2} \left(\frac{\langle \varrho_s \rangle}{\varrho} + \frac{\varrho_s}{\varrho} \right) .$$

Experiments [39] do not have sufficient precision to determine which of eqs. (58) and (59) is more accurate. Using eqs. (59), (57) and (56) we find

$$(60) \qquad u_3 = u_3^0 + \frac{\varrho_s / \varrho}{\langle \varrho_s \rangle / \varrho} \frac{u_3}{y} \left(\frac{3D}{y} - 1 \right) \delta y .$$

Now from eq. (56) we see that the crests of third-sound waves have a particle velocity in the direction of propagation and troughs have a velocity in the opposite direction. From eqs. (35) and (38) we see that, aside from the thickness dependence of ϱ_s/ϱ, the thicker the film the lower u_3. More appropriately the experimental data of Fig. 14 can be used to reach this conclusion with no reservations about the role played by the thickness dependence of ϱ_s/ϱ. Thus, whether crests travel faster than troughs or *vice versa* depends on whether the particle velocity contribution to the additional velocity of third sound exceeds that due to the thickness dependence. From eq. (60) it is clear that for $y > 3D$ crests travel slower than troughs and for $y < 3D$ the reverse is true. When $y > 3D$ the distortion tends to produce shocks on the trailing front and for $y < 3D$ on the leading front. In a deep third-sound ocean the waves break backward! Looking at Fig. 14 one can easily understand why the particle velocity dominates the distortion process for very thin films. At onset u_3 is independent of y.

From eq. (60) we see that when $y = 3D$ there is no distortion and waves propagate without change in shape. Within experimental uncertainty (it is not small) the minimum attenuation occurs at this thickness.

Mechanisms in addition to those considered by BERGMAN, which may be responsible for the excess attenuation shown in Fig. 20, include thermal and quantum fluctuations. CHESTER and MAYNARD [32] have outlined a procedure for accounting for the effect of thermal fluctuations. The striking feature of a maximum attenuation at some small value of y is absent in their result. GOODSTEIN and SAFFMAN [33] obtain a value for the attenuation of saturated films which is close to the observed value but their result has the defect that the frequency and especially the thickness dependence differ from those observed. The most interesting proposal postulates that the excess attenuation is due to quantum fluctuations. The idea is advanced that there is a macroscopic uncertainty principle which operates in the macroscopic quantum system, superfluid helium. This leads to fluctuations in film thickness and it is the coupling of such fluctuations with the wave form distortion process which leads to attenuation (see ref. [1], PUTTERMAN, p. 277). These ideas are speculative and objections to them have been advanced [32, 34]. However, it is possible to calculate values which agree with the experiments.

2˙3.4. The critical velocity of superfluid films. It is a characteristic of hydrodynamic flow that there is a critical velocity above which the character of the flow abruptly changes. For instance, for ordinary fluid flow in pipes, Poiseuille flow is replaced by turbulent flow above a critical Reynolds number. For superfluids there is a critical velocity below which the flow is governed by the two-fluid hydrodynamic equations and above which the flow is dissipative and eqs. (1)-(4) no longer apply. There have been many experiments which determine the critical velocity in films and constricted geometries such as capillaries, and two ideas have dominated in the discussion of the results. On the one hand there was an early observation [35] that in certain experiments the critical velocity of the superfluid component $v_{s.c.}$ was given approximately by

(61) $v_{\text{s.c.}} = \hbar/md$,

where m is the mass of the helium atom and d is the restricted width of the channel. On the other hand, especially in recent years, the thrust has been in the direction of explaining critical velocities in terms of thermal fluctuations which overcome an energy barrier for the creation of excitations such as vortices in the superfluid, thus introducing dissipation in the flow.

In the experiment to be described [36], the flow is that of a film and a unique feature is that the velocity of the film flow is determined directly from the Doppler shift of third sound. In all previous experiments it was inferred from the magnitude of the mass flux, the momentum flux, or the entropy creation.

Consider Fig. 21. The 6 in long glass cylinder has two equal length sections— the bottom section has a diameter of about 4 cm and the top section a diameter

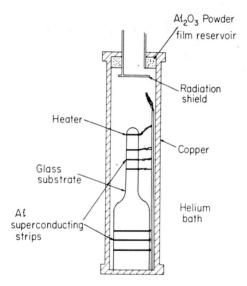

Fig. 21. – Schematic diagram of the apparatus used to measure the critical velocity of films using third sound. The glass substrate is about 15 cm long and is drawn to scale. Taken from ref. [36].

of about 1 cm. Each of these two sections has 3 third-sound transducers wrapped around it. At the very top of the cylinder 20 turns of resistance wire are wrapped and cemented to the cylinder. This is contained within a tight copper container which also contains about 4 g of Al_2O_3 powder (500 Å grain size) which serves as a helium film reservoir. Helium gas is introduced into the container at a pressure p less than the vapor pressure p_0, and consequently an unsaturated film coats everything. When current flows in the heater there is a flow of film up the glass cylinder, evaporation at the heater, a recondensation on the walls of the container and film flow down the walls to the base of the glass cylinder, and so on. The velocity of flow on the glass is determined by measuring the Doppler shift of a pulse of third sound using the outside pair of each triplet of transducers as a source and the middle one as a receiver. The top section of Fig. 22 shows the measured velocities, upstream and downstream, as a func-

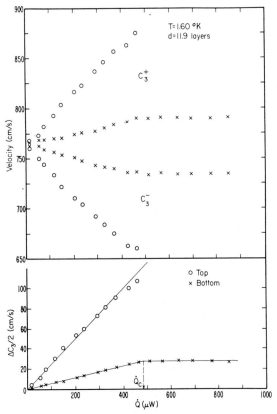

Fig. 22. – Typical measured Doppler-shifted third-sound velocities upstream C_3^- and downstream C_3^+, on both the top (circles) and bottom (crosses) sections of the glass cylinder in Fig. 21. The lower portion shows the net Doppler shift $\Delta C_3/2$ and the dashed line gives the critical heat current $\dot{Q}_c$. Taken from ref. [36].

tion of the heating rate. The lower section shows the net Doppler shift. It is seen that when the convective velocity on the top section of the cylinder is about 100 cm/s the velocity on the bottom section is about 25 cm/s and ceases to increase with further increases in heater power. This is because the critical velocity has been reached on the top section. The third-sound source transducers are heaters and superfluid flows towards them in generating the third-sound signal. When the direction of this flow is opposed to the convective velocity the net flow is less than the critical value, but when the sense of the two flows is the same the net flow remains critical and no third-sound signal is generated. Thus the third-sound signal propagating upstream disappears when the convective velocity approaches its critical value and this accounts for the absence of data on the top section for $\dot{Q} > \dot{Q}_c$. However, the critical

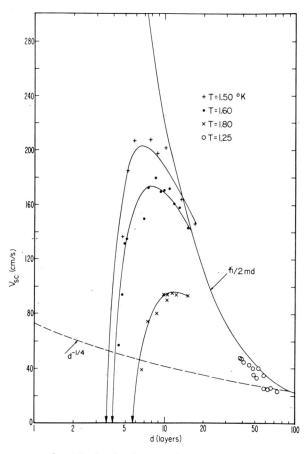

Fig. 23. – The measured critical velocities at three different temperatures. The circles and the solid curves are from ref. [38]. The dashed curve is the LEIDEN proposal for the critical velocity. Taken from ref. [36].

velocity is accurately determined by observations on the bottom section and its value as a function of film thickness and temperature is shown in Fig. 23. It is seen that it reaches a peak at intermediate film thicknesses. For smaller thicknesses it is strongly temperature dependent and has been described in terms of thermal activation processes [37]. For greater thicknesses the values seem to be compatible with values obtained by PICKAR and ATKINS [38] which are shown on the Figure. In this regime the critical velocity looks like it may be determined by the Bijl, de Boer and Michels condition, eq. (61) [35]. The peak velocity at intermediate thicknesses should increase as the temperature is decreased and may become as high as 500 cm/s. There have been no experiments to check this.

2·3.5. The thickness of a moving film. The average of u_3^+ and u_3^- in Fig. 22 gives the velocity of third sound at zero convective velocity (within terms which are first order in v_s/u_3). In the experiments just described it was found that $(u_3^+ + u_3^-)/2$ was a constant, independent of v_s. Equation (38) can be written in the following form:

$$(62) \qquad u_3^2 = \frac{\langle \varrho_s \rangle}{\varrho} \frac{3\alpha}{y^3}$$

when $TS/L \ll 1$. On quite general grounds one expects ϱ_s/ϱ to depend on $(v_n - v_s)^2$ since $(v_n - v_s)^2$ is a Galilean invariant and thermodynamic properties may depend on its magnitude. Also α/y^3 is expected to depend on $(v_n - v_s)^2$ through the Bernoulli equation [39]

$$(63) \qquad -\frac{\alpha}{y^3} + \frac{1}{2} \frac{\varrho_s}{\varrho} (v_n - v_s)^2 = \text{constant}$$

throughout a flowing film. The result that an Euler liquid has its surface depressed at converging sections (where the velocity is high) in canal flow is an analogue of this result. In the film flow experiment the van der Waals potential of the substrate replaces the gravitational potential for canal flow. Thus we can write

$$(64) \qquad \frac{\delta u_3^2}{u_3^2} = \frac{\delta(\langle \varrho_s \rangle /\varrho)}{\varrho_s/\varrho} + \frac{\delta(3\alpha/y^3)}{3\alpha/y^3} .$$

The expected value of the first term on the right is uncertainly known and has never been measured for circulation-free flow because its value is small. However, in high-velocity persistent currents in superleaks where it is known that the flow has vortex motion and is therefore not circulation free, quite large changes in $\delta(\langle \varrho_s \rangle /\varrho)$ have been observed.

Thus the observation that u_3^2 is independent of $(v_n - v_s)^2$ can be taken to mean that (within experimental error) changes in $\delta(3\alpha/y^3)$ balance out changes in $\delta \langle \varrho_s/\varrho \rangle$ at the temperature of the experiments in the range $(1.5 \div 1.9)$ °K. Other experiments designed to detect the changes in α/y^3 conflict in their results and at this writing the situation is not very clear. See ref. [39] for further discussion.

2·4. Fourth sound.

2·4.1. Transducers for fourth sound. The most common transducers used are the condenser transducers used for first sound (see Fig. 24).

But heater, thermometer transducers of the type used for second sound also work. For that matter so should the porous-diaphragm condenser type.

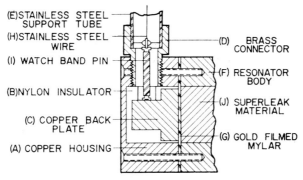

Fig. 24. – Cross-section of transducer used in fourth-sound measurements. Taken from ref. [3].

2˙4.2. The velocity of fourth sound. A plane-wave acoustic cavity is schematically shown in Fig. 25. The superleak must have pores which are small enough that the normal component of the helium is locked. The viscous penetration depth is $\sqrt{2\eta/\varrho\omega}$ (where η is the coefficient of viscosity), and this must exceed the pore radius. It will have its minimum value at the lambda point where it is equal to $76/f^{\frac{1}{2}}$ μm (f is the frequency in Hz). For a

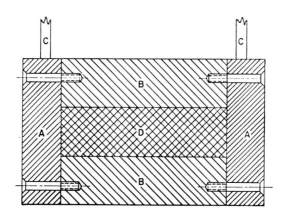

A. Transducers
B. Resonator Body
C. Support Tubes
D. Space For Superleak Material

Fig. 25. – Schematics of the acoustic resonator for fourth-sound measurements. Taken from ref. [3].

frequency of 5 kHz, and this is a typical frequency, the viscous penetration depth is about 1 μm and it becomes clear that the grain size of a powder super-leak should be less than 1 μm if it is to be used near T_λ at kilohertz frequencies. In practice it is found that powder grains 0.3 μm in diameter packed to a porosity P (ratio of open volume to total volume) equal to 0.5 to 0.8 will give good results over the whole temperature range for a resonator a few centimeters long. When the powder grains are 1 μm or greater, the unlocking which occurs near T_λ shows up as reduced Q resonances. In properly designed resonators, Q's as high as 6000 at 1 °K have been obtained.

Figure 26 shows some early measurements of the velocity of fourth sound [3] and compares the results with the value calculated using the best available

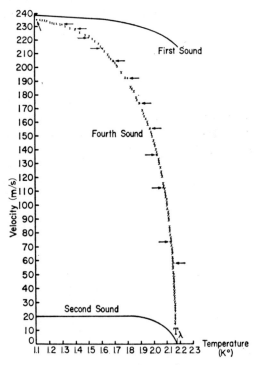

Fig. 26. – The velocity of fourth sound. The × are experimental. The arrows point to values calculated using accepted values of ϱ_s/ϱ. Taken from ref. [3].

values of ϱ_s/ϱ in eq. (29). The agreement is seen to be excellent. In fact a case can be made for using the observed value of the velocity of fourth sound as the primary way of determining ϱ_s/ϱ. It is interesting to note in this connection that the first unambiguous proof that the new phases of ^{3}He found below $2.6 \cdot 10^{-3}$ °K were superfluid came from the observation that fourth sound occurred and the measurements of its velocity gave the superfluid fraction as a function of temperature [40].

2`4.3. Scattering of fourth sound by the superleak. The powder particles act as scattering centers for the fourth-sound wave. Moreover, because of the dense array of grains, multiple scattering occurs. Because the grains are immobile they generate an inertial current (relative to the moving fluid). The best-known instance of this is the result that setting a sphere in motion in a fluid which is initially at rest produces a momentum in the fluid of one-half the mass displaced by the sphere times the velocity of the sphere. As a result dipole scattering occurs. Also because the grains are essentially incompressible there is monopole scattering. Present scattering theory is inadequate for dealing with this extreme case of multiple scattering, and the scattering corrections for fourth sound have been empirically determined. The velocity of fourth sound was measured for a number of powders with porosities ranging from 43% to 94%. When the powder grains were large enough, the theoretical temperature dependence was obtained, but the absolute values were reduced. If the powder grains were too small, further depression, which was temperature dependent, occurred. The origin of this is believed to be the same as that which depresses the velocity of third sound and will be discussed later. Using only the data for superleaks which give the theoretical temperature dependence and defining the index of refraction n as the ratio of the theoretical to the observed values of u_4 the results of Fig. 27 were obtained. Also plotted are values obtained by FERRERO and SACERDOTE [41] for airborne sound in the pores of a matrix of lead shot. It is seen that the equation

(65) $$n^2 = 2 - P$$

fairly describes the results although there is considerable scatter.

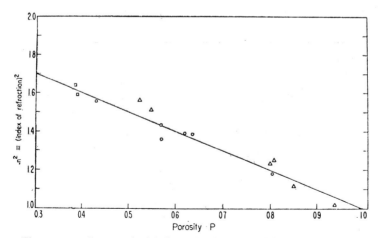

Fig. 27. – The measured acoustical index of refraction of fourth sound due to multiple scattering. The squares are values found by FERRERO and SACERDOTE in measurements on air contained in the pores of a lead pellet matrix (ref. [41]). Taken from ref. [42].

There have been recent attempts to derive the index of refraction by considering the medium as a quasi-continuum, *i.e.* considering volume elements large compared to the powder grains. With this view a description of the motion occurring in the neighborhood of a grain is ignored. The Kelvin inertial current is defined to be the velocity of the superfluid (at low temperatures) which occurs when the superleak is set in motion. Defining λ to be the ratio of the Kelvin inertial current to the velocity of the superleak, REVZEN *et al.* [43] obtain

$$(66) \qquad\qquad n^2 = \frac{P}{1-\lambda}.$$

The denominator is due to the Kelvin current (dipole scattering); the numerator is due to the incompressibility of the grains (monopole scattering). If both eqs. (65) and (66) hold, then $\lambda = (1-P)/(1-P/2)$. There are limited measurements of λ [44] which are in fair agreement with this. At present the most reliable result for the index of refraction of u_4 is eq. (65).

2˙4.4. The reduction of ϱ_s/ϱ, T_λ **and** u_4 **by healing-length effects.** We have already discussed how a finite healing length reduces ϱ_s/ϱ, T_λ and u_3 and how a measurement of u_3 in thin films can lead to a determination of the healing length. A measurement of u_4 can serve the same purpose and historically the use of fourth sound preceded the use of third sound in this connection. The former is inferior to the latter in a very important respect. It is impossible to accurately describe the channel size in practical superleaks and, to the degree it can be done, it can only be statistically. Films, on the other hand, have a satisfying uniformity of thickness. In fact the principal incentive for the development of third-sound techniques for thin unsaturated films comes

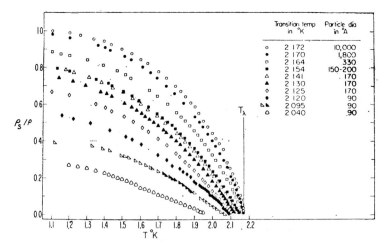

Fig. 28. – Values of ϱ_s/ϱ from fourth-sound measurements. Taken from ref. [42].

from precisely this consideration. It remains true, however, that u_4 can be measured more easily and more accurately than u_3 and fourth sound continues to have interest in connection with healing-length effects.

Figure 28 shows the results of measurements of the velocity u_4 in powders of various grain sizes and porosities. The value of ϱ_s/ϱ calculated from eq. (28), after scattering corrections according to eq. (65) are made, are plotted as a function of temperature. It is clear that T_λ and ϱ_s/ϱ can be seriously depressed. These data can be used to obtain the healing length, but such a determination is inferior to one based on u_3 for the reasons already given. Suffice it to say that, after allowance is made for the statistical distribution of pore sizes, there are no conflicts between healing-length determinations by the two methods. Apart from this, the kind of data shown in Fig. 28 is important in assessing the significance of other u_4 determinations.

2˙4.5. The measurement of persistent current using fourth sound. Just as a superconducting current persists in a superconducting ring, a superfluid current also persists. If the ring has a cross-section dimension of ~ 1 cm the critical velocity is very low according to eq. (61). Accordingly, the ring is generally filled with a superleak with a pore size which typically lies in the $(10^4 \div 10^2)$ Å range and currents of the order of 10^2 cm/s can then be obtained. The persistent currents can be measured by their gyroscopic effects and such measurements have added greatly to our knowledge of the superfluid state [44]. They can also be measured by the Doppler shift they produce in fourth sound and it is our present purpose to discuss this and the information gained by such measurements [45]. We should point out that Doppler shift is loosely used here to mean the change in velocity of fourth sound relative to the superleak produced by the persistent current. The velocity of a persistent current will always be understood to be the relative velocity of the normal and superfluid components, i.e. $v_n - v_s$.

Consider an azimuthal sound mode (no axial dependence) in a cylindrical resonator. The equation for the acoustic pressure distribution δp is

(67)
$$\delta p = \cos m\varphi J_m(k_m r) \cos \omega_a t ,$$

where $\omega_a/u = k_m$ and the value of k_m is determined from the boundary condition at the cylindrical wall, u is the sound velocity, ω_a is the acoustic resonant frequency, and $m = 1, 2, 3 \dots$. Equation (67) can be rewritten

(68)
$$\delta p = \tfrac{1}{2} [\cos (\omega_a t + m\varphi) + \cos (\omega_a t - m\varphi)] J_m(k_m r) .$$

Suppose the fluid undergoes solid-body rotation with a frequency ω_φ. Then eq. (68) is still correct if φ is measured in a frame which rotates with the fluid, but relative to the laboratory frame eq. (65) becomes

(69)
$$\delta p = \tfrac{1}{2} [\cos (\omega_a + m\omega_\varphi) t + \cos (\omega_a - m\omega_\varphi) t] J_m(k_m r) .$$

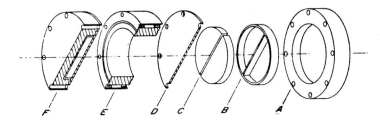

Fig. 29. – Partially exploded view of fourth-sound cylindrical resonator and transducers for investigating persistent currents. A, B, C and D are the elements of the source transducer. A is the case, C are the electrodes driven out of phase so the transducer is a dipole source, B is an insulator. D is metallized mylar. E is the cylindrical resonator. Taken from ref. [46].

When the fluid is stationary, waves which run either clockwise or counterclockwise have the same resonant frequency ω_a. When the fluid rotates this degeneracy is lifted and the resulting doublet has frequencies

$$(70) \qquad\qquad\qquad \omega_a \pm m\omega_\varphi .$$

If the resonator is filled with superleak and He II, then the normal fluid is locked to the superleak (and resonator) and using eq. (43) for the Doppler

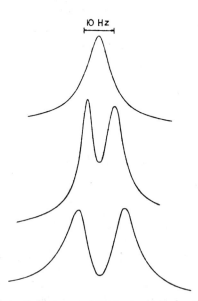

Fig. 30. – Splittings observed from persistent currents in the resonator of Fig. 29. Top curve: cool-down without rotation. Middle curve: cool-down with rotation rate of 5.4 Hz, splitting is 8.7 Hz. Bottom curve: cool-down with rotation rate of 13.5 Hz, splitting is 15.2 Hz. Taken from ref. [46].

shift of fourth sound we get the split modes

$$(71) \qquad \omega_a \pm m(\omega_n - \omega_s)\left(\frac{\varrho_s}{\varrho} + \frac{s[(\partial/\partial T)(\varrho_n/\varrho)]_p}{(\partial s/\partial T)_p} - \frac{\varrho[(\partial/\partial p)(\varrho_n/\varrho)]_T}{(\partial \varrho/\partial p)_T}\right) v_s \, .$$

It is clear then that a measurement of the splitting of the azimuthal modes gives the angular velocity $\omega_n - \omega_s$ of the persistent current. Figure 29 is an exploded view of the first resonator used for this purpose [46]. Condenser transducers were used in a dipole configuration to selectively excite the fundamental azimuthal mode. Figure 30 is a reproduction of the amplitude *vs.* frequency and it is clear that the observed splitting can be measured with considerable accuracy. In subsequent developments the resonator most used was an annulus as seen in Fig. 31 [47]. Most frequently the inside diameter of the annular ring is 10 cm; the cross-section of the ring has dimensions 0.5 cm×0.5 cm. The mode which runs around the annulus will be split by a persistent current according to eq. (68). With the dimensions quoted, there are 32 running modes (*i.e.* $m = 32$) before the first height or radial mode occurs, so there are many easily identifiable modes for use in the measurement. The choice of mode is generally based on the magnitude of the Q which may be as high as 6000. Typically they are $2000 \div 3000$. At low temperatures (say 1.1 °K) the Doppler shift is $(m\varrho_s/\varrho)(\omega_n - \omega_s)$. If $\omega_a = 2\pi \times 15$ kHz ($m = 25$) the bandwidth ($Q = 3000$) is $2\pi \times 5$. The persistent current which produces a splitting equal to the bandwidth has a velocity $\omega_n - \omega_s = 2\pi \times 5/25.2 = 2\pi \times 0.1$. Actually changes in

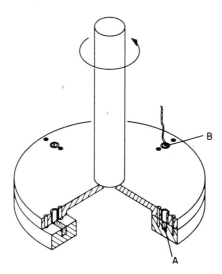

Fig. 31. – A cutaway view of the annular resonator used for persistent-current measurements. *A* is the cavity packed with powder. The inner diameter is 10.02 cm, the width and height are 5 mm. *B* is an electrode of the electrostatic transducer. Only four of the eight used are shown. Taken from ref. [47].

persistent current velocities which are the order of 0.01 rad/s (0.33 cm/s) can be detected. Figure 32 shows an x-y recording of the observed splitting of an $m = 24$ mode (there are 24 wavelengths of fourth sound around the annulus) obtained with the resonator of Fig. 31, packed with a powder whose grain size is nominally 0.05 μm (Linde B 0.05 μm).

Suppose such a resonator is cooled to some low temperature, say 1.3 °K, while at rest. Then both the normal and superfluid component will be at rest. Suppose now that the resonator is rotated at some constant angular velocity. The normal component being locked in the superleak which in turn is locked to the resonator will rotate with this same velocity ω_n. If the superfluid component behaved as a normal liquid, it too would rotate with this velocity. One might suppose that it would remain stationary since it is free of viscosity, but, when the experiment is performed and ω_s is measured by the Doppler shift of fourth sound, it is found that it rotates with a velocity of $\frac{1}{3}$ to $\frac{1}{2}$ of ω_n (depending on the superleak). Moreover, when ω_n is reduced to zero, ω_s goes to zero. The whole process is reversible. This flow of the superfluid has been called the Kelvin inertial current and is produced by the motion of the powder grains through the superfluid as first pointed out by MEHL and ZIMMERMAN [48]. It is a flow which would also occur in an irrotational ideal Euler fluid. There is a well-known result that, when a solid sphere is brought into motion with a velocity v in such a liquid, the liquid acquires a momentum equal to $\frac{1}{2}$ the product of the mass of liquid displaced by the sphere and v. (It follows that when the sphere is brought to rest, the liquid also comes to rest.) The velocity ω_s observed is within a factor of two what one would get assuming the powder grains are spheres and that their effect is additive. Since the superfluid is curl

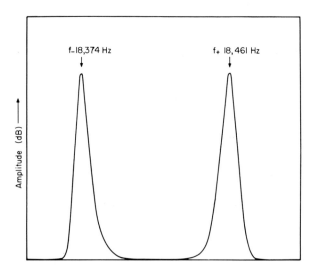

Fig. 32. – Splitting of the 24th azimuthal mode observed from persistent currents in the resonator of Fig. 31.

free $(\nabla \times \boldsymbol{v} = 0)$ and since initially it is at rest, the circulation remains zero at all times (it obeys the Kelvin circulation theorem). Thus we have the interesting situation that the flow of the superfluid component around the annulus has a finite angular momentum although its circulation is zero.

Figure 33 shows the results of measurements in the resonator of Fig. 31. Rotation starts at point A. As long as the angular velocity does not exceed that at point B, ω_{c_1}, the flow is reversible in the sense that the line AB is re-

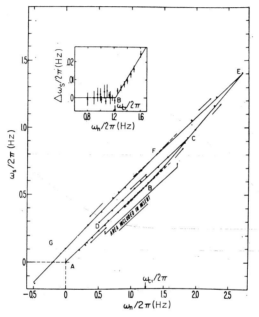

Fig. 33. – A plot of the angular velocities of the normal and superfield components when the resonator is rotated starting from rest. The sequence is in the following order: $ABCDEFG$. AB is the Landau state. $\omega(B) = \omega_{c_1}$ is the angular velocity at which vortices first enter the helium. The insert shows the data in the neighborhood of ω_{c_1}. The difference between the measured value of ω_s and that along the extended line AB is plotted against ω_n. It is apparent that the departure from line AB occurs abruptly at ω_{c_1}. Taken from ref. [49].

traced when the velocity is reduced. AB is called the Landau state. It is vortex free just as the Meissner state in superconductors is flux free. When ω_{c_1} is exceeded, the flow is not reversible and, if at C the rotational speed is reduced, the line CD is followed. Vortex flow enters the fluid in the region BC. Along CD the reduction in ω_s is due to a reduction in the vortex-free flow since it too is reversible (provided $\omega_n(D)$ is not too small). One can thus return to C and then to E by increasing ω_n. EF is also a region of reduction of vortex-free flow. EF, CD and BA are accurately parallel. Finally for $\omega < \omega(F)$ both vortex and vortex-free flow are reduced and when ω_n is finally reduced to zero a finite persistent current will remain. If ω_n is increased well beyond point E, eventually a point will be reached where the curve has a 45° slope. Increments of ω_n are matched by equal increments of ω_s. The persistent current is saturated and cannot be increased any further.

This is particularly evident in Fig. 34 where now $(\omega_n - \omega_s)/2\pi$ is plotted against $\omega_n/2\pi$. Starting from rest ω_n is increased. $\omega_{c_1}/2\pi$ is 1 Hz. The flow saturates at $(\omega_n-\omega_s)/2\pi=1.88$ Hz which occurs at approximately $\omega_n/2\pi=5$ Hz.

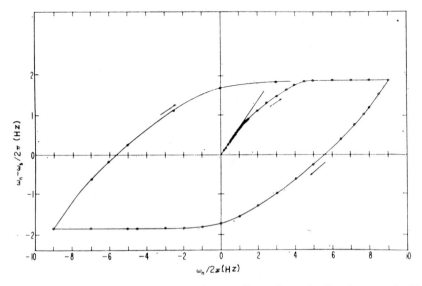

Fig. 34. – A plot from $\omega_n-\omega_s$ for complete cycling of ω_n starting from rest. Taken from ref. [49].

If $\omega_n/2\pi$ is increased to 9 Hz and then reduced to zero, a persistent current close to the saturated value remains. Negative values of $\omega_n/2\pi$ mean that the resonator is rotated in an opposite sense to its original direction of rotation. These results are a close analogue of the hysteresis curves found in highly irreversible, high-field, type-II superconductors when the magnetization is plotted as a function of the applied field. This is evident in comparing Fig. 34 and Fig. 35. The analogous quantities are magnetization $\sim$ persistent current velocity, applied magnetic field $\sim \omega_n$, $H_{c_1} \sim \omega_{c_1}$, $H_{c_2} \sim \omega_{c_2}$, quantized flux lines $\sim$ quantized vortex lines, Landau state $\sim$ Meissner state. ω_{c_2}, whether determined by the coherence length of He II or the lattice spacing of the powder grains, is so high as to be inaccessible.

Another way of producing a persistent current is to rotate the resonator when $T > T_\lambda$. It is then cooled and the rotation is only slowed and stopped at some $T < T_\lambda$. This is the method used in all the gyroscopic methods. Figure 36 shows the results of measurements during the slowdown (and sub-

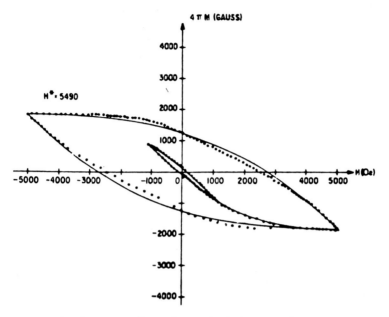

Fig. 35. – Magnetization *vs.* applied field of highly irreversible type-II superconductor made of lead pressed into porous vycor glass. Compare with Fig. 34. Taken from ref. [50].

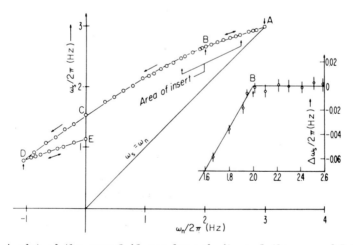

Fig. 36. – A plot of the superfluid angular velocity and the normal-fluid angular velocity when the rotation is started above T_λ then cooled below T_λ keeping ω_i constant, measurements being at A where $\omega_n = \omega_i$. The sequence is then $ABCDE$. The negative values of ω_n correspond to rotation in the opposite sense of those between A and C. AB and DE are strictly parallel and changes are reversible. The insert shows the deviations of $\omega_s/2\pi$ near B from the values along the extended line segment AB. The purpose is to show the abrupt departure from the line AB at B just as in the insert of Fig. 33. Taken from ref. [47].

sequent reversal of rotation before it is finally brought to rest). Note that $\omega_s = \omega_n = \omega_i$ as long as rotation is continued at its original velocity ω_i. In the region AB the change in velocity is due to the generation of a Kelvin drag current and the changes in flow are completely reversible. This is also true of DE which is accurately parallel. The insert shows the departure from the line determined by A and B and it is seen to be abrupt just as in the insert of Fig. 33.

If the velocity of the persistent current $\omega_n - \omega_s$ is plotted against $\omega_i - \omega_n$, then we find a curve independent of ω_i as shown in Fig. 37.

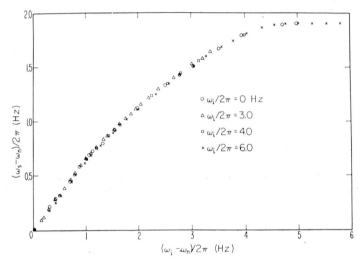

Fig. 37. – A plot of the persistent current $(\omega_s - \omega_n)/2\pi$ against $(\omega_n - \omega_i)/2\pi$ for various values of ω_i. This is taken from the type of data shown in Fig. 36. Taken from ref. [47].

Moderate-velocity persistent currents undergo no observable decay during times of the order of a day. On the other hand gyroscopic measurements showed that initially saturated critical-velocity currents undergo decay and it was found that the decay was proportional to the logarithm of time since their establishment. Superconducting currents in highly irreversible type-II superconductors also behave this way. This logarithmic behavior can be traced to the fact that an activation energy is involved in the process. In the interest of studying this behavior over many decades of time, a method using fourth sound was devised [51].

The frequency of splitting of an azimuthal mode can be determined by measuring the beat rate when the equally excited doublet is allowed to decay. Consequently if the resonator is driven by a band, whose width is slightly more than the splitting, centered on the doublet frequency, when the drive is removed one should see the beating in the decay. Figure 38 shows such a decay. The total decay is 40 db and the ordinate is linear in db. The decay time is

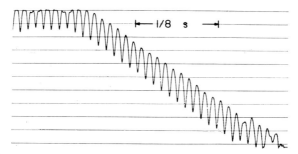

Fig. 38. – A plot of the beats between the Doppler-shifted fourth-sound modes in free decay. The total height of the decay is 40 db. Taken from ref. [51].

approximately $\frac{1}{4}$ s and so measurements could be made every $\frac{1}{2}$ s. Thus decay measurements could be made during the first second after the current was established.

By following the procedure of Fig. 36 and initiating measurements as soon as point C was reached, the data shown in Fig. 39 were collected. In curve A the initial current is saturated. The decay rate is seen to be 0.63 % per decade. An initial response might be that this is large; but after all the age of the Universe is estimated to be 10^{17} s and during this time (if the decay rate is unchanged) the total drop in current is only 12 %. When the initial

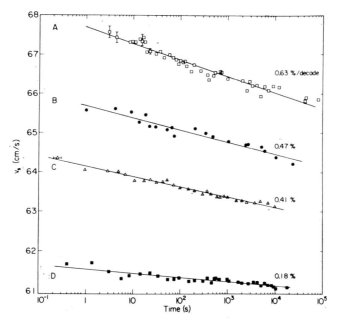

Fig. 39. – Decay of saturated (A) and unsaturated (B, C and D) persistent-current velocities plotted against $\log t$. Vertical bars on some data points in A indicate probable experimental errors in the initial parts of the data. From ref. [51].

current $\left(\omega_s(C) \text{ in Fig. 36}\right)$ is less than saturated the decay rate is reduced as seen in Fig. 39. When the initial current is $\sim 15\%$ less than the saturated value, there is no observable decay in 5 decades of time. The results in Fig. 39 clearly show that the decay rate is not a unique function of the velocity. Thus the rate of decay at the right-hand end of curve A is different from that at the left-hand end of curve B. Clearly the decay rate depends on the history of the persistent current.

It is clear that all the lines of Fig. 39 must intersect if they are extrapolated to the right. If we were able to carry out the experiments for a sufficiently long time is it possible this intersection could be observed? There are difficulties in understanding how it is possible that they intersect. Or is it possible that an intersection signals that other dissipative processes must come into play? Unfortunately (or fortunately) the extrapolated intersections of the curves shown in Fig. 39 occur at about 10^{19} s, which is somewhat greater than the age of the Universe. Surely it is unnecessary to account for the existence or nonexistence of the intersections.

$2 \cdot 4.6$. Application of superfluid acoustic wave guides to prove that persistent currents are caged in superleaks. Suppose the annular cavity of Fig. 31 has only the bottom half or so filled with powder, so that a vertical section has the dimensions shown in Fig. 40. We know that the

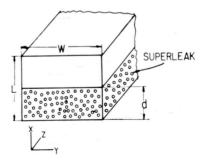

Fig. 40. – Cross-section of a partially packed superfluid acoustic waveguide. From ref. [52].

maximum persistent velocity if $d = 0$ is less than 1 cm/s and that if $d = L$ it is 10^2 cm/s. What can we expect in the co-existing channels? It is difficult to see how a gyroscopic measurement can provide the answer to this question. It is a demonstration of the great utility of acoustic techniques that the answer can be provided by them. But to see how this can be done, we must first discuss the propagating acoustic modes in a straight waveguide whose section is that of Fig. 40.

It has been shown that two sound modes can propagate in the annular channel [52]. One is essentially a modified second-sound mode with a velocity

u_{II}, given by

(72)
$$u_{\mathrm{II}}^2 = u_2^2 \frac{\varrho(L-(1-P)d)}{\varrho(L-d)+\varrho_s Pd},$$

where u_2 is the velocity of second sound and P is the porosity of the powder. Note that $u_{\mathrm{II}} \geqslant u_2$.

The other mode is an interpolated first-fourth sound mode with a velocity u_{14} given by

(73)
$$u_{14}^2 = u_1^2 u_4^2 \frac{\varrho(L-d)+\varrho_s Pd}{u_4^2 \varrho(L-d)+u_1^2 \varrho_s Pd},$$

where u_1 and u_4 are the velocities of first and fourth sound. Note that $u_1 \geqslant u_{14} \geqslant u_4$.

Since both modes have components of first and second sound in the free region and fourth sound in the packed region, first-, second-, or fourth-sound

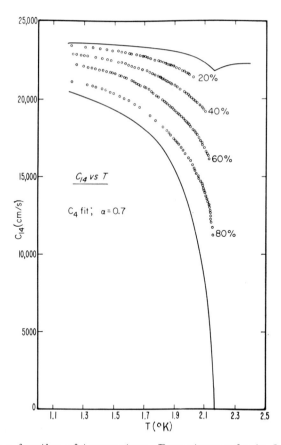

Fig. 41. – C_{14} as a function of temperature. Percentages refer to d, the depth of the powder. For the annulus used in this experiment $L = 0.5$ cm, $P \simeq 0.75$, and the powder size was nominally 0.05 μm. From ref. [53].

transducers may be used to generate and detect both modes. We have experimentally verified that eqs. (72) and (73) are correct to within about 1 % [53]. Figures 41 and 42 show the velocities of C_{14} and C_{II} as functions of temperature for various depths of powder.

When there are persistent current velocities v_f and v_p in the free and packed parts of the channel, the Doppler-shifted velocities of these modes are [54]

$$(74) \qquad u_{\mathrm{II}} = u_{\mathrm{II}0} \pm \frac{\varrho_n}{\varrho} v_f,$$

$$(75) \qquad u_{14} = u_{140} \pm \frac{\varrho_s}{\varrho} \frac{(L-d)\,v_f + P(2-P)\,dv_p}{L-d+P(2-P)\,d},$$

where the zero subscript denotes the value in the absence of current and the plus and minus signs refer to propagation with and against the flow respectively. v_f is determined by measuring the splitting of the azimuthal u_{II}-modes and then v_p by the splitting of the azimuthal u_{14}-modes.

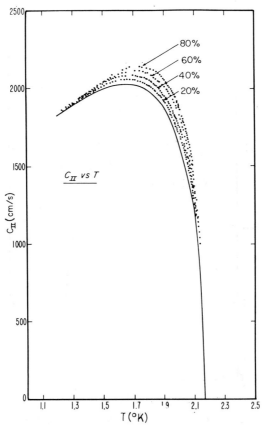

Fig. 42. – C_{II} as a function of temperature for the same annulus as used to find C_{14} in Fig. 41. Percentages refer to d, the depth of the powder. From ref. [53].

The usual procedures for generating persistent currents (epitomized by the results of Fig. 34 and 36) were followed and the outstanding and striking results of extensive measurements [55] with values of d ranging from 0.1 cm to 0.4 cm was that the u_{14}-mode showed considerable splitting and the u_{11}-mode showed no splitting. The maximum persistent current in the superleak was found to be 10^2 cm/s, that in the free region $\leqslant 1$ cm/s. This difference is remarkable since at the same time there is a free exchange of superfluid mass across the boundary separating the two regions. The current in the superleak is stabilized by the presence of a sheath of vortex lines near the boundary and most likely on the superleak side of it. It is this sheath which cages the current in the superleak, and the bars of the cage are the vortex lines. The results clearly imply that persistent currents can exist in a superleak which is not in a container—a bare superleak. The knowledge that this is so may lead to modifications in superfluid gyroscopes—unnecessary mass can be eliminated, and gyroscopic effects can therefore be increased.

2˙4.7. Superfluid Helmholtz resonators. Figure 43 is a schematic diagram of a classical Helmholtz resonator. The volume V is contained within rigid walls. The cylindrical neck has a length l and a cross-sectional area A. In its simplest form the radius of the neck is very much less than its length. When fluid oscillates in the neck the volume V is alternately compressed and expanded. The system can be regarded as a mass (the mass of fluid in the neck) and a spring (provided by the stiffness of the fluid volume V). It will have an angular resonant frequency ω given by

$$(76) \qquad \omega = u \sqrt{A/lV} ,$$

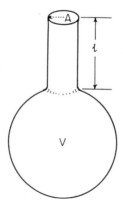

Fig. 43. – Schematic diagram of a Helmholtz resonator. V is the volume of the bulb, l is the length of the neck and A is the cross-sectional area of the neck. Such a vessel would have first- and second-sound resonances. When filled with superleak it would have a fourth-sound resonance. All resonant frequencies are given by $\omega_n = u_n (A/Vl)^{\frac{1}{2}}$, $n = 1, 2, 4$.

where u is the velocity of sound. It is essential that all dimensions be small compared to the sound wavelength.

Suppose, however, that the fluid is He II. Then it can be shown [56] that there are two resonant frequencies given by

$$(77) \qquad\qquad \omega_n = u_n \sqrt{A/lV} ,$$

where $n = 1$ and 2. The $n = 1$ mode is one just like that already discussed and u_1 is the velocity of first sound. The $n = 2$ mode is one in which the motion is characteristic of second sound—u_2 is the velocity of second sound. There is counterflow in the neck which is driven by temperature oscillations in the volume.

Finally if the resonator is packed with superleak, eq. (77) continues to hold with $n = 4$ signifying that this mode is a fourth-sound mode (u_4 is the velocity of fourth sound). All these modes have been observed.

It is clear that a measurement of the resonant frequency of a Helmholtz resonator can be used to determine the velocity of first, second, or fourth sound. In fact it is not difficult to devise a Helmholtz resonator for third sound for which eq. (77) is applicable with $n = 3$. The outstanding characteristic of such a determination of the velocity is that it can be carried out at very low frequencies and yet does not require an apparatus of large dimensions. So far they have had no significant application in this direction. Second sound in superfluid ^{3}He is expected to be highly attenuated because of the large viscosity of the normal component and it may be that these low-frequency resonators will have a real advantage. At this writing, second sound has not been seen in superfluid ^{3}He by any method.

REFERENCES

[1] For a comprehensive introduction to the properties of superfluid helium see
     S. PUTTERMAN: *Superfluid Hydrodynamics* (Amsterdam, 1974); F. LONDON:
     Superfluids (New York, N. Y., 1964); K. R. ATKINS: *Liquid Helium* (Cambridge,
     1959); J. WILKS: *Liquid and Solid Helium* (Oxford, 1967); W. KELLER: *Helium
     3 and Helium 4* (New York, N. Y., 1969); I. M. KHALATNIKOV: *Introduction to the
     Theory of Superfluidity* (New York, N. Y., 1965).
[2] K. R. ATKINS: *Phys. Rev.*, **113**, 962 (1959).
[3] K. A. SHAPIRO and I. RUDNICK: *Phys. Rev.*, **137**, A 1383 (1965).
[4] E. M. LIFSHITZ: *Žurn. Éksp. Teor. Fiz.*, **29**, 94 (1955) (English translation: *Sov.
     Phys. JETP*, **2**, 73 (1956)).
[5] E. S. SABISKY and C. H. ANDERSON: *Phys. Rev. A*, **7**, 790 (1973).
[6] I. M. KHALATNIKOV: *Žurn. Éksp. Teor. Fiz.*, **30**, 617 (1956) (English translation:
     Sov. Phys. JETP, **3**, 649 (1956)).
[7] K. TELSCHOW and I. RUDNICK: *Journ. Low Temp. Phys.*, **18**, 43 (1975).

[8]   H. KOJIMA, W. VEITH, S. J. PUTTERMAN, E. GUYON and I. RUDNICK: *Phys. Rev. Lett.*, **27**, 714 (1971).

[9]   R. D. FINCH, R. KAGIWADA, M. BARMATZ and I. RUDNICK: *Phys. Rev.*, **134**, A 1425 (1964).

[10] R. D. FINCH, T. G. J. WANG, R. KAGIWADA, M. BARMATZ and I. RUDNICK: *Journ. Acoust. Soc. Amer.*, **40**, 211 (1966).

[11] See J. R. SHADLEY and R. D. FINCH: *Phys. Rev. A*, **3**, 780 (1971), and references therein.

[12] M. BARMATZ and I. RUDNICK: *Phys. Rev.*, **170**, 224 (1968).

[13] R. D. WILLIAMS and I. RUDNICK: *Phys. Rev. Lett.*, **25**, 276 (1970).

[14] L. D. LANDAU and I. M. KHALATNIKOV: *Dokl. Akad. Nauk SSSR*, **96**, 469 (1954).

[15] V. L. POKROVSKII and I. M. KHALATNIKOV: *Piz'ma Žurn. Éksp. Teor. Fiz.*, **9**, 255 (1969) (English translation: *JETP Lett.*, **9**, 149 (1969)); I. M. KHALATNIKOV: *Žurn. Éksp. Teor. Fiz.*, **57**, 489 (1969) (English translation: *Sov. Phys. JETP*, **30**, 268 (1970)).

[16] B. GOLDING and M. BARMATZ: *Phys. Rev. Lett.*, **23**, 223 (1969); A. BACHELLERIE, J. JOFFRIN and A. LEVELUT: *Phys. Rev. Lett.*, **30**, 617 (1973).

[17] J. S. IMAI and I. RUDNICK: *Phys. Rev. Lett.*, **22**, 694 (1969); *Journ. Acoust. Soc. Amer.*, **46**, 1144 (1969).

[18] D. COMMINS and I. RUDNICK: *Proceedings of the XIII International Low Temperature Conference*, Vol. **1** (New York, N. Y., 1972), p. 356.

[19] I. M. KHALATNIKOV and C. M. CHERNIKOVA: *Žurn. Éksp. Teor. Fiz.*, **49**, 1957 (1965); **50**, 411 (1966) (English translation: *Sov. Phys. JETP*, **22**, 1336 (1966); **23**, 274 (1966)).

[20] B. D. JOSEPHSON: *Phys. Lett.*, **21**, 608 (1966).

[21] J. R. CLOW and J. D. REPPY: *Phys. Rev. Lett.*, **16**, 887 (1966).

[22] R. WILLIAMS, E. A. BEAVER. J. C. FRASER, R. S. KAGIWADA and I. RUDNICK: *Phys. Lett.*, **29** A, 279 (1969).

[23] D. S. GREYWALL and G. AHLERS: *Phys. Rev. A*, **7**, 2145 (1973).

[24] G. G. IHAS and F. POBELL: *Phys. Rev. A*, **9**, 1278 (1974).

[25] K. R. ATKINS and I. RUDNICK: *Progress in Low Temperature Physics*, Vol. **6**, edited by C. J. GORTER (Amsterdam, 1970).

[26] I. RUDNICK, R. S. KAGIWADA, J. C. FRASER and E. GUYON: *Phys. Rev. Lett.*, **20**, 430 (1968).

[27] R. S. KAGIWADA, J. C. FRASER, I. RUDNICK and D. BERGMAN: *Phys. Rev. Lett.*, **22**, 338 (1968).

[28] J. SCHOLTZ, E. MACLEAN and I. RUDNICK: *Phys. Rev. Lett.*, **32**, 147 (1974).

[29] V. L. GINZBURG and L. P. PITAEVSKII: *Žurn. Éksp. Teor. Fiz.*, **34**, 1240 (1958) (English translation: *Sov. Phys. JETP*, **7**, 858 (1958)).

[30] D. BERGMAN: *Phys. Rev. A*, **3**, 2058 (1971).

[31] T. WANG and I. RUDNICK: *Journ. Low Temp. Phys.*, **9**, 425 (1972).

[32] M. CHESTER and R. MAYNARD: *Phys. Rev. Lett.*, **29**, 628 (1972).

[33] D. L. GOODSTEIN and P. G. SAFFMAN: *Proc. Roy. Soc.*, A **325**, 447 (1971). This paper is criticized by D. BERGMAN: *Proc. Roy. Soc.*, A **333**, 261 (1973).

[34] M. W. COLE and D. L. GOODSTEIN: *Phys. Rev. A*, **9**, 2806 (1974).

[35] A. BIJL, J. DE BOER and A. MICHELS: *Physica*, **8**, 655 (1941).

[36] K. TELSCHOW, T. WANG and I. RUDNICK: *Phys. Rev. Lett.*, **32**, 1292 (1974).

[37] S. V. IORDANSKII: *Žurn. Éksp. Teor. Fiz.*, **48**, 708 (1965) (English translation: *Sov. Phys. JETP*, **21**, 467 (1965)); J. S. LANGER and M. E. FISHER: *Phys. Rev. Lett.*, **19**, 560 (1967); R. J. DONNELLY and P. H. ROBERTS: *Phil. Trans. Roy. Soc.* **271**, 41 (1971).

[38] K. S. PICKAR and K. R. ATKINS: *Phys. Rev.*, **178**, 389 (1969).

[39] K. TELSCHOW, I. RUDNICK and T. WANG: *Journ. Low Temp. Phys.* **18**, 43 (1975).

[40] H. KOJIMA, D. N. PAULSON and J. C. WHEATLEY: *Phys. Rev. Lett.*, **32**, 141 (1974).

[41] M. A. FERRERO and G. G. SACERDOTE: *Acustica*, **1**, 137 (1951).

[42] M. KRISS and I. RUDNICK: *Journ. Low Temp. Phys.*, **3**, 339 (1970).

[43] M. REVZEN, C. KUPER, J. RUDNICK and B. SHAPIRO: *Phys. Rev. Lett.*, **33**, 143 (1974).

[44] J. S. LANGER and J. D. REPPY: *Progress in Low Temperature Physics*, Vol. **6**, edited by C. J. GORTER (Amsterdam, 1970).

[45] H. KOJIMA: Ph. D. Thesis, University of California at Los Angeles (June, 1972).

[46] I. RUDNICK, H. KOJIMA, W. VEITH and R. S. KAGIWADA: *Phys. Rev. Lett.*, **23**, 1220 (1969).

[47] H. KOJIMA, W. VEITH, E. GUYON and I. RUDNICK: *Journ. Low Temp. Phys.*, **8**, 187 (1972).

[48] J. B. MEHL and W. ZIMMERMAN jr.: *Phys. Rev. Lett.*, **14**, 815 (1965); *Phys. Rev.*, **167**, 214 (1968).

[49] H. KOJIMA, W. VEITH, S. PUTTERMAN, E. GUYON and I. RUDNICK: *Phys. Rev. Lett.*, **27**, 714 (1971).

[50] C. P. BEAN: *Rev. Mod. Phys.*, **36**, 31 (1969).

[51] H. KOJIMA, W. VEITH, E. GUYON and I. RUDNICK: *Proceedings of the XIII International Conference on Low Temperature Physics*, Vol. **1** (New York, N. Y., 1974), p. 279.

[52] J. RUDNICK, I. RUDNICK and R. ROSENBAUM: *Journ. Low Temp. Phys.*, **16**, 417 (1974).

[53] J. HEISERMAN and I. RUDNICK: submitted for publication.

[54] These expressions were first derived by J. RUDNICK (private communication).

[55] J. HEISERMAN and I. RUDNICK: submitted for publication.

[56] M. KRISS and I. RUDNICK: *Phys. Rev.*, **174**, 326 (1968).

Absorption des ondes ultrasonores longitudinales et transversales dans les cristaux liquides.

S. Candau et P. Martinoty

Laboratoire d'Acoustique Moléculaire, Equipe de Recherche Associée au C.N.R.S. Université Louis Pasteur - Strasbourg

Introduction.

On a pu constater ces dernières années un développement considérable des recherches sur les cristaux liquides et depuis 1965 des colloques leur sont régulièrement consacrés. Pourquoi cet intérêt aussi spectaculaire que récent? Avant de répondre à cette question nous allons commencer par définir un cristal liquide.

Alors que la plupart des cristaux fondent à une température bien déterminée pour se transformer en un liquide isotrope, il existe des composés organiques pour lesquels cette fusion n'est que partielle. On obtient alors un liquide opaque, fortement diffusant, et ce n'est qu'au-dessus d'une certaine température appelée température de clarification que le composé se transforme en un liquide parfaitement clair.

Entre ces deux températures, le corps a des propriétés qui l'apparentent à la fois à un liquide et à un solide: liquide en raison de sa fluidité, solide en raison de ses propriétés optiques analogues à celles d'un cristal. D'où le nom de « cristal liquide » (ou phase mésomorphe) donné à cet état particulier de la matière.

Sur la Fig. 1, on a représenté les formules de trois molécules constituant des phases mésomorphes qui ont été largement étudiées. Ces molécules peuvent être assimilées à des bâtonnets formés d'une partie centrale rigide, cette rigidité étant assurée par la double liaison reliant les deux noyaux benzéniques et de deux chaînes aliphatiques plus flexibles qui aident à maintenir la fluidité. Ces molécules possèdent toutes, dans la phase cristal liquide, un ordre orientationnel à longue distance; la nature de cet ordre a permis à FRIEDEL [1] de classer les cristaux liquides en 3 catégories bien distinctes schématisées sur la Fig. 2.

Dans un nématique, les grands axes moléculaires sont orientés en moyenne dans une même direction. Les centres de gravité sont distribués aléatoirement comme dans un liquide classique. Ce sont des matériaux fortement biréfringents et optiquement uniaxes. L'axe optique est parallèle à la direction d'alignement des molécules.

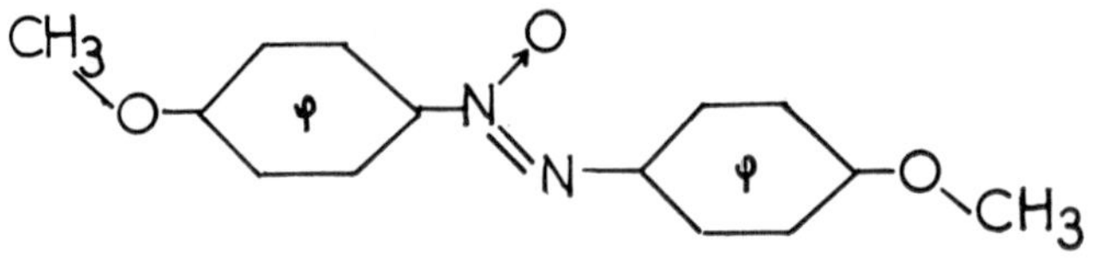

Para Azoxyanisole

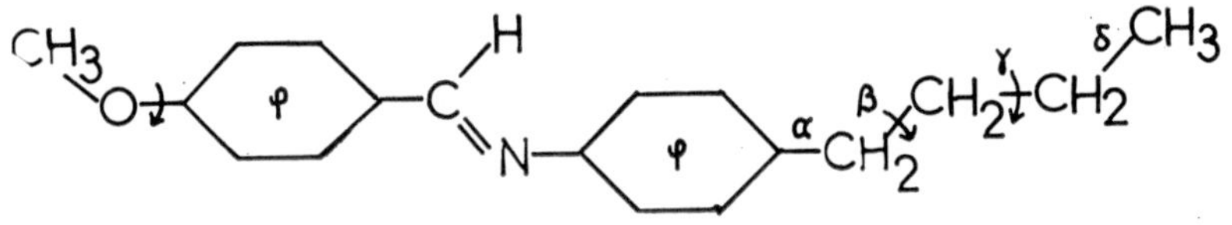

Para Methoxy benzylidene p'n butylaniline (M.B.B.A.)

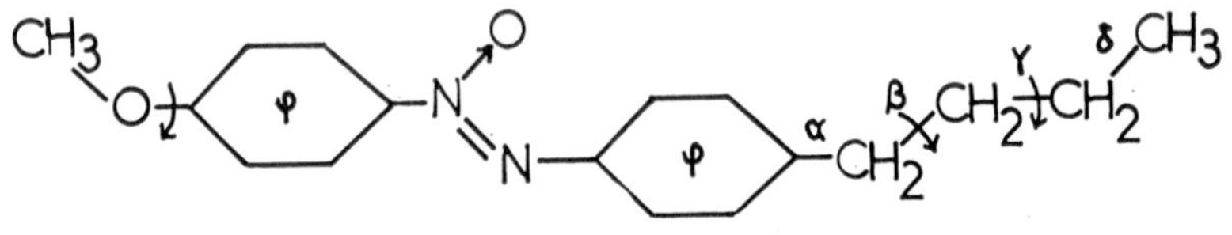

Para Methoxy p'n butylazoxybenzene (Licristal Merck IV)

Fig. 1. – Formules moléculaires de trois cristaux liquides nématiques.

La phase cholestérique est semblable à la phase nématique sur une échelle de quelques distances moléculaires. Cependant, à plus grande échelle, les molécules se répartissent suivant une structure hélicoïdale, l'axe de l'hélice étant perpendiculaire en tout point au grand axe moléculaire.

La phase smectique est caractérisée à la fois par un ordre orientationnel et un ordre de position, les molécules se répartissant en couches d'épaisseur approximativement égale à la longueur de la molécule, soit de l'ordre de 20 à 30 Å. Les phases smectiques sont moins connues que les phases nématiques et cholestériques et toutes ne sont pas encore parfaitement caractérisées. Sur la Fig. 2, on a représenté l'arrangement moléculaire de la phase la plus connue dite A. Les molécules sont dans ce cas perpendiculaires au plan des couches et les centres de gravité répartis aléatoirement au sein de ces couches.

Comme nous l'avons déja souligné, la caractéristique commune aux trois classes de cristaux liquides est l'ordre orientationnel à longue distance (vis à vis des dimensions moléculaires). Il est possible, au moyen de forces relativement faibles fournies par un champ électrique, un champ magnétique ou l'action de parois, d'orienter la direction moyenne des molécules sur des distances macroscopiques. Un exemple en est donné avec le MBBA déjà mentionné (Fig. 1), qui, au contact d'une lamelle de verre recouverte d'un film

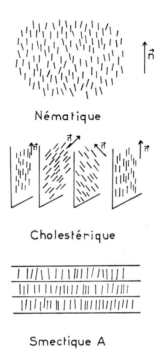

Némaflque

Cholestérique

Smecflque A

Fig. 2. – Arrangement des molécules dans un cristal liquide nématique, cholestérique et smectique *A*.

d'or d'environ 100 Å d'épaisseur, s'oriente spontanément perpendiculairement à la lamelle. (Pour une description détaillée des méthodes d'orientation et de leur interprétation, le lecteur peut se reporter à la réf. [2].)

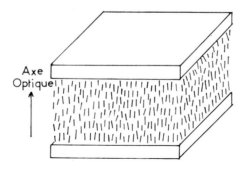

Axe
Optique

Fig. 3. – Lame monocristalline de cristal liquide nématique.

On peut ainsi réaliser des dispositifs tels que celui qui est représenté sur la Fig. 3 et qui se comporte comme un monocristal d'axe optique perpendiculaire aux lames.

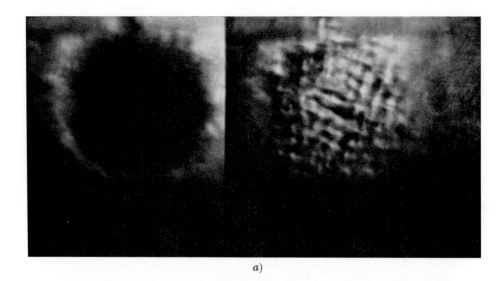

a)

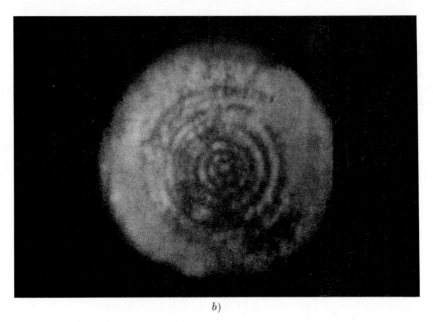

b)

Fig. 4. – Effet d'une irradiation ultrasonore sur une lame monocristalline de MBBA d'axe optique perpendiculaire à la lame (cf. Fig. 3) vue entre nicols croisés. a) Distance à l'émetteur 5 cm. Fréquence 1 MHz. La photo de gauche représente la lame avant irradiation. b) Distance à l'émetteur 40 cm. Fréquence 2 MHz. Intensité ultrasonore 0.5 W/cm². On peut distinguer les anneaux de diffraction du champ acoustique [3].

Comme nous l'avons déjà fait remarquer, la direction de l'axe optique peut, en raison de la fluidité de son support, être facilement modifiée par des agents extérieurs. C'est à partir de cellules de ce type que sont réalisés les dispositifs d'affichage dont le principe est fondé sur la modification des propriétés optiques de la cellule sous l'action d'un champ électrique judicieusement disposé. Un champ ultrasonore peut également avoir les mêmes effets (cf. Fig. 4). Nous voyons ici un aspect des travaux actuellement effectués sur les cristaux liquides et les espoirs qu'ils autorisent, principalement dans le domaine de la visualisation ultrasonore et de l'holographie acoustique [4].

Parallèlement à ces recherches motivées par de possibles applications, l'étude de la propagation d'ondes ultrasonores transversales et longitudinales de faible amplitude fait l'objet de nombreux travaux et apporte une contribution importante à la compréhension des propriétés hydrodynamiques des cristaux liquides.

Il ne peut être question dans ce séminaire de s'étendre trop longuement sur l'ensemble des propriétés physiques des cristaux liquides, le lecteur pouvant se référer aux ouvrages spécialisés [5]. Nous allons simplement nous intéresser aux quelques points fondamentaux nécessaires à l'étude des propriétés acoustiques des cristaux liquides et plus particulièrement aux théories de milieu continu décrivant l'élasticité et l'hydrodynamique de ces matériaux. Nous nous limiterons volontairement par la suite à la phase nématique.

1. – Élasticité.

Lorsqu'on écarte des molécules de leur direction privilégiée, elles ont tendance à y revenir. On rencontre donc ici une élasticité d'un type nouveau. Alors que dans un solide des forces de rappel s'opposent à un déplacement relatif des distances entre les points voisins d'un matériau, dans un cristal liquide déformé ce sont des couples de rappel qui s'opposent aux courbures. La théorie du continuum élastique élaborée par OSEEN [6], ZOCHER [7] et FRANK [8] décrit entièrement les distorsions d'un cristal liquide par un champ de vecteurs $n(r)$. Le vecteur n appelé directeur est caractérisé par une longueur unité et une orientation variable. On admet que n varie continûment avec r.

Lorsqu'il apparaît une distorsion dans l'orientation, elle crée une variation d'énergie libre élastique de la forme

$$F = \frac{1}{2} \int \mathrm{d}^3 r \, K_{ijkl} \frac{\partial n_i}{\partial x_j} \frac{\partial n_k}{\partial x_l},$$

où K_{ijkl} sont des constantes élastiques.

Compte tenu des invariances par symétrie, cette expression se réduit à [8]

(1) $$F = \tfrac{1}{2}\big(K_{11}(\operatorname{div} n)^2 + K_{22}(n \cdot \operatorname{rot} n)^2 + K_{33}(n \wedge \operatorname{rot} n)^2\big).$$

L'élasticité du milieu est donc décrite par trois constantes correspondant à trois déformations bien distinctes. K_{11} n'apparaît par exemple que pour $(\text{div }\boldsymbol{n}) \neq 0$, ce qui traduit une déformation en éventail. Un raisonnement identique permet de montrer que K_{22} est une constante de torsion et K_{33} une constante de flexion (cf. Fig. 5).

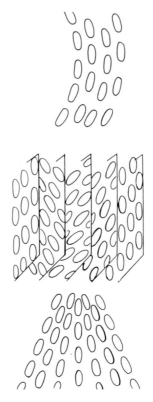

Fig. 5. – Les trois déformations fondamentales — éventail, torsion, flexion — d'un cristal liquide nématique.

Pour caractériser les conditions d'équilibre, il est pratique d'introduire un « champ moléculaire » $\boldsymbol{h}$ défini comme la dérivée fonctionnelle de l'énergie libre de déformation par rapport à $\boldsymbol{n}$ [5, 9]:

$$(2) \qquad h_a = -\frac{\delta F}{\delta n_a} = -\frac{\partial F}{\partial n_a} + \frac{\partial}{\partial_\beta}\frac{\partial F}{\partial n_{a\beta}}.$$

Le champ moléculaire qui a les dimensions d'une énergie peut être assimilé à un couple généralisé. En écrivant que l'énergie totale de déformation doit être minimum par rapport à toute variation du directeur, on obtient la condition d'équilibre

(3) $$h = \lambda n \, ,$$

où λ est un multiplicateur de LAGRANGE.

En tout point r, le directeur doit donc être parallèle à h. $(n \wedge h)$ représente alors le couple élastique exercé sur une molécule par celles qui l'entourent.

2. – Hydrodynamique [5, 9-14].

L'hydrodynamique des nématiques est assez complexe car il existe, par rapport à un liquide ordinaire, un degré de liberté supplémentaire, le directeur qui est couplé à l'écoulement. Ce couplage va se traduire par l'existence d'un couple élastique $n \wedge h$ exercé sur le directeur par l'écoulement et d'un couple visqueux de frottement Γ qui dépend de l'orientation du directeur par rapport à l'écoulement (cf. Fig. 6).

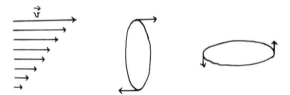

Fig. 6. – Couples de frottement exercés sur les molécules dans un écoulement non uniforme.

Historiquement la théorie hydrodynamique des cristaux liquides nématiques a été conçue par ERICKSEN et LESLIE [10, 11] puis raffinée par PARODI [12].

Sept variables hydrodynamiques sont nécessaires pour décrire les propriétés hydrodynamiques d'un cristal liquide nématique: les cinq variables d'un liquide ordinaire (densité ϱ, quantité de mouvement g et énergie ε) auxquelles il faut ajouter les deux composantes indépendantes du directeur (avec la condition $|n|^2 = 1$).

Ces variables obéissent à un système de sept équations différentielles

(4a) $$\frac{d\varrho}{dt} = - \operatorname{div}(\varrho v) \, ,$$

(4b) $$\frac{d}{dt}(\varrho v_i) = \frac{\partial \sigma_{ji}}{\partial x_j} \, ,$$

(4c) $$\Gamma = n \wedge h \, ,$$

(4d) $$\frac{d\varepsilon}{dt} = - \operatorname{div} j \, .$$

L'équation (4a) traduit la conservation de masse.

L'équation (4b) est l'équation de conservation de quantité de mouvement. σ_{ij}, tenseur des contraintes, s'écrit

$$\sigma_{ij} = -p\delta_{ij} + \sigma'_{ij} + \sigma^0_{ij} ,$$

p est la pression hydrostatique, σ^0_{ij} un terme d'origine élastique qui a été calculé par ERICKSEN [10] et qui, dans le cas des mouvements de faible amplitude, est négligeable, σ'_{ij} est un terme dissipatif.

L'équation (4c) exprime l'égalité entre le couple visqueux Γ et le couple élastique $\boldsymbol{n} \wedge \boldsymbol{h}$ agissant sur le directeur.

L'équation (4d) décrit la conservation d'énergie, $\boldsymbol{j}$ étant la densité de courant relatif au flux d'énergie.

Par ailleurs les formes les plus générales des lois de frottement compatibles avec la symétrie du milieu s'écrivent

$$\sigma'_{ij} = \alpha_4 A_{ij} + \alpha_1 n^0_i n^0_j n^0_k n^0_l A_{kl} + \alpha_5 n^0_j n^0_k A_{ki} + \alpha_6 n^0_i n^0_k A_{kj} + \alpha_2 n^0_j N_i + \alpha_3 n^0_i N_j ,$$

$$\Gamma = \boldsymbol{n} \wedge (\gamma_1 N + \gamma_2 \overline{\overline{A}} \boldsymbol{n}) ,$$

N est la vitesse de rotation du directeur par rapport à celle du fluide

$$N = \frac{\mathrm{d}\boldsymbol{n}}{\mathrm{d}t} - \frac{1}{2} (\mathbf{rot}\, \boldsymbol{v} \wedge \boldsymbol{n}) ,$$

$\overline{\overline{A}}$ est la partie symétrique du tenseur des déformations

$$A_{ij} = \frac{1}{2}\left(\frac{\partial v_i}{\partial x_j} + \frac{\partial v_j}{\partial x_i}\right).$$

Les coeficients α_i, γ_1 et γ_2 ont la dimension d'une viscosité.

Récemment une théorie hydrodynamique généralisée à toutes les phases mésomorphes a été mise au point par plusieurs écoles [13-15]. Cette théorie, bien que conceptuellement différente, conduit aux mêmes résultats que la théorie d'Ericksen-Leslie-Parodi. Elle a cependant l'avantage de tenir compte explicitement de la compressibilité du milieu et c'est la raison pour laquelle nous l'avons adoptée dans ce qui suit.

Dans cette nouvelle formulation σ_{ij} se décompose en une partie réactive σ^R_{ij} et une partie dissipative σ^D_{ij}

$$(5a) \qquad \sigma^R_{jk} = p\delta_{jk} + \left\{ -\frac{\gamma_2}{\gamma_1} K_{kpqr}(\nabla_p \nabla_r n_q) n^0_j + \right.$$
$$\left. + \frac{1}{2} (K_{jkqr} n^0_p \nabla_p \nabla_r n_q - K_{pkqr} n^0_j \nabla_p \nabla_r n_q) + (jk \to kj) \right\}.$$

Les coefficients K_{ijkl} étant reliés aux trois constantes élastiques précédemment définies par

$$K_{ijkl} = K_{33}(\delta_{ik} - n_i^0 n_k^0) n_j^0 n_l^0 + (K_{22} - K_{11})\varepsilon_{ijp} n_p^0 \varepsilon_{klq} n_q^0 + K_{11}(\delta_{ik} - n_i^0 n_k^0)(\delta_{jl} - n_j^0 n_l^0) .$$

La partie dissipative de σ_{ij} s'écrit

$$(5b) \qquad \sigma_{ij}^{\mathrm{D}} = -2\nu_2 A_{ij} - 2(\nu_3 - \nu_2)(A_{ik} n_k^0 n_j^0 + A_{jk} n_i^0 n_k^0) -$$

$$- (\nu_4 - \nu_2)\delta_{ij} A_{kk} - 2(\nu_1 + \nu_2 - 2\nu_3) n_i^0 n_j^0 n_k^0 n_l^0 A_{kl} -$$

$$- (\nu_5 - \nu_4 + \nu_2)(\delta_{ij} n_k^0 n_l^0 A_{kl} + n_i^0 n_j^0 A_{kk}) ,$$

les coefficients ν_i étant des viscosités. Nous donnons ci-dessous la correspondance entre les coefficients ν_i et α_i:

$$(6) \quad \begin{cases} 2\nu_2 = \alpha_4 , \qquad 2\nu_3 = \alpha_4 + \alpha_5 - (\gamma_2 \alpha_2 / \gamma_1) , \qquad \gamma_1 = \alpha_3 - \alpha_2 , \\[2mm] \gamma_2 = \alpha_6 - \alpha_5 , \qquad 2\nu_1 = \alpha_1 + \alpha_4 + \alpha_5 + \alpha_6 , \\[2mm] \alpha_6 - \alpha_5 = \gamma_2 = \alpha_2 + \alpha_3 . \end{cases}$$

Les coefficients $\nu_4 - \nu_2$ et ν_5 sont des coefficients de viscosité de volume qui n'apparaissent pas dans la théorie d'Ericksen-Leslie, cette dernière étant relative à un nématique incompressible ($A_{kk} = 0$).

Ces coefficients sont susceptibles d'être couplés aux degrés internes du milieu et de donner naissance à des mécanismes de relaxation.

Modes propres. L'aspect des solutions du système d'équations (4), (5) dépend du paramètre sans dimensions $\mu = K\varrho/\eta^2$ où η est une viscosité moyenne. μ est très petit devant 1 et on obtient les modes propres suivants (ω fréquence circulaire, q vecteur d'onde):

2 modes sonores:

$$(7) \qquad \omega_s = \pm C^* q ,$$

1 mode de relaxation thermique:

$$(8) \qquad \omega_t = i\varkappa q^2 / \varrho .$$

Ces trois modes sont tout à fait identiques à ceux que l'on observe dans un liquide ordinaire à la différence près que la vitesse de propagation complexe C^* et la conductibilité thermique $\varkappa$ sont des fonctions simples de l'angle entre le vecteur d'onde et le directeur:

2 modes rapides de cisaillement qui, pour certaines géométries expérimentales, peuvent s'écrire sous la forme simple

(9) $\omega_r = i\eta q^2/\varrho$,

où η est un coefficient de viscosité.

2 modes lents de fluctuation d'orientation:

(10) $\omega_l = \dfrac{iK}{\gamma} q^2$,

où γ est une viscosité effective de rotation du directeur tenant compte du couplage avec les gradients de vitesse.

Les modes hydrodynamiques auxquels nous allons nous intéresser sont, d'une part, les modes sonores que l'on peut étudier au moyen de techniques de propagation d'ondes ultrasonores longitudinales et les modes de cisaillement, d'autre part, pour l'étude desquels on utilise des ondes transversales.

Pour décrire les propriétés acoustiques des cristaux liquides nématiques, nous allons diviser la suite de cet exposé en deux parties:

La première sera consacrée à la comparaison entre les résultats de mesures d'atténuation d'ondes ultrasonores transversales et longitudinales et la théorie hydrodynamique classique des nématiques qui ne tient pas compte des phénomènes de relaxation.

Dans la deuxième partie, on discutera des mécanismes de relaxation observés dans les phases nématique et isotrope.

On concluera avec quelques perspectives sur les études en cours sur les phases smectiques.

3. – Attenuation d'ondes ultrasonores et théorie hydrodynamique des cristaux liquides nématiques.

3`1. *Ondes ultrasonores transversales*. L'équation de dispersion pour des ondes transversales de vecteur d'onde q et de fréquence circulaire ω s'écrit pour un cristal liquide nématique incompressible ([16] éq. (9))

$$\omega = i\eta q^2/\varrho .$$

Ces modes sont caractérisés par une fréquence purement imaginaire et ne peuvent donc se propager à travers un fluide. Cependant dans un liquide visqueux des ondes transversales très amorties peuvent prendre naissance à partir d'une surface solide plane oscillant périodiquement. L'amortissement de ces ondes est caractérisé par un paramètre appelé «longueur de pénétration» et qui est défini comme la distance à laquelle l'amplitude de l'onde n'est plus que $1/e$ fois sa valeur initiale. Cette longueur de pénétration δ est donnée par

$$\delta = (\eta/\varrho\omega)^{\frac{1}{2}} \text{ soit } \delta \simeq 1\mu\text{m pour } \eta = 1 \text{ Poise et } \omega = 10^8 \text{ s}^{-1}.$$

Par suite de cette très faible valeur de la longueur de pénétration, il n'est guère possible de déterminer l'atténuation d'ondes ultrasonores transversales au moyen de techniques de transmission. Il existe cependant des techniques permettant d'atteindre ce paramètre, fondées sur la mesure du coefficient de réflexion d'une onde de cisaillement sur un interface solide-liquide. Cette méthode est particulièrement adaptée à l'étude des cristaux liquides puisque nous avons vu qu'une surface plane d'un solide est susceptible d'orienter ces matériaux.

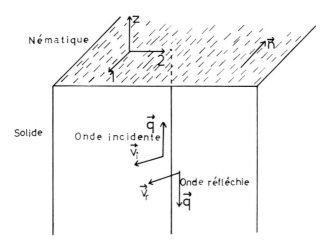

Fig. 7. – Réflexion d'une onde transverse à l'interface solide-cristal liquide nématique, $\boldsymbol{v}_i$ vitesse hydrodynamique de l'onde incidente, $\boldsymbol{v}_r$ vitesse hydrodynamique de l'onde réfléchie.

Le principe de l'expérience est schématisé sur la Fig 7. Un cristal liquide nématique déposé à la surface d'un solide est orienté selon une direction donnée (dans l'exemple choisi, parallèle à la surface). Choisissons un référentiel O_{12z} tel que le directeur soit orienté selon 1. On crée dans le solide une onde transversale polarisée parallèlement à l'interface solide-nématique. Cette onde donne naissance, à l'interface, à une onde transmise qui est très rapidement amortie. Le rapport entre les vitesses de particules associées aux ondes respectivement incidente et réfléchie définit un coefficient de réflexion. Nous allons voir comment la mesure de ce coefficient de réflexion conduit à la détermination de coefficients de viscosité.

Modes propres. Pour la géométrie expérimentale considérée le système d'équations (4), (5) et (6) conduit, lorsqu'on admet que le milieu est incompressible et isotherme, à l'équation de dispersion suivante:

$$
\begin{vmatrix}
\omega q \dfrac{\gamma_1 + \gamma_2}{2} & i\omega\varrho - q^2\left(\nu_3 + \left(1 + \dfrac{\gamma_2}{\gamma_1}\right)^2 \dfrac{\gamma_1}{4}\right) & 0 \\[3mm]
i\omega\gamma_1 - Kq^2 & -\dfrac{iq}{2}(\gamma_1 + \gamma_2) & 0 \\[3mm]
0 & 0 & i\omega\varrho - \nu_2 q^2
\end{vmatrix} = 0 \; .
$$

Les deux modes propres, comparables aux modes transverses des liquides ordinaires, sont donnés par

$$(11) \qquad q_1 = (1-i)\left(\frac{\omega\varrho}{2\nu_3}\right)^{\frac{1}{2}} ,$$

$$(12) \qquad q_2 = (1-i)\left(\frac{\omega\varrho}{2\nu_2}\right)^{\frac{1}{2}} .$$

La troisième solution de l'équation de dispersion décrit le mode lent de relaxation d'orientation

$$(13) \qquad q_0 = (1-i)\frac{\omega\nu_3\gamma_1}{2K\left(\nu_3 + (1 + \gamma_2/\gamma_1)^2\,(\gamma_1/4)\right)} \; .$$

La longueur de pénétration de ce mode est environ cent fois plus faible que celle des modes de cisaillement.

Impédance mécanique de cisaillement. L'impédance mécanique, qui dans le cas d'un liquide istrope est un scalaire, défini comme le rapport de la contrainte à la vitesse, est de nature anisotrope pour un nématique. Elle prend alors une forme matricielle Z définie par la relation

$$\bar{\sigma} = \overline{\overline{Z}}\boldsymbol{v} , \qquad \sigma_{z1} = Z_{11}v_1 + Z_{12}v_2 , \qquad \sigma_{z2} = Z_{21}v_1 + Z_{22}v_2 \; .$$

Les expressions de σ_{z1} et σ_{z2} à la surface ($z = 0$) sont obtenues à partir du système d'équations (4)-(6). En annulant dans ces expressions successivement v_1 et v_2 avec la condition supplémentaire d'une orientation rigide du directeur à l'interface, on déduit

$$(14) \qquad Z = \begin{vmatrix} Z_{11} = (1-i)\left(\dfrac{\omega\varrho\nu_3}{2}\right)^{\frac{1}{2}} & Z_{12} = 0 \\[3mm] Z_{21} = 0 & Z_{22} = (1-i)\left(\dfrac{\omega\varrho\nu_2}{2}\right)^{\frac{1}{2}} \end{vmatrix} \; .$$

Coefficient de réflexion à l'interface solide-nématique. On le détermine en écrivant les conditions de continuité des vitesses et des contraintes à l'interface.

Son expression prend une forme simple lorsque la vitesse est *a*) parallèle à 1, *b*) parallèle à 2.

a) $\boldsymbol{v}\|1$.

Dans ce cas $v_2 = 0$ et le coefficient de réflexion complexe s'écrit

$$r_1 \exp\left[-i\varphi_1\right] = \frac{v_{r1}}{v_{i1}} = \frac{Z_s - Z_{11}}{Z_s + Z_{11}} \simeq 1 - 2\frac{Z_{11}}{Z_s} \qquad (Z_s \gg Z_{11}),$$

r_1 et φ_1 sont respectivement la variation d'amplitude et le déphasage subis par l'onde à la réflexion.

Dans l'hypothèse généralement justifiée où φ est petit

$$r_1 \simeq 1 - 2\frac{R_{11}}{Z_s},$$

où R_{11}, partie réelle de Z_{11}, est égal à

$$R_{11} = \left(\frac{\omega\varrho\nu_3}{2}\right)^{\frac{1}{2}}.$$

b) $\boldsymbol{v}\|2$.

Dans ce cas $v_1 = 0$ et on a de la même façon que précédemment

$$r_2 \simeq 1 - 2\frac{R_{22}}{Z_s}$$

avec

$$R_{22} = \left(\frac{\omega\varrho\nu_2}{2}\right)^{\frac{1}{2}}.$$

Ainsi la mesure de r_1 et r_2 permet de déterminer deux coefficients de viscosité ν_3 et ν_2.

Une troisième géométrie expérimentale est facilement réalisable; c'est celle où le directeur est perpendiculaire à la surface

c) $\boldsymbol{n}\|0z$, $\boldsymbol{v}$ dans le plan 0 1 2 .

Le calcul montre que dans ce cas

$$r \simeq 1 - 2\frac{R_{11}}{Z_s}.$$

Dans cette géométrie on mesure donc de nouveau le coefficient ν_3. La Fig. 8 représente les trois géométries expérimentales et les coefficients de viscosité correspondants.

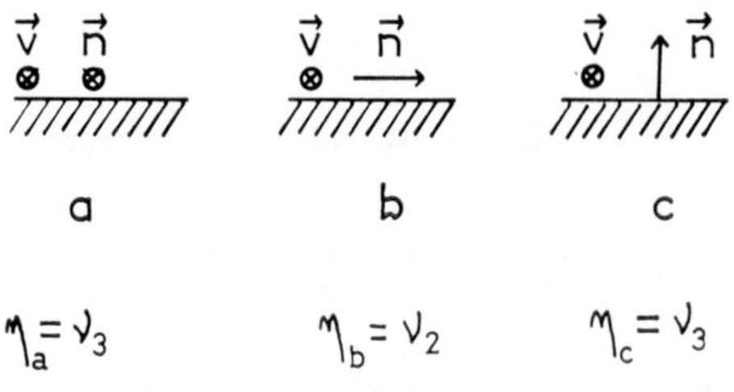

Fig. 8. – Les trois géométries expérimentales.

Remarque. Il est souvent avantageux en pratique d'utiliser une incidence oblique des ondes transversales sur l'interface, car on augmente ainsi la précision des mesures [17]. On peut montrer que dans ce cas le coefficient de réflexion est donné par [18]

$$r \simeq 1 - 2\,\frac{R_{ii}}{Z_s \cos\theta}\,,$$

où θ est l'angle d'incidence.

Viscosité capillaire. Il peut être intéressant de comparer les viscosités déterminées par les techniques ultrasonores avec celle que l'on mesure par écoulement dans un tube capillaire. L'écoulement impose un couple au directeur qui est donné par l'équation (4c). LESLIE [11] a montré que si $\gamma_1 + \gamma_2 < 0$ (ce qui est généralement le cas) les molécules du nématique s'orientent sous l'effet du cisaillement, l'angle δ entre le directeur et l'écoulement étant donné par $\cos 2\delta = -\gamma_2/\gamma_1$ pour des distances supérieures à la portée des parois.

Habituellement l'angle δ est très faible ($\gamma_2/\gamma_1 \sim -1$) et dans ce cas la viscosité capillaire calculée à partir des éq. (4) est donnée par

$$\eta_{\text{cap}} \simeq \nu_3\,,$$

soit la même viscosité que celle que l'on mesure par les techniques ultrasonores dans les géométries *a*) et *c*) (cf. Fig. 8).

Résultats expérimentaux. La vérification des prédictions de la théorie hydrodynamique et la mesure des coefficients de viscosité ν_2 et ν_3 ont été effectuées sur le MBBA par les auteurs [16] et ultérieurement par LEE *et al.* [19].

Les techniques de mesure utilisées sont conventionnelles et ont été décrites en détail dans la littérature [20].

Le cristal liquide à étudier est déposé sur la surface plane d'un barreau de silice fondue, préalablement traité de manière à obtenir une orientation macroscopique de l'échantillon (Fig. 9). Pour le MBBA les deux orientations *a*) et *b*) (cf. Fig. 8) ont été obtenues par frottement de la surface, l'orientation *c*) étant réalisée par dépôt de détergent.

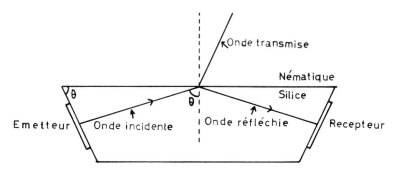

Fig. 9. – Barreau de mesure.

Aux deux extrémités du barreau sont fixés deux quartz piezoélectriques taillés de manière à vibrer parallélement à la surface du barreau (perpendiculairement au plan de la Figure).

Le quartz émetteur est excité au moyen d'un générateur d'impulsions. Ces impulsions sont réfléchies sur la surface du barreau et détectées par le deuxième quartz. Pour mesurer le coefficient de réflexion et le déphasage on opère en deux temps: une première mesure dans laquelle la surface du barreau est à l'air et une deuxième après dépôt du cristal liquide. (Les détails de la technique de mesure peuvent être trouvés dans [16, 20].)

Pour les cristaux liquides nématiques, en fait, les parties réelle et imaginaire de l'impédance acoustique sont égales et la mesure du coefficient de réflexion est suffisante pour la détermination des coefficients de viscosité.

Pour le MBBA à $T = 25$ °C [16]

$$\text{géométrie } b) \quad v_2 = (42 \quad \pm 2 \quad) \text{ cp},$$

$$\text{géométrie } a) \quad v_3 = (27.0 \pm 1.5) \text{ cp},$$

$$\text{géométrie } c) \quad v_3 = (27.1 \pm 1.5) \text{ cp}.$$

On retrouve bien l'égalité prévue par la théorie entre les viscosités mesurées dans les géométries *a*) et *c*).

Sur la Fig. 10 on a reporté les variations thermiques de v_2 et v_3 et de la viscosité η_{cap} mesurée par écoulement capillaire. L'égalité entre η_{cap} et v_3 que l'on peut constater loin de la température de transition nématique-isotrope montre que l'angle δ entre le directeur et l'écoulement est faible. Effectivement des mesures de biréfringence ont montré qu'il était de l'ordre de 5° [18]. Par contre au voisinage de la transition cet angle devient beaucoup plus grand (∼20°) ce qui peut expliquer la différence observée entre η_{cap} et ν_3.

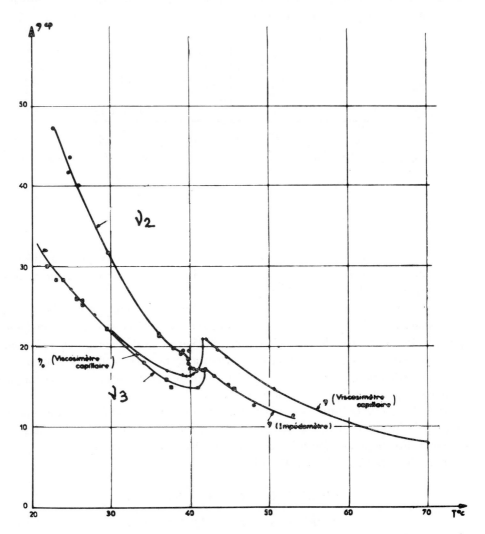

Fig. 10. – MBBA. Variation thermique des coefficients de viscosité mesurés par écoulement capillaire et par absorption d'ondes de cisaillement dans les phases nématique et isotrope [16].

On peut aussi remarquer que dans la phase isotrope la viscosité statique capillaire est supérieure à la viscosité haute fréquence mesurée par les techniques ultrasonores. Cette différence traduit la présence d'effets relaxationnels que nous considérerons ultérieurement.

Remarque. Il existe également pour les cholestériques et les smectiques des géométries expérimentales simples pour lesquelles on peut mesurer des coefficients de viscosité effectifs. C'est le cas par exemple d'un cholestérique dont l'axe hélicoïdal est dirigé parallèlement au gradient de cisaillement. Si le pas de l'hélice est très supérieur à la longueur de pénétration des ondes transversales, la viscosité dépend de l'orientation des molécules à l'interface vis à vis de l'écoulement [21]. La variation de ces coefficients de viscosité avec le pas hélicoïdal a été étudiée pour des mélanges MBBA-propionate de cholestérol [22]. De même, lorsqu'on oriente un smectique avec le plan des couches parallèles à l'interface on mesure le coefficient de frottement visqueux associé au mouvement des molécules dans le plan des couches [23].

3˙2. *Ondes ultrasonores longitudinales.* – Lorsqu'on tient compte de la compressibilité du cristal liquide, 7 coefficients de viscosité sont nécessaires pour décrire la dissipation par frottements (cf. éqs. (5)-(6)).

L'expression complète de l'atténuation ultrasonore α s'écrit [13]

$$(15) \qquad \frac{\alpha(\theta)\varrho V^3}{2\pi^2 f^2} = \nu_1 \cos^2\theta + (\nu_2 + \nu_4)\sin^2\theta - \frac{1}{4}(\nu_1 + \nu_2 - 4\nu_3 + \nu_4 - 2\nu_5)\sin^2 2\theta +$$
$$+ (\varkappa_\perp \sin^2\theta + \varkappa_\parallel \cos^2\theta)(c_v^{-1} - c_p^{-1}),$$

θ est l'angle entre le vecteur d'onde sonore et le directeur,

f est la fréquence sonore,

V la vitesse de propagation,

$\varkappa_\perp$ et $\varkappa_\parallel$ sont les conductibilités thermiques respectivement pour $\theta = \pi/2$ et $\theta = 0$.

En pratique le terme de conductibilité thermique est négligé. Les deux viscosités de volume $\nu_4 - \nu_2$ et ν_5 sont susceptibles d'être couplées aux degrés internes. Ce couplage qui n'est pas pris en considération dans la théorie hydrodynamique classique donne naissance à des phénomènes de relaxation. Ces derniers peuvent être mis en évidence par une étude de l'absorption ultrasonore en fonction de la fréquence. Sur la Fig. 11 on a reporté la variation thermique de la quantité α/f^2 pour des orientations respectivement parallèle et perpendiculaire du vecteur d'onde par rapport au directeur, pour une fréquence relativement basse (6 MHz) et des fréquences élevées (> 100 MHz). On a également reporté les valeurs de α/f^2 mesurées dans le domaine de température où la phase est isotrope.

On sait que pour un liquide non relaxant la quantité α/f^2 est indépendante de la fréquence. La différence observée sur la Fig. 11 entre les courbes obtenues à haute fréquence et basse fréquence est significative de la présence de processus relaxationnels.

  L'analyse des résultats d'absorption ultrasonore en termes de la théorie hydrodynamique classique ne peut donc être effectuée que pour des fréquences nettement plus élevées que les fréquences de relaxation. Cette analyse a été

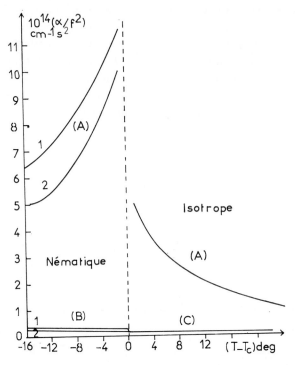

Fig. 11. – MBBA. Variations de α/f^2 en fonction de $T - T_c$. Courbes 1: $\boldsymbol{n}$ parallèle à $\boldsymbol{q}$. Courbes 2: $\boldsymbol{n}$ perpendiculaire à $\boldsymbol{q}$. A) Résultats de KEMP et LETCHER [33] $f = 6$ MHz. B) Courbes estimées à partir des résultats de BACRI [24, 26] $f = 200$ MHz. C) Résultats des auteurs [28, 34] $f = 155$ MHz.

réalisée par BACRI [24] qui a mesuré l'atténuation ultrasonore de divers cristaux liquides et notamment du MBBA dans un domaine de fréquences compris entre 200 et 560 MHz.

  Revenons à l'expression (15) de l'atténuation ultrasonore. Elle peut être écrite sous une forme légèrement différente

$$(16) \qquad \Delta\alpha(\theta) = (\alpha_\parallel - \alpha_\perp) \cos^2 \theta - \alpha_c \sin^2 2\theta \,,$$

où

$$\Delta\alpha(\theta) = \alpha(\theta) - \alpha(\pi/2)$$

et

$$\alpha_c = \frac{2\pi^2 f^2}{4\,V^3} \left(\nu_1 + \nu_2 - 4\nu_3 + \nu_4 - 2\nu_5 \right).$$

Ainsi la fonction $f(\theta) = \Delta\alpha(\theta) - (\alpha_{\parallel} - \alpha_{\perp}) \cos^2\theta$ doit être représentée par une courbe en $\sin^2 2\theta$. Ceci a été vérifié expérimentalement par BACRI comme l'illustre la Fig. 12.

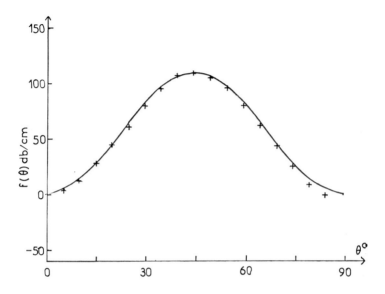

Fig. 12. – MBBA. $T = 24°$ C. Variation de $f(\theta)$ en fonction de θ, angle entre le vecteur d'onde ultrasonore et le directeur pour $f = 200$ MHz. (D'après BACRI [26].) + Résultats expérimentaux, ———— courbe en $\sin^2 2\theta$.

En combinant les résultats d'absorption longitudinale haute fréquence et ceux que nous avons obtenus au moyen des techniques d'ondes transversales, il est possible de déterminer cinq coefficients de viscosité.

Pour le MBBA à 25 °C

$$\nu_1 = 96 \text{ cp}, \quad \nu_2 = 42 \text{ cp}, \quad \nu_3 = 27 \text{ cp}, \quad \nu_4 = 43 \text{ cp} \quad \text{et} \quad \nu_5 = 18 \text{ cp}.$$

Remarque. Les mesures ultrasonores peuvent permettre également de déterminer des constantes élastiques de cristaux liquides nématiques. Ces dernières sont en effet généralement obtenues à partir de la mesure d'un champ magnétique critique au-dessus duquel on induit un changement d'orientation dans une lame monocristalline (transition de FREEDERICKZ [25]). L'atténuation ultrasonore, qui est fonction de cette orientation, peut servir de détecteur. Ces expériences ont été réalisées par BACRI qui a pu par cette méthode étudier notamment le comportement critique de la constante élastique de déformation en éventail, au voisinage d'une transition nématique-smectique A [26].

4. – Processus relaxationnels.

Nous allons discuter maintenant de la fraction de α/f^2 qui dépend de la fréquence; on peut voir sur la Fig. 11 que sa contribution est significative sur un grand domaine de températures aussi bien dans la phase nématique que que dans la phase isotrope.

Auparavant nous rappellerons brièvement le principe de la relaxation ultrasonore.

Les mécanismes relaxationnels décrivent des systèmes qui répondent à une perturbation en réalisant un nouvel équilibre après un certain temps de réponse. Aux basses fréquences, la période est suffisamment longue pour que l'énergie puisse être transférée de l'onde à l'équilibre interne. Lorsque la fréquence sonore augmente, un déphasage apparaît entre les variations sinusoïdales de pression et de température d'une part et le réajustement de l'équilibre d'autre part. Ce déphasage conduit à un maximum de l'absorption par longueur d'onde et à une décroissance de α/f^2. Tout mécanisme moléculaire mettant en jeu une variation de volume, d'entropie ou d'énergie interne du système peut contribuer à absorber de l'énergie ultrasonore par relaxation.

Lorsque le mécanisme est décrit par un seul temps de relaxation, la variation de α/f^2 avec la fréquence est donnée par

$$(17) \qquad \frac{\alpha}{f^2} = \frac{A}{1 + (f/f_r)^2} + B \,,$$

f_r est la fréquence de relaxation, A et B étant deux constantes. Lorsque le processus est caractérisé par plusieurs constantes de temps, la variation de α/f^2 avec la fréquence est plus graduelle que celle décrite par la loi (17).

Parmi les cristaux liquides ayant donné lieu à des travaux sur les propriétés acoustiques, c'est le MBBA qui a été le plus étudié [24, 25, 29-33]. Les mesures d'absorption ultrasonore en fonction de la température et de la fréquence font apparaître deux comportements différents, suivant le domaine de température considéré.

Au voisinage de la transition nématique-isotrope, la variation de l'absorption avec la fréquence n'obéit pas à une simple loi à un seul temps de relaxation [31]. Le temps de relaxation moyen présente un maximum aigu à la transition, de même que l'absorption basse fréquence [31].

A des températures très inférieures à la température de transition, le processus relaxationnel est caractérisé par un temps unique, la contribution relaxationnelle demeurant encore très importante.

Dans la discussion des phénomènes de relaxation, nous allons donc envisager séparément d'une part le domaine des températures $T \ll T_c$ et d'autre part le domaine critique de part et d'autre de la transition T_c.

a) $T \ll T_c$.

Dans ce domaine de températures, l'origine de l'absorption ultrasonore en excès a été attribuée à l'existence d'équilibres intramoléculaires. Les molécules de MBBA comme du reste celles de tous les composés mésomorphes sont constituées d'une partie centrale rigide et de deux extrémités flexibles. Dans ces extrémités, plusieurs conformations sont possibles, qui ne sont pas toutes équivalentes en ce qui concerne leur énergie libre. On sait que, lorsque les molécules d'un composé sont réparties dans des états d'énergie libre différente, le passage de l'onde sonore perturbe la distribution d'équilibre, subissant de ce fait une atténuation [27].

Nous avons tout d'abord envisagé la possibilité d'un équilibre entre isomères cis et trans du groupe méthyl [34]. Ce choix avait été dicté par l'analogie avec le cas des esters et des vinyléthers pour lesquels on observe des relaxations uniques dans le domaine de 1 à 100 MHz [27, 35]. Dans ces substances, la barrière de rotation autour de la liaison C-O est due au caractère partiellement double de cette liaison, provenant de la délocalisation des électrons π. Dans le MBBA, bien que le carbone de la liaison C-O fasse partie du noyau benzénique très stable, les électrons π sont également délocalisés par suite de la conjugaison.

Il paraît cependant peu probable que ce mécanisme soit à l'origine de l'absorption ultrasonore du MBBA, car les états cis et trans, par suite de l'absence d'interactions stériques, doivent posséder des énergies libres comparables, ce qui entraîne une très faible atténuation ultrasonore.

Une autre possibilité a été envisagée par JÄHNIG [35] qui a suggéré que le processus de relaxation provient de transitions trans-gauche de la chaîne butyle, la rotation pouvant s'effectuer autour des liaisons (C-C) β ou γ (Fig. 1). De tels changements conformationnels ont déjà été étudiés par les techniques ultrasonores pour de petites molécules. L'étude de la variation thermique de la fréquence de relaxation fournit notamment les paramètres cinétiques de l'équilibre.

Pour un système à deux états, la théorie cinétique prévoit la loi suivante pour la fréquence de relaxation f_r [27]:

$$(18) \qquad f_r = \frac{k_{\mathrm{B}} T}{2 \pi h} \exp\left[\Delta S_b / R\right] \exp\left[-\Delta H_b / RT\right],$$

k_{B} est la constante de Boltzmann, h la constante de Planck, ΔS_b et ΔH_b sont les entropie et enthalpie d'activation.

D'après l'éq. (18), en portant $\log(f_r/T)$ en fonction de $1/T$ on doit obtenir une droite dont l'ordonnée à l'origine et la pente fournissent ΔS_b et ΔH_b.

En général, dans les cristaux liquides tels que le MBBA, la température de fusion (22 °C) est trop proche de T_c (47 °C) en sorte que, dans tout le domaine de température, l'effet critique prédomine. C'est la raison qui a motivé une étude récente [36] sur le *p*-methoxy *p'n*-butylalkoxybenzène (Licristal

Merck IV), qui est nématique dans une assez grande gamme de températures (16 °C÷76 °C). Ce composé possède les même groupes terminaux que le MBBA (Fig. 1).

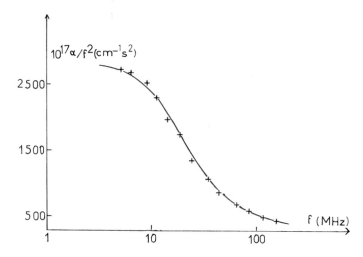

Fig. 13. – Licristal Merck IV. $T = 20$ °C. Variation de α/f^2 en fonction de f [36]. La courbe en traits pleins a été calculée à partir de l'éq. (17).

Pour des températures suffisamment éloignées de la transition ($T < 47$ °C) la variation de l'absorption ultrasonore avec la fréquence obéit à une loi de relaxation unique (Fig. 13). D'autre part la variation de $\log(f_r/T)$ avec $1/T$ est bien représentée par une droite de pente négative à partir de laquelle on peut déterminer ΔH_b et ΔS_b (Fig. 14).

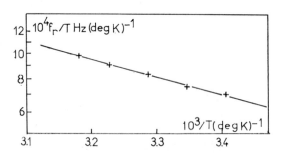

Fig. 14. – Licristal Merck IV. Variation de $\log(f_r/T)$ en fonction de $10^3/T$.

On obtient $\Delta H_b = 3$ kcal/mole, $\Delta S_b = -11$ cal/mole·°K. La valeur de ΔH_b ainsi trouvée est en bon accord avec la valeur mesurée pour le pentane (3.3 kcal/mole) par spectroscopie infrarouge [37]. Cet accord est un sérieux argument en faveur de l'hypothèse de changements conformationnels dans la chaîne hydrocarbonée. En ce qui concerne la barrière entropique, elle n'est pas connue pour le pentane. Les mesures ultrasonores permettent cependant de montrer qu'elle est nettement plus faible (en valeur absolue) que la valeur obtenue pour le nématique. En effet, les fréquences de relaxation des paraffines pures sont nettement plus élevées à température ambiante et en dehors du domaine expérimental analysé dans cette étude. Il faudrait donc admettre que l'origine de l'entropie d'activation mesurée réside dans l'ordre des chaînes paraffiniques dans la phase nématique [38, 39].

D'autres études sont cependant nécessaires pour établir sans ambiguité la nature de l'équilibre donnant naissance au mécanisme de relaxation.

b) $T \sim T_c$.

La variation thermique de l'absorption ultrasonore basse fréquence au voisinage de la température de transition T_c ressemble à celle qui est obtenue pour des phénomènes critiques au voisinage d'une transition de phase de 2ème espèce. En fait, on sait que la transition nématique-isotrope est une transition de première espèce dans laquelle le paramètre d'ordre à longue distance décrivant la corrélation orientationelle des molécules passe d'une valeur finie dans la phase nématique à une valeur nulle dans la phase isotrope (*).

Cependant, il subsiste dans la phase isotrope un ordre à courte distance très important qui se manifeste notamment par des valeurs anormales de la biréfrigence magnétique [40], de la biréfringence d'écoulement [41, 42] et de l'intensité de lumière diffusée [40]. En réalité le matériau se comporte comme s'il allait subir une transition de 2ème espèce à une température très légèrement inférieure (~ 1 °K) à T_c. DE GENNES a développé une théorie phénoménologique dans laquelle l'ordre local de la phase isotrope est décrit au moyen d'un tenseur symétrique $Q_{\alpha\beta}$ défini à partir d'une propriété macroscopique du matériau, par exemple la partie anisotrope du tenseur de susceptibilité diamagnétique [43].

L'énergie libre de la phase isotrope peut s'écrire sous la forme d'un développement de LANDAU:

$$(19) \qquad F = F_0 + \tfrac{1}{2}AQ_{\alpha\beta}Q_{\beta\alpha} + \tfrac{1}{3}BQ_{\alpha\beta}Q_{\beta\gamma}Q_{\gamma\alpha} + O(Q^4)\,,$$

où le terme cubique assure l'existence d'une transition de première espèce. Cette transition a cependant un caractère fortement de deuxième espèce. Par analogie avec ce que l'on connaît des propriétés acoustiques au voisinage d'autres transitions de deuxième espèce, on peut envisager d'interpréter les processus relaxationnels observés au moyen des techniques d'ondes longitudinales et transversales, à partir du couplage de ces ondes avec le paramètre d'ordre à longue distance pour la phase nématique, et avec les fluctuations de l'ordre à courte distance pour la phase isotrope.

(*) Le paramètre d'ordre S est défini per $S = \tfrac{3}{2} \langle \cos^2 \theta - \tfrac{1}{3} \rangle$, où θ est l'angle entre le grand axe de la molécule et l'axe du nématique.

On discutera successivement des résultats obtenus au-dessous et au-dessus de T_c.

Phase nématique $(T \sim T_c)$. KAWAMURA *et al.* [32] ont tenté d'interpréter le comportement critique dans la phase nématique à partir de la théorie de LANDAU et KHALATNIKOV [44] sur l'atténuation ultrasonore au voisinage d'une transition de deuxième espèce.

L'anomalie de l'absorption ultrasonore à la transition est attribuée à l'amortissement critique du paramètre d'ordre à longue distance. Dans cette théorie cependant, la nature exacte du couplage entre l'onde ultrasonore et le paramètre d'ordre n'est pas précisée. D'autre part, il ne semble pas possible d'effectuer une comparaison significative entre théorie et expérience du fait de la présence d'une distribution de temps de relaxation dans ce domaine de températures, cette distribution pouvant provenir de la superposition des mécanismes intramoléculaire et critique [35].

Phase isotrope. Ainsi que nous l'avons mentionné précédemment, les corrélations orientationnelles dans la phase isotrope sont décrites par un paramètre d'ordre tensoriel $Q_{\alpha\beta}$. Le paramètre d'ordre subit des fluctuations d'origine thermique dont l'amplitude et l'amortissement augmentent d'une manière critique lorsqu'on s'approche de la température de transition.

DE GENNES a montré que le temps de relaxation Γ^{-1} des fluctuations de $Q_{\alpha\beta}$ varie comme le rapport v/A, où v est un coefficient de transport et A le coefficient du terme quadratique dans le développement de LANDAU de la densité d'énergie libre [43] (éq. (19)). Dans le cadre d'une théorie de champ moyen A varie comme $(T - T^*)^{-1}$, où T^* est une température légèrement inférieure à T_c. Les résultats expérimentaux se sont avérés en parfait accord avec la théorie de DE GENNES [40, 42]. En particulier des mesures de diffusion inélastique de la lumière réalisées par STINSON et LITSTER ont permis de déterminer le temps de relaxation des fluctuations de $Q_{\alpha\beta}$ et de préciser la température critique T^* $(T^* = T_c - 1°)$.

Par ailleurs, le paramètre d'ordre $Q_{\alpha\beta}$ est couplé aux gradients de vitesse. Il existe notamment un couplage direct entre les contraintes de cisaillement et les composantes non diagonales de $Q_{\alpha\beta}$, couplage qui conduit à une relaxation de la viscosité dynamique de cisaillement [43]:

$$(20) \qquad \eta(\omega) = \eta_0 - \frac{2\mu^2}{v} \frac{i\omega}{\Gamma + i\omega},$$

η_0 est la viscosité statique, μ un coefficient de transport et Γ le temps de relaxation des fluctuations de $Q_{\alpha\beta}$.

L'équation (20) permet de comprendre la différence observée sur la Fig. 10 entre les valeurs de la viscosité statique η_0 et de la viscosité dynamique $\eta(\omega)$. Les mesures de $\eta(\omega)$ ayant été effectuées dans le cas de la Fig. 10 à des fréquences ($\omega \sim 10^8 \, \text{s}^{-1}$) très supérieures à la fréquence de relaxation Γ^{-1}, la différence entre η_0 et $\eta(\omega)$ représente la quantité $2\mu^2/\nu$ [42, 45].

La relaxation de viscosité de cisaillement mise ainsi en évidence ne contribue que très faiblement à la variation avec la fréquence de l'absorption des ondes ultrasonores longitudinales. Ces dernières en fait sont couplées aux composantes diagonales de $Q_{\alpha\beta}$ et par suite aux fluctuations de densité du milieu; on ne s'attend donc pas a priori à un comportement critique.

Récemment, IMURA et OKANO [46] et KAWAMURA et al., reprenant un idée de EDMONDS et ORR [47], ont interprété le comportement anormal de l'absorption ultrasonore au voisinage de la transition au moyen d'une théorie analogue à celle qu'avait développée FIXMAN [48] pour le point critique.

Dans cette théorie, le couplage prédominant entre l'onde sonore et la fonction de corrélation orientationelle s'effectue par l'intermédiaire de la variation thermique du coefficient A. Au voisinage de la transition, les fluctuations du paramètre d'ordre ont une forte corrélation spatiale et la réponse de la fonction de corrélation à une oscillation locale de température provoquée par les ondes ultrasonores ne peut être instantanée. Ce retard dans la réponse du système produit une atténuation des ondes ultrasonores. La fréquence caractéristique à laquelle l'atténuation par longueur d'onde est maximum obéit à la loi

$$(21) \qquad\qquad f_c = \frac{2.1}{\pi} \frac{A}{\xi},$$

où ξ est un coefficient de transport.

On devrait donc observer une variation en $T - T^*$ de la quantité $f_c \xi$. On ne connaît pas la variation thermique de ξ, mais on peut en première approximation admettre qu'elle est analogue à celle de ν. Il est cependant impossible actuellement d'effectuer une comparaison significative entre la théorie et le résultats expérimentaux étant donnée la dispersion considérable dans les résultats de différents auteurs [29, 31, 46].

Cette dispersion peut s'expliquer par le fait que le MBBA est un produit instable et que les mesures ont été effectuées sur des échantillons présentant des températures de transition différentes. Il serait intéressant de refaire des études précises sur des composés nématiques beaucoup plus stables que l'on sait actuellement synthétiser [49].

5. – Conclusion et perspectives.

Dans cet exposé, nous avons principalement axé la discussion sur les nématiques car ces matériaux se prêtent bien à une introduction sur les propriétés acoustiques des phases mésomorphes ainsi que sur la nature des problèmes soulevés par leur étude.

Cependant, le champ d'investigation qui s'avère actuellement le plus riche concerne les phases smectiques auxquelles la structure en couches confère des propriétés tout à fait particulières. Dans le domaine de l'acoustique deux caractéristiques essentielles de ces matériaux sont à dégager, sur lesquelles nous donnerons quelques éléments d'information en guise de conclusion:

    a) l'existence de modes transverses propagatifs,

    b) l'utilisation de ces matériaux comme modèles d'études de transition de phases.

Rappelons tout d'abord que dans un smectique, les molécules se répartissent en couches équidistantes et que, dans chacune des couches, il n'existe pas d'ordre de position des centres de gravité. Les grands axes des molécules sont parallèles entre eux et, dans le cas du smectique A, perpendiculaires aux planes des couches (cf. Fig. 2).

Le directeur $\boldsymbol{n}$ étant astreint à demeurer perpendiculaire aux couches, la nouvelle variable hydrodynmamique indépendante (par rapport à un liquide ordinaire) est le déplacement unidimensionnel u des couches perpendiculairement à elles-mêmes (dans la direction $0z$). Une distorsion $\delta\boldsymbol{n}$ du directeur s'exprime en fonction du déplacement u; $\delta\boldsymbol{n}$ a pour composantes $-\partial u/\partial x$ et $-\partial u/\partial y$ et l'énergie libre de distorsion s'écrit [50]

$$F = \frac{1}{2} K_{11} \left(\frac{\partial^2 u}{\partial x^2} + \frac{\partial^2 u}{\partial y^2} \right)^2 + \frac{B}{2} \left(\frac{\partial u}{\partial z} \right)^2,$$

où le terme $B(\partial u/\partial z)^2$ tient compte d'une modification éventuelle de la distance entre couches. Dans cette description, la variation de densité à l'intérieur des couches n'est pas prise en considération. Une expression générale tenant compte de cet effet a été donnée par DE GENNES [50]:

$$F = \frac{1}{2} A\theta^2 + \frac{1}{2} B\gamma^2 + C\theta\gamma + \frac{1}{2} K_{11} \left(\frac{\partial^2 u}{\partial x^2} + \frac{\partial^2 u}{\partial y^2} \right)^2,$$

θ est la dilatation en volume, $\gamma = \partial u/\partial z$ et A, B, C sont des modules isothermes de rigidité.

Le couplage entre les deux variables θ et γ a une incidence très remarquable sur l'élasticité des smectiques en donnant naissance à des modes transverses propagatifs correspondant à une oscillation de couches à densité presque constante.

La signification physique de ce « deuxième son » peut être dégagée de la Fig. 15 où l'on a représenté les deux polarisations possibles pour des modes hydrodynamiques transverses de vecteur d'onde $\boldsymbol{q}$ oblique par rapport au directeur.

La polarisation 1 perpendiculaire au plan de la Figure correspond à un mode visqueux amorti ($\omega = i\eta q^2/\varrho$) comparable à ceux que l'on met en évidence en phase nématique.

La polarization 2 parallèle au plan de la Figure est associée à deux modes propagatifs ($\omega = \pm\, C_2 q$) qui déforment les couches en introduisant une légère variation de densité. C'est le deuxième son des smectiques.

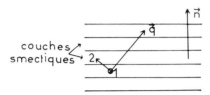

Fig. 15. – Modes transverses dans les smectiques.

Ces modes s'évanouissent lorsque le vecteur d'onde est parallèle ou perpendiculaire aux plans des couches; ils sont alors remplacés par des modes amortis [14, 50].

Aux modes transverses décrits ci-dessus, il faut ajouter deux modes longitudinaux propagatifs ($\omega = \pm\, C_1 q$) dont la vitesse de propagation et l'amortissement sont fonction de l'angle entre le directeur et le vecteur d'onde, et un mode de relaxation thermique ($\omega = i\varkappa q^2/\varrho$).

L'étude de la variation de la vitesse de propagation et de l'atténuation des ondes longitudinales avec l'orientation du directeur par rapport au vecteur d'onde a été effectuée pour différents smectiques [51, 52].

La mise en évidence du deuxième son au moyen de techniques acoustiques paraît difficile, mais elle a pu être réalisée par des expériences de diffusion Brillouin dues à LIAO, CLARK et PERSHAN [53]. Ces résultats sont reportés sur la Fig. 16.

Qu'advient-il maintenant lorsqu'un composé passe d'une phase mésomorphe à une autre phase mésomorphe?

Ces transitions sont en général discontinues (transitions du premier ordre), mais ceci n'est nullement imposé par la symétrie du système. Ces transitions peuvent s'effectuer de façon plus continue (transitions du 2ème ordre). Dans ce cas les fonctions thermodynamiques (entropie, énergie, volume) restent continues lors du passage au point de transition. En particulier, contrairement aux transitions du 1er ordre, les transitions du 2ème ordre ont une chaleur latente de transformation nulle. Ces transitions sont intéressantes car elles sont annoncées par des phénomènes de prétransition. Nous en avons vu un exemple avec la transition nématique-isotrope qui est « presque du 2ème ordre ». La transition nématique-smectique A peut également être du 2ème ordre. Quand la température diminue vers la température de transition la phase smectique « germe » progressivement dans la phase nématique. Ceci se traduit par l'apparition de petits domaines smectiques (groupes cybotactiques) en phase nématique. Un groupe cybotactique peut être caractérisé par sa dimension ξ et son temps de vie τ. Lorsque l'on s'approche de la transition la divergence de ξ et de τ provoque des accroissements anormaux de certains paramètres (constantes élastiques, coefficients de viscosité).

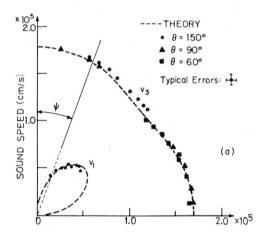

Fig. 16. – Vitesse des deux modes propagatifs dans un smectique A en fonction de la direction de propagation. V_3 vitesse de propagation des ondes longitudinales. V_1 vitesse de propagation du deuxième son. ψ angle entre $\boldsymbol{n}$ et $\boldsymbol{q}$. D'après [53].

Les théoriciens d'Orsay suivant une ligne de pensée due à DE GENNES ont développé des théories fondées sur des analogies avec d'autres systèmes déjà connus [54, 58]. Par exemple la transition nématique-smectique A est analogue à la transition λ de l'hélium. Comme pour l'hélium, le paramètre d'ordre des smectiques ψ est une quantité complexe, dont le module définit la densité de couches et la phase, la position des couches. Néanmoins la transition nématique-smectique A est plus complexe à cause du couplage entre ψ et le paramètre d'ordre S du nématique. Les propriétés associées au paramètre d'ordre ψ au voisinage de la transition sont décrites en détail dans les articles dont les références vont du no. 54 au no. 58.

Nous allons énumérer simplement les principales prédictions théoriques dans le domaine de l'acoustique:

la vitesse du deuxième son s'annule comme $(T-T_c)^{\frac{1}{3}}$,

l'atténuation du deuxième son diverge comme $((T-T_c)/T_c)^{-\frac{1}{3}}$,

le temps de relaxation τ des fluctuations du module du paramètre d'ordre diverge comme $(T-T_c)^{-1}$ et peut être en principe déterminé à partir des mesures d'atténuation d'ondes longitudinales, l'absorption ultrasonore basse fréquence ($\omega \ll \tau^{-1}$) diverge également comme $(T - T_c)^{-1}$.

Parmi ces différentes prédictions seul le dernier point a fait l'objet de vérifications expérimentales. SETTE et ses collaborateurs ont en effet mis en évidence un comportement critique de l'absorption ultrasonore pour le N-p–cyano-benzylidène p-n–octyloxyaniline au voisinage de la transition smectique A-nématique [59].

Une autre transition qui présente également des propriétés intéressantes a été étudiée sur le plan théorique, c'est la transition smectique A-smectique C [60].

Un smectique C ne diffère d'un smectique A que par la direction des molécules par rapport à la normale aux couches. Alors que pour un smectique A le directeur est confondu avec la normale aux couches, dans le smectique C elle fait un angle fini avec celle-ci. La transition smectique A-smectique C peut être du deuxième ordre. Dans ce cas, lorsque partant de la phase A on s'approche de la transition, les fluctuations du directeur autour de la normale aux couches deviennent très importantes et se ralentissent. Les fluctuations sont caractérisées par un temps de relaxation τ' qui diverge à la transition. Comme pour le nématique, le directeur est couplé à l'écoulement. Une mesure en fonction de la fréquence de l'absorption d'une onde de cisaillement, au moyen du dispositif précédemment décrit, permet théoriquement de déterminer τ'. Il est également possible d'obtenir ce paramètre à partir de l'atténuation basse fréquence ($\omega \ll \tau'^{-1}$) du deuxième son.

L'ensemble des prédictions théoriques que l'on vient d'énoncer, prédictions non encore vérifiées expérimentalement, permet de donner un aperçu de la richesse des informations que l'on peut obtenir de l'étude des matériaux mésomorphes, notamment dans le domaine des transitions de phase. On peut remarquer au passage que ces transitions ont lieu dans des domaines de températures assez facilement accessibles, ce qui n'est pas le cas pour les supraconducteurs par exemple.

Notons également qu'il existe d'autres phases mésomorphes dont nous n'avons pas parlé dans cet exposé, et qui sont constituées par des systèmes à deux composantes chimiques indépendantes. On peut citer par exemple les phases lamellaires lipide-eau pour lesquelles on retrouve les modes caractéristiques d'un smectique A et un mode nouveau correspondant à un glissement des lames lipidiques par rapport aux lames d'eau [61].

$$* \, * \, *$$

Les auteurs désirent exprimer leurs remerciements à J. C. BACRI et P. S. PERSHAN pour les avoir autorisés à reproduire leurs résultats expérimentaux.

Une partie du travail exposé ici a été réalisée sous contrat D.R.M.E. 71/034.

BIBLIOGRAPHIE

[1] G. FRIEDEL: *Ann. de Phys.*, **18**, 273 (1922).
[2] E. GUYON, P. PIERANSKI et M. BOIX: *Lett. Appl. Engin. Sci.*, **1**, 19 (1973);
     J. L. JANNING: *Appl. Phys. Lett.*, **21**, 173 (1973).
[3] La photo *a*) est due à l'obligeance de M. PETERS. La photo *b*) est due à l'obligeance
     de MM. S. NAGAÏ et K. IIZUKA: travail non publié.
[4] Voir, par exemple, P. GREGUSS: *Acustica*, **29**, 52 (1973).
[5] Voir, par example, P. G. DE GENNES: *Liquid Crystal Physics* (London, 1974).
[6] C. W. OSEEN: *Trans. Farad. Soc.*, **29**, 883 (1933).
[7] H. ZOCHER: *Trans. Farad. Soc.*, **25**, 1 (1958).
[8] F. C. FRANK: *Disc. Farad. Soc.*, **29**, 945 (1933).
[9] GROUPE D'ETUDE DES CRISTAUX LIQUIDES D'ORSAY: *Journ. Chem. Phys.*, **51**,
     2816 (1969).
[10] J. L. ERICKSEN: *Arch. Rat. Mech. Anal.*, **4**, 231 (1960); **28**, 265 (1968).
[11] F. M. LESLIE: *Quart. Journ. Mech. Appl. Math.*, **19**, 357 (1966); *Arch. Rat.
     Mech. Anal.*, **28**, 265 (1968).
[12] O. PARODI: *J. Physique*, **31**, 581 (1970).
[13] D. FORSTER, T. C. LUBENSKY, P. C. MARTIN, J. SWIFT et P. S. PERSHAN: *Phys.
     Rev. Lett.*, **26**, 17 (1971).
[14] P. C. MARTIN, O. PARODI et P. S. PERSHAN: *Phys. Rev. A*, **6**, 2401 (1972).
[15] F. JÄHNIG et H. SCHMIDT: *Ann. of Phys.*, **71**, 129 (1972).
[16] P. MARTINOTY et S. CANDAU: *Mol. Cryst. Liq. Cryst.*, **14**, 243 (1971).
[17] H. T. O'NEIL: *Phys. Rev.*, **75**, 928 (1949).
[18] C. H. GÄHWILLER: *Phys. Lett.*, **36** A, 311 (1971).
[19] Y. S. LEE, S. L. GOLUB et G. H. BROWN: *Journ. Phys. Chem.*, **76**, 2409 (1972).
[20] Voir, par exemple, H. J. MC SKIMIN: *Physical Acoustics*, edité par W. F. MASON,
     Vol. 1, Part A (New York, N. Y., 1965).
[21] F. BROCHARD: *J. Physique*, **32**, 685 (1971).
[22] P. MARTINOTY et S. CANDAU: *Phys. Rev. Lett.*, **28**, 1361 (1972).
[23] F. KIRY et P. MARTINOTY: a paraître.
[24] J. C. BACRI: *J. Physique Lett.*, **35**, L143 (1974).
[25] V. FREEDERICKS et V. ZOLINA: *Trans. Farad. Soc.*, **29**, 919 (1933).
[26] J. C. BACRI: *J. Physique*, **35**, 601 (1974).
[27] Voir, par example, K. F. HERZFELD et T. A. LITOVITZ: *Absorption and Dispersion
     of Ultrasonic Waves* (New York, N. Y., 1960); A. J. MATHESON: *Molecular Acoustics*
     (New York, N. Y., 1970).
[28] P. MARTINOTY et S. CANDAU: *Compt. Rend.*, **271** B, 107 1970).
[29] G. G. NATALE et D. E. COMMINS: *Phys. Rev. Lett.*, **28**, 1439 (1972).
[30] M. E. MULLEN, B. LÜTHI et J. M. STEPHEN: *Phys. Rev. Lett*, **28**, 799 (1972).
[31] D. EDEN, C. W. GARLAND et R. C. WILLIAMSON: *Journ. Chem. Phys.*, **58**, 1861
     (1973).
[32] Y. KAWAMURA, Y. MAEDA, K. OKANO et S. IWAYANAGI: *Japan Journ. Appl.
     Phys.*, **12**, 1510 (1973).
[33] K. A. KEMP et S. V. LETCHER: *Phys. Rev. Lett.*, **27**, 1634 (1971).
[34] P. MARTINOTY: Thèse, Strasbourg (1972).
[35] F. JÄHNIG: prétirage.
[36] S. CANDAU, P. MARTINOTY et R. ZANA: *J. Physique Lett.*, a paraître.

[37] N. Sheppard et G. J. Szasz: *Journ. Chem. Phys.*, **17**, 86 (1969).

[38] S. Marcelja: *Journ. Chem. Phys.*, **60**, 3599 (1974).

[39] B. Deloche, J. Charvolin, L. Liebert et L. Strzelecki: *Fifth International Liquid Crystal Conference, Stockholm, June 1974.*

[40] T. W. Stinson et J. D. Litster: *Phys. Rev. Lett.*, **25**, 503 (1970).

[41] V. Zvetkov: *Acta Physicochemica USSR*, **19**, 86 (1944).

[42] P. Martinoty, S. Candau et F. Debeauvais: *Phys. Rev. Lett.*, **27**, 1123 (1971).

[43] P. G. de Gennes: *Mol. Cryst. Liq. Cryst.*, **12**, 193 (1971).

[44] L. D. Landau et E. M. Khalatnikov: *Collected Papers of L. D. Landau*, edités par D. ter Haar (New York, N. Y., 1965), p. 626.

[45] S. Nagai et K. Iizuka: *Japan Journ. of Appl. Phys.*, a paraître.

[46] H. Imura et K. Okano: *Chem. Phys. Lett.*, **19**, 387 (1973).

[47] P. D. Edmonds et D. A. Orr: *Mol. Cryst. Liq. Cryst.*, **2**, 135 (1966).

[48] M. Fixman: *Journ. Chem. Phys.*, **36**, 1957 (1962).

[49] G. W. Gray, K. J. Harrison et J. A. Nash: *Electron. Lett.*, **9**, 130 (1973).

[50] P. G. de Gennes: *J. Physique*, colloque C4, supplément au n. 11-12, **30**, 65 (1969).

[51] K. Miyano et J. B. Ketterson: *Phys. Rev. Lett.*, **31**, 17 (1973).

[52] A. E. Lord jr.: *Phys. Rev. Lett.*, **29**, 1366 (1972).

[53] Y. Liao, N. A. Clark et P. S. Pershan: *Phys. Rev. Lett.*, **30**, 14 (1973).

[54] P. G. de Gennes: *Sol. Stat. Comm.*, **10**, 753 (1972).

[55] P. G. de Gennes: *Compt. Rend.*, **274** B, 758 (1972).

[56] P. G. de Gennes: *Mol. Cryst. Liq. Cryst.*, **21**, 49 (1973).

[57] F. Brochard: *J. Physique*, **34**, 411 (1973).

[58] F. Jähnig et F. Brochard: *J. Physique*, **35**, 299 (1974).

[59] R. Sette, F. Scudieri et R. Bartolino: *Fifth International Liquid Crystal Conference, Stockholm, June 1974.*

[60] F. Brochard: *Compt. Rend.*, **276** B, 87 (1973).

[61] F. Brochard et P. G. de Gennes: prétirage.

Acoustical Properties of Solids.

W. P. MASON

Columbia University - New York, N. Y.

Glossary of Symbols.

a = Hertz radius of contact, separation of atoms in a unit cell, separation between Peierls' troughs

a_0 = cut-off radius $= \frac{3}{4} b$

A_1, B_1 = constants

b = Burgers vector

B = drag coefficient for dislocations

$\partial c / \partial x$ = concentration gradient

Δc = increase in elastic modulus

c_{66}^{C} = shear modulus at constant field

c_{66}^{D} = shear constant at constant P

C_V = specific heat at constant volume

d = distance between equivalent sites

D = diffusion constant

$\bar{D}$ = average grain diameter

e = electron charge

E = electric field

E_i = thermal energy in i-th mode

ΔE = difference between magnetically saturated and demagnetized rod, difference between elastic constant with dislocations and without

f = frequency

F = flux of atoms, force on a dislocation $= \tau b$

$\hbar$ = Planck's constant h divided by 2π

h_{36} = piezoelectric constant

H = activation energy in calories per mole

I = magnetic intensity

k = Boltzmann's constant

K = thermal conductivity, relaxation strength

K_1 = anisotropic energy

l_e = mean free electron path

l = domain thickness

l_A = average separation of impurity atoms

l_n = network length

m = mass of electrons

M = average atomic mass, mass of dislocation per unit length $= \pi \varrho b^2$

N = number of kinks crossing barrier, number of electrons per cm³

N_0 = total number of kinks

$\bar{N}$ = total number of dislocations per cm³

$N(l)$ = number of dislocations of length l

P = electrical polarization

q = wave vector = ω/V

Q^{-1} = internal friction

r = radius of solvent atom

r' = radius of impurity atom

R = resistivity, gas constant $\doteq 2$ cal·mole, orientation factor $\doteq \frac{1}{4}$ for longitudinal waves

S_S = shearing strain including dislocation effects

S_E = shearing strain without dislocations

t_1 = thickness of contact layer

t_2 = thickness of noncontact layer

T_1 = longitudinal stress, sum of applied pressure

T_{10} = stress holding rock together

T = absolute temperature

v = activation volume = bA^*, A^* being activation area

V = sound velocity

V_0 = Fermi velocity

$\bar{V}$ = Debye average velocity

w = maximum energy stored

w_k = energy of one kink

ΔW = energy loss per cycle

W = width of dislocation kink

x = displacement along dislocation loop

$\bar{x}$ = average displacement of dislocation loop

y = distance along dislocation loop

Y_0 = Young's modulus

z = number of atoms per unit cell

α, β = orientation factors of order $\frac{1}{4}$

α = attenuation per cm in neper

α_{12} = forward numbers of jumps

α_{21} = backward numbers of jumps

β = ratio of τ_{db}/σ_k

γ_j = Grüneisen constant

γ_i = Grüneisen numbers

ε = $(r-r')/r$ = ratio of radii differences over r

$\dot{\varepsilon}$ = velocity of end of tin crystal

η_p = phonon viscosity

η_{ijkl} = viscosity tensor

λ = wavelength, total magnetostrictive expansion

μ = magnetic permeability, shearing modulus

ν, σ = Poisson's ratio

θ = Debye temperature

ϱ = density

σ = electrical conductivity

σ_p = Peierls' stress

σ_k = kink stress

$\sigma_x, \sigma_y, \tau_{xz}, \tau_{\theta z}$ = stresses around a dislocation

τ = shearing stress, relaxation time

τ_{db} = dynamic Peierls' stress

τ_0 = inverse of attempt frequency
τ_r = rate constant determining diffusion strength
τ^* = $\tau(1-j\beta)$ = complex shearing stress
χ_0 = magnetic susceptibility
ω = angular frequency

1. – Introduction.

The transmission of sound waves in solids has been used to study many solid-state motions such as domain wall motions, point-imperfection motions, dislocation motions and many others. The techniques used involve the measurement of velocities and attenuations occurring in the solids and their variation with temperature and frequency. Often the effects manifest themselves in the form of relaxation effects for which the attenuation rises to a maximum for a definite frequency at a definite temperature. The peak maximum will usually increase in frequency with an increase in temperature which allows an activation energy for the process to be determined. For other effects such as the conversion of acoustic-wave energy to electrons and phonons, the attenuation usually increases proportionally to the square of the frequency except at very low temperatures. Hence measurements of the two quantities, attenuation and velocity, and their variation with frequency, temperature and amplitude can give considerable information on possible solid-state motions occurring in the materials.

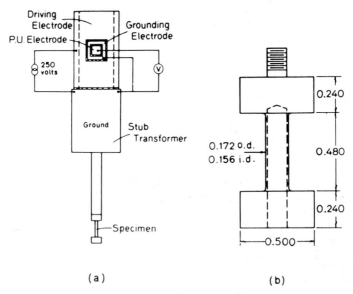

(a)   (b)

Fig. 1. – Transducer, transformer and sample for measuring modulus defect and internal friction at high strain amplitudes: *a*) schematic lay-out of ultrasonic generator, *b*) specimen shape.

The methods of measurement depend on the frequency range to be covered. From very low frequencies up to 100 kHz the usual method is to excite some resonant mode of motion such as a flexural vibration, a longitudinal vibration, or a torsional mode of vibration by means of a driving transducer and to measure the resonant frequency and the width of the resonant curve. These two measurements will determine the velocity—or elastic modulus—and the internal friction Q^{-1} which is related to the attenuation of waves in the solid. Figure 1 shows a recent apparatus [1] for measuring low- and high-amplitude effects in solids.

For higher frequencies it is usual to send pulsing or steady-state waves into the body of the material and to pick up transmitted or reflected pulses from the sample. If a continuous wave is used the resonance properties of the sample are measured. Figure 2 shows a typical pulsing system. Velocities are essentially measured by determining the transmission time of the pulses while attenuations are measured by observing the rate for which successive pulses die down. All these measurements require careful consideration of spreading losses, parallelism of samples, etc. A very complete account of these considerations has been given in a recent book [2].

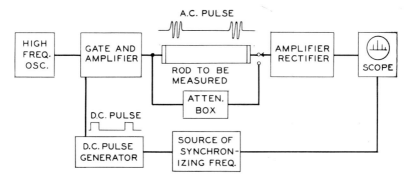

Fig. 2. – Pulse system for measuring velocity and attenuation in a solid material (after McSkimin).

2. – Attenuation of sound in a perfect crystal.

2'1. *Effect of energy conversion to thermal phonons.* – Even if there are no imperfections in a crystal, there still is a background attenuation caused by the conversion of acoustic energy to the energy of thermal phonons or electrons in a metal. This loss usually becomes so large that it overshadows other types of loss above about 200 MHz and usually makes determinations of imperfection's motion impossible above such frequencies. A possible exception is at very low temperatures in dielectric solids or in metals under high magnetic fields. Hence before other effects are mesured it is usually necessary to evaluate these two background losses.

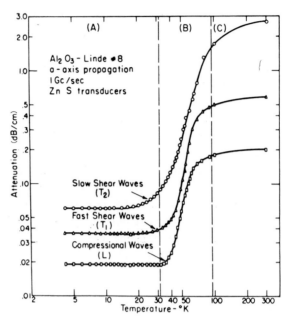

Fig. 3. – Hypersonic attenuation for a longitudinal and two shear waves in Al_2O_3. Measured at 1 GHz. Regions (A), (B) and (C) are regions for which different phonon processes are taking place (after DeKlerk).

Figure 3 shows one of the best measurements [3] of the attenuation in a dielectric crystal Al_2O_3. As can be seen there are three regions, a low-temperature region (A) from 0 °K to about 30 °K in which the attenuation is independent of the temperature, a rapidly rising region (B), and a relatively flat region (C) for which the attenuation is again independent of the temperature. The lowest-temperature region is assumed to be controlled by the deviation from a perfect crystal. The attenuation in this region depends on the degree of perfection of the crystal. The region (B), which occurs when the product of ω (2π times the measuring frequency) by the relaxation time τ, for the interchange of energy between the various modes, is greater than unity, is usually considered to be due to the direct interaction of acoustic waves with the thermal phonons. The slow shear wave has an increase in attenuation proportional to the fourth power of the temperature which is in agreement with the theory of Landau and Rumer [4] which considers a three-phonon collision between the acoustic phonon and two thermal phonons. This equation takes the form

$$(1) \qquad \alpha = \frac{60\gamma^2 kT}{M\overline{V}^2}\left(\frac{T}{\theta}\right)^3 \frac{2\pi}{\lambda},$$

where α is the attenuation in neper per cm, γ is the Grüneisen constant, k the Boltzmann constant, M is the average atomic mass, $\overline{V}$ is the average sound velocity, T the absolute temperature, θ the Debye temperature and λ the acoustic wavelength. The agreement with the formula is quite good. The fast shear wave and the longitudinal wave behave in a different manner with slopes proportional to T^7 and T^9 respectively. Explanations [5] for these results usually lie along the line that collinear collisions can occur between acoustic waves and phonons even though ordinarily these collisions would be forbidden.

When $\omega\tau < 1$ the individual collisions cannot be followed. In this region two effects cause the attenuations. These are 1) the thermoelastic effect which depends on the propagation of heat from the compressed part of the wave to the expanded part and 2) the Akheiser effect which causes an instantaneous separation of the phonon modes followed by an equilibration of their temperatures which occurs with a relaxation time τ. This effect produces a loss about 40 times the thermoelastic loss for insulators. A very general treatment using the Boltzmann integral was given by WOODRUFF and EHRENREICH [6] in the form

$$(2) \qquad \alpha = \frac{C_v T \gamma_{av}^2 \omega^2 \tau}{2\varrho V^3},$$

where C_v is the specific heat per unit volume, T the absolute temperature, γ_{av} is some average Grüneisen constant, ϱ is the density, V the ultrasonic velocity and τ a time to readjust to the new equilibrium values. γ_{av} was an adjustable parameter and τ a readjustment time which could not be calculated from any measurements of physical quantities.

A more recent and restricted theory has been given by the writer [7] in which the istantaneous separation of the phonon modes resulted in a change in the elastic modulus ΔC connected with the thermal energy and the nonlinear third-order elastic moduli. For a shearing mode the equation takes the form

$$(3) \qquad \Delta C = 3 \sum_i E_i (\gamma_i^j)^2,$$

where E_i is the thermal energy of mode i and γ_i^j are the Grüneisen numbers associated with the particular mode and strain. These are related to the second- and third-order moduli as discussed in [7].

The energy storage connected with an increase in the elastic modulus is relaxed down to the thermal-equilibrium temperature with a relaxation time τ which for shear waves has been taken equal to the thermal-relaxation time

$$(4) \qquad \tau = 3K/\varrho C_v \bar{V}^2,$$

where K is the lattice thermal conductivity, ϱC_v is the specific heat per unit volume, and $\bar{V}$ is the Debye average velocity. This value is chosen since the low-frequency shear acoustic waves can satisfy energy and momentum conservation by communicating energy directly to the thermal phonons.

For longitudinal waves one has to substract the thermoelastic loss from the equivalent expression of eq. (2) which becomes

(5)
$$\Delta C = \left[3 \sum_i E_i (\gamma_i^j)^2 - \gamma^2 \varrho C_v T \right],$$

where γ is the average Grüneisen constant for the material. Longitudinal acoustic waves do not couple directly to the thermal phonons since to satisfy momentum and energy conservation they interact only with shear and longitudinal waves in the same frequency range. Hence the relaxation time should consist of two parts, one part expressing the time required to convert longitudinal waves into low-frequency shear waves and the other for the thermal phonons to equilibrate their energy. Experimentally the total relaxation time appers to be about twice the thermal-relaxation time.

So far it has been assumed that all of the thermal modes have the same energy and that 39 pure modes existing for a cubic crystal or the 57 modes of a trigonal [8] crystal have been used to evaluate the expression for ΔC. Neither of this assumptions is necessary since one can take account of the greater energy in shear modes by assigning larger values to E_i for them and a computer program could be written which takes account of all directions. However, since the theory is valid only in the region (C), these extensions do not seem justified. Hence for these cases we can define a nonlinearity constant

(6)
$$D = 3 \left[3 \sum_i (\gamma_i^j)^2 / n - \gamma^2 \varrho C_v T / E_0 \right]$$

for longitudinal waves with the last term dropping out for shear waves. Here n is the total number of modes and E_0 is the total thermal energy.

The rate for which wave energy is converted into thermal energy is determined by the product

(7)
$$\eta_p = \Delta C \tau,$$

where, in analogy with the expression for the viscosity of an ordinary gas or liquid, the product $\Delta C \tau$ is equal to a viscosity. In terms of the nonlinearity constant D, the « phonon » viscosity term is

(8) $\eta_l = 2D(E_0 K / \varrho C_v \overline{V}^2)$ longitudinal, $\eta_s = D(E_0 K / \varrho C_v \overline{V}^2)$ shear,

and the attenuation is given by

(9)
$$\alpha = \omega^2 \eta_p / 2 \varrho V^3,$$

where V is the velocity of the appropriate wave.

This derivation of the equation is somewhat empirical although when these formulae are applied [9a] to six cubic crystals for which third-order moduli

have been measured and to one trigonal crystal [8] (quartz) the agreement as
to magnitude and temperature variation is quite good.

Recently LAMB and RICHTER [10] have produced experimental evidence
that the attenuation in a solid is determined by a phonon viscosity having the
same number of terms as the elastic tensor. This would require an attenuation
of the form

(10) $$\alpha = \omega^2 \eta_{ijkl}/2\varrho V^3 .$$

For a cubic crystal the number of constants would be 3, *i.e.*

(11) $$\eta_{11} , \quad \eta_{12} \quad \text{and} \quad \eta_{44} .$$

In this form one has to measure the attenuations and velocities for each crystal.
However the expressions in (8) are in the form of a viscosity and by evaluat-
ing η_l and η_s along a cube axis, one can determine the values of η_{11} and η_{44}.
A calculation for a longitudinal wave along the $\langle 110 \rangle$-axis [9b], given recently,
can be used to evaluate η_{12}. The results have been used to compare measured
and calculated results for LiF with fair agreement.

2.2. *Effect of electrons on acoustic attenuation.* – When electrons are present
in a metal it is found that there is an attenuation caused by the communication
of energy to the electrons by the sound wave. At very low temperatures for
which the mean free path of the electrons is large the attenuation can be quite
high. This effect was first discovered by BÖMMEL [11] and Fig. 4 shows his

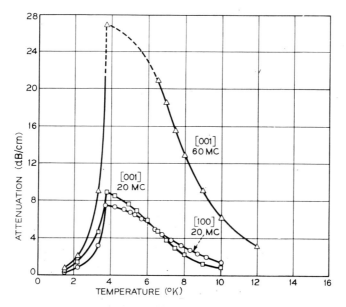

Fig. 4. – Shear wave attenuation measurements for a single tin crystal along the
[001]-axis and along the [100]-axis (after BÖMMEL).

measurements for a 60 MHz shear wave in a very pure sample of single-crystal tin. The superconducting temperature is 3.7 °K and the fall-off of attenuation below this temperature shows that the free electrons are disappearing in accordance with the Bardeen, Cooper, Schrieffer [12] theory of superconductivity.

For the region above the superconducting temperature and in the absence of a magnetic field the attenuation is quite well accounted for by theories that assume that the electrons lie on spherical Fermi surfaces. The most complete theory is due to PIPPARD [13], who showed that in the absence of collisions of electrons with the lattice, the separation of the positive ions by the acoustic vibration would produce a large electric field. In order to maintain charge neutrality the sum of the lattice current and the electron current must equal zero. Hence an acoustic wave can cause the electrons to move and energy is communicated to them. These considerations have been used by PIPPARD to show that the attenuation for longitudinal waves is given by

$$(12) \qquad \alpha_{l \mathrm{nep}(m)} = \frac{N m V_0 \omega}{V^2} \left[\frac{q l_e}{3} \frac{A}{1-A} - \frac{1}{q l_e} \right],$$

where N is the number of electrons per cm³, m their mass, V_0 is the Fermi velocity, V the acoustic velocity, $q = \omega/V$, l_e the mean free electron path and

$$(13) \qquad A = (\mathrm{tg}^{-1} q l_e)/q l_e .$$

If one expands this in a power series and inserts the value of

$$(14) \qquad l_e = \frac{\sigma m V_0}{N e^2},$$

the series becomes

$$(15) \qquad \alpha_l = \frac{2 \pi^2 f^2}{\varrho V_l^3} \left[\frac{4}{3} \frac{\hbar^2 (3 \pi^2 N)^{\frac{2}{3}} \sigma}{5 e^2} \left[1 - \frac{9}{35} (q l_e)^2 + \frac{23}{175} (q l_e)^4 \right] \right],$$

and the attenuation is determined by an electron viscosity [14] given by the term multiplying the bracket. At high frequencies Pippards relation indicates that the attenuation approaches a limit

$$(16) \qquad \alpha_l = \frac{2 \pi^2 N m \overline{V} f}{6 \varrho V_l^2},$$

and the attenuation increase is only in proportion to the frequency.

A similar expression occurs for shear waves with the low-frequency term being given by

$$(17) \qquad \alpha_s = \frac{2 \pi^2 f^2}{\varrho V_s^3} \left[\hbar^2 \frac{(3 \pi^2 N)^{\frac{2}{3}} \sigma}{5 e^2} \right].$$

Many measurements are found to agree approximately with these formulae. When the Fermi surface differs substantially from a spherical surface deviations from these values occur [15]. Measurements of the attenuation drop in the superconducting regions have been used to test the Bardeen, Cooper, Schrieffer theory of superconductivity. In many materials a good confirmation is obtained but in some there are deviations [15]. When magnetic fields are applied to the crystal, oscillatory values of the attenuation [16] are obtained as a function of the inverse applied field. These have been applied in delineating the shape of the Fermi surface, in studying open orbit transmission of electrons on the Fermi surface, and many other effects.

3. – Effect of structure in a solid.

The two background losses discussed in Sect. 2 usually produce measurable effects at quite high frequencies. These can be eliminated by measurement or calculation leaving a remainder due to other sources. These sources can usually be divided into solid-state structures in the solid and imperfections in the solid.

3˙1. *Effect of grain structure in a solid.* – The simplest structure in a solid is the polycrystalline formation consisting of individual crystals with different orientations separated by grain boundaries. If the orientations of different grains do not differ much, the boundaries can be made up of a series of dislocations [17]. If they differ by a larger amount, the grain boundaries are regions of vacancies and considerable strain in the lattice.

These grain boundaries are the origin of two effects, the grain boundary rotation occurring at low frequencies and high temperatures and sound scattering from the difference of the mechanical impedance of adjacent grains. This last effect severely limits the frequency range that can be used to investigate sound transmission in these polycrystalline solids.

Figure 5 shows a measurement of the grain boundary relaxation in a polycrystalline aluminum rod. The other curves show that it does not occur in a single crystal. These measurements were made using a torsional vibration of about 0.8 Hz. By measuring the shift in the temperature of the maximum attenuation as the driving frequency is changed, Ké [18] found an average activation energy of 31 000 calories per mole but a spread from a single relaxation equation of the form

$$(18) \qquad\qquad \alpha = \frac{K\omega^2\tau}{1 + \omega^2\tau^2}, \qquad\qquad \tau = \tau_0 \exp\left[-H/RT\right],$$

where K is a constant determined by the relaxation strength, $\tau_0 = 1/\omega_0$ is the inverse of the angular attempt frequency of the process. It requires an activation energy range of from 27 kcal to 35 kcal to account for the width. The

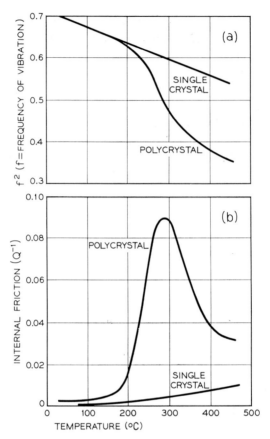

Fig. 5. – Elastic constants and internal dissipation in polycrystal and single-crystal aluminum showing grain boundary effect (after KÉ).

measurements indicate that the grain boundary can only move by its various parts diffusing under a stress bias.

The other effect of grain boundaries is the scattering of sound waves by the difference of impedances of differently oriented crystals adjacent to each other. This effect is an important one since it produces sound scattering at low enough frequencies to prevent polycrystals from being used in such studies as the high-frequency damping of dislocations. For such high-frequency studies single-crystal specimens are required.

The scattering type of loss begins to be important when the grain size approaches one-third of the acoustic wavelength. As in the Rayleigh-type scattering of light waves, the scattering loss is proportional to the fourth power of the frequencies. While it is not a true loss in the sense that the energy is converted into heat, nevertheless it prevents any coherent pulses from being propagated in the solid and hence velocities and true attenuations cannot be measured when this loss becomes high enough.

   While the origin of the effect is obvious, if account is taken of mode conversion as well as impedance differences the formulae become complex and the reader is referred to a recent review [19]. Figure 6 shows the attenuation in db/μs for shear waves in α brass. The dashed lines correspond to the theory when the mid frequency f_B is taken as $2\pi\overline{D}$, where $\overline{D}$ is the average grain size. The triangles correspond to measured points. As can be seen the initial scattering loss is proportional to the fourth power of the frequency. At higher frequencies the slope decreases to 2. The scattering loss becomes so high that other effects are lost.

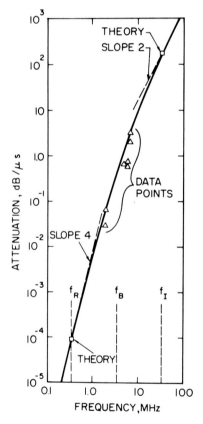

Fig. 6. – Attenuation *vs.* frequency of shear waves in brass (after PAPADAKIS).

   3˙2. *Domain wall motion in ferromagnetic material.* – One of the earlier effects of structure noted in a solid was the effect of domain wall motion on the elastic properties and the attenuation properties of ferromagnetic materials. The elastic effect is called the ΔE effect, while the dissipative properties are connected with eddy currents generated by the motion of the domain walls and by hysteresis loops in the flux curve caused by the applied stress.

   A domain is a region inside a polycrystal or single crystal for which all the polarization is in a single direction. Adjacent to these regions are other regions or domains for which the polarization is directed either 90° or 180° from the first region. These domains are separated by domain walls which are regions in the order of 1000 Å thick in which the direction of magnetization changes by small steps from one domain to the other. The 90° and 180° difference in orientation apply to such materials as iron which has its direction of easy magnetization along the ⟨100⟩-direction. For nickel or other materials with ⟨111⟩ being the direction of easy magnetization, the corresponding angles for the 90° wall are 71° and 109°. For a positive magnetostrictive material the magnetic-flux direction is also accompanied by a slight expansion, while for a negative magnetostrictive material a slight compression takes place. Hence along 90° domains there is a strain, while for 180° boundaries there is no strain. Since an applied stress will cause the same strain in 180° domains, there is no tendency for a motion of the magnetic flux. However for 90° domains a stress along one domain may cause a compression in the direction of the flux, while it will cause an expansion in the direction of the flux—by the Poisson-ratio effect—for the other domain. The result is a movement of the domain wall as shown by Fig. 7 for a positive magnetostrictive material.

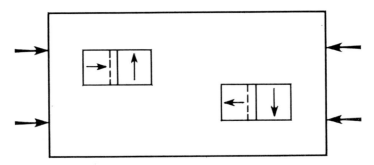

Fig. 7. – Change in domain wall positions in a ferromagnetic material as a function of the stress.

The ΔE effect, or the change in the elastic modulus, has its origin in the fact that a motion of the domain wall will include more material in domains perpendicular to the compressive stress. Since, for a positive magnetostrictive material, the dimension perpendicular to the polarization is less than that parallel to the polarization, a shortening of the rod occurs not only because of the elastic strain but also because of the magnetostrictive shortening. Since the elastic Young's modulus E is the ratio of the stress to the total strain, the value of Young's modulus decreases. If now the material is saturated magnetically, all the domains go into one domain and a stress can no longer move domain walls. The result is that the Young's modulus increases, and the fraction of difference between the magnetized and unmagnetized Young's moduli divided by the unmagnetized Young's modulus is called the ΔE effect. The microeddy current effect is due to the fact that, when a domain wall moves along, the direction of magnetization changes and eddy currents are generated in the material. The microhysteresis effect is connected with the fact that the displacements of the domain walls are not quite reversible but occur in the form of a small hysteresis loop.

Theoretical expressions for these effects were derived many years ago. The ΔE effect depends on the square of the total magnetostrictive expansion λ, the magnetic susceptibility for the unmagnetized condition

$$(19) \qquad\qquad \chi_0 = \frac{\mu_0 - 1}{4\pi},$$

μ_0 being the initial permeability, E_s the saturated Young's modulus and the square of I_s the saturation magnetization. For iron or any material in which the direction of easy magnetism lies along the $\langle 100 \rangle$-direction, an approximate formula due to AKULOV and KONDORSKY [20] is

$$(20) \qquad\qquad \frac{\Delta E}{E_0} = \frac{3}{5} \frac{\chi_0 \lambda_{100}^2 E_s}{I_s^2}.$$

When the direction of easy magnetism lies along the $\langle 111 \rangle$-direction, as in nickel, a calculation due to DÖRING [21] gives the value

$$(21) \qquad\qquad \frac{\Delta E}{E_0} = \frac{\mu_0 \lambda_{111}^2 E_s}{5\pi I_s^2} \left[\frac{5c_{44}}{c_{11} - c_{12} + 3c_{44}} \right]^2,$$

where c_{11}, c_{12} and c_{44} are the elastic constants of nickel, and λ_{111} the total fractional change along the $\langle 111 \rangle$-axis in going from a magnetized to an unmagnetized state.

The microeddy current effect, which results from the generation of eddy currents as the domain walls move, results in a relaxation-type loss having the form

$$(22) \qquad Q^{-1} = \frac{\Delta E}{E_0} \frac{f/f_0}{1 + (f/f_0)^2}, \qquad \text{where} \qquad f_0 = \frac{R}{96\chi_0 l^2}.$$

Here R is the electrical resistivity and l the thickness of the magnetic domain.

For the microhysteresis effect it is assumed that a Rayleigh-type loop occurs for which the area of the loop is proportional to the third power of the driving stress T_{11}, for a longitudinal bar. Hence the loss per cycle ΔW divided by the maximum energy stored is

$$(23) \qquad\qquad Q^{-1} = \frac{1}{2\pi} \frac{\frac{4}{3} b T_{11}^3}{\frac{1}{2} T_{11}^2 / Y_0} = \frac{4b Y_0 T_{11}}{3\pi},$$

where the constant $b = \mathrm{d}(1/Y_0)/\mathrm{d}T_{11}$, which is the slope of the compliance-stress curve. As long as the domain wall follows the stress, the loss is independent of the frequency, but proportional to the applied stress.

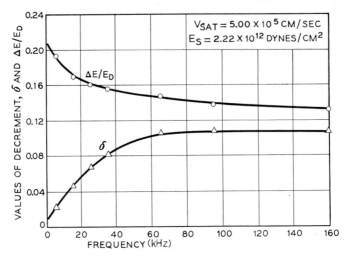

Fig. 8. – $\Delta E/E$ and decrement of a polycrystalline nickel rod as a function of the frequency (after Bozorth, Mason and McSkimin).

Figure 8 shows measurements of the modulus change $\Delta E/E$ and the decrement $\delta = Q^{-1}/\pi$ plotted as a function of the frequency for a well-annealed nickel rod. The microhysteresis effect for this amplitude is $Q^{-1} = 0.01/\pi = 0.003$. The eddy current loss increases initially in proportion to the frequency but then reaches a maximum and starts to decrease. This is a consequence of the relaxation effect of eq. (22). The breadth of the curve is the results of a distribution of domain sizes in agreement with a photograph of the domain structure. The $\Delta E/E_D$ curve starts out with a value of 0.21 which is in agreement with eq. (21) since $\lambda_{111} = 25 \cdot 10^{-6}$, $E_s = 2.22 \cdot 10^{12}$ dyn/cm², $I_s = 484$ CGS units of polarization, the elastic-constant ratio squared is 1.76 and the measured value of $\mu_0 = 340$. With these values the calculated value of $\Delta E/E_D = 0.225$ is in good agreement with the measured value. Due to domain relaxation $\Delta E/E_D$ decreases with increased frequency. Another effect occurs at much higher frequencies when all the domain motion is relaxed out. Under these circumstances rotation of the magnetization can still occur with a rotational susceptibility

$$(24) \qquad \chi_R = \frac{\mu_0 - 1}{4\pi} = \frac{I_s^2}{2K_1} = \frac{235\,000}{2 \cdot 59\,000} = 1.99 \text{ for nickel,}$$

where K_1 is the anisotropic energy. This is only about 7% of the domain boundary susceptibility. It introduces another plateau in the $\Delta E/E_D$ curve at higher frequencies.

These effects are important for magnetostrictive transducers. Also if other effects, such as dislocation motions, are to be investigated by internal-friction methods, this type of loss has to be eliminated by saturating the ferromagnetic rod.

3'3. *Phase transitions and critical points.* – Another structural change in a solid is one for which a phase transition occurs in the solid. Examples of these changes are liquid-vapor critical points, binary-liquid phase separations, ferroelectric and antiferroelectric transitions, ferromagnetic and antiferromagnetic transitions and order-disorder transitions. These have been investigated by ultrasonic means and a very complete review of them has recently been given [23]. Two simple examples will be used for illustrative purposes.

3'3.1. Ferroelectric crystal KH_2PO_4 (potassium dihydrogen phosphate). One well-known system is the ferroelectric crystal potassium

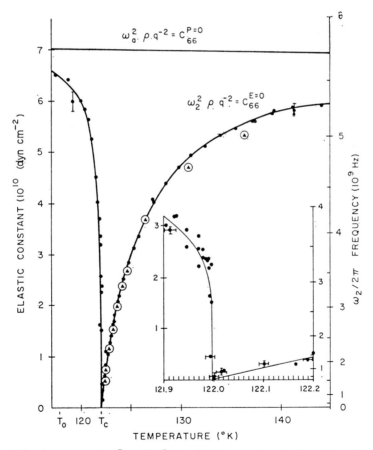

Fig. 9. – Elastic constants c_{66}^E and c_{66}^P for KDP near its Curie point. Solid points were obtained from Brillouin shifts and the open circles from ultrasonic measurements (after GARLAND).

dihydrogen phosphate (KDP) which is in the paraelectric phase above 122 °K, but suddenly changes to a ferroelectric phase below this temperature. In this phase a spontaneous polarization occurs and the mechanical and dielectric properties are entirely changed. Due to the piezoelectric effect there is a strong coupling between the mechanical and dielectric behavior in all ferroelectric materials. Figure 9 shows measurements of the elastic constant c_{66}^E—which is the one affected by the piezoelectric coupling—as a function of the temperature [24]. The constant c_{66}^E is the one measured when the crystal is fully electroded and the resonant frequency ω is measured at the lowest impedance point for the crystal. If an air gap holder or a very small electrode is used to drive the crystal, the elastic constant c_{66}^P becomes independent of the temperature. The insert shows the value of the elastic modulus closer to the ferroelectric transition temperature. Attenuations have been measured in the isomorphous crystal potassium dideuterium phosphate $(T_c = 222 °K)$ and is shown since similar measurements have not been made for KDP. As shown by Fig. 10, these measurements [25] were made with the low field of 1.5 kV per cm in order to eliminate domain structure in the ferroelectric region. The attenuations increase in proportion to the square of the frequency.

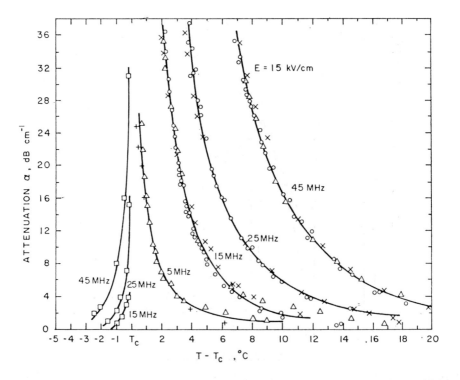

Fig. 10. – Attenuation in KD*P *vs.* temperature at several ultrasonic frequencies (after Litov and Uehling).

The elastic and attenuation properties are connected by the equation

$$(25) \qquad C^* = c_{66}^P - h_{36}^2 \chi_{33}^s(\omega) \,,$$

where h_{36} is the piezoelectric strain constant $(\partial E_3 / \partial S_6)_P$ and $\chi_{33}^s(\omega)$ is the complex linear susceptibility of the clamped crystal. For ultrasonic frequencies for which the product $\omega\tau \ll 1$, the susceptibility can be presented in the form

$$(26) \qquad \chi(\omega) = \chi(0) - i\big(\chi(0)\big)\omega\tau \,,$$

where τ is the relaxation time associated with a polarization reversal. Solving for the real and imaginary parts

$$(27) \qquad \begin{cases} c_{66}^P - c_{66}^E = h_{36}^2 \chi^s(0) \,, \\ \text{Im}\,(C^*) = 2\varrho V^3 \alpha/\omega = h_{36}^2 \chi^s(0)\omega\tau^s \,, \end{cases}$$

where these quantities are understood to be adiabatic. Hence the attenuation in neper per cm can be written in the form

$$(28) \qquad \alpha = \frac{c_{66}^P - c_{66}^E}{2\varrho V^3}\omega^2\tau^s \,,$$

where ϱ is the density and V the acoustic-wave transmission velocity.

From direct mesurements the dielectric susceptibilities are known to follow a Curie-Weiss law above T_c the Curie temperature. This takes the form

$$(29) \qquad \chi^s = C/(T - T_0) \,,$$

where T_0 is the temperature for which the clamped crystal would become ferroelectric. This is lower by 3.5 °K than T_c the actual temperature for which spontaneous polarization appears in the unstrained crystal. The Curie constant C has a value of 259 °K for KDP. If we combine (25) with (29), the elastic-modulus difference also obeys a Curie-Weiss law of the form

$$(30) \qquad c_{66}^P - c_{66}^E = \frac{Ch_{36}^2}{T - T_0} \,.$$

Plots of the elastic-moduli difference from Fig. 9 are in good agreement with this equation with T_0 equal to 118.5 °K and h_{36} equal to $3.04 \cdot 10^4$ stat. volt per unit strain. Hence the difference between the two types of elastic moduli satisfies a Curie-Weiss law.

From eq. (28) it is obvious that the attenuation is going to peak at the Curie temperature in agreement with Fig. 10. Ferroelectric materials are usually interpreted in terms of a « soft » transverse optical mode for which the frequency ω_0 of polarization oscillation goes to zero at the ferroelectric temperature. The relaxation time τ^T is equal to

$$(31) \qquad\qquad \tau^T = 2L/\omega_0^2 \,,$$

where L is a constant and τ^T is the relaxation time at constant stress. This relaxation time is related to the one at constant strain by $\tau^T/\tau^S = c^P/c^E$ and hence the relaxation time τ^S of eq. (28) is finite at T_0 and varies according to the equation

$$(32) \qquad\qquad \tau^S = CL^{-1}/(T - T_0) \,.$$

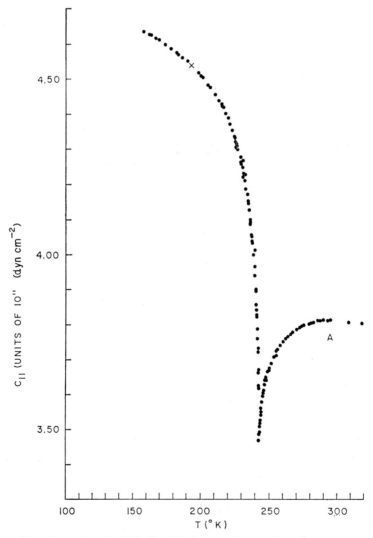

Fig. 11. – Variation of c_{11} in NH$_4$Cl with temperature (after GARLAND).

The attenuation data of Fig. 10 show that τ^s varies from $9 \cdot 10^{-11}$ s to much smaller values far from the temperature T_0.

3·3.2. Order-disorder transition in ammonium chloride [23].

« Ammonium chloride undergoes an order-disorder transition which involves the relative orientations of the tetrahedral ammonium ions in a CsCl-type structure. The most stable orientation of the NH_4^+ ion in the cubic unit cell is for the hydrogen atoms to point toward the nearest-neighbor Cl^- ions. Thus there are two possible positions for the ammonium ion. In the completely ordered state all NH_4^+ tetrahedra have the same relative orientation with respect to the crystallographic axes; in the completely disordered state the orientations are random with respect to these two positions. »

This change in order affects the longitudinal and transverse velocities and attenuations and the temperature of ordering is a function of the pressure. Figure 11 shows the elastic modulus c_{11} plotted as a function of the temperature for atmospheric pressure. The value takes a dip but does not go to zero as in the case of the shear modulus of KDP. However the bulk modulus $(c_{11} + 2c_{13})/3$ does go to zero since there is a change in volume for this material. The shear constant shown in Fig. 12 is a confirmation of this result since it shows a discontinuity at the transformation temperature. The large dip in the longitudinal

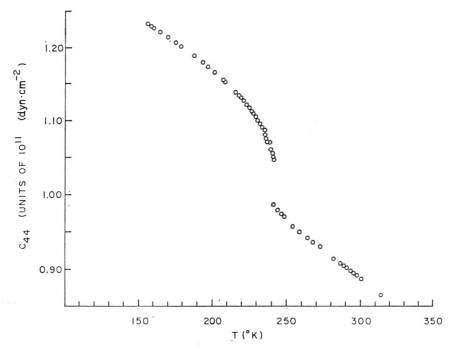

Fig. 12. – Variation of c_{44} in NH_4Cl with temperature on cooling. Open circles are experimental data at 1 atmosphere.

constant c_{11} is accompanied by a peak in the ultra-sonic attenuation as in the shear mode of KDP. There is no increase in attenuation for the shear mode since this is acting as a probe for the phase change. Hence ultrasonic methods are very useful for studying such transitions as well as all the other types investigated. An interpretation of this effect in terms of a compressible Ising model has been discussed by GARLAND [23].

4. – Effect of imperfections.

4'1. *Introduction.* – Imperfections in a material can also affect the velocity and attenuation of acoustic waves in the material. The origin of these effects is that stresses can change the positions of the defects in the crystal lattice and this change affects the elastic moduli of the crystal and the internal friction. It often requires thermal energy as well as mechanical stresses to cause the change in position. The thermal energy takes the imperfection over an energy barrier which is biased by the applied stress. Since the number of jumps in position depends on the height of the barrier and the attempt frequency in crossing the barrier, ultrasonic measurements are one way of studying these quantities. The acoustic methods usually cover a lower-temperature range than thermal methods for studying these properties, but for some materials such as carbon and nitrogen in iron a join between the two methods has been made.

It has been shown by NOWICK and HELLER [26, 27] that the macroscopic behavior of such a crystal depends on the type of symmetry of the defect. This defect symmetry determines whether or not the crystal will undergo anelastic or dielectric relaxation and which coefficients of elastic compliance or dielectric susceptibility show the relaxation effect.

For linear defects, such as dislocations, the motion is much more complicated since motions of parts of the dislocation—such as kinks—can occur in two dimensions. Several models of dislocation motion have been proposed which give results agreeing with experiment for many conditions. There appear to be two sources of dissipation connected with dislocation motion, one involving lattice vibrations generated when dislocation kinks cross kink energy barriers and the other connected with the drag on dislocations as they communicate energy to phonons and electrons. The first type depends on the number of kink displacements and results in an internal friction Q^{-1} independent of the frequency. The second type initially produces an internal friction proportional to the frequency which reaches a maximum when the ratio of the angular frequency of measurement to the relaxation frequency ω_0 is 0.33, where

$$(33) \qquad\qquad \omega_0 = \mu b^2 / B l_A^2 .$$

Here μ is the elastic modulus in the glide plane, b the Burgers distance, B the drag coefficient per unit length of the dislocation and l_A the average dislocation loop length. Above this frequency the internal friction falls off inversely proportional to the frequency. The rate of fall-off is given by

(34)
$$\frac{Q^{-1}}{\bar{N}Rl_A^2} = \frac{\omega_0}{\omega} = \frac{\mu b^2}{\omega Bl_A^2} \quad \text{or} \quad B = \frac{\bar{N}R\mu b^2}{(\omega Q^{-1})_L}.$$

This formula has been used to evaluate the drag coefficient B which is determined when the limiting value of $(\omega Q^{-1})_L$ is measured. The number $\bar{N}$ of dislocations can be determined by etch pattern techniques, R the orientation factor which relates the stress in the glide plane to the stress applied to the crystal can be calculated from the elastic moduli of the crystal. μ the shear modulus and b the Burgers distance are known from the crystal constants.

4'2. Effect of point defects. – Most phenomena in which diffusion of atoms is detected depend on randon motion of atoms from one interstitial position to another. If there is a stress or field applied, there is a tendency for the atoms to diffuse in some particular direction connected with the gradient of the stress or electric field. If D is the diffusion constant, it is usually defined in terms of Fick's equation

(35)
$$F = -D\frac{\partial c}{\partial x},$$

where F = flux of atoms and $\partial c/\partial x$ = concentration gradient at that point. This coefficient can be defined [28] in terms of the frequency f with which an atom moves from one crystal site to another equivalent site and the distance d between them, in the form

(36)
$$D = \frac{fd^2}{6} = \frac{d^2}{6\tau},$$

where τ is the mean time of stay in a given site and is equal to $1/f$.

Most chemical or radioactive measurements determine the diffusion constant directly, but anelastic measurements determine the time τ. The anelastic measurements determine an internal friction Q^{-1} or a modulus change $\Delta M/M$. The damping peak is usually used and this internal friction follows the form

(37)
$$Q^{-1} = \Delta \frac{\omega \tau_r}{1 + \omega^2 \tau_r^2},$$

where τ_r is a rate constant determined by the frequency of the maximum loss and Δ is called the relaxation strength.

It is generally assumed that τ_r is proportional to τ, i.e.

(38)
$$\tau = \alpha \tau_r,$$

and WERT has discussed methods for relating these two. For the simple case of the Snoek peak

(39) $\tau = \frac{3}{2}\tau_r$.

Hence anelastic measurements can be used to determine the diffusion constant. The temperature for which the diffusion constant is determined is usually lower than that determined by radioactive methods. However LORD and BESHERS [29] by determining the internal friction at 7 MHz have closed the gap between the two methods and for nitrogen in iron have shown that both sets of data lie on a smooth curve.

The first and most completely studied point-defect relaxation is the Snoek [30] peak which results from the solution of nitrogen or carbon in iron. Nitrogen and carbon enter the iron—a body-centered crystal—as an interstitial atom half-way between the iron atoms in the positions marked x, y, z in Fig. 13. Since the structure is cubic, each of these positions is equally favored. Thermal agitation can cause the atoms to interchange positions, and hence equilibrium is established among the x, y and z positions. The effect of a stress along x is to stretch out the iron atoms along the x-direction and compress the atoms along the y- and z-direction. To go from x to y or z requires a thermal energy equal to the activation energy for the jump.

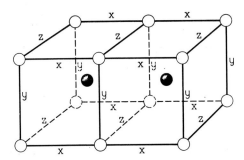

Fig. 13. – Proposed mechanism for stress-induced diffusion of nitrogen in iron (after SNOEK).

In order to measure this effect, it is necessary to saturate the iron to remove the ΔE and microeddy current effects discussed in Subsect. **3**$\cdot$**2**. When this is not done peaks connected with the Snoek effect are still seen but at a different temperature. These have been called Maringer [31] peaks. In order to measure the Snoek peaks the samples are first wet hydrogen-treated at 720 °C to remove the nitrogen and carbon impurities. Nitridizing was then accomplished at 580 °C by using a 5 % NH_3 gas in H_2. The most complete measurements are those of LORD and BESHERS [29] which have been carried up to 7 MHz. Figure 14 shows a plot of the diffusion constant D measured by various techniques cover-

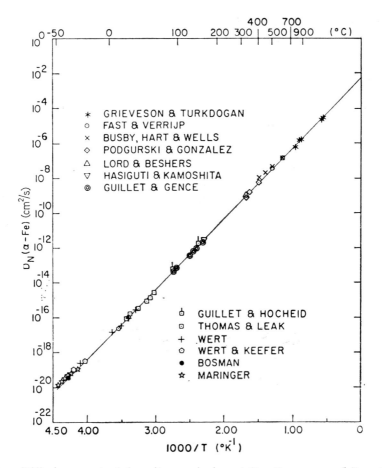

Fig. 14. – Diffusion constant for nitrogen in iron (after BESHERS and LORD).

ing a temperature range from 220 °C to 1470 °C. The low-temperature values are obtained by anelastic methods, while the higher values are obtained by radioactive methods. As can be seen the two techniques give a good join. From the slope of the curve it is found that the activation energy is 18 350 calories per mole. The equation for the diffusion constant of nitrogen in iron is

$$(40) \qquad D = 4.88 \cdot 10^{-3} \exp\left[-18\,350/RT\right] \text{ cm}^2/\text{s}.$$

For carbon in iron the diffusion constant is measured to be

$$(41) \qquad D = 3.94 \cdot 10^{-3} \exp\left[-19\,160/RT\right] \text{ cm}^2/\text{s}.$$

The initial constant, which is nearly the same for the two materials, indicates an attempt angular frequency $\omega_0 \doteq 10^{14}$ angular hertz. This follows from (36) and (39) with $d = a/\sqrt{2} = 1.65$ Å and $\omega_0 = 1/\tau_r$. Many other point-defect motions have been studied and complete reviews have been given [26, 28, 31].

Point defects can interact with dislocations and produce relaxation-type peaks. These peaks were first reported for many metals by HASIGUTI [32] and his co-workers and subsequently were seen by others. These occur after deformation and the frequency factor of 10^{12} s^{-1} and an activation energy of 0.27 eV suggest that they are determined by the interaction of point defects and dislocations. Annealing and cold-work have confirmed this interpretation.

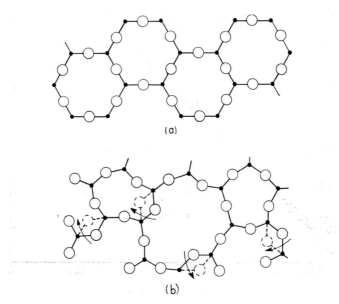

(a)

(b)

Fig. 15. – a) Regular network of a repeating crystal, b) the random network of glass (after WARREN).

4˙3. *Internal friction in fused silica.* – Another type of internal friction due to the motion of single atoms appears in fused silica. In this case the oxygen atom in the random network of fused silica has two positions having nearly the same energy with an energy barrier between the two positions as shown by Fig. 15. For a single crystal all the atoms have definite positions with respect to each other and stresses cannot cause a relative motion. The random network of Fig. 15 results in a series of relaxation curves as shown by Fig. 16. From the change in position of the peaks with frequency the angular frequency *vs.* temperature follows an equation

$$(42) \qquad \omega = \omega_0 \exp\left[-1300/RT\right], \qquad \text{where } \omega_0 = 5 \cdot 10^{13}.$$

The activation energy is an average one since the relaxation curve is much broader than a single one. By assuming a series of relaxation mechanisms with

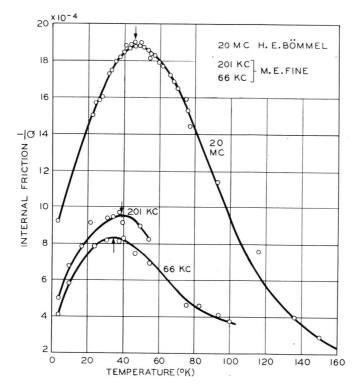

Fig. 16. – Internal friction in fused silica as a function of the temperature at three frequencies (after FINE and BÖMMEL).

activation energies from a few calories per mole to 3500 with a decreasing relaxation strength [33] $\Delta G/G$, the form of the attenuation curve can be matched. This distribution is consistent with the model of Fig. 15 since from X-ray measurements [34] most of the bonds are nearly straight and only a few of them have angles as large as 5°. Losses due to this flop mechanism are the principal losses in the room-temperature range. As shown by LEWIS [8], the Akheiser mechanism produces a loss only about 10% of that measured.

4'4. *Effect of dislocations on acoustic attenuation.*

4'4.1. Introduction. The imperfection that produces the largest effect on the attenuation and modulus defect in a crystal is the dislocation. This is partly because it can move rapidly and also because it is closely coupled with the applied stress. Some of the properties of dislocations are shown by Fig. 17. An edge dislocation is a region in the crystal for which there is one more plane of atoms above a plane—called the glide plane—than there is below this glide plane. When a shearing stress is applied to the crystal with its forces parallel

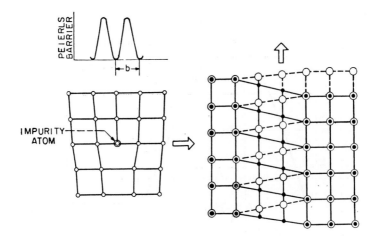

Fig. 17. – Edge and screw dislocations, Burgers vectors and Peierls-Nabarro barriers. Effect of impurity atom is to pin dislocation.

to the glide plane there is a force on the dislocation equal to

$$(43) \qquad\qquad\qquad F = \tau b$$

tending to move the dislocation along the glide plane. Here b is the Burgers' distance which is equal to the separation between atom planes. This motion is resisted by several forces. If the dislocation lies along a straight line it is most stable when the extra plane of atoms is midway between the two adjacent planes. When the force is applied to the dislocation, it has to overcome a barrier—known as the Peierls-Nabarro barrier—formed by the force required for the atoms to break one bond and form another. This is shown by the energy-distance diagram at the top. When the extra plane of atoms has moved over by one Burgers' distance b the dislocation is again in its minimum-energy position. The dislocation on the right is known as a screw dislocation since all the atoms lie in the form of a screw thread. For this type of a dislocation, a shearing stress causes the atoms on the right side to displace from the solid line positions to the dashed-line positions. Here the Burgers vector, shown by the arrow, is along the dislocation, rather than perpendicular to it as in the edge dislocation.

The edge dislocation figure shows another property of the dislocation, namely that it can interact with impurity atoms or vacancies. The dislocations are regions of stress in the crystal with a compression above the glide plane, a tension below the glide plane and a shearing stress across the glide plane. It is shown in books on dislocation theory [35] that the three stresses in the

xy-plane are

$$(44) \quad \begin{cases} \sigma_x = -\dfrac{\tau_0 b}{r} \sin\theta\,(2 + \cos 2\theta)\,, \\[2ex] \sigma_y = \dfrac{\tau_0 b}{r} \sin\theta \cos 2\theta\,, \\[2ex] \tau_{xy} = \dfrac{\tau_0 b}{r} \cos\theta \cos 2\theta \end{cases}$$

for an edge dislocation where $\tau_0 = \mu/2\pi(1-\nu)$, μ being the shearing modulus and ν the Poisson's ratio. For a screw dislocation the only stress is $\tau_{\theta z} = \mu b/2\pi r$, r being in both cases the distance from the center of the dislocation. When an impurity atom of a different size than the solute atom is introduced into the crystal, the energy will be smaller if a large atom occurs below the glide plane and also smaller if a smaller atom occurs above the glide plane. According to COTTRELL [36] the binding energy of an impurity is

$$(45) \qquad V_B = \frac{4}{3}\frac{1+\nu}{1-\nu}\frac{\mu b r^3 \varepsilon \sin\theta}{R}\,,$$

where r is the radius of the solvent atom, $\varepsilon = (r^1 - r)/r$, where r^1 is the radius of the impurity atom, and θ is the angle connecting the impurity atom from the center of the dislocation as measured from the glide plane. R is the distance of the impurity from the center of the dislocation. Hence impurity atoms can act as pins for the dislocation and it requires a high stress to cause the dislocation to break away from pinning points.

Except for materials with very high Peierls-Nabarro barriers—such as materials in the diamond-type lattice (for example silicon and germanium)—dislocations do not lie in the troughs between barriers but cross them at an angle. One theory of dislocation damping, the Granato-Lücke theory [37], assumes that they lie in an essentially straight line between pinning points and have no reaction with the Peierls-Nabarro barrier. The only source of dissipation in this model is a drag on the dislocation proportional to the velocity which represents a communication of energy to the thermal phonons. This type of theory gives good agreement with experiment for pure materials at high frequencies. However it is generally believed [38] that a more likely disposition of the dislocation is for part of it to be in the potential minimum and part to cross the barrier in the form of a « kink » of width w as shown in Fig. 18. The motion of the dislocation consists of a jump from one position of minimum energy across a « kink barrier » to the next position. The height of the kink barrier and the stress σ_k required to cause the kink to cross the barrier have been calculated by SCHOTTKY [39] to be

$$(46) \qquad \sigma_k = \frac{192}{\pi^2\sqrt{3}}\frac{1-\nu}{1+\nu}\frac{b}{w}\left(\frac{\sigma_P}{\mu}\right)^2 \mu\,, \qquad V_k = \sigma_k b^3/5\,,$$

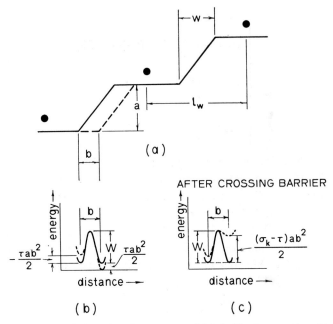

Fig. 18. – *a*) Motion of dislocation by means of kink displacements. *b*) Kink energy barrier and effect of an applied stress. *c*) Energy barrier showing effect of energy stored after kink crosses the barrier.

where σ_p is the Peierls stress, b/w is the inverse width of the kink, and V_k is the kink energy barrier. Experiments at very low temperatures [40], which determine the conditions for the internal-friction and modulus defect disappearance, indicate that the kink energy barrier for copper is $1.5 \cdot 10^{-16}$ erg or $1.25 \cdot 10^{-4}$ eV. With this activation energy the kink stress for copper becomes $4.5 \cdot 10^7$ dyn/cm². Since most internal-friction measurements are made with stresses from $4 \cdot 10^3$ to $4 \cdot 10^5$ dyn/cm², it is evident that the stresses alone are not sufficient to cause the kinks to cross the barriers and it requires thermal energy to do so. Figure 18*b*) shows the effect of a stress applied to the energy barrier. It lowers one side by an amount

$$(47) \qquad\qquad\qquad \tau a b^2 / 2$$

and raises the other minimum by the same amount. From reaction rate theory it is well known that the number of jumps per second in the forward direction is

$$(48) \qquad\qquad \alpha_{12} = \omega_0 \exp\left[-(V - \tau a b^2 / 2)/kT\right],$$

while the number in the reverse direction is

$$(49) \qquad\qquad \alpha_{21} = \omega_0 \exp\left[-(V + \tau a b^2 / 2)/kT\right].$$

Here ω_0 is the attempt angular frequency which is taken to be $2\pi \cdot 10^{13}$ in agreement with the values found in eqs. (40) and (42) for single-element attempt frequencies. Since $\tau ab^2/kT$ is much smaller than unity, even for the highest stresses and the lowest temperatures considered, then the net rate can be written

$$(50) \qquad \alpha_{12} - \alpha_{21} = \omega_0 \frac{\tau ab^2}{kT} \exp\left[-V/kT\right].$$

Since from the measurements of ALERS and SALAMA [40] the modulus defect and the decrement have dropped down to about half their values at 0.1 °K for a frequency of 10 MHz and a strain of 10^{-7} (stress of $4 \cdot 10^4$ dyn/cm²) and assuming that it takes 2π jumps to complete a cycle, we have determined the value of V to be $1.5 \cdot 10^{-16}$ erg or $1.25 \cdot 10^{-4}$ eV. For temperatures above this the exponential term can be dropped. At room temperature—300 °K—and under the assumption of 2π jumps per cycle, the kink motion will follow the applied stress up to a frequency of 125 MHz for a strain of 10^{-7}. This is as high a frequency as dislocation effects have been measured at this strain. One test of this thermal mechanism would be to measure dislocation effects at lower strains to see if the modulus change and internal-friction values are independent of the amplitude.

After the kink has crossed the barrier it acquires an energy $\sigma_k ab^2/2$ from the negative slope. If the dislocation kink is unpinned, one might expect that this energy would take it over the next barrier, etc. However this neglects the fact that lattice vibrations are generated on the negative-slope side which takes away some of the stored energy. Calculations by WEINER [41] and by ATKINSON and CABRERA [42] show that it takes a dynamic Peierls stress τ_{dP} of from 0.01 to 0.1 of the kink stress to keep the dislocation kink moving. Experimentally a value of 0.03 has recently been found [43].

When the kink is in a pinned system it cannot keep moving over barriers and the energy from the negative slope is stored in pushing the kinks closer together. The energy loss per kink is still the same and hence if this is the only source of loss, the Q^{-1} of the dislocation system alone is

$$(51) \qquad Q^{-1} = \frac{\text{energy dissipated}}{\text{energy stored}} = \frac{\tau_{dP}}{\sigma_k} = \beta.$$

Since an applied stress not only moves the dislocation but also stores elastic energy, the Q^{-1} of the complete specimen is

$$(52) \qquad Q^{-1} = \beta(\Delta S/S),$$

where ΔS is the stiffness of the system due to dislocation motion and S the stiffness due to elastic forces.

For slowly varying stresses the rate equation (50) would indicate much too fast a rate of kink displacements. However when a kink has crossed the barrier and stored an energy $\sigma_k ab^2/2$, it will have a higher energy as shown by Fig. 18c) Hence these kinks will tend to flow back to their original position at a faster rate than the lower-energy kinks flow in the forward direction. An equilibrium value is attained when

$$(53) \qquad\qquad N_0 \tau = N \sigma_k \,,$$

where N_0 is the total number of kinks, τ the applied stress and N the number of kinks that have crossed the barrier. This is obvious since the potentials balance when

$$(54) \qquad\qquad (N_0 - N) \frac{\tau ab^2}{2kT} = \frac{N(\sigma_k - \tau)}{2kT} \, ab^2 \,,$$

which results in eq. (53). The time required to reach this equilibrium condition is still determined by the jump time of eq. (50).

4˙4.2. Granato-Lücke theory. The most widely used theory of dislocation damping is the Granato-Lücke theory of dislocation damping which assumes that dislocations lie along straight lines between pinning points. The pinning points are of two types, impurity pinning points and network pinning points which are formed when three dislocations come together with the sum of their Burgers vectors equal to zero. The network pinning points are very

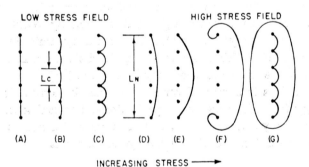

Fig. 19. – Dislocation displacements under various conditions of the applied stress (after Granato and Lücke).

stable and cannot be broken. At very high amplitudes they can act as centers for Frank-Read sources for the generation of new dislocations. For medium amplitudes, as shown by Fig. 19, the dislocations can break away from their impurity pinning points out to the network pins. This process introduces a nonlinear effect with stress for both the internal friction and the modulus defect. At lower strain amplitudes the stress causes the dislocation to bow out about pinning points of all types. The bowing-out is resisted by a tension

T of the dislocations, by a drag coefficient B which results in a force proportional to the velocity, and a mass reaction proportional to the dislocation acceleration which is determined by the mass M of the dislocation per unit length. This is usually taken to be equal to $\pi \varrho b^2$, where ϱ is the density of the crystal. Hence the motion of the dislocation is determined by the equation of a stretched string

$$(55) \qquad M \frac{\partial^2 x}{\partial t^2} + B \frac{\partial x}{\partial t} - T \frac{\partial^2 x}{\partial y^2} = \text{force} = \tau b .$$

For all the measurements that have been made the dislocations are over-damped and hence the acceleration term can be neglected. The drag coefficient B has been given several derivations. One calculation due to ESHELBY, which considers only the thermoelastic damping, has been found to be quite small and is usually neglected. LIEBFRIED [44] calculated the reaction of the dislocation against the radiation pressure of the phonon field and came up with the expression

$$(56) \qquad B = \frac{a}{10 V_s} 3kT \frac{Z}{a^3} ,$$

where a is the lattice constant, V_s the shear velocity and Z the number of atoms per unit cell. A second source, proposed by the writer [45], was based on the Akheiser-type loss occurring in a solid. The dislocation is surrounded by a strain field and as this is moved through the solid an acoustic loss occurs through the phonon viscosity effect. The expression for the damping constant B becomes for screw dislocations

$$(57) \qquad B = \frac{b^2 \eta_s}{8 \pi a_0^2} ,$$

where η_s is the shear phonon viscosity of eq. (8), and a_0 is a cut-off radius below which the strain due to the dislocation becomes so large that the concept of a phonon as an elastic wave no longer is valid. On the basis that the energy of the third-order terms are in the order of 20 percent of the energy of the second-order terms, the result for copper is $a_0 = 3b/4$. From eq. (8) the temperature variation of the phonon viscosity depends on the product $E_0 K/\varrho C_v$. For a number of materials for which the thermal conductivity varies as $1/T$ the value decreases as the temperature decreases, but for some materials such as lithium fluoride [46] for which the thermal conductivity increases as the temperature decreases, with a factor greater than $1/T$, the drag coefficient remains constant with temperature. For a metal the same expression holds, but one has to use the thermal conductivity due to the lattice rather than the total thermal conductivity which is mostly due to electrons. This follows since the interchange of energy stored in the lattice by the suddenly applied strain has to be equilibrated by means of the lattice thermal conductivity. The electronic and lattice thermal conductivity can be separated by means of thermal-conductivity measurements for crystal swith differing amounts of impurity content. This seems to have been done only for copper [47], with the result that the lattice thermal conductivity is inversely proportional to the temperature.

Various calculations have been given for the line tension T [48], which indicate that the screw dislocation is stiffer than the edge dislocation. As an approximation, the tension is taken to be $\mu b^2/2$. Solutions of the string equation have been widely discussed [49] and will not be repeated here. Two types of solutions, neglecting the mass M, have been obtained [50]. These are for an equal spacing of the impurities along the dislocation line and for an exponential distribution of the form

$$(58) \qquad N(l)\,\mathrm{d}l = \frac{\bar{N}}{l_A^2}\exp[-l/l_A]\mathrm{d}l\,,$$

where $\bar{N}$ is the total number of dislocations per cm², l_A is the average loop length and $N(l)\,\mathrm{d}l$ the number of loops lying between l and $l+\mathrm{d}l$. The results for the two cases are shown by Fig. 20. These are plotted in terms of a normalized frequency

$$(59) \qquad \omega_0 = \frac{\mu b^2}{B l_A^2}$$

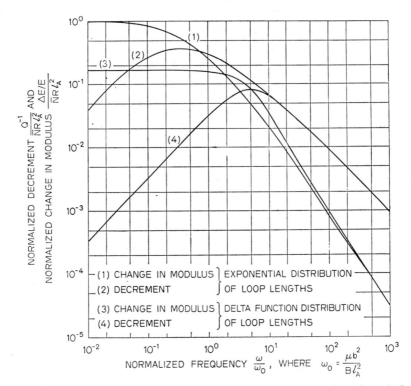

Fig. 20. – Normalized internal-friction and modulus change as a function of the normalized frequency (after OEN, HOLMES and ROBINSON).

and a normalized internal-friction and modulus change

(60) $$\frac{Q^{-1}}{\bar{N}Rl_A^2} \quad \text{and} \quad \frac{\Delta E/E}{\bar{N}Rl_A^2}.$$

Here ΔE is the change in elastic modulus due to dislocations, E the complete elastic and dislocation modulus and R is an orientation factor which determines the ratio of the stress in the glide plane to the applied stress.

At high frequencies the internal friction decreases inversely proportional to the frequency with a value determined by eq. (34). By measuring the limiting value of the product ωQ^{-1} at high frequencies and the number of dislocations per cm^2 $\bar{N}$ by etch pattern techniques an evaluation of the drag coefficient B can be made. For copper the results vary from 2 to $8 \cdot 10^{-4}$ dyn cm/s^2. Other measurements by the increase in velocity of single dislocations as a function of the applied stress—measured by the introduction of a stress pulse with displacements measured by etch patterns—give results in the same order of magnitude.

It has been shown [51] that if one neglects the low-frequency source of dissipation inherent in the motion of kinks and assumes that the kinks move freely over the kink barriers—which will be true for all temperatures above 5 °K, as can be seen from eq. (50)—the kink formulation gives results very similar to the string model. If the dislocations lie mostly in Peierls-Nabarro troughs as appears to be the case for freshly generated dislocations, thermal agitation can generate double kinks which give rise to an attenuation peak known as the Bordoni peak—which has an activation energy in the order of 0.1 eV, *i.e.* much higher than the kink barrier. For undeformed crystals the Bordoni peak does not appear and all the dislocation motion is by the motion of geometrical kinks. In any case these form the background attenuation for all crystals.

The motion of geometrical kinks is controlled by the mass of the kink, the damping of the kinks and the interaction force between the kinks. By writing out a series of partial differential equations for each kink and neglecting interactions other than nearest-neighbor interactions, SEEGER and SCHILLER [48] and SUZAKI and ELBAUM [51] have shown that the principal results obtained for the string equation may also be derived from the kink picture apart from slight differences in the numerical values of the parameters. Slight differences occur in the onset of nonlinearity which will not be discussed here. However the kink model predicts a source of dissipation connected with the number of kink displacements, whereas the string model does not. This results in a « low frequency » component of internal friction with an internal friction Q^{-1} independent of the frequency. This source is discussed in the next Subsection.

4˙4.3. Low-frequency loss connected with kink motion. As discussed in the Introduction the height of the kink barrier is large enough to require thermal energy to overcome the barrier to the motion. The rate equation is given by (50) and the equilibrium number of kinks is given by (53). This number of kinks is under the assumption that the only interaction is the pushing together of adjacent kinks by the energy acquired from the negative slope of the potential barrier. Since the string model gives essentially the same result as the kink model—when the displacement dissipation is neglected—we can obtain an equivalent string model by equating the area swept out by the kinks to the area determined by the displacement of the dislocations assumed initially to be straight lines. For the low-frequency case considered here we can neglect damping and inertial terms. For $N(l)$ dislocations of length l, we can set

$$(61) \qquad\qquad nab = \frac{\tau}{\sigma_k} n_0 ab = N(l) \bar{x} l,$$

where n is the number of kink jumps, a the kink height, n_0 the total number of kinks and $\bar{x}$ the average displacement of a string of length l. With M and B set equal to zero and with a loop of length l pinned on both ends, a solution of eq. (55) is

$$(62) \qquad\qquad x = \frac{\tau}{\mu b} [ly - y^2],$$

where y is the distance from one pin and l the total distance between pins. Hence the average displacement $\bar{x}$ is

$$(63) \qquad\qquad \bar{x} = \frac{1}{l} \int\limits_0^l x \, \mathrm{d}x = \frac{\tau l^2}{6\mu b}.$$

If we put this value into (61) and equate the stress τ applied to the kink to the stress τ applied to the loop, we have

$$(64) \qquad\qquad n_0 = \frac{N(l)\sigma_k l^3}{6\mu ab^2}.$$

When the kink crosses the barrier under an applied stress and thermal agitation, it stores an energy $\sigma_k ab^2/2$ from the negative slope of the kink barrier. Hence the energy stored in obtaining the maximum extension of the dislocation is

$$(65) \qquad\qquad \tfrac{1}{2} n_0 \sigma_k ab^2 = \frac{N(l)\sigma_k^2 l^3}{12\mu}.$$

The energy stored in the crystal is

$$(66) \qquad\qquad \frac{\sigma_k^2}{2\mu'} = \frac{\sigma_k^2}{2\mu^E[1 - \Delta\mu/\mu]} \doteq \frac{\sigma_k^2}{2\mu^E}\left[1 + \frac{\Delta\mu}{\mu}\right],$$

where μ' is the elastic constant including the effect of dislocations and μ^E is the elastic constant without dislocation effects. Hence

$$(67) \qquad \frac{\Delta\mu}{\mu} = \frac{N(l)l^3}{6},$$

a well-known result for the string model as shown by Fig. 20 with the orientation factor R equal to unity.

This derivation does not take account of the dissipation connected with the motion over the kink barriers which, as shown by eq. (52), results in an internal friction proportional to β and $\Delta\mu/\mu$. For sinusoidal motion this effect can be obtained by applying a stress with a dissipation component

$$(68) \qquad \tau^* = \tau(1-j\beta)$$

to the string model. Then the average displacement becomes

$$(69) \qquad \bar{x} = \frac{\tau(1-j\beta)l^2}{6\mu b}.$$

To obtain the effect of this term we have that the total strain S_s consists of

$$(70) \qquad S_s = S_s^E + N(l)\bar{x}lb = S_s^E + \frac{N(l)\tau(1-j\beta)l^3}{6\mu}.$$

Dividing through by the applied shear stress τ, we have

$$(71) \qquad \frac{1}{\mu'} = \frac{1}{\mu^E} + \frac{N(l)l^3}{6\mu}(1-j\beta),$$

where μ' is the complex shearing modulus. Since the last term is small compared to unity

$$(72) \qquad \mu'[1+jQ^{-1}] = \mu\left[1 - \frac{N(l)l^3}{6}(1-j\beta)\right].$$

Separating out the real and imaginary parts we have

$$(73) \qquad \frac{\mu-\mu'}{\mu} = \frac{\Delta\mu}{\mu} = \frac{N(l)l^3}{6}, \qquad Q^{-1} = \frac{\Delta\mu}{\mu}\beta.$$

Hence the use of a complex stress reproduces for sinusoidal vibrations the same considerations as given in [44] since $\Delta S/S = \Delta\mu/\mu$.

The effect of the acceleration and drag terms can be obtained by applying the stress term τ^* to the complete string model equation (55). Since the method is similar to the case already discussed, the results obtained by neglecting M can be written as

(74) $$\frac{\Delta\mu}{\mu} = \frac{N(l)\,l^3}{6[1 + (\omega l^2 B/6\mu b^2)^2]}, \qquad Q^{-1} = \frac{\Delta\mu}{\mu}\left[\beta + \frac{\omega l^2 B}{6\mu b^2}\right].$$

For any other type of stress, such as a longitudinal vibration, the same solution holds if we introduce an orientation factor R which relates the average shear stress in the glide plane to the applied stress. This results in

(75) $$\frac{\Delta E}{E} = \frac{N(l)\,Rl^3}{6[1 + (\omega l^2 B/6\mu b^2)^2]}, \qquad Q^{-1} = \frac{\Delta E}{E}\left[\beta + \frac{\omega l^2 B}{6\mu b^2}\right].$$

For a distribution of loop lengths of the exponential type one can use the calculations shown by Fig. 20. Since the modulus defect is not affected by the low-frequency component, the curve labelled (1) is still valid. For the internal friction eq. (75) shows the there are two parts, one proportional to $\Delta E/E$ and the other part which is given by the curve labelled (2). Since β has been evaluated as being equal to 0.03 for an alloy for which the drag coefficient term is negligible, the complete internal-friction curve will be given by adding 0.03 times curve (1) to curve (2). The cross-over point occurs when $\omega/\omega_0 = 7.8 \cdot 10^{-3}$. Below this frequency the internal friction will be practically independent of the frequency, while above this frequency it will follow curve (2) of Fig. 20, as shown by Fig. 21.

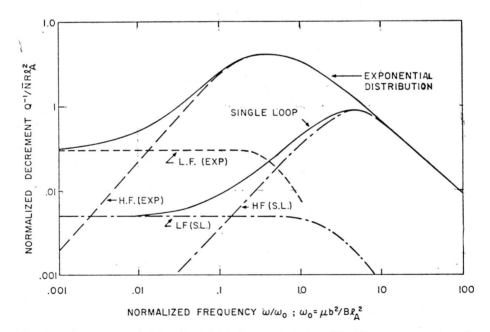

Fig. 21. – Combination of low- and high-frequency internal-friction components under the assumption that $\beta = 0.03$. Both single loop and an exponential distribution are shown.

For very pure copper GRANATO and STERN [52] have shown that the internal friction at 15 kHz is of the order of the high-frequency component. This requires an average loop length of $1.6 \cdot 10^{-4}$ cm or larger, not an unreasonable value. There is considerable evidence for the low-frequency component in the work of WEGEL and WALTHER [53], ROUTBORT and SACK [54] and MASON and WEHR [43]. For a metal the only measurements found that cover both low- and high-frequency ranges are measurements on annealed 30-70 brass [55]. As shown by Fig. 22, measurements have been made from 10^5 Hz to 10^6 Hz,

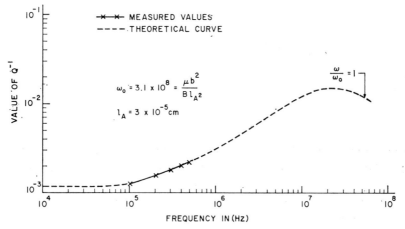

Fig. 22. – Measured attenuation coefficient for brass tube 10 cm in diameter (after SAVAGE and HASEGAWA).

which show a rise in internal friction. If we fit the curve of Fig. 21 to these measurements, the dashed curve of Fig. 22 results. This indicates a value of $\omega/\omega_0 = 1$ at $4.95 \cdot 10^7$ Hz giving $\omega_0 = 3.1 \cdot 10^8$. As shown on the Figure, with $\mu = 4.6 \cdot 10^{11}$ dyn/cm², $b = 2.55 \cdot 10^{-8}$ cm, $B = 8 \cdot 10^{-4}$ (similar to the value for copper) we find that l_A in of the order of $3 \cdot 10^{-5}$ cm, which is a reasonable value for an alloy. A check on this value is found from the high-amplitude internal-friction measurements discussed in Subsect. 4'4.6.

4'4.4. Internal friction in rocks. A more complete confirmation of the two sources of internal friction present for metals and rocks is given by the measurements made for three rocks taken over a frequency range from 32 kHz to 5 MHz.

The first measurements were made on Westerley granite [56] by two techniques. The low-frequency measurements were made by cementing two PZT-4 ceramics, having a resonant frequency of 64.4 kHz, to a block of Westerley granite adjusted to the same resonant frequency. The results are shown by Fig. 23. By measuring the Q of the composite resonator and the two drivers

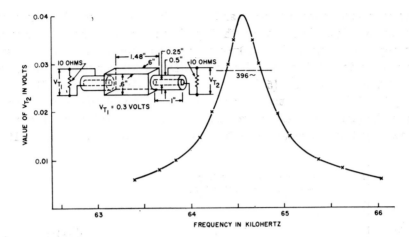

Fig. 23. – Response of two PZT transducers driving a half-wave section of Westerley granite.

separately, the internal friction Q^{-1} can be determined and is plotted in Fig. 24. Using different-length transducers, three frequencies were measured in this manner. Pulsing methods were used up to 2 MHz. Above this the grain structure of the rock was large enough to cause scattering.

Two other rocks with smaller grain size were also measured. Figure 25 shows measurements of Pennsylvania slate [57] which has a grain size of about

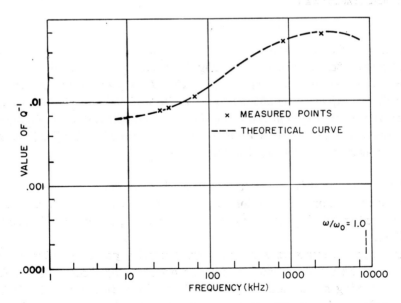

Fig. 24. – Measurement of Q^{-1} of fine-grained Westerley granite over a wide frequency range (after MASON, BESHERS and KUO).

$4 \cdot 10^{-4}$ cm, which is small enough not to cause scattering losses up to 15 MHz. The shape of this curve is very similar to that of Fig. 21. The most complete measurements have been made on Solenhofen limestone [58] which has a grain

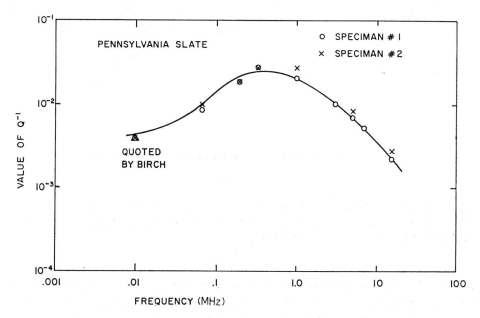

Fig. 25. – Internal friction for Pennsylvania slate. Points show measurements of several samples measured by low- and high-frequency measurements. Circles show specimen 1 and crosses specimen 2 (after MASON and KUO).

size of $9 \cdot 10^{-4}$ cm. The squares and crosses of Fig. 26 show measurements by resonance and pulsing methods. The two techniques have covered the same frequency range with similar results. The dashed line is a replot of the theoretical curve of Fig. 21. Up to 5 MHz the agreement is good. Other measurements are also plotted in the Figure. Hence all three rocks give good agreement with the theoretical curve which considers the two sources of internal friction. Most seismic measurements are made at frequencies less than a Hz, and for these measurements only the low-frequency source of internal friction is operative and the internal friction is independent of the frequency.

Besides the dislocation mechanism considered here two other mechanisms have received consideration for producing an internal friction independent of the frequency. The first mechanism considered was a distribution of relaxation times placed so that the peaks of the relaxation times produced a Q independent of the frequency. This mechanism has been discussed by KNOPOFF [59] and has been shown to be unlikely.

Another suggestion [60] for an internal friction of this type was that it was due to the Coulomb friction of rock grains rubbing on each other. A theoretical

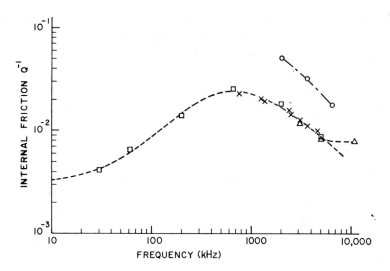

Fig. 26. – Internal friction for Solenhofen limestone. o: KRISHNAMURTHI and BALA-
KRISHNA; ▵: PESELNICK and ZEITZ; ◻: continuous-wave measurement; ×: pulsing
measurement; dashed line: theoretical.

solution for the tangential force-displacement curve for two spherical lenses
pressed together by a normal force was first given by MINDLIN [61]. The elastic
solution indicated that slip should occur over a circular ring and that it should
produce a hysteresis loop whose area should vary as the cube of the amplitude
of the motion. The ring was first verified by MINDLIN, MASON, OSMER and
DERESIEWICZ [62] for amplitudes near that of gross slide. However, as the
amplitude decreased, the area went from a cube law to a square law which corresponded to a Q independent of the amplitude. This is indicative of a plastic
flow due to dislocations. Similar results were also obtained by JOHNSON [63]
for the motion of a steel ball on steel flats.

Since the internal friction in the experiments of MINDLIN *et al.* except at the
highest amplitudes was due to plastic flow, it is probable that the internal friction of rocks at low amplitude is not due to Coulomb friction. In any case the
variation of Q^{-1} with frequency found in Fig. 24 to 26 is not in agreement with
the Coulomb friction model.

Discussion of the measurements. The constant value of Q^{-1} at low frequencies
followed by an increasing and then a decreasing value of internal friction found
for all three rocks is consistent with a dislocation mechanism. By matching
the measured curves to the theoretical curves it is found that there is a good
agreement when $\omega/\omega_0 = 1$ at $6.0 \cdot 10^6$ for Pennsylvania slate and $9 \cdot 10^6$ for
Solenhofen limestone. Setting

(59) $$\omega_0 = \mu b^2 / B l_A^2$$

and noting that

(60) $$Q^{-1}/\bar{N}Rl_A^2 = 0.37$$

at the top of the curve, we obtain a method for eliminating l_A^2. Then we have

(76) $$\frac{\bar{N}}{B} = \frac{Q^{-1}\omega_0}{0.37R\mu b^2}.$$

For these rocks the shear moduli are $2.2 \cdot 10^{11}$ dyn/cm² (P.S.) and $3.25 \cdot 10^{11}$ (S.L.). b for many materials is $3 \cdot 10^{-8}$ cm, and R the orientation factor is often found to be 0.25 for longitudinal waves. Hence

(77) $$\frac{\bar{N}}{B} = 8 \cdot 10^9 \text{ (P.S.)} \quad \text{and} \quad 8.3 \cdot 10^9 \text{ (S.L.)}.$$

For a single-crystal metal the number of dislocations is determined by an etching process or by introducing a known amount by compressing the material and determining the increase in attenuation and the change in velocity. These processes cannot be followed for a rock since it would crumble before slip occurred. Hence to obtain any numerical values one has to assume a value for $\bar{N}$ or B. The measured range of B is much less than that of $\bar{N}$. The best determination of B is for copper ($8 \cdot 10^{-4}$ dyn·s/cm²). Using a round figure of $B = 10^{-3}$ dyn·s/cm² we find

(78) $$\bar{N} = 8 \cdot 10^6 \text{ (P.S.)} \quad \text{and} \quad 8.3 \cdot 10^6 \text{ (S.L.)},$$

and from (62)

(79) $$l_A = 1.82 \cdot 10^{-4} \text{ cm (P.S.)} \quad \text{and} \quad 1.75 \cdot 10^{-4} \text{ cm (S.L.)}.$$

These values are consistent with rather pure materials rather than alloys such as brass which have a smaller average length l_A and higher dislocation densities. As will be shown here, these values are associated with the numbers expected for free surfaces where the loop length is effectively longer and $\bar{N}$ smaller than for a continuous medium.

For a rock most of the internal friction is connected with the grain boundaries and cracks, as can be seen from the fact that hydrostatic pressure reduces the internal friction by a factor of 10 or more [60, 64]. A model is used here for which a grain boundary is represented by a series of spherical surfaces having a shearing modulus $\mu(1 + j/Q_1)$ pressing on each other, while the interior is represented by a solid having a stiffness $\mu(1 + j/Q_2)$. Making use of the Hertz theory of contacts and Mindlin's [61] calculations of the displacement of a set of convex surfaces, for an applied alternating force T we find for the sidewise displacement dx_2

(80)
$$dx_2 = \frac{(2-\sigma)\,T}{8a\mu(1+j/Q_1)}.$$

The tangential force T is related to the applied shearing stress T_6 by

(81)
$$T_6 = T(\pi a^2)\,n_0,$$

where a is the Hertz radius of contact and n_0 the total number of contacts per cm². The Hertz radius of contact is determined by

(82)
$$a = \sqrt[3]{\frac{3}{4}\,Nr\,\frac{1-\sigma}{u}},$$

where r is the radius of the spherical contact surfaces and N the normal force on each contact. The product Nr can be eliminated by considering the thickness t_1 of the layer $r\theta^2$, where θ is half the angle determined by the radius of curvature r. Then

(83)
$$Nr = T_1/n_0\pi t_1,$$

where T_1 is the sum of the pressure holding the rock together $T_{1_\bullet}$ plus the applied hydrostatic pressure. Introducing all these values the sidewise displacement is

(84)
$$dx_2 = \frac{2-\sigma}{6(1-\sigma)}\,\frac{T_6\,t_1}{T_1(1+j/Q_1)}.$$

To the displacement of the intermediate layer we have to add that in the two outside layers which results in

(85)
$$dx_1 = \frac{T_6\,t_2}{\mu(1+j/Q_2)}.$$

If we make use of the fact that Q_1^{-1} and Q_2^{-1} are small, the ratio between the applied shearing stress T_6 and the shearing strain

$$S_6 = \frac{2(dx_1 + dx_2)}{t_1 + t_2}$$

is

(86)
$$\mu_T = \mu(1+t_1/t_2)/(1+A), \qquad Q_T^{-1} = Q_2^{-1} - \frac{A(Q_2^{-1} - Q_1^{-1})}{1+A},$$

where

$$A = \frac{2-\sigma}{6(1-\sigma)}\,\frac{\mu}{T_1}\,\frac{t_1}{t_2}.$$

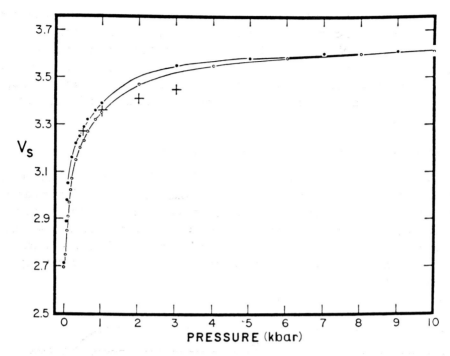

Fig. 27. – Shear stiffness of Westerley granite (after SIMMONS).

By means of measurements [64] of the shear stiffness of Westerley gran-
ite (Fig. 27) it is found that $T_{1_0} = 3 \cdot 10^8$ dyn·cm^{-2}, $t_1/t_2 = 1.7 \cdot 10^{-3}$ and
$\mu = 3.57 \cdot 10^{11}$ dyn·cm^{-2}. Hence for no external pressure the internal friction is

$$(87) \qquad\qquad Q_T^{-1} = 0.545 Q_2^{-1} + 0.455 Q_1^{-1},$$

while for high pressures it is equal to Q_2^{-1}, the value for the medium. By means
of the above values and Q_1^{-1} equal to $1.1 \cdot 10^{-2}$ to agree with the torsional wave
measurements of BIRCH and BANCROFT [64] on Rockport granite ($550 \cdot 10^{-5}$ at
200 bar and $60 \cdot 10^{-5}$ at 4 kbar) the value due to the boundary is less than $1.2 \cdot 10^{-4}$
at 10 kbar, while Q_2^{-1}, the internal friction of rocks without flaws, is less than
$2 \cdot 10^{-4}$. Hence granite under pressure has internal-friction values of the order
of that found for Moon rocks, *i.e.* 1000 to 3000 [65]. Recently TITTMANN [65]
has shown that the amount of moisture trapped in the rock has a large
effect on the internal friction. By evaluating specimens to 10^{-7} Torr and
outgassing them, Q's as high as 1500 have been obtained on terrestrial basalt.
Since viscosity effects of the absorbed water would give a Q^{-1} proportional to
the square of the frequency, it appears that the effect of the moisture is to increase
the number of cracks and decrease the area of contact. Both of these effects
would increase Q_1^{-1} and decrease T_{1_0} and hence increase the value of A in eq. (86).

Q_1^{-1} will increase when the number of cracks increases, since its value is proportional to the number of contacting surfaces per unit volume. Also since the dislocations are localized on the surface, moisture may increase the value of β in eq. (74). A decrease in T_{1_0} and an increase in t_1/t_2—which will occur for a layer of moisture—will increase A which tends to cause Q_T^{-1} to approach Q_1^{-1}. Eliminating the moisture would increase the Q to values found for high hydrostatic pressures.

4'4.5. Relaxations due to dislocations.

4'4.5.1. Bordoni peak. – The types of internal friction discussed in the previous Sections are connected with the motion of geometrical kinks across energy barriers and with the viscous drag of the kinks due to phonons and electrons. When a material is compressed, new dislocations are introduced which lie mostly in the Peierls-Nabarro troughs. Under these conditions it is possible to have double-kink generation, by thermal agitation, as shown by Fig. 28.

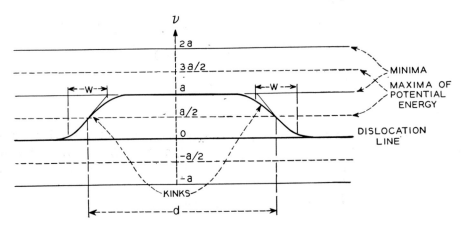

Fig. 28. – Double kinks which form the basis for the Bordoni peak.

This type of a kink motion is accompanied by a relaxion peak which occurs at a given temperature for a given frequency. The first measurements of such peaks were made by BORDONI [66], with the results shown by Fig. 29. Many measurements have been made since, particularly for deformed and annealed copper. One of the results which shows the effect of deforming and annealing copper, due to MECS and NOWICK [67], is shown by Fig. 30. The Bordoni peak is accompanied by a satellite peak, occurring at a lower temperatue, which is known as the Niblett-Wilks peak. Both of these peaks, as well as the background, decrease for successive anneals. This suggests the close relation between the two peaks. The Figure also shows a third peak P_1, which has been called the Hasiguti peak [68] after its discoverer.

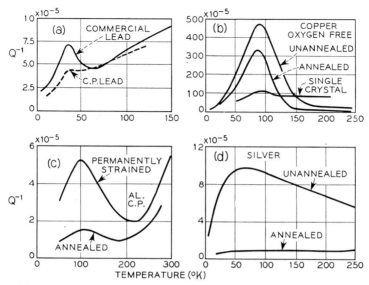

Fig. 29. – Attenuation peaks at low temperatures for four face-centered metals (after BORDONI).

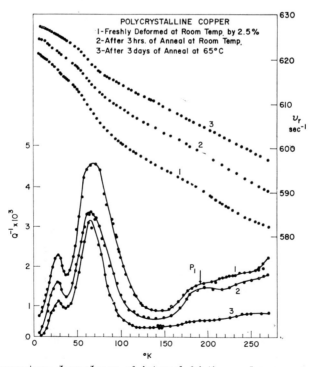

Fig. 30. – Temperature dependence of internal friction and resonant frequency for deformed polycrystalline copper immediately after deformation by 2.5% (curve 1) and after recovery anneals (curve 2: after 3 h at room temperature and curve 3: after 3 days anneal at 65 °C). (From MECS and NOWICK.)

As one varies the frequency, the temperature positions of all these peaks change. By plotting the logarithm of the peak frequency against $1/T$ an Arrhenius-type plot is obtained. For the Bordoni peak, for pure copper, the activation energy is about 0.14 eV and the frequency factor is about $4 \cdot 10^{12}$ Hz. Since this is lower than the attempt frequency for a point imperfection—about $1.6 \cdot 10^{13}$ Hz from eq. (41)—it is thought that a larger imperfection, such as a dislocation kink, is involved in the relaxation. The peak below the Bordoni peak has the same attempt frequency but an activation energy of about $(0.04 \div$ $\div 0.05)$ eV. Both peaks are wider than a single-relaxation peak. The height and position are practically independent of the strain amplitude in contrast with what one finds for the internal friction for geometrical kinks. Bordoni peaks have been found for a number of face-centered metals. For body-centered and hexagonal close-packed metals, peaks have been obtained but the identification as Bordoni peaks is rather elusive.

The most widely accepted theories of the Bordoni effect consider that the peak is connected with the thermal generation of double kinks, as shown by Fig. 28, but modified by the presence of internal stresses in the material and by the diffusion of the kinks. The kinks have to be separated by a critical distance d_{cr}. For a separation closer than d_{cr} the pair will attract each other and will rapidly come together and be annihilated, while for distances greater than d_{cr} the applied stress will drive the two kinks apart, thus producing a displacement of the entire dislocation segment by one interatomic distance.

In his first treatment SEEGER [69] took the activation energy to be twice the kink energy W_k or

$$(88) \qquad\qquad 2W_k = 4a^{\frac{3}{2}} b^{\frac{1}{2}} \sqrt{\frac{\mu \sigma_P}{\pi}},$$

where a is the separation of the Peierls-Nabarro troughs, and σ_P is the Peierls-Nabarro stress. The theory has been extended to include the effect of the applied stress on the activation energy, but according to a criticism presented by LOTHE [70] the activation energy of eq. (88) is still nearly correct if the ratio of the applied stress σ to σ_P is not too large. If we use eq. (88), the values of σ_P/μ range from 3 to $10 \cdot 10^{-4}$, which is consistent with theoretical estimates. The lower-temperature relaxation peak results from the fact that the two types of dislocations may lie along a $\langle 110 \rangle$ close-packed direction, namely a pure screw and those for which the Burgers vector lies at $\pm 60°$ to the direction of the dislocation line. The latter will have a lower relaxation energy.

Seeger's theory presents a reasonable qualitative picture but does not explain the breadth of the relaxation peak which is much wider than a single relaxation. PARÉ [71] has modified the theory to take account of the effect of internal stresses existing in the material. These have the effect of widening the activation energy distribution and explaining why more than half the dislocations enter into the effect since the stress serves to sweep geometrical kinks out to the ends of the dislocation loops leaving a segment in the center which in fact does lie parallel to the Peierls-Nabarro valley. Strong support for the Seeger-Paré model comes from the work of ALEFELD *et al.* [72] that the Bordoni peak can be observed in slightly cold-worked materal—for which there would be no appreciable peak in the low-amplitude internal-friction experiment— when a high static stress is used in addition to a small alternating stress.

The Hasiguti-type peaks, shown by P_1 of Fig. 30, and other peaks at higher temperatures, have been shown to be due to the interaction of impurities with dislocations. A number of different theories have been given but none is generally accepted. A connection between the Bordoni peak and the Hasiguti peak is shown by the correlation between the flow strain [79] and the height of both the Bordoni peak and the Hasiguti peak as shown by Fig. 31. This confirms the Seeger-Paré theory for the Bordoni peak and shows that dislocations as well as impurities are involved in the Hasiguti peak.

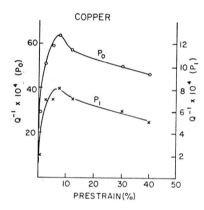

Fig. 31. – Height of Bordoni (P_0) and Hasiguti (P_1) peaks for copper *vs.* prestrain (after DE FOUQUET *et al.*).

4˙4.5.2. Simpson-Sosin peak. – Recently another relaxation peak has been found which is associated with the geometrical kinks rather than the double-kink generation of the Bordoni peak. This peak, known as the Simpson-Sosin peak [74], is connected with the effects of irradiation by 1.0 MeV electrons on polycrystalline copper. The irradiation produces internal-friction and elastic-modulus changes. Since the frequency of measurement is around 500 Hz, the internal friction should be in the low-frequency range for which the internal friction is proportional to the modulus change, in agreement with the measured results. Figure 32 shows two curves showing the ratio of the internal friction to the initial internal friction for 99.999 pure polycrystalline copper plotted as a function of the irradiation time. The initial decrement $\delta_0 = \pi Q^{-1}$ is determined by eliminating dislocation effects by holding the sample at 100 °C for 10 minutes after an electron irradiation. This thermal

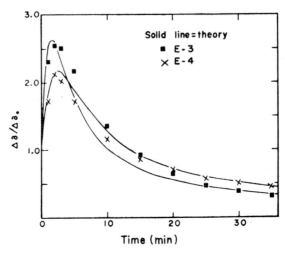

Fig. 32. – Change in internal friction of copper as a function of the radiation time of 10^6 MeV electrons (after SIMPSON and SOSIN).

procedure is sufficient to guarantee that the irradiation-produced point defects arrive in sufficient numbers to eliminate essentially all dislocation contribution to internal-friction and modulus values. The internal-friction and modulus changes are then determined by observing the changes from the initial values. For the two samples E-3 and E-4 Fig. 32 the internal-friction and modulus change—after annealing the samples to 500° and 600 °C—are for zero irradiation:

$$
(89) \quad
\begin{cases}
\begin{array}{lcccc}
 & \delta_0 & Q_0^{-1} & (\Delta E/E)_0 & Q_0^{-1}/(\Delta E/E)_0 \\
\text{E-3} & 0.00359 & 0.00113 & 0.047 & 0.024 \\
\text{E-4} & 0.0026 & 0.000825 & 0.033 & 0.025
\end{array}
\end{cases}
$$

Since the measurements were made at 500 Hz, the unirradiated values should satisfy the ratio $Q_0^{-1} = (\Delta E/E)\beta$ of eq. (75). The last column gives the ratio for β as being about 0.025 in fair agreement with the value of 0.03 found for a titanium alloy.

After irradiation the internal friction first rises and then falls. SIMPSON and SOSIN ascribe the rise as being due to a very large drag coefficient produced by the radiation-produced impurity. Since the effect is to increase the internal friction in a ratio of 2.0 to 2.5 times the initial value, an idea of the drag coefficient can be obtained by substituting typical values for $l = 10^{-4}$ cm, $\mu = 4.6 \cdot 10^{11}$ dyn/cm² and $b = 2.55 \cdot 10^{-8}$ cm. For the E-3 sample with $\beta = 0.024$ we find

$$
(90) \quad \frac{\omega l_B^2}{6\mu b^2} = 1.50 \times 0.024 \quad \text{or} \quad B \doteq 2.0 ,
$$

which is a thousand times larger than that for the dislocation alone. As the irradiation time increases more impurities settle on the dislocation and the length l decreases so that the value in the bracket approaches β. At the same time since $\Delta E/E$ is proportional to l^2 (since $N(l)$ is inversely proportional to l) the modulus defect decreases and the internal friction becomes smaller.

4'4.6. High-amplitude internal friction.

4'4.6.1. Internal friction due to the break-away of dislocations from pinning points. – When the strain amplitude becomes larger, the internal friction is no longer independent of the stress but increases above a limiting value. For example Fig. 33 [75] shows the value of the internal friction Q^{-1} plotted

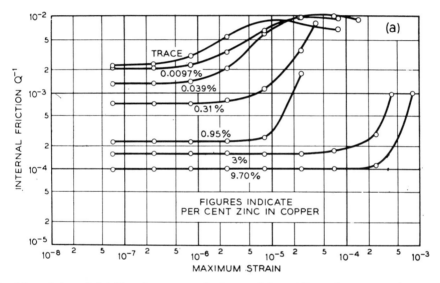

Fig. 33. – Internal friction in copper single crystals, with various amounts of zinc added, as a function of the strain amplitude (after TOKAHASHI).

against the maximum strain for copper with certain percentages of zinc atoms. For traces the internal friction begins to rise for strains as small as $5\cdot10^{-7}$, while for large amounts of zinc atoms in the copper the internal friction is lower and is independent of the amplitude up to strains of $2\cdot10^{-4}$. Figure 34, for 99.995 % pure aluminum single crystal [76], shows that the internal friction is also a function of the temperature.

The first suggestion as to the cause of this amplitude-dependent internal friction was made by GRANATO and LÜCKE [77]. They suggested that the effect was due to the break-away of dislocations from their pinning points as shown by Fig. 19 D and E. As shown by Fig. 35, if we subtract out the linear strain, the dislocation component of strain will be as shown by the successive letters

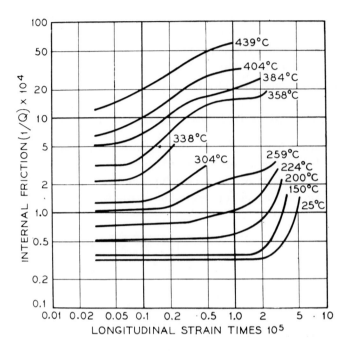

Fig. 34. – Internal friction as a function of temperature and strain for 99.995% pure aluminum single crystal (after BAKER).

a to *f*. Before break-away occurs, the relation between stress and strain is determined by the successive points *a, b, c*. When break-away has occured the dislocation increases suddenly in displacement to the point *d*. Any further displacement increases the line *d, e, f*. When the stress is reversed the dislocation will follow the line *d, e, f* and will be repinned when it reaches the zero-stress position. Hence it traverses a hysteresis loop, shown by the shaded area, which extracts energy from the vibration.

The amount of energy lost, under the assumption that all the impurity pinning lengths l_A and all the Frank-Read distances—*i.e.* the network pinning lengths—are the same, can be calculated as follows. The potential energy in each loop is the force times the displacement or

(91)
$$\tfrac{1}{2} \int_0^{l_A} Fx(y)\, \mathrm{d}y \,,$$

where $x(y)$ is the shape of the displacement curve. This from eq. (55) is

$$x = \frac{\tau}{\mu b}\,(ly - y^2) \,, \qquad \bar{x} = \frac{1}{l} \int_0^l x\, \mathrm{d}x = \frac{\tau l^2}{6\mu b} \,,$$

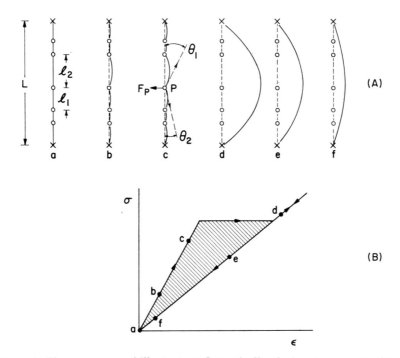

Fig. 35. – *A*) The sequence *a-f* illustrates schematically the break-away of a dislocation segment from a number of pinning points during a loading cycle. *B*) The corresponding stress-strain curve and hysteresis loop. The lettered points match the sequence *a-f* in *A*) (after NOWICK and BERRY).

where $\bar{x}$ is the average displacement. The force is one-half τb since the force goes from zero to the final value τ. Hence the energy stored in each subloop is

$$(92) \qquad \frac{\tau^2 l_A^3}{12\mu} = \frac{S_{13}^2 \mu l_A^3}{12} ,$$

where S_{13} is the shearing strain in the glide plane.

For n loops of length l_A, where $n l_A = l_n$ the network length, this energy is multiplied by n.

When one dislocation is pulled off the pinning point, the resulting loop becomes larger and more force is put on the next pinning point, so that the whole loop will be torn off up to the nodes. Since the full stress is acting on the loop, the total potential energy is equal to

$$(93) \qquad \frac{S_{13}^2 \mu l_n^3}{6} .$$

This energy will be converted into heat and is lost to the motion. We arrive at the same result if we consider the area of the hysteresis loop covered by the sudden displacement as compared to the smooth return, which occurs without interruption until it is pinned again by the impurities. The internal friction for the single-loop length l_n can be determined from (93) if we neglect l_A^3 compared to l_n^3. Since this loss occurs twice for each cycle, the sum over all the loops will be

$$(94) \qquad Q^{-1} = \frac{2\delta W}{2\pi W} = \frac{\bar{N} l_n S_{13n}^2}{6\pi S_{13}^2},$$

since $N_n l_n = \bar{N}$ the number of dislocations per unit volume in the sample. The strain S_{13_n} is required to break the dislocation away from the pinning point. It has been shown by COTTRELL [78] that the energy of pinning a dislocation to an impurity is

$$(95) \qquad U_B = \frac{1}{3} \frac{1+\sigma}{1-\sigma} \frac{\mu b r^3 \varepsilon}{R},$$

where $\varepsilon = (r'-r)/r$, where r is the radius of the solute atom and r' is the radius of the solvent atom, σ is Poisson's ratio, and R is the distance of the dislocation center from the impurity atom. The force with which the solute atom is bound to the dislocation is given by

$$(96) \qquad F = \frac{-\partial U_B}{\partial R} = \frac{+\sqrt{2}}{3} \frac{1+\sigma}{1-\sigma} \frac{\mu b^4 \varepsilon}{R^2}$$

when we introduce the value of $r = b/\sqrt{2}$ for a face-centered crystal. To break the dislocation away from the impurity atom requires a stress

$$(97) \qquad \tau bl = \frac{\sqrt{2}}{3} \frac{1+\sigma}{1-\sigma} \frac{\mu b^4 \varepsilon}{R^2}.$$

Under the assumption that $R = Kr$ or that the impurity atom settles at a distance K atomic radii r from the dislocation, the limiting shearing strain becomes

$$(98) \qquad S_{13_n} = \frac{\tau}{\mu} = \frac{2\sqrt{2}}{3} \frac{1+\sigma}{1-\sigma} \frac{b\varepsilon}{K^2 l_A} \doteq \frac{1.75 b\varepsilon}{K^2 l_A}.$$

Hence the internal friction of eq. (94) is zero until the strain equals the critical strain S_{13_n}. It then increases to the value $\bar{N} l_n^2/6\pi$. Any further increase in strain causes the value to decrease in the ratio of $S_{13_n}^2/S_{13}^2$.

Actually the subloops are not equal but probably follow an exponential law of the type

$$(99) \qquad N(l) = \frac{\bar{N}}{l_A^2} \exp[-l/l_A] \, dl,$$

where l_A is now the average pinned-loop length.

To determine the form that the summation takes with a distribution of this sort, one has to find the distribution in sizes of adjacent subloops of lengths l_1 and l_2 which will produce the most force on a pinning atom and start the break-away process. This problem has been considered by GRANATO and LÜCKE [78] and they have derived an internal friction vs. strain amplitude in the form of an equation

(100)
$$Q^{-1} = \frac{C_1 C_2}{S_{13}} \exp\left[-C_2/S_{13}\right],$$

where

$$C_1 = \beta \left(\frac{2}{\pi}\right)^4 \frac{\bar{N} l_n^3}{l_A}, \qquad C_2 = \frac{\alpha 1.75 \varepsilon b}{K^2 l_A},$$

where α and β are orientation factors determining the ratio of the force along the glide plane to the longitudinal force. These factors will ordinarily have values about $\frac{1}{4}$.

This calculation neglects the effect of thermal agitation which adds to the applied stress. This has the effect of increasing the loop length l_A for which break-away can occur. FRIEDEL [79] and PEQUIN [80] have given a discussion of this case and have shown that the binding energy deduced from plotting the slopes of the inverse of the strain amplitude against the strain amplitude times the internal friction results in reasonable values. As an example [81] Fig. 36 shows a Granato-Lücke plot for the internal-friction values of Fig. 33 for four of the alloys shown. Up to strains of 2 to $5 \cdot 10^{-6}$, depending on the impurity content, the measured values diverge from the straight-line relation.

This indicates the approach of another source of dislocation damping as discussed in Subsect. 4˙4.6.2. The increasing slope as the impurity content increases shows that the constant C_2 of eq. (100) is increasing. This indicates that the average separation l_A between impurity atoms is decreasing. While reasonable values have been obtained from an analysis of C_1 and C_2 constants of eq. (100), there are so many unknowns that no definitive values result.

4˙4.6.2. Internal friction due to unstable Frank-Read loops. – At strains larger than the Granato-Lücke region, the internal friction increases much faster than that due to break-away, as shown by Fig. 36. From Fig. 19, the next dislocation event is the generation of a Frank-Read source which can send a pile of dislocations out to a grain boundary for a polycrystaline metal or out to cross-dislocations or impurity barriers for a single crystal. This effect often occurs at strains nearly the same as that associated with the break-away of dislocations from pinning points and the break-away curve cannot be separated from the plastic curve. For example Fig. 37 shows the internal friction for a brass polycrystalline sample measured at 17.5 kHz by means of the transducer-transformer shown by Fig. 1. The internal friction is constant to strains of 10^{-4} after which it increases with strain. At a strain of $3 \cdot 10^{-4}$ another source of

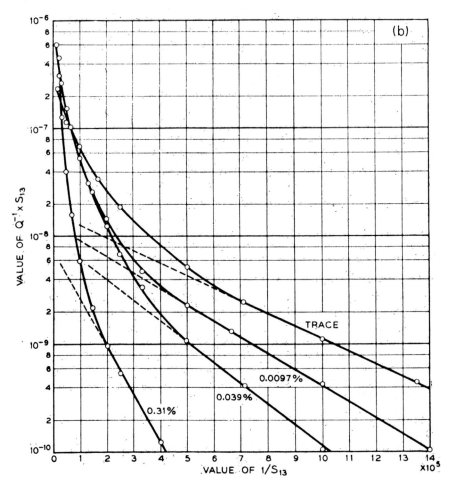

Fig. 36. – Granato-Lücke plot of $\log{(Q^{-1}S_{13})}$ against $1/S_{13}$ for various amounts of zinc content.

internal friction takes over and measured values rise rapidly with strain. It is to be noted that the modulus change $\Delta S/S$—shown by the dashed line—agrees well with the increment of internal-friction change for the high-amplitude effect. It seems that the region from 10^{-4} to $3\cdot10^{-4}$ is predominantly due to break-away loss. Above that the internal friction is due to plastic flow from Frank-Read pile-ups.

It appears that the best mathematical account of this phase has been given by PEQUIN et al. [79] Starting with the well-known expression for the strain velocity, valid for $\varepsilon > \varepsilon_p$,

(101)
$$\dot{\varepsilon} = A_1 \sinh{(B_1\sigma)},$$

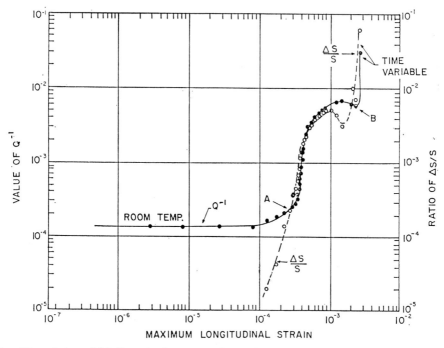

Fig. 37. – Internal friction of α-brass sample, of the form shown by Fig. 1, as a function of the maximum longitudinal strain.

where $\sigma = \sigma_0 - \sigma_p$, $\sigma_0 =$ applied stress, ε_p is the strain connected with the start of the plastic-strain region, $\sigma_p = \mu \varepsilon_p$, where μ is the shear modulus, and A_1 and B_1 are constants and using the equation

$$(102) \qquad Q_p^{-1} = \frac{\Delta W}{2\pi W} = \frac{\mu}{\sigma_0^2} \int_0^T \sigma_0 \dot{\varepsilon} \, dt \, ,$$

where ΔW is the energy loss over the time of a cycle T and W is the maximum energy stored, PEQUIN et al. [79] show that the internal friction due to plastic strain is approximated by the formula

$$(103) \qquad Q_p^{-1} = \frac{2A_1 \mu \Omega}{\sigma_0^2 \omega} \sinh B_1(\sigma_0 - \sigma_p) \, ,$$

where Ω is a constant factor which lies between 0.5 and 1.

Equation (101) is formally identical to the expression obtained if we suppose that the dislocation lines can pass over obstacles with the help of thermal activation [82]. The plastic velocity can be written as

$$(104) \qquad \dot{\varepsilon} = 2\varrho b v_a s \exp \left[- U/kT \right] \sinh \frac{\alpha V}{kT} (\sigma_0 - \sigma_p) \, ,$$

where ϱ is the dislocation density, v_a the natural dislocation frequency in its potential well, s the activation distance, $i.e.$ the distance the dislocation jumps under the influence of thermal and mechanical energy, U the activation energy associated with dislocation jumps, k Boltzmann's constant, T the absolute temperature, α an orientation factor generally taken to be of the order of 0.5 and V the activation volume, $i.e.$ the product of bla, where l is the loop length and a the forward motion of the dislocation per jump.

To conserve the validity of this equation it is assumed that the strain is less than 10^{-3} so that a noticeable variation of the dislocation density does not occur. Also if the value of the hyperbolic sine function is 1.5 or greater, it can be replaced by the exponential function. Then the expression for Q_p^{-1} can be written as

(105)
$$Q_p^{-1} = \frac{A}{2\varepsilon_0 \Omega} \exp\left[B(\varepsilon_0 - \varepsilon_p)\right],$$

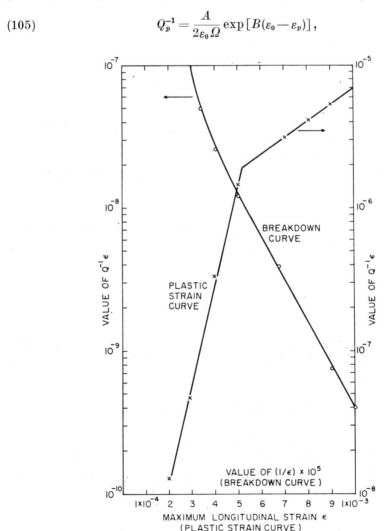

Fig. 38. – Plot of $Q^{-1}\varepsilon$ $vs.$ $1/\varepsilon \cdot 10^5$ (break-away) and $Q^{-1}\varepsilon$ $vs.$ $\varepsilon \cdot 10^{-4}$ for plastic-strain curve.

where

$$A = \frac{4 \varrho b v_a s \exp[-U/kT]}{\pi \omega}, \qquad B = \alpha V \mu / kT .$$

This equation can be written

(106) $$\ln(Q_p^{-1} \varepsilon_0) = B \varepsilon_0 + C ,$$

where $C = \ln(A/2\Omega - B\varepsilon_p)$.

Hence if $\log(Q_p^{-1} \varepsilon_0)$ is plotted against ε_0 in the plastic range, one can expect to obtain a straight line, the slope B of which gives the activation volume V. If we perform this operation for the measurement of Fig. 37, we find that there are two activation volume ranges as shown by Fig. 38. In the first range from $3 \cdot 10^{-4}$ to $5.2 \cdot 10^{-4}$, the activation volume is $3.4 \cdot 10^{-21}$ cm³. Above $\varepsilon = 5.2 \cdot 10^{-4}$ a different slope prevails with an activation volume of $5.5 \cdot 10^{-22}$ cm³. In the lower strain range from $\varepsilon = 10^{-4}$ to $2.5 \cdot 10^{-4}$ the Granato-Lücke plot shows that the internal friction is connected with dislocation break-away. For both types of curves the low-amplitude value of the internal friction has been subtracted from the amplitude-dependent values.

The change in slope above a strain of $5.2 \cdot 10^{-4}$ indicates that a change has occurred for which thermal agitation has a reduced importance. It is believed that the lower range is connected with Frank-Read pile-ups which are limited by grain boundaries. In this region the pile-up returns to its equilibrium position and repeats itself cycle for cycle. In the higher strain range slip occurs within the grain as shown by the optical photograph—Fig. 39a)—of the chemically polished surface [83]. In this region thermal agitation is not as important as direct strain and hence activation volume decreases. At the higher stress levels microcracks are generated as shown by Fig. 39b). These extend through the specimen as has been verified by observation.

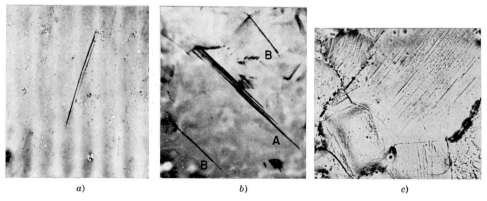

a)   b)   c)

Fig. 39. – a) Isolated slip band in brass peculiar to high-frequency strain. b) Slip band A in brass degenerating into microcrack. Note isolated bands at B. c) Slip bands at low frequencies (room temperature) and at high frequencies at high temperature. Cracks now form in grain boundaries.

The plastic region in aluminum has been extensively studied by Pequin *et al.* [80]. When aluminum is water-quenched from 420° to 600° no break-away region is observed but two plastic ranges are observed. The lower region is the region of depleted loops, while the higher region is ascribed to the interaction of the loops and the moving dislocation. No slip occurs in either region. One material for which the break-away loss and plastic loss can be separated is aluminum with three percent of zinc [80]. Well-defined break-away regions and plastic regions are found. The effects of cold-working, annealing and other variables are also found to be represented in terms of the two types of internal-friction curves.

5. – Fatigue in metals at ultrasonic frequencies.

Considerable work has been done on studying fatigue at ultrasonic frequencies for which a large number of cycles can be obtained in a short time. In fact when a large number of cycles is required, it appears to be the only feasible way of making measurements.

Since most applications are for a much slower rate than given by ultrasonic measurements, the question arises of whether the two rates give similar results. The general opinion is that ultrasonic measurements give higher fatigue stresses than for the ordinary rate of testing. For iron, however, the result is the opposite [84]. For brass, measurements have been made for the complete stress for both low frequencies and high frequencies. Figure 39a) and 39b) show that at ultrasonic frequencies the initial effects are slip bands which degenerate into microcracks for higher strains. These occur only over isolated grains. On the other hand, for low frequencies, as shown by Fig. 39c), slip bands occur throughout the grains.

Similar results can be obtained at high frequencies if the temperature of the specimen is raised. Hence it appears that at high frequencies and room temperatures, slip occurs on single planes, and cannot spread over the grain. For higher temperatures dislocations can climb out of their slip planes and spread over the whole grain as shown by Fig. 39c). This can happen at room temperature, at the low-frequency rate, since climb is a thermally activated process which requires an incubation time. For the high frequencies the time of the oscillation is smaller than the incubation time and climb does not occur. For the high-frequency measurements fatigue occurs by isolated slip bands joining up, as shown by Fig. 40a), and microcracks become macrocracks. On the other hand, for higher temperature ultrasonic frequencies, of for low-frequency measurements, as shown by Fig. 40b), fatigue occurs in the grain boundaries. The strain for ultrasonic frequencies is slightly higher than that for lower frequencies. Fatigue is also accompanied by a very rapid increase in internal friction and modulus change as shown by Fig. 37. For strains greater than $2 \cdot 10^{-3}$

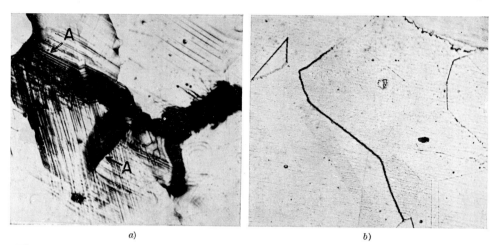

Fig. 40. – *a*) Joining-up of several microcracks to form a fatigue failure. *b*) Section of high-temperature brass specimen confirms grain-boundary cracking.

both of these quantities become time variable, and fatigue occurs in a matter of minutes. Fatigue can also occur at lower amplitudes for a large number of cycles, *i.e.* greater than 10^8.

A number of metals has been investigated by ultrasonic techniques [85]. Generally the results are similar for other face-centered material, but for the

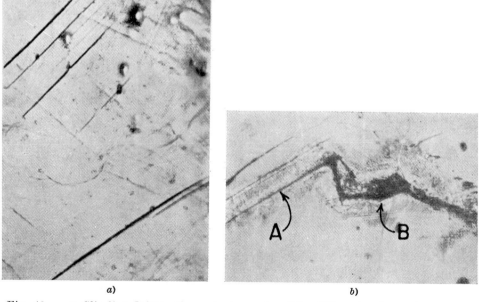

Fig. 41. – *a*) Slip-line deformation only in produced by HF cycles, here after amplitude as small as $2 \cdot 10^{-4}$ (500×). *b*) HF slip line A starts microcrack B which results in fatigue in iron at strains as small as $4 \cdot 10^{-4}$.

body-centered material iron the high-frequency strain is less than that for the low-frequency values [85]. For high-frequency measurements the first effect is slip in single grains, as shown by Fig. 41a), which may occur at strains as small as $2 \cdot 10^{-4}$. The grain slip can join up, as shown by Fig. 41b), at strains as low as $4 \cdot 10^{-4}$. On the other hand, for low frequencies the effect of the strain is taken up by deformation in the boundary and it is not till strains of $13 \cdot 10^{-4}$ are reached that fatigue occurs. Hence the ultrasonic vibrations are much more damaging than the lower-frequency vibrations.

The damaging effect of ultrasonic vibrations in steel is confirmed by the work of WERNER [86], who has studied the pitting of steel railway wheels. Ultrasonic vibrations, generated by the impacts between wheel and rail, have been measured in the range above 35 kHz. The pitting point distribution appears to be connected with ultrasonic Rayleigh waves, transmitted around the perifery, setting up standing waves at certain intervals connected with the speed of motion. WERNER concludes that « this leaves hardly any doubt that the periodic pittings on the wheel are in fact fatigue phenomena due to ultrasonics ».

A different type of fatigue mechanism is found for titanium [87] and some of the high-strength materials such as the nickel-base superalloys [88]. Titanium and titanium alloys such as 90Ti-6Al-4V have an internal-friction curve as shown by Fig. 42. The internal friction is about $4 \cdot 10^{-5}$, independent of the strain up to values of $3 \cdot 10^{-3}$, after which a sudden rise occurs for both the in-

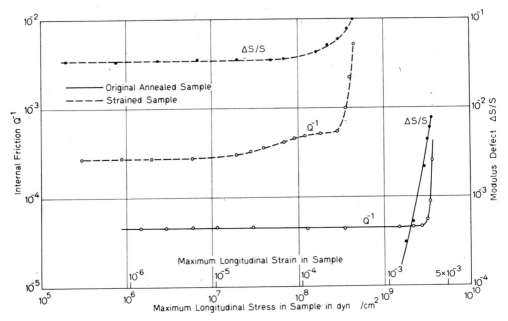

Fig. 42. – Internal-friction curve for 90Ti-6Al-4V alloy as a function of strain. Dashed line shows repeat of sample after break-away of dislocations from pinning points.

ternal-friction and the modulus change. It is believed that dislocations are tightly bound by an atmosphere approximating a Cottrell-type pinning [78]. There is no indication of a plastic-strain internal friction and the increase is due to dislocation break-away. This is confirmed by the measurements shown by the dashed lines, which are a repeat after break-away. The internal-friction and modulus changes are both higher and go into the rapidly rising range at much lower strains.

The fatigue strains are the same for both low and high frequencies, *i.e.* about $3 \cdot 10^{-3}$. This is due to the fact that dislocations break away from pinning atmospheres at this strains and go through their fatigue-producing motions at this strain. For low frequencies this is transgranular slipless shear cracking across grains which turns into a tensile mode of cracking, which then fractures the specimen. For high-frequency vibrations slip-band microcracks extend the length of the grain as shown by Fig. 43a). These microcracks are in the maximum

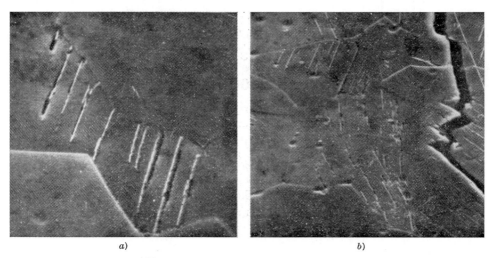

a)   b)

Fig. 43. – a) Slip band microcracks extend through grain in direction of maximum shear. b) Tensile fatigue cracks caused by joining of fatigue cracks (after WOOD and MACDONALD).

shear direction. These cracks extend in adjacent grains until macrocracks form which turn into tensile cracks causing fatigue as shown by Fig. 43b). Titanium specimens have been tested at strain amplitudes of $2 \cdot 10^{-3}$ for $2 \cdot 10^8$ cycles and did not fatigue; and on these specimens no microscopic slip was observed. Thus the specimens exhibited a safe limit for both low and high frequencies.

For nickel-base superalloys, having strengths up to 200000 lbs per square inch up to temperatures as high as 1600 °F, fatigue occurs by crack propagation along slip planes as in titanium. These is a large difference between fatigue

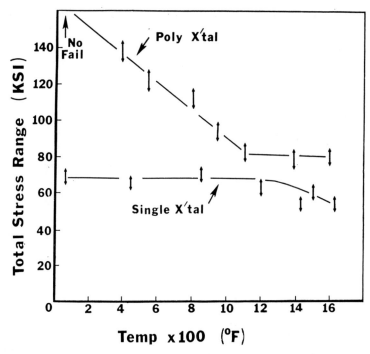

Fig. 44. – $5 \cdot 10^4$ cycle endurance limit of superalloy single crystals (lower curve) and superalloy polycrystals (upper curve) as a function of the temperature (after PURUSHO-THAMAN et al., ref. [87]).

strengths in polycrystalline samples and single crystals as shown by Fig. 44. With an average grain size of 100 μm, the no fail stress is twice as high at room temperature for the polycrystalline form than for the single-crystal form.

6. – Acoustic emission.

An effect related to internal friction and the motion of dislocations is the noise in the specimens produced by strain. Historically [90] it appears that the first work was done for rocks. This work started in the 1930's and was under-taken to test mine areas which were near danger regions for slide. Since the attenuation of sound waves in rocks is quite large at high frequencies—see Fig. 24, 25 and 26—most of the measured frequencies were in the 150 to 10 000 Hz range. Later work has attempted to relate acoustic emission and earthquake properties. A relation was found by SCHOLZ [90] that acoustic emission and microfracturing in rocks could be directly related to the ane-lastic part of the stress-strain behavior, i.e. the internal friction Q^{-1}. This suggests that the acoustic emission in rocks is connected with dislocation motion as has been established for metals.

For metals one of the first measurements—which correlates with the burst-type acoustic emission—was the work of MASON, McSKIMIN and SHOCKLEY [91] on the effect of a stress on a tin polycrystal with large grain sizes. The experimental arrangement and the type of response measured in shown by Fig. 45.

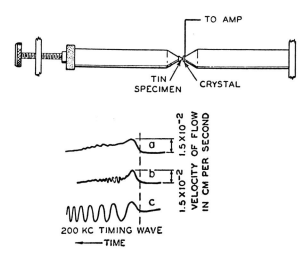

Fig. 45. – Specimen arrangement, pick-up crystal and response of a tin polycrystal with large grains.

The voltage response is in a direction to indicate a relief of the force applied by the turning of the screw thread on the end of the left-hand rod. The tin sample was 1/8 inch in diameter on the end connected to the glass rod and 1/16 inch on the side pressed against the quartz crystal used to pick up the response. The thickness of the quartz crystal was 0.5 mm and when backed on both sides by the glass rods—which were three feet long—the sensitivity of the system was uniform from a few kilohertz to 5 MHz. This type of pick-up device is essential if a replica of the dislocation event is to be obtained. By calibration it was found that a total force of 1000 dyn would produce a displacement of 3/8 inch on the oscillograph. The force is related to the displacement velocity by

$$(107) \qquad\qquad \dot{\varepsilon} = F/\varrho v A \,,$$

where ϱ the density is about 7.1 for a polycrystal, $v = 2.7\cdot10^5$ cm/s for a Young's modulus velocity and A the area is 0.02 cm². Hence a velocity of yield scale can be put on Fig. 45. The total volume change is obtained by integrating the curve with respect to time and is $1.2\cdot10^{-8}$ cm³ for curves of type a and half that for type b. The fine structure shown on the curves indicates that the process is one first proposed by FRANK [92] for the dynamic generation of dislocations. In this process a dislocation starts across the glide plane with the shear velocity, reaches the edge of the crystal, and a return dislocation is generated on an adjacent slip plane by the momentum associated with the motion of the first dislocation. This process keeps up until the dissipation associated with the motion brings it to a halt. This appears to be about 10 oscillations for a and 15 for b. The speed of dislocations can be approximated by taking the distance at 45° from the length since this is the direction of greatest shearing stress. The longest distance from one edge to the opposite side is 0.26 cm. The sound picked up by the crystal will be less when the dislocation is half the time of a complete cycle or $1.25 \cdot 10^{-6}$ s for a and $0.5 \cdot 10^{-6}$ s for b. Taking the complete length of 0.26 cm this gives a velocity of $2.08 \cdot 10^5$ cm/s. It is probable that the effective length is somewhat less than 0.26 cm on account of the curved surface and hence the velocity of the dislocation is close to the shear velocity, which is the limiting speed for a dislocation. The forward motion, which determines the area under the curve, will be about twice the number of complete cycles—since a motion forward occurs for each half cycle times the Burgers distance, which for tin is $b = 3.18 \cdot 10^{-8}$ cm. For 10 cycles the forward motion should be $6.4 \cdot 10^{-7}$ cm, which agrees reasonably well with the measured area $1.2 \cdot 10^{-8}$ cm³, which is the forward motion times the area of the tin specimen, 0.02 cm². For the trace b it appears that the dislocation starts near the middle of the specimen and runs to the edge. If we take the velocity as $1.75 \cdot 10^5$ cm, the distance to agree with the measured timing is 0.0875 cm. The forward motion should be about 30—twice the number of cycles—times b times the relative area of the two path lengths, which is about 0.4. Hence the area under the curve should be

$$(108) \qquad\qquad 30 \cdot 0.4 \cdot 3.18 \cdot 10^{-8} \cdot 0.02 = 7.7 \cdot 10^{-9} \text{ cm}^3 \,,$$

in good agreement with the measured result. Hence all the measurements are in good agreement with a Frank source as the cause of the burst.

The first serious work on the application of acoustic emission is credited to KAISER [90], who studied emission in polycrystalline zinc, steel, aluminum, copper and lead. He applied a steadily increasing stress and observed the acoustic emission which occurred in the form of pulses which could be counted. KAISER also observed that emissions are not generated during the reloading of a material until the stress exceeds its previous high value, an effect known as the « Kaiser effect ». This effect applies to most metals, but not generally to other materials.

SCHOFIELD (1961) [90] performed some very important early acoustic-emission experiments. He showed that single crystals were important sources of emission, showing that grain boundaries are not the only sources of emission. In this work SCHOFIELD was the first to make a distinction between burst-type (discrete) and continuous-type emissions. By means of simple dislocation-deformation arguments, it was estimated that each acoustic-emission pulse resulted from the motion of some (5÷50) dislocations. His later work showed that the major contribution to acoustic emission was from volume effects and not from surface effects.

This fundamental work has received wide practical application in such subjects as structural integrity of metal and rock structures, flaws in metals, composite materials, concrete, ceramics, ice, soils, wood, integrity of welds, martensitic transformations, nuclear-power reactors, leaks in pressure systems and many other uses. A very good description of these applications is given by LORD [90].

Attempts [90] have been made to account for the different types of emissions, namely continuous emissions and burst emissions. Continuous emissions are ascribed to dislocations moving through the crystal and possibly to slip movements. Burst emissions are ascribed to failures such as twinning, microcracks, blocks of dislocations breaking through obstacles (Frank generation source) and with the growth of an existing crack.

By using a high-amplitude system such as that shown in Fig. 1, the writer [93] and associates (BESHERS and JON) have studied the acoustic emission produced in single and polycrystalline metals and rocks. For a thin metal layer—in this case a layer of lead-tin solder superposed on a stiffer metal α-

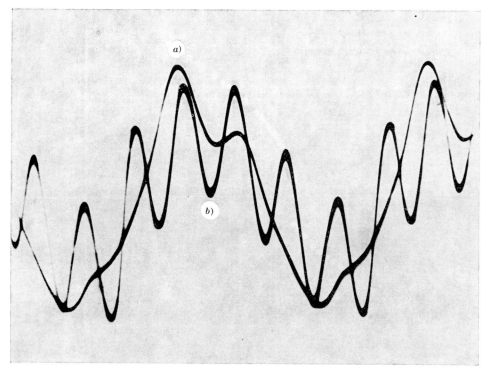

Fig. 46. – *a*) Harmonic generated in a copper single crystal at a strain of $3 \cdot 10^{-4}$, *b*) strain of $5.8 \cdot 10^{-4}$.

brass—a single pile-up from a Frank-Read source was observed. Calculations on the break-away time—about 10^{-6} s—and the energy pick-up are in agreement with Cottrell's [94] calculation for a dislocation pile-up. For a thicker sample many pile-ups occur in the specimen resulting in the generation of harmonics of the driving system. Since the plastic strain reverses itself on each half-cycle, it is an odd function of the strain which results in odd-harmonic generation. An example of such a harmonic generation is hown by Fig. 46. The curve labelled a) is the pick-up from the center of the specimen using a wide-band pick-up device attached to a small hole in the specimen. The curve labelled b) is the pick-up on the strain-measuring electrode at the center of the PZT transducer. Both show a predominant fifth harmonic. This measurement is for a copper single crystal cemented into brass end pieces for a strain of $5.8 \cdot 10^{-4}$. For higher strains—up to $2 \cdot 10^{-3}$—a nonharmonic burst-type response is seen as shown by Fig. 47. From the timing of the bursts it is found that the velocity across the grain in near the shear velocity and it is probable that the response is a Frank-type dynamic generation as for the tin specimen of MASON, McSKIMIN and SHOCKLEY. This is one type of burst emission. This type of emission has also been found for rocks.

Since the continuous-type emission observed in most acoustic-emission experiments with a continuously increasing stress is probably due to the same

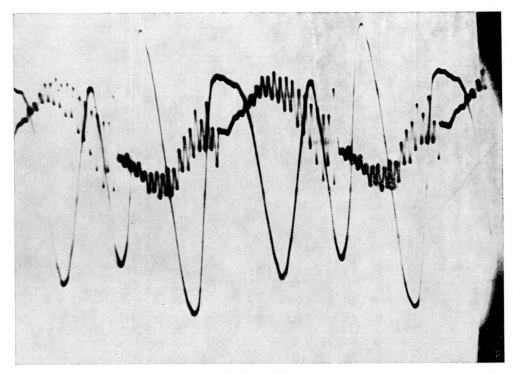

Fig. 47. – Burst-type emission for single-crystal copper.

mechanism as that producing the harmonics in the alternating-stress system, the harmonic production has been analyzed as a function of the applied strain. JON and the writer have made a number of measurements of the harmonic by using a spectrum analyzer. Since the odd harmonics are connected with dislocation motion, the third harmonic has been analysed in the same manner as the internal friction of α-brass shown by Fig. 38. Plotting the product $V_{th}\varepsilon$ against $1/\varepsilon$ and ε as shown in Fig. 38, three ranges are obtained as shown by Fig. 48. These can be analyzed as being due to break-away and the first and second plastic ranges. When two adjacent ranges come together, there is a dip which shows that the phases of the voltages of the different regions are not the same. However, whenever one region predominates, the harmonic voltage is proportional to Q^{-1} in that region.

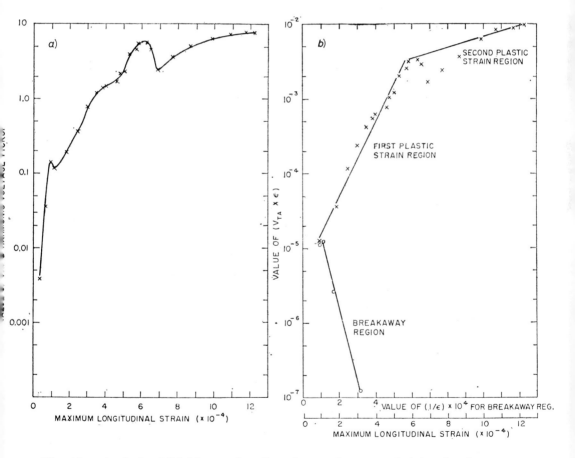

Fig. 48. – Analysis of third harmonic voltage for an α-brass sample into a break-away region and two plastic-flow regions. By comparison of this with the analysis of the internal friction of Fig. 38, it is seen that the voltage is proportional to Q^{-1} in any region.

REFERENCES

[1]   W. P. MASON: *Low and high-amplitude internal-friction measurements in solids and their relation to imperfection motions*, in *Microplacticity*, edited by C. J. MC-MAHON jr. (New York, 1968), p. 287.
[2]   R. TRUEL, C. ELBAUM and B. B. CHICK: *Ultrasonic Methods in Solid State Physics* (New York, N. Y., 1968).
[3]   J. DEKLERK: *Fabrication of vapor-deposited thin film piezoelectric transducers for the study of phonon behaviour in dielectric materials at microwave frequencies*, in *Physical Acoustics*, Vol. 4-A, Chap. V (New York, N. Y., 1966).
[4]   L. LANDAU and G. RUMER: *Phys. Zeits. Sowjet.*, **11**, 18 (1937).
[5]   H. J. MARIS: *Phil. Mag.*, **9**, 901 (1964).
[6]   R. O. WOODRUFF and H. EHRENREICH: *Phys. Rev.*, **123**, 1553 (1961).
[7]   W. P. MASON: *Physical Acoustics*, Vol. **3** B, Chap. VI (New York, N. Y., 1965).
[8]   M. F. LEWIS: *Journ. Acoust. Soc. Amer.*, **44**, 713 (1968).
[9]   a) W. P. MASON and T. B. BATEMAN: *Journ. Acoust. Soc. Amer.*, **40**, 1343 (1966);
      b) W. P. MASON: *Journ. Acoust. Soc. Amer.*, **42**, 253 (1967).
[10] J. LAMB and J. RICHTER: *Proc. Roy. Soc.*, A **293**, 479 (1966).
[11] H. BÖMMEL: *Phys. Rev.*, **96**, 220 (1954).
[12] J. BARDEEN, L. N. COOPER and J. R. SCHRIEFFER: *Phys. Rev.*, **108**, 1175 (1966).
[13] A. B. PIPPARD: *Phil. Mag.*, **46**, 1104 (1955).
[14] This relation was first pointed out by the writer, W. P. MASON: *Phys. Rev.*, **97**, 557 (1955).
[15] See J. RAYNE and C. K. JONES: in *Physical Acoustics*, Vol. **7**, edited by W. P. MASON, Chap. II (New York, N. Y., 1970).
[16] See B. W. ROBERTS: *Oscillatory magnetoacoustic phenomena in metals*, in *Physical Acoustics*, Vol. 4-B, Chap. I (New York, N. Y., 1968).
[17] W. T. READ: *Dislocations in Crystals*, Chap. 11 to 14 (New York, N. Y., 1953).
[18] T. S. KÉ: *Phys. Rev.*, **70**, 105 (1946); **71**, 533 (1947).
[19] E. P. PAPADAKIS: *Physical Acoustics*, Vol. 4-B, Chap. 15 (New York, N. Y., 1968), p. 269.
[20] N. AKULOV and E. KONDORSKY: *Zeits. Phys.*, **78**, 801 (1933).
[21] W. DÖRING: *Zeits. Phys.*, **114**, 579 (1939).
[22] R. M. BOZORTH, W. P. MASON and H. J. MCSKIMIN: *Bell Syst. Tech. Journ.*, **30**, No. 4, Part 1, 970 (1951).
[23] C. W. GARLAND: *Use of ultrasonic methods to study phase transitions*, in *Physical Acoustics*, Vol. **7**, Chap. III (New York, N. Y., 1970).
[24] C. W. GARLAND and D. B. NOVETNEY: *Phys. Rev.*, **177**, 9711 (1969).
[25] E. LITOV and E. A. UEHLING: *Phys. Rev. Lett.*, **21**, 809 (1968).
[26] A. S. NOWICK and W. R. HELLER: *Adv. Phys.*, **14**, 101 (1965).
[27] A. S. NOWICK: *Adv. Phys.*, **16**, 1 (1967).
[28] C. WERT: *Determination of the diffusion coefficient of impurities by anelastic methods*, in *Physical Acoustics*, Vol. **3**-A, edited by W. P. MASON, Chap. 2 (New York, N. Y., 1966).
[29] A. E. LORD jr. and D. N. BESHERS: *Acta. Met.*, **14**, 1659 (1966).
[30] J. L. SNOEK: *Physica*, **8**, 711 (1941).
[31] B. S. BERRY and A. S. NOWICK: *Anelasticity and internal friction due to point defects in crystals*, in *Physical Acoustics*, Vol. **3**-A, Chap. I (New York, N. Y., 1966).
[32] R. R. HASIGUTI, N. IGATA and G. KAMOSHITA: *Acta Met.*, **10**, 442 (1962).

[33] W. P. MASON: *Physical Acoustics and the Properties of Solids*, Chap. X (Amsterdam, 1958).

[34] B. E. WARREN: *Journ. Appl. Phys.*, **8**, 645 (1937).

[35] W. T. READ: *Dislocations in Crystals* (New York, N. Y., 1953), p. 115.

[36] A. H. COTTRELL: *Dislocations and Plastic Flow in Crystals* (Oxford, 1953), p. 134.

[37] See A. GRANATO and K. LÜCKE: *The vibrating-string model of dislocation damping*, in *Physical Acoustics*, Vol. 4-A, Chap. 6 (New York, N. Y., 1966).

[38] See A. SEEGER and P. SCHILLER: *Kinks in dislocation lines and their effects on the internal friction in crystals*, in *Physical Acoustics*, Vol. 3-A, Chap. 8 (New York, N. Y., 1966).

[39] *Loc. cit.*, p. 426.

[40] G. ALERS and K. SALAMA: *Dislocation Dynamics*, edited by A. R. ROSENFIELD, G. T. HAHN, A. L. BEMENT jr. and R. I. JAFFEE (New York, 1968), p. 211.

[41] J. H. WEINER: *Phys. Rev.*, **136**, A 863 (1964).

[42] W. ATKINSON and N. CABRERA: *Phys. Rev.*, **138**, A 763 (1965).

[43] W. P. MASON and J. WEHR: *Journ. Phys. Chem. Sol.*, **31**, 1925 (1970).

[44] G. LEIBFRIED: *Zeits. Phys.*, **127**, 344 (1950).

[45] *Physical Acoustics*, Vol. 3-B, edited by W. P. MASON, Chap. VI (New York, N. Y., 1965).

[46] O. M. M. MITCHELL: *Journ. Appl. Phys.*, **36**, 2083 (1965).

[47] G. K. WHITE and S. B. WOODS: *Phil. Mag.*, **45**, 1343 (1954).

[48] See A. SEEGER and P. SCHILLER: *Physical Acoustics*, Vol. 3-A, Chap. 8 (New York, N. Y., 1966), p. 413.

[49] See A. V. GRANATO and K. LÜCKE: *Physical Acoustics*, Vol. 4-A, Chap. VI (New York, N. Y., 1966).

[50] O. S. OEN, D. K. HOLMES and M. T. ROBINSON: U.S. Atomic Energy Commission, Report ORNL-3017, 3.

[51] T. SUZUKI and C. ELBAUM: *Journ. App. Phys.*, **35**, 1539 (1964).

[52] A. V. GRANATO and R. J. STERN: *Journ. Appl. Phys.*, **33**, 2880 (1962).

[53] R. L. WEGEL and H. WALTHER: *Physics*, **6**, 141 (1935).

[54] J. L. ROUTBORT and H. S. SACK: *Journ. Appl. Phys.*, **37**, 4803 (1966).

[55] J. C. SAVAGE and H. HASEGAWA: *Geophysics*, **32**, 1003 (1967).

[56] W. P. MASON, D. N. BESHERS and J. T. KUO: *Journ. Appl. Phys.*, **41**, 5206 (1970).

[57] W. P. MASON and J. T. KUO: *Journ. Geophys. Res.*, **76**, 2084 (1971).

[58] W. P. MASON: *Nature*, **234**, 461 (1971).

[59] See L. KNOPOFF: *Physical Acoustics*, Vol. 3-B, edited by W. P. MASON, Chap. VII (New York, N. Y., 1965).

[60] R. B. GORDON and L. A. DAVIS: *Journ. Geophys. Res.*, **73**, 3917 (1968).

[61] R. D. MINDLIN: *Journ. Appl. Mech.*, **71**, 259 (1949).

[62] R. D. MINDLIN, W. P. MASON, T. F. OSMER and H. DERESIEWICZ: *First National Congress of Applied Mechanics* (1951); also *Bell Syst. Tech. Journ.*, **31**, 469 (1952).

[63] K. L. JOHNSON: *Proc. Roy. Soc.*, A **230**, 531 (1955).

[64] F. BIRCH and D. BANCROFT: *Journ. Geology*, Pt. I, 59, Pt. II, 113 (1938).

[65] B. R. TITTMAN: *IEEE Group on Sonics and Ultrasonics Meeting in Boston* (1973), p. 130; and Monograph No. 923, North American Rockwell Science Center, Thousand Oaks, Cal. 91360.

[66] P. G. BORDONI: *Journ. Acous. Soc. Amer.*, **26**, 495 (1954).

[67] B. MECS and A. S. NOWICK: *Acta Met.*, **13**, 771 (1965).

[68] R. R. HASIGUTI: *Journ. Phys. Japan, Suppl.*, **118**, 567 (1963).

[69] A. SEEGER: *Phil. Mag.*, **1**, 651 (1956).

[70] J. LOTHE: *Phys. Rev.*, **117**, 704 (1960).

[71] V. K. PARÉ: *Journ. Appl. Phys.*, **17**, 271 (1966).

[72] G. ALEFELD, J. FILLOUX and H. HARPER: in *Dislocation Dynamics*, edited by A. R. ROSENFIELD, G. T. HAHN, A. L. BEMENT jr. and R. J. JAFFEE (New York, N. Y., 1968).

[73] J. DE FOUQUET, P. BOCH, J. PETIT and G. RIEU: *Journ. Phys. Chem. Solids*, **31**, 1901 (1970).

[74] H. M. SIMPSON and A. SOSIN: *Phys. Rev. B*, **5**, 705 (1974).

[75] S. TOKAHASHI: *Journ. Phys. Soc. Japan*, **1**, 1253 (1956).

[76] G. S. BAKER: *Journ. Appl. Phys.*, **28**, 734 (1957).

[77] A. GRANATO and K. LÜCKE: *Journ. Appl. Phys.*, **27**, 789 (1956).

[78] A. H. COTTRELL: *Dislocations and Plastic Flow in Crystals* (Oxford, 1953), p. 134.

[79] J. FRIEDEL: *Dislocations* (London, 1964), p. 359.

[80] P. PEQUIN, J. PEREZ and P. GOBIN: *Trans. Metal. Soc. AIME*, **239**, 438 (1967).

[81] W. P. MASON: *Physical Acoustics and the Properties of Solids* (Amsterdam, 1958), p. 250.

[82] J. FRIEDEL: *Dislocations* (London, 1964), p 211.

[83] W. P. MASON and W. A. WOOD: *Journ. Appl. Phys.*, **39**, 5581 (1968).

[84] W. A. WOOD and W. P. MASON: *Journ. Appl. Phys.*, **40**, 4514 (1969).

[85] *First International Symposium on High Power Ultrasonics, Gras* (1970); also *Second Symposium London* (1973).

[86] K. WERNER: *ETR Eisenbahntechnische Rudschau*, Vol. **4** (1973), p. 1.

[87] D. E. MacDONALD and W. A. WOOD: *Journ. Appl. Phys.*, **42**, 5531 (1971).

[88] S. PURUSHOTHAMAN, J. P. WALLACE and J. TIEN: *Second International Congress of High Power Ultrasonics, London, March 1973*, in press.

[89] W. P. MASON and J. WEHR: *Journ. Phys. Chem. Sol.*, **31**, 1925 (1970).

[90] A. E LORD: *Review article on acoustic emission*, in *Physical Acoustics*, Vol. **11**, edited by W. P. MASON and R. N. THURSTON (New York, N. Y., 1975).

[91] W. P. MASON, H. J. McSKIMIN and W. SHOCKLEY: *Phys. Rev.*, **73**, 1213 (1948).

[92] F. FRANK: *Report on Strength of Solids, London Phys. Soc.*, **46**, 46 (1948).

[93] W. P. MASON: *Acoustic emission produced by large alternating stresses, Society of Engineering Science* (1973), in press.

[94] A. H. COTTRELL: *Dislocations and Plastic Flow in Crystals* (Oxford, 1953), p. 102.

Superconducting Transducers for Use in the 50 to 1000 GHz Range.

E. F. CAROME

Physics Department, John Carroll University - Cleveland, O.

1. – Introduction.

When one considers the application of conventional acoustic techniques for studies of materials in the frequency range in excess of 20 GHz one notes that the required surface finish and parallelism exceed that attainable with ordinarily available methods since the acoustic wavelength becomes shorter than optical wavelengths. Thus it is necessary to turn to other techniques. In recent years generators and detectors of incoherent phonons have been developed utilizing superconducting elements and these have been applied for various types of studies in the frequency range from 50 to 1000 GHz [1-5]. For the past two years we have been conducting preliminary studies of the characteristics of these transducers and in this paper I would like to present a qualitative review of some basic concepts involved in their operation. I also would like to briefly mention several applications investigated by various workers.

2. – Tunnel junctions as electrical elements.

To begin let us consider in detail the physical make-up of one type of these transducers, a tunnel junction. Referring to Fig. 1, suppose we have a cylindrical sample of the material to be studied, for example a 1 cm long, 0.5 cm diameter sapphire rod, with polished and parallel ends suitable for ultrasonic studies in the megahertz range. On the end we first deposit four conducting pads to which electrical leads may be soldered. Then, using standard thin-film evaporation techniques, a two millimeter wide metallic strip approximately 1000 Å thick is deposited between two of the pads. Next this film is oxidized to produce a thin insulating layer and a second 2 mm wide metallic strip, possibly 3000 Å thick, is deposited at right angles to the first. The 4 mm² metal-metal oxide-metal element at the center of the crossed strips forms the transducer.

Let us first consider the behavior of this element as an electrical device.

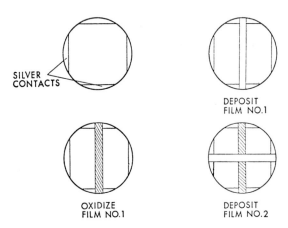

Fig. 1. – Sketches indicating successive steps of depositing a tunnel junction on the end of a cylindrical delay rod.

If both strips are normal metals at a temperature slightly above absolute zero the conduction electrons are distributed over the available energy levels as sketched in Fig. 2. By applying a voltage V between the strips quantum-mechanical tunneling of electrons through the oxide barrier can be produced and the resulting current I is, to a high degree of approximation, a linear function of the applied voltage V.

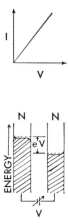

Fig. 2. – Sketches of the density of states and IV characteristic curve for a tunnel junction consisting of two normal-metal films.

On the other hand, if one of the strips is superconducting, by means of the single-particle model of a superconductor the density of states may be shown to be as sketched in Fig. 3. A gap of magnitude 2Δ develops in the allowed energy spectrum, this being the amount required to separate a Cooper pair into two unpaired electrons or quasi-particles. There is also a sharp increase in the density of allowed states at each edge of the energy gap. If a voltage smaller than Δ/e (where e is the electronic charge) is applied across the junction little or no current will flow because electrons on the left face either filled or forbidden states on the right. (In reality a small current will flow at temperatures above absolute zero due to thermal excitations.) When the voltage is increased through $V = \Delta/e$, a large current increase occurs due to the large number of available

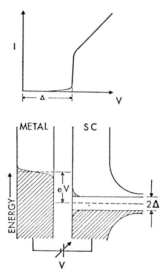

Fig. 3. – Sketches of the density of states and IV characteristic curve for a tunnel junction consisting of a superconducting and a normal metal film.

unoccupied states in the superconductor. Further increases in V produce a less sharply increasing current because of the reduction and flattening out of the density-of-states function of the superconductor and the IV characteristic curve for higher voltages approaches that of normal-normal junction.

A similar analysis of the behavior of a junction consisting of two films of the same superconducting material indicates that the IV characteristic curve is similarly shaped to that in Fig. 3. However, from a consideration of the density-of-states function, it may be shown that the sharp current increase would occur at $V = 2\Delta/e$. A typical experimental IV characteristic curve that we have obtained for a tin-tin oxide-tin junction at 1.74 °K is shown in Fig. 4. Note that the sharp current increase occurs for $V = 1.2$ mV. The normal-to-superconducting transition for tin occurs at approximately $T_c = 3.7$ °K. A set of IV curves for this same junction at various temperatures below T_c are shown in Fig. 5. Note that the origin is shifted upward for each successive curve, all of which begin at $I = 0$. It may be seen that the current steps occur at lower voltages and become less sharp as $T_c - T$ decreases. This

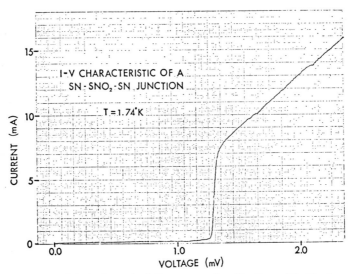

Fig. 4. – Experimental IV characteristic curve of a tin-tin oxide-tin tunnel junction obtained at $t = 1.74\,°\mathrm{K}$.

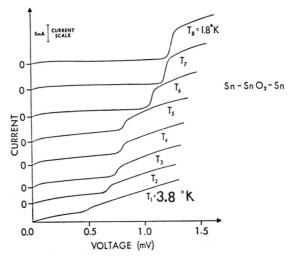

Fig. 5. – Sketches of experimental IV characteristic curves of a tin-tin oxide-tin tunnel juction at various temperatures between 1.8 °K and approximately 3.8 °K.

follows from the fact that the energy gap decreases and the density-of-states discontinuity becomes less sharp as T approaches T_c.

3. – Tunnel junctions as acoustic transducers.

We now inquire how such superconducting tunnel junctions may be used as phonon generators and detectors for acoustic-propagation studies. Consider

a 1 cm long cylindrical sapphire delay rod with tin-tin oxide-tin junctions centered on the opposite flat ends. Assume that this is at 1.7 °K and that 0.1 μs
wide voltage pulses several millivolt in amplitude are applied cyclically at
1 ms intervals across the junction on one end. Since for an appropriate junction
the linear portion of the IV characteristic curve has a slope $\Delta E/\Delta I = R_{normal} \simeq$
$\simeq 50 \text{ m}\Omega$, a tunneling current pulse of several hundred milliampere peak
amplitude would flow. Suppose next that the junction on the opposite end is
connected to a constant-current source, with the current set to a value corresponding to the bottom edge of the step increase shown in Fig. 4, and that the
voltage across the junction is monitored with a sensitive pulse amplifier and
oscilloscope system. With the oscilloscope trace triggered each time the voltage
pulse is applied to the first junction, the resulting voltage signal across the second
junction might look like the typical experimental trace sketched in Fig. 6,

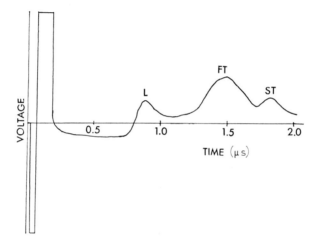

Fig. 6. – Sketch of oscilloscope trace of the output signal *vs.* time from a receiver
tunnel junction on one end of a 1 cm long sapphire delay rod when a voltage pulse is
applied at $t = 0$ to a source tunnel junction on the opposite end of the rod.

i.e. a « main bang », electrically fed through signal initiated at the same time
as the applied voltage pulse, followed by a group of three or more pulses occurring at delay times corresponding roughly to the acoustic-travel times through
the sapphire rod for longitudinal (L), fast shear (FT) and slow shear (ST) acoustic waves.

Let us consider the presently well-established interpretation of this experiment. When voltage pulses of the proper polarity and $V > 2\Delta/e$ are applied
to the first tin-tin oxide-tin junction or « source transducer » an electron or
quasi-particle tunneling current pulse is produced from the outermost to the
inner tin film. In the latter the excited electrons lose energy in collisions by the
emission of phonons until they reach the top of the energy gap. Then two such

quasi-particles combine to form a Cooper pair, dropping to an unoccupied state below the gap with the emission of a phonon of energy 2Δ. One speaks of phonons emitted in the « relaxation » process as the quasi-particles lose energy and drop to the gap edge, and of those emitted in the « recombination » process in the formation of Cooper pairs. If in the relaxation process of a relatively high-energy quasi-particle a phonon of energy greater than 2Δ is emitted in a material such as tin, where the electron-phonon interaction is strong, there would be a high probability that the phonon would be absorbed in the breaking of a Cooper pair in the tin film. On relaxation and recombination the resulting quasi-particles would emit lower-energy relaxation phonons and 2Δ recombination ones. Thus a continuous spectrum of relaxation phonons of energy less than 2Δ, together with a large number of recombination phonons that form a peak at 2Δ, would be produced. For tin $2\Delta = h\nu$ corresponds to phonons of energy 1.2 meV with $\nu = 280$ GHz: in the case of a lead-lead oxide-lead junction $2\Delta = 2.8$ meV, corresponding to a frequency of 650 GHz.

In the above-discussed process an « acoustic » signal consisting of pulses of incoherent phonons of various polarities would be produced and some of these would propagate into the sapphire rod. Those of energy 2Δ that reach the similar tunnel junction on the opposite end of the rod could trigger « phonon-assisted » tunneling processes, i.e. on being absorbed in the first tin film they reach, they could break Cooper pairs, increasing the population of excited quasi-particle states above the energy gap and, with a constant bias current, lead to a drop in resistance and in voltage across the junction. From another viewpoint, this might be looked on as an effective increase in the temperature of the first film of this « receiving transducer ». Thus, referring to Fig. 7, a pulsed shift

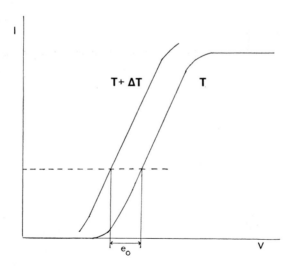

Fig. 7. – Sketch of two adjacent IV characteristic curves for T and $T + \Delta T$, indicating how an output voltage of magnitude e_0 may be produced when a tunnel junction receiver absorbs a phonon pulse.

from the T to the $T + \Delta T$ characteristic curve would occur and a voltage pulse of amplitude e_0 would be produced.

4. – Superconducting fluorescent phonon generators.

In the above treatment an attempt has been made to outline qualitatively the operation of tunnel junctions as acoustic sources and detectors in the high GHz range. We have taken this approach since historically they were the first elements used as narrow-band transducers in this frequency range. However, we have overlooked some of the practical problems encountered in the production of pairs of low-resistance junctions suitable for propagation studies, and also the fact that the receiver output signals e_0 are frequently in the low microvolt range. Fortunately other types of transducers have been developed for work in this same range [4, 5]. A relatively simple generator of large-amplitude phonon signals is shown in Fig. 8. It is referred to as a phonon

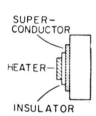

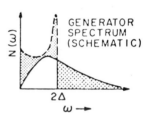

Fig. 8. – Sketches indicating the structure of a phonon fluorescent generator and the manner in which the high-frequency region of the heater's black-body phonon spectrum is down shifted to produce a semi-monochromatic spectrum.

fluorescent generator and consists of a superconducting film, roughly 500 Å thick, an insulating layer and a heater such as a constantan thin film strip. When a current pulse is passed through the heater a « black-body » spectrum phonon pulse is produced, as indicated in Fig. 8. Those of energy greater than 2Δ that propagate into the superconducting film again have high probability of breaking Cooper pairs, and through the relaxation and recombination processes discussed in the previous Sections 2Δ phonons and ones of energy less than 2Δ are produced. Thus the black-body phonon spectrum is transformed to the cross-hatched peaked spectrum shown in Fig. 8, *i.e.* a « semi »-monochromatic one.

Fewer problems are encountered in the fabrication of this type of transducer since the insulating layer is relatively simple to deposit and is much more stable than the low-resistance oxide layer of suitable tunnel junctions. The resistance of the constantan heater may easily be made of the order of 50 Ω and the im-

pendance-matching problems encountered in the use of 50 mΩ junctions may
be avoided. In addition the level of acoustic signals attainable with the fluorescent generators appears to be of the order of several watt so that signal-to-
noise problems are greatly reduced [5, 6].

5. – Superconducting bolometer receivers.

It is also possible to avoid some of the difficulties encountered in the use of
tunnel junctions as detectors by employing superconducting bolometers as
receiving elements [7]. Again these consist of 500 to 1000 Å thick metallic
strips deposited in the same way as the first strip of a tunnel junction on the
end of a sample delay rod. The low-temperature normal-state resistance of
such a strip may easily be made of the order of 100 Ω by scribing a number
of interlaced scratches partially across the width of the strip to increase its
effective length. Then in the normal-to-superconductor transition region the
strip exhibits a very sharp decrease in resistance. The temperature at which
this change takes place may be decreased by passing a bias current through
the bolometer or by applying a magnetic field. Typical curves of resistance
vs. temperature for a tin bolometer are shown in Fig. 9 for bias currents of 1
and 20 mA.

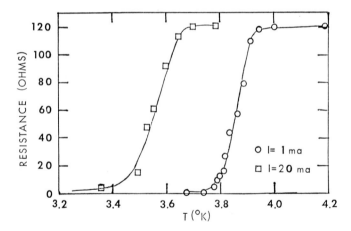

Fig. 9. – Graph of the resistance vs. temperature of a tin bolometer for bias currents
of 1 and 20 mA.

To employ such an element as phonon detector the bias current is set at
a value that brings the resistance of the bolometer to the middle of the transition region at the desired operating temperature. The voltage across the element
is then monitored with a sensitive amplifier; an incident phonon pulse produces
an increase in the temperature of the strip through absorption of phonons and

thus an increase in voltage is produced. In this way one may obtain a sensitive detector of phonons that is simple to fabricate and operate. It should be emphasized, however, that such detectors are not frequency selective, and respond effectively only to changes of temperature.

6. – Applications.

Several applications of these superconducting transducers have been made by various workers. DYNES and NARAYANAMURTI have developed a phonon spectrometer, as sketched in Fig. 10, and have used it to study the spectrum

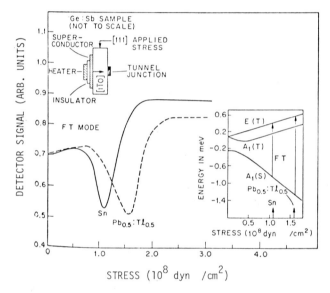

Fig. 10. – Sketches indicating the structure and behavior of a phonon spectrometer. The tracing is of the peak intensity of the FT mode propagating along the $[1\bar{1}0]$-axis as a function of stress for Sn and $Pb_{0.5} Tl_{0.5}$ fluorescent generators. (Reproduced with permission [5].)

of the phonons produced by tunnel junctions and fluorescent generators [6, 8]. They employed a single-crystal germanium delay rod doped with antimony that is known from infra-red studies to possess a set of energy states whose separation may be scanned through the millielectronvolt range by the application of a uniaxial mechanical stress. As indicated in Fig. 10 (lower curves), they have been able to confirm that both tunnel junctions and fluorescent generators produce strongly peaked phonon spectra. They have observed the stress-dependence of the peak amplitude of the fast shear (FT) portion of received signals such as that shown in Fig. 6. As indicated by the solid curve in Fig. 10 of the FT peak signal amplitude *vs.* applied stress obtained with tin sources and receivers, they found that this peak has a relatively sharp minimum. This occurs for a stress value that produces an antimony donor energy level spacing of 1.2 meV, confirming that the received signal is peaked at 280 GHz as previously discussed.

HUET, MANEVAL and ZYLBERSZTEJN have reported measurements of acoustic-wave velocity dispersion in indium-antimonide single crystals [9]. They examined the arrival times of thermal phonon pulses generated by a constantan heater and detected simultaneously by tunnel junctions and an aluminum bolometer. Since the junctions were sensitive only to phonons of particular high frequencies, while the bolometer was sensitive to the entire phonon spectrum and indicated the arrival time of low-frequency, low-velocity phonons, these workers were able to obtain data on the dispersion of both longitudinal and fast shear phonons.

NARAYANAMURTI, ANDRES and DYNES have used a time-of-flight system employing superconducting transducers to study the dispersion and attenuation of high-frequency phonons in helium II at $T = 0.1$ °K [10]. In this case the phonon energy $h\nu \gg kT$ and from their results they concluded that such « superthermal » phonons propagate with negligible dispersion and that their lifetimes are unexpectedly long. To obtain velocity data over a range of frequencies they employed the fact that the energy gaps of their aluminum fluorescent generator and tunnel junction detector may be reduced by applying a magnetic field parallel to the metal films. In this way they obtained data in the frequency range 0.15 to 0.91 GHz, the latter being the frequency available at zero field when $2\varDelta$ for aluminum has its largest value at 0.1 °K, of approximately 0.32 meV.

In a recent publication FORKEL, WELTE and EISENMENGER have shown that it is possible to generate even higher frequencies with aluminum junctions [11]. As pointed out earlier, in the relaxation and recombination processes for quasi-particles in tin relaxation phonons with energies greater than $2\varDelta$ have a high probability of breaking Cooper pairs. Thus with tin sources recombination phonons of energy $2\varDelta$ and mainly lower-frequency relaxation phonons may propagate from the source. Using aluminum tunnel junctions, however, for which the phonon-electron interaction is less strong, these authors have been able to detect phonons of energies up to $24\varDelta$, corresponding to a frequency of 870 GHz. These have propagated through a 2.3 mm long silicon crystal and were detected with a second aluminum junction.

7. – Conclusion.

It is hoped that this brief discussion of the behavior and recent applications and developments of superconducting phonon transducers will bring to mind other possible uses for such elements. Their availability greatly extends the upper frequency limit for physical acoustic studies of materials and they should have many applications in studies of the properties and behavior of materials at low temperatures.

REFERENCES

[1] W. EISENMENGER and R. C. DYNES: *Phys. Rev. Lett.*, **18**, 125 (1967).
[2] J. M. ANDREWS jr. and M. W. P. STRANDBERG: *Journ. Appl. Phys.*, **38**, 2660 (1967).
[3] A. H. DAYEM, B. I. MILLER and J. J. WIEGAND: *Phys. Rev. B*, **3**, 2949 (1971).
[4] V. NARAYANAMURTI and R. C. DYNES: *Phys. Rev. Lett.*, **27**, 410 (1971).
[5] R. C. DYNES and V. NARAYANAMURTI: *Phys. Rev. B*, **6**, 143 (1972).
[6] J. P. D'IPPOLITO and W. E. BRON: *Phys. Lett.*, **47** A, 309 (1974).
[7] J. M. ANDREWS jr. and M. W. P. STRANDBERG: *Journ. Appl. Phys.*, **38**, 2660 (1967).
[8] R. C. DYNES, V. NARAYANAMURTI and M. CHIN: *Phys. Rev. Lett.*, **26**, 181 (1971).
[9] D. HUET, J. P. MANEVAL and A. ZYLBERSZTEJN: *Phys. Rev. Lett.*, **29**, 1092 (1972).
[10] V. NARAYANAMURTI, K. ANDRES and R. C. DYNES: *Phys. Rev. Lett.*, **31**, 687 (1973).
[11] W. FORKEL, M. WELTE and W. EISENMENGER: *Phys. Rev. Lett.*, **31**, 215 (1973).

Production and Detection
of Very-High-Frequency Sound Waves (*).

K. Dransfeld

Max-Planck-Institut für Festkörperforschung - Grenoble

1. – Introduction.

We will discuss in this short review the new methods which were developed within about the last ten years for the generation and detection of acoustic waves at frequencies above 1 GHz.

Perhaps the most remarkable development is the discovery by Eisenmenger and Dayem [1, 2] that superconducting tunnel diodes can be used for the generation and detection of phonons up to almost 10^{12} Hz [3]. Kinder [4] has developed a special modulation technique, by which he has been able to emit and to observe phonons up to presently 870 GHz, having a frequency resolution of better than 3 % [3]. In this context one should also mention the demonstration by Narayanamurti [5] that superconducting films are excellent phonon filters, transmitting only phonons whose energy is lower than the gap energy $2\varDelta$. The whole exciting development is the subject of Carome's review (*) in this same volume and will therefore only be mentioned here.

We will restrict ourselves here to a treatment of 3-dimensional bulk waves. Surfaces waves [6-8] and their generation will be discussed by de Klerk in this volume.

In contrast to ultrasonic waves at MHz frequencies the *acoustic wavelegth* is for waves above 1 GHz only of the order of 10 000 Å or less. Consequently the diffraction losses due to *beam spreading* become negligible in the high-frequency regime. On the other hand, the slightest local inhomogeneities of the velocity of sound—due to dislocations or mosaic structure—often transform a perfectly plane *coherent* wave after a travelled distance of 1 cm into an *incoherent* wave showing a strongly corrugated wave front. This disturbing transformation—which can simply be considered as small-angle scattering—

(*) This review is a supplement to the article by E. F. Carome published in this same volume. Surface waves are also treated separately by J. de Klerk in this volume.

becomes so severe above 100 GHz that it is appropriate to speak only of incoherent waves at higher frequencies.

Another consequence of the short wavelength λ is the correspondingly shorter distance needed for the transformation from a perfectly sinusoidal density wave (density variation $\Delta\varrho/\varrho$) into a shock wave. This transformation takes place (see Berktay's contribution [9]) after the wave has travelled the distance $\sim\lambda/(\Delta\varrho/\varrho)$. Particularly in high-frequency surface waves shock formation is reached—even at moderate acoustic power levels—after the wave has propagated only a few mm. Joffrin's lecture (see this volume [10]) has given another good example for the increased importance of acoustic nonlinear processes at elevated frequencies.

High-frequency phonons have many possibilities to *decay spontaneously* into low-frequency phonons, even at zero temperature. This is a consequence of the increased density of phonon states at higher frequencies, caused by their short wavelength. Therefore we do not know whether, for example, 10^{12} Hz phonons can propagate in high-purity germanium at $T = 0$ for one centimeter.

At frequencies above 1 GHz the acoustic wavelength becomes *comparable to the wavelength of visible light*. Therefore one can use Brillouin scattering to switch light beams acoustically fast from one direction to another—both by bulk or surface waves. QUATE [11] has built the first acoustic scanning microscope using microwave ultrasonic waves in water. High-frequency surface waves can also be used to read out or display images.

High-frequency acoustic waves above, for example, $1000 \text{ GHz} = \omega/2\pi$ have a phonon energy $\hbar\omega$ which is larger than kT at helium temperatures. Therefore the equipartition principle is no longer valid and it is no longer appropriate to describe the acoustic absorption by a classical viscosity. In the acoustic transmission through, for example, a superconductor or a paramagnetic crystal it is rather becoming necessary to consider the *quantum-mechanical phonon-induced transitions* between two different energy states in the particular material.

Summing up: The presently available frequencies from 10^9 to 10^{12} Hz are still well below the Debye frequency of 10^{13} Hz. Nevertheless their short wavelengths and high frequencies make it interesting both from a theoretical and practical point of view to consider the generation and detection of these waves.

2. – Coherent phonons.

The piezoelectric and magnetostrictive methods for the generation of ultrasonic waves have been used for a long time in order to generate waves in the MHz frequency range. However, these methods were not extended to frequencies above about 1 GHz until 1959. For the historical development see BEYER and LETCHER [12].

2ʹ1. Surface excitation. – BARANSKII [13], BÖMMEL *et al.* [14] and JACOB-
SEN [15] adopted a new method to excite high-frequency sound waves in a large
piezoelectric crystal whose dimensions were much larger than the acoustic
wavelength. When a polished surface of such a crystal is exposed to an electrical
microwave field, acoustic waves of the same frequency are emitted from the
surface into the interior of the sample even though the electric field may be
uniform across the sample. ILUKOR and JACOBSEN [16] have used this method
for ultrasonic frequencies of up to 114 GHz. In view of the availability of strong
molecular vapour lasers in the submillimeter spectral range it does not seem
impossible to use this principle of phonon generation at even higher frequencies.

2ʹ2. Thin-film piezoelectric transducers. – Nowadays one can deposit oriented
films of CdS, ZnS, ZnO, AlN or LiNbO$_3$ on almost any crystalline or amorphous
substrate making excellent acoustic contact. These techniques which make
it possible to emit efficiently high-frequency sound waves also into nonpiezoelectric materials have been described by FOSTER [17], DE KLERK [18] and more
recently also by SITTIG [19]. The different authors use various techniques of
deposition: evaporation in vacuum, epitaxial growth or sputtering. Mostly
the *c*-axis seems to grow oriented perpendicular to the substrate surface, thereby
rendering the film piezoelectric.

Longitudinal waves can be emitted into the substrate if the r.f. electric
field is parallel to the *c*-axis (*i.e.* normal to the surface). If one wants to excite
transverse waves by the same film, one needs only to incline the direction of
the electric field (at best 39° for CdS) relative to the *c*-axis, as has recently
been described by ARNOLD *et al.* [20].

2ʹ3. Piezoelectric semiconducting transducers. – Another interesting proposal
has come from WHITE [21]. He suggested using the *depletion layer* of a piezoelectric semiconductor (see Fig. 1*a*)) as a transducer. For example, in GaAs
the depletion layer may have a thickness of (100÷800) Å.

If the biasing voltage is increased further an *inversion layer* is formed. Inside
the potential minimum of the inversion layer a two-dimensional electron gas
is formed; the momentum of the electron is quantized normal to the surface.
By applying a magnetic field perpendicular to the surface (see Fig. 1*b*)) also

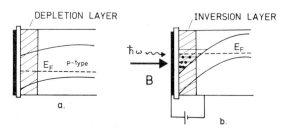

Fig. 1. – *a*) Depletion layer transducer and *b*) inversion layer in a magnetic field.

the electronic motion parallel to the surface is quantized. By changing the biasing voltage and/or the magnetic field the separation of the energy levels from each other and from the Fermi energy can be tuned in wide ranges. Phonon absorption may populate levels above the Fermi energy and leads to phonon-induced conductivity of the inversion layer. Photon absorption has already been demonstrated very recently by KOCH et al. [22].

2'4. *Piezoelectric high polymers.* – A strong piezoelectric effect of a high polymer has been discovered 1969 by KAWAI [23] and FUKADA [24] (deformation $\Delta l/l$ in an electric field E) (Fig. 2).

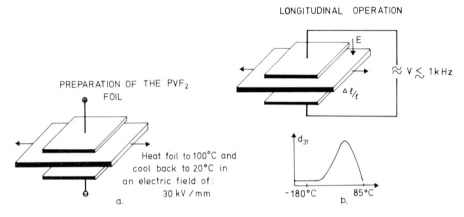

Fig. 2. – Preparation and longitudinal operation of a piezoelectric high-polymer film.

The piezoelectric coefficient $(\Delta l/l)/E$ is three times higher than d_{31} for quartz. Although the origin of this strong piezoelectric effect is not well understood, VON SCHICKFUSS and SUSSNER [25] have used this flexible plastic material to build a spherical loudspeaker shaped like a balloon which has a good high-frequency response and is a truly spherical radiator of sound (see Fig. 3).

SUSSNER et al. [26] have also demonstrated that PVF_2 foils can be operated

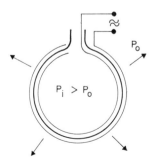

Fig. 3. – Spherical loudspeaker using a PVF_2 foil.

in the piezoelectric thickness vibration as transducers up to 800 MHz, particularly at low temperatures as indicated in Fig. 4.

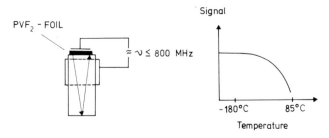

Fig. 4. – Excitation of microwave ultrasonic waves by a high-polymer film.

2ʹ5. *Magnetostrictive transducers.* – If a Ni sphere at a temperature well above the Curie temperature T_c is cooled through T_c, it acquires a magnetic moment and thereby it deforms because of the magnetostrictive effect: the sphere becomes an ellipsoid of revolution whose axis of revolution is parallel to the magnetization M. Let us now consider a Ni film deposited onto a dielectric and let us excite ferromagnetic resonance; the strain ellipsoid will precess with the magnetic moment thereby inducing a shear motion of the Ni film and thus emitting circularly polarized phonons into the dielectric (Fig. 5). For further details see BÖMMEL *et al.* [27] and BIALAS [28].

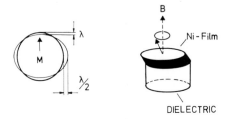

Fig. 5. – Excitation of ultrasonic waves by a Ni film.

2ʹ6. *Electromagnetic generation of microwave sound.* – Usually ultrasonic waves are generated by applying an a.c. electric field to a piezoelectric or magnetostrictive transducer bonded or deposited on a substrate, if the latter is not piezoelectric itself.

For the generation of ultrasonic waves in metals there is, however, the possibility to excite ultrasonic waves directly by applying an a.c. electric field to the surface of a pure metal as reviewed by DOBBS [29]. The fact that electromagnetic waves can excite sound waves directly was accidentally discovered by GANTMAKER [30] on bismuth and by LARSEN *et al.* [31] on aluminium. All these initial experiments were performed at MHz frequencies.

The first direct generation of microwave sound by this method was achieved by ABELES [32] and is indicated in Fig. 6. He used an indium film less than one micron thick on a single crystal of germanium. The microwave power was pulsed (10 W peak power) and he could receive several acoustic echoes. The insertion loss is much larger than for piezoelectric excitation, but the coupling between acoustic waves and electromagnetic radiation is interesting to study for its own sake.

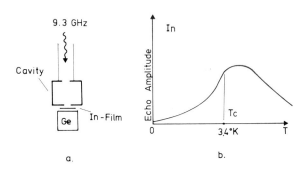

Fig. 6. – Abeles' experiment: *a*) experimental set-up, *b*) results.

The tangential a.c. electric fields penetrate the metal up to the skin depth (normal or anormal) which in pure metals is of the same order as the Meissner penetration depth in superconductors, *i.e.* of the order of 500 Å or a bit more. The electric fields act in opposite directions and independently firstly on the ions and secondly on the mobile electrons. If the scattering rate of the electrons is very rapid (mean free path $\ll$ acoustic wavelength)—either because of impurities or because of thermal phonon scattering—there is no net momentum transfer to the lattice. But if the mean free path increases on cooling there remains a net force acting on the ions, because the momentum transferred to the ions and to the electrons does not cancel locally to better than one part in 1000 any more. In this view COHEN [33] predicts the largest phonon generation in the purest metals and highly « pure » superconductors. However, in « dirty » superconductors, where the mean free path l is smaller than the coherence length ξ, the conversion efficiency drops by the factor $(l/\xi)^2$ as indicated by Abeles' results. It now seems that these phenomena are reasonably well understood and the prevention of these processes of energy conversion may prove important for the construction of high-quality superconducting cavities. It may furthermore be quite interesting to study this conversion in very high magnetic fields which act predominantly on the electrons.

In this context it is interesting also to consider an experiment of BELLESSA [34] at 150 MHz shown in Fig. 7.

When the nuclear magnetic resonance condition is fulfilled for the ^{3}He nuclei (at ~ 46 kOe), one observes an *acoustic free decay* (AFD). The AFD

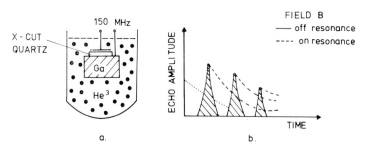

Fig. 7. – Experiment of BELLESSA: *a*) experimental arrangement, *b*) acoustic signals.

of successive acoustic echoes superimpose to interferences. If gallium is replaced by germanium, no anomalies are observed at all: apparently acoustic waves in pure metals generate a sufficiently strong r.f. magnetic field reacting with the spinning ³He nuclei in the vicinity of the surface (Fig. 8).

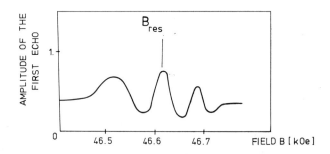

Fig. 8. – Experiment of BELLESSA: field dependence of acoustic signals.

This process of coupling between lattice vibrations in pure metals and liquid ³He may be of great practical importance to account for the observed strong reduction of the thermal Kapitza resistance at millidegree temperatures between metals and liquid ³He.

3. – Scattering of light and X-rays.

3˙1. *Spontaneous and stimulated Brillouin scattering.* – Let us consider (see Fig. 9) a monochromatic light beam (wave vector $k = 2\pi/\lambda$) incident under an angle α with a receding sound wave (wave vector $K = 2\pi/\Lambda$).

Reflection of light occurs (Bragg condition) if $2\Lambda \cos\alpha = \lambda$.

This is equivalent to the conservation of momentum. Since $\lambda_1 \approx \lambda_2$, $\frac{1}{2}(2\pi/\Lambda) = (2\pi/\lambda)\cos\alpha$.

The reflected light suffers a Doppler shift to smaller frequencies since the reflecting object, *i.e.* the sound wave, is receding at a speed v ($\ll$ velocity of light c). The frequency shift is

$$\Delta v = -2 \frac{v}{c} v \frac{\lambda}{2\Lambda} = \frac{v}{\Lambda} = -\text{ sound frequency } \Omega \, .$$

$$\uparrow$$
$$\cos \alpha$$

Quantum mechanically this implies the transfer of the energy $\hbar\Omega$ from the primary light beam to the sound wave.

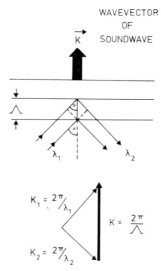

Fig. 9. – Brillouin scattering.

The detection of microwave sound waves by measuring the frequency shift of the scattered light is nowadays a standard procedure. One of the most sensitive Fabry-Perot spectrometers has been built by SANDERCOCK [46].

As we have seen above, each photon reflected by the sound wave loses a quantum $\hbar\Omega$ to the acoustic wave. If one increases the intensity of the incident light, soon a condition is reached such that the acoustic wave gains more energy from the light than it loses by dissipative processes. The acoustic wave gets amplified and grows at the expense of the laser beam energy. Since the reflection coefficient for a given acoustic energy is largest for backscattering, where the incident and reflected light beams largely overlap, this stimulated process (*stimulated Brillouin scattering*) occurs nearly always in the backscattering geometry. The growth rate of the acoustic waves is so rapid that after a fraction of a nanosecond almost all incident light is reflected. Details can be found in many laser textbooks. For us it is only important to remember that SBS is an important tool for the generation of very-high-frequency sound waves of high intensities in almost any transparent medium. By probing the population a little bit later using a second delayed pulse we have measured the phonon lifetimes of 30 GHz phonons in quartz crystal, in glass, liquid and solid helium (1 GHz). For more detailed information see [35, 36].

3`2. *Phonon detection by X-rays.* – If one wants to detect in perfect dielectric crystals phonons of shorter wavelength, scattering of light becomes impossible. In this Subsection we will see that phonon detection by X-ray scattering becomes however possible.

At first one might think—as indicated in Fig. 10—that X-*rays traversing*

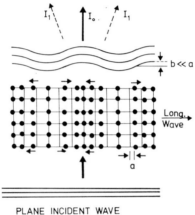

Fig. 10. – Raman-Nath scattering of X-rays.

a longitudinal wave in liquids or solids at right angles could be deflected as is well known to occur for the Raman-Nath scattering of light by an ultrasonic wave in a standard liquid [37]. But it turns out that—due to the small refractive index $n-1$ for X-rays—the modulation amplitude b (see Fig. 10) is much smaller than the acoustic amplitude a and consequently the intensity scattered into the first order $(I_0(b^2/\lambda^2))$ is in general *very small*.

As indicated in Fig. 11 it is more advantageous to observe the X-ray intensity in the *vicinity of a Bragg peak*. For simplicity we will only consider in Fig. 11 *backscattering* for which the Bragg condition is $2d = n\lambda$. A periodic deformation of the X-ray wave front occurs only if a transverse acoustic wave, polarized in the plane of the X-ray beam, crosses this beam at right angles. Now the modulation amplitude equals twice the acoustic amplitude and therefore the intensity scattered into the first order is considerably larger

$$\frac{I_1}{I_0} \simeq \frac{4a^2}{\lambda^2},$$

where a is the acoustic amplitude and λ the X-ray wavelength. (The acoustic strain is a/Λ with Λ being the acoustic wavelength.)

Phonon detection in the vicinity of a Bragg peak is—as just shown— most useful. However, it is unfortunately not possible to use this well-established

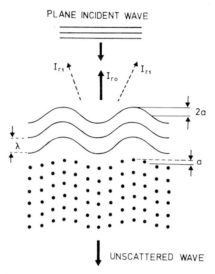

Fig. 11. – Scattering of X-rays near a Bragg peak by a transverse acoustic wave.

technique for the observation of phonons in liquids. The classical literature on phonon-induced X-ray scattering has been reviewed by LAVAL [38], CRIBIER [38], LE CORRE [38], CURIEN [38] and JACOBSEN [38]. More recently microwave phonons excited by the acousto-electric effect have been directly observed and studied by X-ray scattering [39].

3˙3. *Raman scattering by acoustic phonons.* – Recently PETICOLAS *et al.* [40] have observed by means of Raman scattering the thickness vibrations of the crystalline lamellae of polyethylene. (Practically all high polymers form these characteristic $(150 \div 300)$ Å thick folded lamellae upon crystallization.) The thickness vibration caused an inelastic energy loss of the scattered light by about 20 cm^{-1}. In view of the possible importance of similiar acoustic vibrations in biological membranes of the same thickness there is great interest in finding these vibrations by light or by neutron scattering. But, at present, no results have been published yet.

3˙4. *Microwave-induced Raman scattering.* – The Raman signals described in the last Subsection arising from the thermally excited acoustic vibrations of macromolecules are rather weak. It might therefore be advantageous to make use of the fact that these vibrations are often also piezoelectric, and therefore to strongly excite them well above the thermal background by means of the powerful molecular vapour lasers (at $\sim 10^{12}$ Hz) now available. Simultaneous Raman scattering may serve as an indicator for the response of the macromolecules.

It appears also promising to adopt this technique to the study of the acoustic absorption in powdered piezoelectric materials.

Small particles of diameter $d \gtrsim \lambda$ are Raman active also as far as their *acoustic* phonons are concerned, because in a surface layer of thickness λ the momentum conservation is broken. Therefore the fraction $\lambda d^2/d^3 = \lambda/d$ of each particle is fully Raman active for its acoustic phonons. The general statement that acoustic phonons are not Raman active is only true for large particles $(d \gg \lambda)$.

4. – Incoherent phonons.

4`1. *Heat pulses.* – The study of heat pulses of short duration was originally developed in the field of superfluid helium for the investigation of *second sound*, a temperature wave which propagates through the liquid at a speed lower than the velocity of sound [41].

Second sound has also been discovered more recently in highly pure crystals of ^{4}He, NaI and Bi [42]. Second sound can be thought of as a compressional wave in the phonon gas. Thus collisions between thermal phonons are important.

However, at lower temperatures collisions become rare and phonons propagate ballistically (in pure crystals). The study of phonon pulses in this regime has been greatly advanced by VON GUTFELD and NETHERCOT [42].

In their experiments 100 ns heat pulses are generated at one end of a sapphire or quartz rod and detected after some delay by a fast bolometer on the opposite side of the rod. The detected signal showed clearly with sufficient time resolution first the arrival of the longitudinal and somewhat later the transverse phonons.

What remains less clear is the exact frequency distribution of the phonons emitted by the heater. If one assumes for all phonons a temperature equilibrium with the electron gas of the heater, the phonons emitted by the heater during the heating pulse will have a Planck distribution corresponding to the transient heater temperature. On the other hand—as has been pointed out by PERRIN *et al.* [43]—the low-frequency phonons of the metal heater escape to the substrate without much interaction with the overheated electrons, which communicate their energy gain therefore mainly to the high-frequency phonons. To sum up, there seems to be serious doubt as to whether the phonons emitted from an electrically heated metal film really have a Planck distribution. It seems therefore highly desirable to *measure* the frequency distribution of phonons propagating in heat pulse experiments.

4‘2. *Monochromatic detection.* – A monochromatic detection of phonons generated by heat pulses up to frequencies of 10^{12} Hz has become possible by two optical methods, both making use of impurities whose optical behaviour is in a resonant way affected by phonons of a discrete energy [44, 45].

RENK [44] studied the propagation of 10^{12} Hz phonons in ruby—generated by a heat pulse—in the following way. The crystal was pumped by a mercury lamp and the intensity of the R_2-line was compared with that of the R_1-line. The ratio of both intensities depends only on their relative population and this —in turn—is only determined by the density of 10^{12} Hz phonons in the crystal; the splitting between the R_1- and the R_2-level is 29 cm^{-1} $\sim 10^{12}$ Hz. Renk's technique allows a study of the 10^{12} Hz phonon population with a high spatial resolution and a time-recording sufficiently rapid for pulsed observation. By using this technique RENK has demonstrated that phonons of 10^{12} Hz propagate in sapphire at 4 °K at least for one centimeter.

ANDERSON and SABISKY [45] had used earlier the energy levels of another impurity also for the purpose of monochromatic phonon detection. Mostly they used a paramagnetic crystal—CaF$_2$ doped with Tm^{++}—and exposed it to a magnetic field. The two lowest-energy levels are split by the magnetic Zeeman energy $\hbar\omega$ and, if the crystal is kept cool enough, mainly the lowest-energy level is populated. Therefore the magnetization and the magnetic dichroism of the sample is strong. As soon as phonons at the resonance energy $\hbar\omega$ are generated in the sample, the optical dichroism (and the magnetization) can be seen to decrease. The advantage of this method, which has been described in detail elsewhere [31], lies in the fact that phonon energies can be tuned by simply varying the magnetic field. A certain disadvantage is perhaps the relatively slow response which makes its use difficult in pulsed operation. By adopting this method for the study of liquid helium the authors have been able to investigate the acoustic properties of helium films at frequencies of 55 GHz and the acoustic reflection of such high-frequency sound waves from the interface between a crystal and superfluid helium.

REFERENCES

[1]   W. EISENMENGER and A. H. DAYEM: *Phys. Rev. Lett.*, **18**, 125 (1967).
[2]   W. EISENMENGER et al.: *Tunnelling Phenomena*, edited by E. BURSTEIN and LUNDQUIST (New York, N. Y., 1969).
[3]   W. DIETZSCHE and H. KINDER: to be published.
[4]   H. KINDER: *Phys. Rev. Lett.*, **28**, 1564 (1972).
[5]   V. NARAYANAMURTI and R. C. DYNES: *Phys. Rev. Lett.*, **27**, 410 (1971).
[6]   E. SALZMANN et al.: *Physical Acoustics*, **7**, 220 (1970).
[7]   H. ÜBERALL: *Physical Acoustics*, **10** (1973).
[8]   J. DE KLERK: this volume, p. 437.

[9]   H. O. BERKTAY: this volume, p. 369.
[10] J. JOFFRIN: this volume, p. 291.
[11] C. F. QUATE: *Appl. Phys. Lett.*, **24**, 163 (1974).
[12] BEYER and S. V. LETCHER: *Physical Acoustics* (New York, N. Y., 1969).
[13] K. N. BARANSKII: *Sov. Phys. Doklady*, **2**, 237 (1957).
[14] H. BÖMMEL and K. DRANSFELD: *Phys. Rev. Lett.*, **1**, 234 (1958); **3**, 83 (1959).
[15] E. H. JACOBSEN: *Phys. Rev. Lett.*, **2**, 249 (1959).
[16] J. LUKOR and E. H. JACOBSEN: *Physical Acoustics*, **5**, 221 (1968).
[17] N. F. FOSTER: *Handbook for Thin Film Technology*, edited by L. F. MAISSEL and R. GLANY, Chap. 15 (New York, N. Y., 1970).
[18] J. DE KLERK: *Physical Acoustics*, **4**, 195 (1969).
[19] SITTIG: *Physical Acoustics*, **11** (1972).
[20] W. ARNOLD et al.: *Appl. Phys.*, submitted (1974).
[21] D. L. WHITE: *Physical Acoustics*, **1** B, 321.
[22] A. KAMGAR, P. KNESCHAUREK, G. DORDA and J. F. KOCH: *Phys. Rev. Lett.*, **32**, 1251 (1974).
[23] H. KAWAI: *Japan Journ. Appl. Phys.*, **8**, 975 (1969).
[24] E. FUKADA and S. TAKASHITA: *Japan Journ. Appl. Phys.*, **8**, 960 (1969).
[25] M. V. SCHICKFUSS and H. SUSSNER: Pat. application.
[26] H. SUSSNER, D. MICHAS, A. ASSFALG, S. HUNKLINGER and K. DRANSFELD: *Phys. Lett.*, **45** A, 475 (1973).
[27] H. BÖMMEL et al.: *Phys. Rev. Lett.*, **3**, 83 (1959).
[28] A. BIALAS and O. WEIS: *Appl. Phys. Lett.*, **13**, 81 (1968).
[29] E. R. DOBBS: *Physical Acoustics*, **10**, 127 (1973).
[30] V. F. GANTMAKER and V. T. DOLGOPOLOV: *JETP Lett.*, **5**, 12 (1967).
[31] P. K. LARSEN and K. SAERMARK: *Phys. Lett.*, **24** A, 374, 668 (1967).
[32] B. ABELES: *Phys. Rev. Lett.*, **19**, 1181 (1967).
[33] R. W. COHEN: *Phys. Lett.*, **29** A, 85 (1969).
[34] G. BELLESSA: *Phys. Rev. Lett.*, **32**, 46 (1974).
[35] W. HEINICKE, G. WINTERLING and K. DRANSFELD: *Phys. Rev. Lett.*, **22**, 170 (1969).
[36] P. LEIDERER, P. BERBERICH and S. HUNKLINGER: *Rev. Sci. Inst.*, **44**, 1610 (1973).
[37] L. BERGMANN: *Ultraschall* (Berlin, 1954).
[38] J. LAVAL: *Rev. Mod. Phys.*, **30**, 222 (1958); D. CRIBIER: *Rev. Mod. Phys.*, **30**, 228 (1958); Y. LE CORRE: *Rev. Mod. Phys.*, **30**, 229 (1958); H. CURIEN: *Rev. Mod. Phys.*, **30**, 232 (1958); E. H. JACOBSEN: *Rev. Mod. Phys.*, **30**, 234 (1958).
[39] D. G. CARLSON and A. SEGMÜLLER: *Phys. Rev. Lett.*, **27**, 195 (1971).
[40] W. L. PETICOLAS, G. W. HIBLER, J. L. LIPPERT, A. PETERLIN and H. OLF: *Appl. Phys. Lett.*, **18**, 87 (1971).
[41] J. WILKS: *Liquid Helium* (Oxford, 1970).
[42] R. J. VON GUTFELD et al.: *Physical Acoustics*, **6** A, 233 (1968).
[43] N. PERRIN and H. BUDD: *Phys. Rev. Lett.*, **28**, 1701 (1972).
[44] K. F. RENK and J. DEISENHOFER: *Phys. Rev. Lett.*, **26**, 764 (1971).
[45] CH. ANDERSON and E. S. SABISKI: *Physical Acoustics*, **8**, 2 (1971).
[46] J. R. SANDERCOCK: *Opt. Comm.*, **2**, 73 (1970); *Proceedings of the International Conference on Light Scattering in Solids* (Paris, 1971).

Les échos de phonons.

J. Joffrin et A. Levelut

Laboratoire d'Ultrasons (*), *Université Paris VI - Paris*

1. – Introduction.

Depuis deux ou trois ans, on a vu se développer en U.R.S.S., en France et aux Etats-Unis, toute une série d'études qu'on peut regrouper sous le vocable d' « échos de phonons ». Ce nom n'est pas fortuit. Comme il l'indique, ces études ont d'abord en commun de mettre en jeu des phonons; ceux-ci ont n'importe quelle fréquence: on connait en effet des expériences qui couvrent une gamme allant de quelques mégahertz jusqu'a 36 GHz. Ces expériences ne couvrent une si large bande que parce que le spectre des vibrations élastiques des cristaux est lui-même fort étendu. Par ailleurs, et c'est une différence avec l'acoustique habituelle, elles ne nécessitent aucune propriété de résonance mécanique des échantillons, bien que la résonance favorise l'effet.

D'autre part, le mot « écho » doit s'entendre d'une manière tout-à-fait différente de celle utilisée en acoustique. Les spécialistes de cette technique savent comment on fait parcourir un certain nombre d'allers et retours à un train d'onde ultrasonore dans une cellule d'expérience; à chaque fois que le train d'onde revient à l'émetteur, on obtient un signal ou « écho ultrasonore ». Au sens des échos de phonons, le mot « écho » doit être pris tel qu'il est appliqué aux systèmes de spins car l'explication du phénomène observé conduit à utiliser les mêmes concepts. C'est un autre exemple physique où le résultat de l'action de deux impulsions n'est pas simplement la somme des actions de chacune des impulsions; il y a donc nécessairement une non-linéarité dans l'évolution du système.

Le but de cet article est donc triple; dans un premier temps, il vise à expliquer à un public d'acousticiens et de physiciens en quoi consiste une expérience d'échos de phonons, quelles en sont les caractéristiques essentielles nouvelles; il espère souligner ce qui, *a priori*, les rend séduisantes et fixer des ordres de grandeur.

(*) Associé au Centre National de la Recherche Scientifique.

Dans un deuxième temps, son but est de fournir des explications microscopiques sur les concepts qui permettent de comprendre ce phénomène et sur les mécanismes élémentaires mis en jeu. Quoi qu'ils soient encore relativement obscurs, on en sait suffisamment à l'heure actuelle pour décrire deux mécanismes importants et pour se fixer des règles empiriques dans le but d'augmenter la grandeur de l'écho.

Dans un troisième temps enfin, cet article cherche à montrer quelle peut être l'utilité de cette nouvelle technique aussi bien du point du vue du physicien que du point de vue de l'ingénieur; paradoxalement, on verra que ses applications semblent plus prometteuses en ce moment pour le deuxième que pour le premier.

2. – Une expérience d'échos de phonons: les échos à 2τ.

Une expérience standard d'échos de phonons comporte les étapes suivantes:

Production et excitation d'une population de phonons de fréquence centrale ω_1 et de largeur $\Delta\omega_1$ à l'instant de référence zéro; pour cela tous les moyens d'excitation sont bons: transducteurs résonnants à basse fréquence, peignes interdigités pour ondes de surface, cavités résonnantes pour les plus hautes fréquences. Cette population, suivant les modes de production, comprend des ondes de polarisation et de vecteur d'onde bien définis ou au contraire des ondes qui n'ont en commun que leur fréquence. Mais dans tous les cas, la phase des différentes ondes est imposée par celle du générateur électrique qui leur donne naissance: elle est fixée pour $t = 0$ et en un point (ou une surface) où toutes ces ondes sont engendrées. Cette remarque sur la phase est essentielle; elle signifie encore que la première impulsion de champ électrique a introduit une cohérence entre les diverses ondes.

Deuxième étape: application à l'instant τ sur tout ou partie de l'échantillon dans lequel se propagent les ondes incidentes d'une seconde impulsion de champ électrique uniforme dans l'espace, de fréquence ω_2 et de largeur $\Delta\omega_2$. S'il existe dans le cristal un mécanisme d'interaction non linéaire entre le champ électrique et le carré de la déformation élastique, alors il y aura production d'ondes ultrasonores réfléchies, de vecteurs d'ondes opposés à ceux des ondes créées par la première impulsion de champ électrique. C'est une opération de renversement du vecteur d'onde, ou de renversement du sens du temps. Ces ondes réfléchies vont cheminer en sens inverse dans le cristal, et suivre un trajet identique à celui parcouru par les ondes incidentes.

Troisième étape: on détecte un écho à l'instant 2τ. La population réfléchie retourne en effet vers les sources qui ont donné naissance aux ondes incidentes et tous les modes excités se focalisent, en phase, à l'instant 2τ en produisant un écho généralement détecté sous forme de signal électrique (Fig. 1).

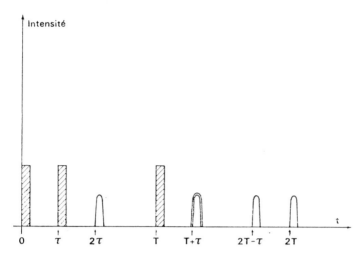

Fig. 1. – Séquence à trois impulsions. En hachuré on a représenté les trois impulsions, avec un trait simple les échos de genre 2τ, avec trait double l'écho de genre $T + \tau$.

Il est tout-à-fait remarquable que cet écho, et pour les raisons précédentes, soit indépendant de la forme de l'échantillon et de la localisation des sources sonores; cela est dû à la réversibilité des propriétés élastiques de l'échantillon, à son charactère markovien tant que l'on néglige les interactions phonons-phonons, et enfin à la cohérence de la population incidente et donc de la population réfléchie.

Toujours en ignorant le détail du mécanisme d'interaction non linéaire, on peut rendre l'explication précédente plus convaincante en calculant la réponse en termes classiques tout en la formulant dans un langage qui sera généralisable à des situations plus compliquées. On développera un formalisme faisant intervenir les modes propres élastiques; ils correspondent aux ondes progressives repérées par un vecteur d'onde dans le cas où l'on ne tient pas compte des effets de bord et des impuretés.

Pour un cristal harmonique, l'hamiltonien élastique est

$$(1) \qquad \mathscr{H}_{\text{élast}} = \sum_k (\dot{e}_k^* \dot{e}_k + \Omega_k^2 e_k^* e_k) \lambda_k \,,$$

k est un indice de mode. e_k est le vecteur propre associé avec le mode k; c'est une grandeur dépendant du temps. Pour un cristal de forme quelconque, la décomposition sous la forme (1) n'est pas immédiate; mais, en principe, on peut toujours diagonaliser l'hamiltonien, car il est hermitique. λ_k est de l'ordre c/Ω_k^2, où c est une constante élastique. Les e_k constituent un ensemble de vecteurs orthogonaux, mais non normés. On écrira qu'en un point R la déformation locale $\varepsilon(R, t)$ peut se développer sur les modes e_k, soit

$$(2) \qquad \varepsilon(R, t) = \sum_k a_k(R) e_k(t)$$

avec

(2′)
$$\int \mathrm{d}^3 R \, a_k(R) a_{k'}(R) = \delta_{kk'} \, .$$

L'équation du mouvement libre du mode k est

$$\ddot{e}_k(t) + \Omega_k^2 e_k(t) = 0 \, ,$$

et la solution de Green $G_k(t)$ de cette équation est telle que

(3)
$$\ddot{G}_k(t) + \Omega_k^2 G_k(t) = \delta(t) \, .$$

La solution causale de (3) est

(4)
$$G_k(t) = \frac{1}{2} \frac{Y(t)}{\Omega_k} \sin \Omega_k t \, .$$

La fonction de Green complète du problème élastique est donc

$$G(R, R' ; t) = \sum_k a_k(R) a_k^*(R') G_k(t) \frac{1}{\lambda_k} \, .$$

On écrira que le champ électrique de la première impulsion est

$$E_1(t) = E_1 \exp\left[-i\omega_1 t\right] f_1(t) \, ,$$

où $f_1(t)$ est la fonction de forme. On supposera que $f_1(t)$ est à support borné et que $f_1(t) \neq 0$ pour $0 < t < \Delta t_1$ (on a donc $\Delta t_1 \simeq 1/\Delta\omega_1$):

$$f_1(t) = \int \mathrm{d}(\delta\omega_1) \exp\left[i \, \delta\omega_1 t\right] \tilde{f}_1(\delta\omega_1) \, .$$

On appellera S le point courant de la surface d'émission.

L'interaction non linéaire du champ électrique uniforme $E_2(t)$ de la seconde impulsion et des ondes acoustiques est écrite sous la forme

(5)
$$\mathscr{H}_{\text{int}} = \int \mathrm{d}^3 R \, \beta\big(E_2(t)\varepsilon^*(Rt)\varepsilon^*(Rt) + \text{c.c.}\big) \, .$$

En portant (2) dans (5) et en utilisant les relations d'orthogonalité des a_k, on arrive à l'expression qui servira de point de départ:

(5′)
$$\mathscr{H}_{\text{int}} = \beta\Big(E_2(t) \sum_k e_k^*(t) e_k^*(t) + \text{c.c.}\Big) \, .$$

On prendra

(6)
$$E_2(t) = E_2 \exp\left[-i\omega_2 t\right] f_2(t) \, ,$$

$f_2(t)$ est la fonction de forme de la deuxième impulsion; par convention elle est non nulle pour $\tau < t < \tau + \Delta t_2$:

$$f_2(t) = \int \mathrm{d}(\delta\omega_2) \exp\left[i\,\delta\omega_2\,t\right]\tilde{f}_2(\delta\omega_2)\,.$$

L'expression (5') de $\mathscr{H}_{\mathrm{int}}$ tient compte du fait que le champ électrique appliqué est presque uniforme dans l'espace, tout au moins vis à vis de l'échantillon. β est un coefficient de couplage non linéaire que, pour simplifier, on a pris indépendant de k. On reviendra sur cette approximation ultérieurement.

Ceci fait, en présence du champ électrique, l'équation du mouvement de la grandeur $e_k(t)$ comporte un second membre qui joue la rôle de source. Soit

(7) $$\ddot{e}_k(t) + \Omega_k^2 e_k(t) = -\beta E_2\big(e_k^*(t)\exp\left[-i\omega_2 t\right]f_2(t) + \mathrm{c.c.}\big)\frac{1}{\lambda_k}\,.$$

Puisque β est petit, on résoudra cette équation par approximation en écrivant que, au premier ordre, dans le second membre $e^* = e^*_{\mathrm{incidente}}$ et que dans le premier $e = e_{\mathrm{réfléchie}}$. Connaissant la fonction de Green $G(R, R'; t)$, on a

(8) $$\varepsilon_{\mathrm{réfl}}(Rt) = -\beta \int_{-\infty}^{t}\mathrm{d}t' \int \mathrm{d}^3R'\; G(R, R'; t-t')\varepsilon_{\mathrm{inc}}^*(R't')E_2(t')\,.$$

Par ailleurs, on sait calculer $\varepsilon_{\mathrm{inc}}(Rt)$ à partir des contributions de chaque point S de la surface d'émission:

(9) $$\varepsilon_{\mathrm{inc}}(R't') = \alpha \int_{-\infty}^{t'}\mathrm{d}t'' \int \mathrm{d}^3S\; G(R', S; t'-t'')E_1(t'')\,,$$

α est un coefficient piézoélectrique que par simplification on a pris indépendant de S.

Après avoir remplacé (9) dans (8) et utilisé les propriétés d'orthogonalité des modes, on arrive à la solution suivante:

$$\varepsilon_{\mathrm{réfl}}(Rt) = -\frac{\beta\alpha}{4}\int_{-\infty}^{t}\mathrm{d}t'\int_{-\infty}^{t'}\mathrm{d}t''\int \mathrm{d}^3R'\int \mathrm{d}^3S \sum_{k,k'}Y(t-t')\,Y(t'-t'')\cdot$$

$$\cdot\, a_k(R)a_k^*(R')a_{k'}^*(R')a_{k'}(S)\frac{1}{\lambda_k}\frac{1}{\lambda_{k'}}\frac{1}{\Omega_k\Omega_{k'}}\sin\Omega_k(t-t')\sin\Omega_{k'}(t'-t'')E_2(t')E_1^*(t'')\,.$$

Le résultat est simple à obtenir: l'intégration d^3R' entraîne $k = k'$. Par ailleurs, si les deux impulsions sont bien séparées dans le temps, les intégrales $\int \mathrm{d}t'$ et $\int \mathrm{d}t''$ s'effectuent simplement et les bornes peuvent être étendues à l'infini en comparaison de la largeur de f_1 et de f_2.

On tire

$$\varepsilon_{\text{réfl}}(R, t) = + \beta\alpha \int \mathrm{d}^3S \sum_k a_k(R) a_k^*(S) \frac{1}{\Omega_k^2} \int \mathrm{d}(\delta\omega_2) \int \mathrm{d}(\delta\omega_1) \int \mathrm{d}t' \int \mathrm{d}t'' \cdot$$

$$\cdot \exp\left[-i(\omega_2 - \delta\omega_2)t'\right] \exp\left[+i(\omega_1 - \delta\omega_1)t''\right] \tilde{f}_1^*(\delta\omega_1) \tilde{f}_2(\delta\omega_2) \cdot$$

$$\cdot \frac{1}{\Omega_k^2} \frac{1}{\lambda_k^2} E_1 E_2 \sin \Omega_k(t - t') \sin \Omega_k(t' - t'') .$$

Les termes non nuls qui subsistent après intégration en t' et t'' sont tels que

$$\Omega_k = \omega_1 - \delta\omega_1 ,$$

$$2\Omega_k = \omega_2 - \delta\omega_2 ,$$

$$(10) \qquad \varepsilon_{\text{réfl}}(R, t) = \frac{\beta\alpha}{4} \sum_k \int \mathrm{d}^3S \, a_k(R) a_k^*(S) \frac{1}{\Omega_k^2 \lambda_k^2} \cdot$$

$$\cdot \exp\left[-i\Omega_k t\right] \tilde{f}_1^*(\omega_1 - \Omega_k) \tilde{f}_2(\omega_2 - 2\Omega_k) E_1 E_2 .$$

La formule (10) contient toutes les informations physiques dont on a besoin pour expliquer la formation d'un écho. On a calculé ici la déformation élastique engendrée en R, t par application des deux impulsions successives; en pratique, c'est plutôt un champ électrique que l'on détecte et le signal $s(t)$ collecté sur la surface de réception de point courant R est la somme linéaire, à un coefficient de conversion près, des différents $\varepsilon_{\text{réfl}}(R, t)$; $s(t) = \int \mathrm{d}^3R \, \varepsilon_{\text{réfl}}(R, t)$. Le coefficient de conversion $\varepsilon \to E$ est proportionnel au même coefficient piézoélectrique α que celui qui joue à l'émission.

Les propriétés d'orthogonalité des $a_k(R)$ imposent, par ailleurs, que pour maximiser le signal on prenne $R \simeq S$; la surface où l'on détecte l'écho doit donc être la même que celle où l'on a excité les modes de vibration au moment de la première impulsion.

Il reste encore dans (10) une somme $\sum_k$ à effectuer que l'on transforme en une intégrale sur l'écart $\delta\Omega_k$ par rapport à la fréquence centrale. En prenant $2\omega_1 = \omega_2$, les dépendances en temps donnent un signal dont la fréquence centrale est ω_1 et le facteur de forme est

$$(11) \qquad f_s(t) \simeq \int \delta\Omega_k \, \tilde{f}_k^*(\delta\Omega_k) \tilde{f}_2(2\,\delta\Omega_k) \exp\left[i\,\delta\Omega_k t\right] n(\delta\Omega_k) ,$$

$n(\delta\Omega_k)$ densité de modes de fréquence Ω_k.

Soit, en définitive,

$$(12) \qquad s(t) \simeq \omega_1^2 n(\omega_1) \exp\left[-i\omega_1 t\right] f_s(t) E_1 E_2 \frac{\beta}{2\varrho v^2} \frac{\alpha}{\varrho v^2} \frac{\alpha}{\chi_L} .$$

Ce résultat qui conclut les calculs appelle les commentaires suivants.

1) La forme du signal de sortie donnée dans (11) correspond à la convolution

$$(11') \qquad f_1(-t) * f_2\left(\frac{t}{2}\right) \simeq f_s(t) \,.$$

Avec les conventions adoptées, $f_1(t)$ est localisée autour de $t = 0$, $f_2(t)$ autour de $t = \tau$; en conséquence $f_s(t)$ est située près de $t = 2\tau$. Si f_1 et f_2 sont des fonctions carrées de largeurs temporelles Δt_1 et Δt_2, f_s aura une forme trapézoïdale de largeur au pied $\Delta t_1 + 2 \Delta t_2$; c'est une propriété connue de la convolution.

2) β est un tenseur de rang cinq. Le cristal sur lequel s'effectue l'expérience doit au moins en permettre l'existence: à cet effet, il doit être dénué de centre de symétrie; en mettant à part la seule classe de symétrie 432 il doit donc être piézoélectrique. On note dans la formule (12) qu'il intervient le terme α^2, carré du coefficient piézoélectrique effectif; un acousticien se convaincra aisément qu'en réalité il doit intervenir le carré du coefficient de couplage électro-mécanique K^2.

Par ailleurs, α est un tenseur; en différents points source S, le rendement de l'interaction piézoélectrique n'est pas le même car ce ne sont pas les mêmes composantes du tenseur α qui interviennent. Dans l'éq. (10), nous avons négligé cet effet.

3) La formule (5) montre que βE_2 est une mesure de la variation de la constante élastique du cristal c sous l'influence du champ électrique E_2. Une évaluation du coefficient de couplage associé au phénomène non linéaire est donc

$$(13) \qquad \frac{\beta E_2}{2c} = \frac{\beta E_2}{2\varrho v^2} = r \,,$$

où v est une vitesse acoustique moyenne pour le cristal.

4) La présentation précédente a focalisé l'attention sur l'aspect cohérence et phase de l'écho; en s'écartant de l'éq. (8), où l'on a séparé les évolutions des ondes incidentes et réfléchies, on peut, comme dans tous les phénomènes paramétriques, résoudre simultanément les équations d'évolution des amplitudes des deux ondes pour une amplitude de pompe (E_2) fixe.

L'onde réfléchie a, pendant la durée de la deuxième impulsion, une amplitude qui évolue comme

$$(14) \qquad \varepsilon_{\text{réfl}}(t) = \varepsilon_{\text{inc}} \sinh{(r\omega_1 t)} \,.$$

On peut en principe rendre ce terme aussi grand que l'on veut.

En appelant Γ le facteur qui donne l'absorption par unité de temps des modes acoustiques excités (incidents ou réfléchis), (14) reste encore valable mais avec r défini comme il suit:

$$(15) \qquad r^2 = \left(\frac{\beta E_2}{2\varrho v^2}\right)^2 - \left(\frac{\Gamma}{2\omega_1}\right)^2.$$

(15) est une condition de seuil.

5) L'expression (5′) repose sur l'uniformité du champ électrique. Ainsi le renversement du temps, ou du vecteur d'onde, si l'on peut parler en ces termes, se fait mode à mode; pour employer le langage des opticiens, on dira qu'il y a conservation du vecteur d'onde dans le processus paramétrique $(E_2, \varepsilon_{\text{inc}}, \varepsilon_{\text{réfl}})$.

On se rend aisément compte en analysant (10) que toute inhomogénéité du champ électrique appliqué E_2 mélangera les modes propres du cristal et diminuera le signal.

6) L'interaction en $\beta E_2(t)\varepsilon\varepsilon$ (éq. (5)), n'est pas la seule possible. Il est clair qu'un couplage de la forme

$$(16) \qquad \mathcal{H}_{\text{int}} = \int \mathrm{d}^3 R\, \gamma \big(E_2(t)\,E_2(t)\,\varepsilon^*(R,t)\,\varepsilon^*(R,t) + \text{c.c.}\big)$$

a les mêmes propriétés en ce qui concerne le renversement du temps. Toutefois, si ω_2 est la fréquence centrale de la seconde impulsion, la condition à satisfaire dans ce cas est $\omega_2 = \omega_1$. γ est un tenseur de rang six, qui existe quelle que soit le symétrie du cristal.

7) Dans le cas de l'interaction β (éq. [5]), la génération de l'écho a lieu si la seconde impulsion est à la fréquence $\omega_2 = 2\omega_1$ ou bien si la fréquence ω_1 de la deuxième impulsion peut être doublée par un processus non linéaire interne au cristal. Ceci peut être réalisé par l'anharmonicité diélectrique du cristal, représentée par un tenseur χ_{NL}; il est de rang trois, donc compatible avec l'absence de centre de symétrie. Dans ce dernier cas, la fonction de forme du signal est

$$(17) \qquad f_s(t) = f_1(-t) * f_2^2\left(\frac{t}{2}\right).$$

8) Nous avons vu que l'existence de l'interaction β entraînait la présence d'un terme de source au second membre de l'éq. (7). Tout se passe comme si on avait une source piézoélectrique en volume, tout au moins dans le volume du cristal occupé par l'onde incidente. Cela peut se voir directement si l'on écrit la densité d'énergie électroacoustique de la façon suivante:

$$u(R) = \alpha E\varepsilon + \beta E\varepsilon\varepsilon = (\alpha + \beta\varepsilon)E\varepsilon.$$

L'effet a lieu en volume car l'interaction conserve le vecteur d'onde (en supposant qu'on a affaire à des ondes progressives), au contraire de l'interaction piézoélectrique $\alpha E\varepsilon$ qui n'est génératrice d'ondes élastiques que sur une surface de discontinuité: la surface du cristal.

9) En pratique, l'amplitude du signal s dépend de l'intervalle de temps τ: il y a une relaxation, c'est-à-dire que le signal, détecté à l'instant 2τ, varie comme

$$(18) \qquad s(2\tau) = s_0 \exp\left[-\frac{2\tau}{T_2}\right].$$

L'origine du temps de relaxation T_2 devient un nouveau problème posé au physicien.

3. – Une expérience d'échos de phonons: les échos à $T + \tau$.

Il est toujours tentant, après avoir appliqué deux impulsions de champ électrique, d'essayer d'en appliquer une troisième. Ceux qui l'ont fait ont immédiatement repéré l'existence d'une nouvelle sorte d'écho aux caractéristiques très différentes de celui baptisé précédemment de type 2τ. Si 0, τ, T sont les temps correspondant à une séquence d'impulsions de champ électrique de même fréquence (Fig. 1), il apparaît un nouvel écho à l'instant $T + \tau$: celui-ci présente la caractéristique d'avoir une amplitude qui varie avec T et τ comme

$$(19) \qquad s(t) \simeq \exp\left[-\frac{2\tau}{T_2}\right]\exp\left[-\frac{T-\tau}{T_1}\right].$$

Les propriétés essentielles de cet écho sont les suivantes. Dans tous les cas où il a été mesuré, T_1 est toujours beaucoup plus long que T_2; les trois impulsions de champ électrique doivent avoir la même fréquence; si $f_1(t)$, $f_2(t)$, $f_3(t)$ sont les fonctions de forme des trois impulsions, le signal a la fonction de forme suivante:

$$(20) \qquad f_s(t) = f_1(-t) * f_2(t) * f_3(t) ;$$

il est proportionnel au produit des trois champs électriques $E_1 E_2 E_3$; émission et détection impliquant une conversion électromécanique, il apparaît encore le coefficient de couplage K^2 dans l'expression de $s(t)$; enfin, cet écho, comme celui à 2τ, ne peut être détecté que sur la surface d'émission.

Mais il reste à expliquer, phénoménologiquement au moins, ce qui se passe dans l'échantillon pendant l'intervalle $(0, T + \tau)$. On peut présenter les choses ainsi: la première impulsion a pour objet de produire une population cohérente de phonons de fréquence ω_1. La seconde impulsion, uniforme dans l'espace, de fréquence centrale ω_1, « fige » l'onde acoustique dans le cristal qui possède un mécanisme d'inscription de l'information. Il en résulte une modulation spatiale statique de la constante piézoélectrique effective du milieu; cette modulation reproduit, au moment où on la fige, la distribution de l'amplitude de vibration de chacun des modes excités par la première impulsion. En principe, si rien ne vient perturber le cristal, cette modification de la constante piézoélectrique peut persister très longtemps; c'est la raison pour laquelle T_1 est grand devant T_2. T_1 traduit cette persistance, alors qu'au contraire T_2 est lié à l'absorption des phonons dans l'intervalle $(0, \tau)$. La troisème impulsion, appliquée à l'instant T, doit être uniforme dans l'espace et de fréquence centrale ω_1; associée à la modulation spatiale de la constante piézoélectrique, elle produit une source de tension élastique qui remplit très exactement les conditions d'accord de phase pour engendrer une onde acoustique; dans l'intervalle de temps $(T, T + \tau)$, une onde «réfléchie» chemine en sens inverse dans le cristal et rejoint la surface d'émission où la population se retrouve focalisée en phase à l'instant $T + \tau$. Une explication plus mathématique du processus de formation de cet écho suivrait les mêmes lignes que celles développées pour celui de type 2τ.

On peut, à ce stade, souligner quelques similitudes entres les deux types d'échos: surface de détection et surface d'émission doivent être identiques; la conversion se fait mode à mode; la forme des échantillons, leur coupe, n'ont pas d'importance; la cohérence des populations excitées par la première impulsion est essentielle.

Inversement, les différences tiennent surtout à cet état figé intermédiaire et à l'origine du coefficient qui couple paramètriquement l'onde sonore, le champ électrique et l'état figé.

Mais le schéma d'explication précédent a déjà hypothéqué le processus microscopique qui assure l'existence de cet état figé intermédiaire et du coefficient du couplage. Deux explications ont successivement été fournies pour ces échos à $T + \tau$. La première utilisait encore le coefficient électroacoustique β, valable pour les échos à 2τ; la deuxième, liée à un processus d'interaction acoustoélectrique entre porteurs libres et ondes acoustiques, conduisait à l'existence de temps de vie des états figés beaucoup plus longs; on a appelé β' le coefficient qui donne dans ce cas l'intensité du couplage non linéaire. On reviendra ultérieurement sur ces coefficients β et β'; mais c'est en raison de la grande efficacité de ce deuxième processus qu'on a choisi le schéma d'explication précédent. On se souviendra seulement qu'il n'est peut-être pas unique.

4. – Les expériences d'échos de phonons.

Avant que les concepts précédents soient précisés, on avait vu apparaître dans la littérature quelques expériences dont on sait maintenant qu'elles constituent autant d'expériences qui portent sur les échos de phonons de type 2τ.

Généralement effectuées sur des cristaux ferroélectriques tels que SbSI [1-4], par des physiciens orientés vers des études de résonance quadrupolaire, elles impliquaient des basses fréquences (2 à 70 MHz) et s'étendaient de ce fait sur une grande gamme de température.

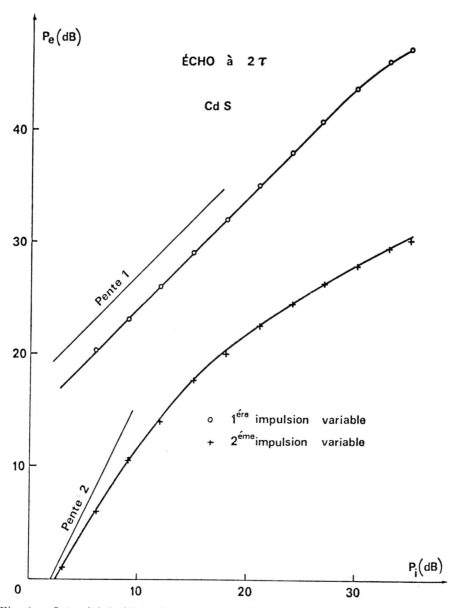

Fig. 2. – Intensité de l'écho à 2τ dans un cristal de CdS, en fonction des intensités de deux impulsions. L'expérience est faite à 4.2 °K, avec une seule cavité résonnant à $\nu = 9.07$ GHz. Les variations attendues, indiquées par les droites de pentes 1 et 2, sont observées à bas niveau.

Dans la gamme des hautes fréquences (10 GHz), le groupe de KAZAN, qui avait étudié depuis fort longtemps les échos de spins, les échos cyclotroniques …, publiait les premières observations sur KDP et des composés ferroélectriques analogues [5]; il suggérait que les échos étaient liés à l'excitation cohérente d'une assemblée d'oscillateurs [6].

Ces résultats furent repris et étendus au même moment dans une suite de trois articles où les auteurs s'efforcèrent de vérifier [7] les relations (11) et (20), de montrer que les phonons jouaient un rôle essentiel dans le phénomène [8] et de fournir une explication à l'origine de ces échos [9] de type 2τ ou $T + \tau$ en même temps qu'ils les observaient dans de nombreux cristaux (Fig. 2 et 3).

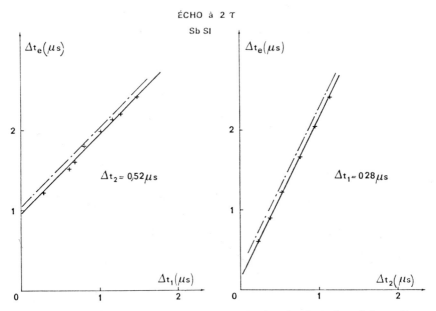

Fig. 3. – Durée de l'écho à 2τ en fonction des durées des deux impulsions. La variation attendue $\Delta t_e(2\tau) = \Delta t_1 + 2\,\Delta t_2$ est représentée par les droites en traits et points. Les conditions expérimentales sont température 4.2 °K, fréquence $\nu = 9.30$ GHz, cristal de SbSI, une seule cavité.

Ils exploraient à la suite toute une série d'aspects des échos de phonons tels que les auto-échos [10], le processus paramètrique de couplage [11] et le temps de relaxation T_2 dans les poudres cristallines [12]. Cette série d'articles orientait la compréhension du phénomène et l'origine du coefficient β vers l'existence d'un terme non linéaire dans le développement de l'énergie du cristal en fonction de la déformation et du champ électrique:

$$(21) \qquad \beta_{ijklm}\varepsilon_{ij}\varepsilon_{kl}E_m \,.$$

C'était introduire un terme électroacoustique.

Peu de temps après, une autre origine au coefficient de couplage de (21) était suggérée [13]; elle tenait à l'existence de charges plus ou moins libres dans le cristal. Le couplage résultait d'une interaction non linéaire entre cette densité de charges et les ondes sonores au sens où elle était connue depuis une dizaine d'années dans le cadre plus général des phénomènes acoustoélectriques. Cette deuxième explication, sans contredire la première, avait l'avantage de fournir un coefficient de couplage, que dans la suite nous appelons β', beaucoup plus grand que β en même temps qu'elle donnait une explication physique à la grande valeur de T_1.

Nous aurons l'occasion de revenir plus en détail sur ces mécanismes.

5. – Aperçu physique sur les mécanismes d'interaction.

Nous voudrions donner maintenant des informations supplémentaires concernant les deux mécanismes qui conduisent à la formation d'un écho. Pour le processus électroacoustique, la présentation sera phénoménologique; pour le processus acoustoélectrique, la description donnera seulement un rapide aperçu de cet effet qui nécessite encore des recherches.

5`1. *Processus électroacoustique: coefficient β.*

5`1.1. Développement au troisième ordre de l'équation d'état. – Le coefficient β s'introduit naturellement dans le développement de l'énergie interne (par unité de volume) en fonction des variables E et ε, lorsque l'on va au delà du deuxième degré.

En effet, aux termes bien connus $U = \frac{1}{2}c\varepsilon^2 + \alpha E\varepsilon + \frac{1}{2}\chi_L E^2$ qui représentent les énergies élastique, piézoélectrique et électrique, s'ajoutent ceux du troisième degré $+ \mathrm{d}\varepsilon^3 + \beta E\varepsilon^2 + QE^2\varepsilon + \chi_{\mathrm{NL}}E^3$. On reconnaît les termes d'anharmonicité élastique $(\mathrm{d}\varepsilon^3)$, diélectrique $(\chi_{\mathrm{NL}}E^3)$, d'électrostriction $(QE^2\varepsilon)$ auxquels s'ajoute $\beta E\varepsilon^2$ qui nous intéresse au premier chef. On peut d'ailleurs poursuivre ce développement et faire apparaître $\gamma EE\varepsilon\varepsilon$ à l'ordre suivant [10, 14].

β est apparu jusqu'à maintenant sous la forme d'un simple coefficient; en réalité, c'est un tenseur de rang 5 dont les éléments sont définis par

$$(22) \qquad\qquad \beta_{ijklm}\varepsilon_{ij}\varepsilon_{kl}E_m \,.$$

Il n'est donc non nul que pour les cristaux ne possédant pas de centre de symétrie.

Mais une question se pose: la forme (22) admise est-elle légitime? Autrement dit, cette écriture qui, comme pour l'énergie élastique $\frac{1}{2}c\varepsilon^2$, ne prend en compte que la partie symétrique ε du tenseur des déformations est-elle correcte? La réponse est non.

On peut s'en convaincre de deux manières.

L'argument pour justifier l'expression $\frac{1}{2}c\varepsilon^2$ est le suivant. Une rotation du cristal ne change pas son énergie interne; cela résulte de l'isotropie de l'espace. Mais pour le terme en β, qui peut être considéré comme une contribution dépendant de E à la constante élastique, l'argument ne s'applique plus puisque le champ électrique $\boldsymbol{E}$ brise l'isotropie de l'espace.

L'autre argument est calqué sur un résultat connu qui concerne la photoélasticité [15-16] et pour lequel on sait que la variation de l'indice fait intervenir les déformations symétriques et les rotations (déformations antisymétriques). La raison en est qu'une rotation du cristal fait tourner l'ellipsoïde des indices; un observateur fixe y est sensible, sauf dans des cas particuliers dont le plus évident est celui d'une sphère des indices. Dans notre cas, une rotation change, pour un observateur fixe, le tenseur piézoélectrique, sauf, ici aussi, dans quelques cas particuliers.

On doit donc définir le tenseur β par

$$(23) \qquad\qquad U = \ldots + \beta_{ijklm}\frac{\partial u_i}{\partial x_j}\frac{\partial u_k}{\partial x_l}E_m$$

et la seule symétrie indicielle qu'il possède, dans la limite des petits vecteurs d'onde, est $\beta_{ijklm} = \beta_{klijm}$.

Les quatre termes, où $u_{i,j} = \partial u_i/\partial x_j$,

$$\beta_{ijklm}u_{i,j}u_{k,l}E_m\,, \qquad +\,\beta_{jiklm}u_{j,i}u_{k,l}E_m\,,$$

$$+\,\beta_{ijlkm}u_{i,j}u_{l,k}E_m\,, \qquad +\,\beta_{jilkm}u_{j,i}u_{l,k}E_m$$

peuvent se mettre sous une autre forme si l'on utilise la notation (ij) couple symétrique, $[ij]$ couple antisymétrique. Il vient

$$\beta_{(ij)(kl)m}E_{ij}\varepsilon_{kl}E_m\,,$$

$$\left\{ \begin{array}{l} +\,\beta_{(ij)[kl]m}\varepsilon_{ij}R_{kl}E_m\,, \\[2mm] +\,\beta_{[ij](kl)m}R_{ij}\varepsilon_{kl}E_m\,, \\[2mm] +\,\beta_{[ij][kl]m}R_{ij}R_{kl}E_m\,. \end{array} \right. \qquad \text{avec} \quad \left\{ \begin{array}{l} \varepsilon_{kl} = \dfrac{1}{2}\left(\dfrac{\partial u_k}{\partial u_l} + \dfrac{\partial u_l}{\partial u_k}\right), \\[3mm] R_{kl} = \dfrac{1}{2}\left(\dfrac{\partial u_k}{\partial u_l} - \dfrac{\partial u_l}{\partial u_k}\right), \end{array} \right.$$

Le premier terme est celui que nous avions utilisé au départ. Les deux suivants s'expriment à partir des constantes piézoélectriques du cristal.

Toutefois, dans la suite de cet article, pour plusieurs raisons (souci de simplicité, absence de mesures précises), nous conserverons la notation $\beta\varepsilon\varepsilon E$. Les quelques mesures faites ont été exploitées dans cette approximation qui est d'ailleurs exacte pour les coefficients $\beta_{ii,jj,m}$.

5'1.2. Formation d'échos par mécanisme électroacoustique. –

L'existence de l'interaction $\beta\varepsilon\varepsilon E$ suffit pour expliquer le phénomène d'écho. Elle permet même de prévoir des échos de deux natures différentes et nous devrons distinguer entre échos dipolaires et échos quadrupolaires. Ces dénominations sont prises aux sens suivants. De même qu'un opérateur de spin S_x est un moment dipolaire magnétique et $S_x S_y + S_y S_x$ est un moment quadrupolaire magnétique, nous dirons que ε_{ij} est un moment dipolaire élastique et que $\varepsilon_{ij}\varepsilon_{kl}$ est un moment quadrupolaire élastique.

5'1.2.1. Échos dipolaires.

Un moment dipolaire élastique est une déformation ε, telle qu'on peut l'engendrer par effet piézoélectrique. Les échos de type 2τ dont nous avons traité dans les paragraphes précédents sont donc des échos dipolaires.

Nous noterons seulement que l'instant où apparaît l'écho et l'instant où on excite la population cohérente sont symétriques par rapport à celui où l'on renverse le temps. Ainsi dans une séquence de trois impulsions $(0, \tau, T)$ — voir Fig. 1 —, trois échos existent aux instants 2τ, $2T - \tau$ et $2T$; ils sont de nature dipolaire élastique. Mais, dans une telle séquence, il apparaît un écho supplémentaire à $T + \tau$; il est inexplicable dans le modèle dipolaire qui ne fait intervenir que le coefficient électroacoustique β.

5'1.2.2. Échos quadrupolaires.

De même que l'interaction piézoélectrique $\alpha\varepsilon E$ permet d'engendrer un moment dipolaire ε proportionnel à αE, le couplage non linéaire $\beta\varepsilon\varepsilon E$ permet de créer un moment quadrupolaire $\varepsilon\varepsilon$ proportionnel à βE et de même fréquence ω que celle de E. Ce moment est évidemment détectable par effet inverse.

Dans le moment quadrupolaire, chacun des ε est à fréquence $\omega/2$; l'interaction paramétrique $\beta\varepsilon(\omega/2)\varepsilon(\omega/2)E(\omega)$ renverse le sens du temps pour l'un des ε et donne un moment quadrupolaire $\varepsilon(\omega/2)\varepsilon(-\omega/2)$ qui est un état figé. L'application à l'instant T d'une troisième impulsion de fréquence ω renverse le sens du temps pour $\varepsilon(\omega/2)$ qui n'avait pas été affecté par la deuxième impulsion; elle conduit à l'apparition d'un écho quadrupolaire à $T + \tau$. Celui-ci résulte donc de l'application successive de deux impulsions où le terme d'interaction β est traité au premier ordre.

Une seule impulsion pendant laquelle cette interaction est traitée au deuxième ordre renverse le temps pour le moment quadrupolaire et donne une contribution aux échos de type 2τ. Dans la référence [9], cet aspect des échos de phonons a été traité dans un langage quantique; ce formalisme met en évidence l'analogie avec les échos de spins.

5˙1.3. Mesure du coefficient β.

5˙1.3.1. Principe de la mesure. On a vu que l'on pouvait considérer le terme $\beta E \varepsilon \varepsilon$ comme une contribution à la piézoélectricité dépendant de la déformation $\beta \simeq \partial\alpha/\partial\varepsilon$. Mais en écrivant

$$u(R, t) = \tfrac{1}{2} c \varepsilon\varepsilon + \beta E \varepsilon\varepsilon = \tfrac{1}{2}(c + 2\beta E)\varepsilon^2$$

il apparaît aussi comme une contribution, dépendant du champ électrique, à la constante élastique $\beta = \tfrac{1}{2}(\partial c/\partial E)$. Ce résultat induit le principe d'une méthode de mesure: c'est la modification linéaire de la constante élastique c_{ijkl} en présence d'un champ électrique de composantes E_m. Pratiquement ceci est réalisé par la mesure du changement de vitesse d'une onde élastique progressive ou bien par la modification de la fréquence d'un résonateur acoustique.

Cependant, l'autre aspect ($\beta = \partial\alpha/\partial\varepsilon$) n'a pas été négligé et quelques expériences de mesure de la polarisation P sous contrainte extérieure ont été faites. Elles fournissent une valeur de β par les relations

$$P(\sigma) = a\sigma + b\sigma^2 \,,$$
$$= \alpha\varepsilon + \beta\varepsilon^2 \,.$$

5˙1.3.2. Ordre de grandeur. Des arguments généraux indiquent que deux coefficients thermodynamiques, dont l'un est la dérivée de l'autre par rapport à la déformàtion, sont du même ordre de grandeur. L'expérience montre que ceci est vérifié pour la constante élastique anharmonique $d = \tfrac{1}{2}(\partial c/\partial\varepsilon)$ où l'on a $d \simeq c$. De même, ici $\beta = \partial\alpha/\partial\varepsilon$ entraîne *a priori* $\beta \simeq \alpha$; ou, autrement dit, $(\partial/\partial\varepsilon)\log\alpha$ est peu dépendant du matériau considéré [17].

En conséquence, on en déduit la règle empirique que, pour trouver des matériaux ayant des grandes valeurs de β, on doit les sélectionner parmi ceux présentant déjà un grand coefficient piézoélectrique.

5˙1.3.3. Résultats des mesures.

5˙1.3.3.1. Quartz. Pour la classe de symétrie $3\,2$ à laquelle appartient le quartz, le tenseur $\beta_{(ij)(kl)m}$ possède huit composantes indépendantes non nulles. Ces huit composantes on été mesurées [18], dont certaines avec une très grande précision [19]. Les valeurs numériques s'étalent entre 0.25 et $4\,\mathrm{NV^{-1}\,m^{-1}}$ qui sont à comparer à $\alpha_{111} \simeq 0.2\,\mathrm{NV^{-1}\,m^{-1}}$. La règle grossière énoncée plus haut est donc vérifiée dans ce cas.

5˙1.3.3.2. Niobate de lithium. LiNbO$_3$ est un bon piézoélectrique: par exemple, $\alpha_{222} = 2.5\,\mathrm{NV^{-1}\,m^{-1}}$. Quelques composantes de β ont été évaluées [17-20]; ainsi on a $\beta_{22222} = 18\,\mathrm{NV^{-1}\,m^{-1}}$ et $B_{21111} \simeq 40\,\mathrm{NV^{-1}\,m^{-1}}$. Les mesures de polarisation sous contrainte dynamique ont donné $\beta_{33333} = (18 \pm 6)\,\mathrm{NV^{-1}\,m^{-1}}$. Ces résultats expliquent pourquoi l'interaction $\beta E \varepsilon^2$ a été mise en évidence d'abord sur ce composé [21].

5'1.3.3.3. Molybdate de gadolinium. Le molybdate de gadolinium présente une transition de phase ferroélectrique vers 159 °C. Le coefficient β_{31111} a été mesuré en fonction de la température [22]. Dans la phase paraélectrique, loin de la transition $\beta_{31111} \simeq 2.5 \, \mathrm{NV^{-1} \, m^{-1}}$, valeur qui n'est pas grande. Mais le résultat important obtenu est le comportement critique de ce coefficient: $\beta = A + B(T - T_0)^{-1}$. Près de la température critique, β peut donc devenir très grand; des valeurs supérieures à $200 \, \mathrm{NV^{-1} \, m^{-1}}$ ont été notées.

L'intérêt de cette observation est clair: dans un cas comme celui-ci, l'intensité de l'écho peut être considérablement augmentée en opérant au voisinage de T_0. Toutefois, un raccourcissement critique du temps de vie des phonons [23] peut contrebalancer le bénéfice d'une telle opération.

5'1.3.3.4. Autres matériaux. L'effet du champ électrique sur la constante élastique du phosphate di-acide de rubidium (RbDP) a été observé [24]. Dans la phase ferroélectrique, le résultat est spectaculaire ($\beta \simeq 700 \, \mathrm{NV^{-1} \, m^{-1}}$). Mais il n'est pas sûr que l'origine du phénomène soit microscopique et il est possible qu'il s'agisse d'un basculement de domaines sous l'influence du champ.

Enfin, la méthode de la polarisation sous contrainte, appliquée au tantalate de lithium (LiTaO$_3$) a donné $\beta \simeq 4 \, \mathrm{NV^{-1} \, m^{-1}}$ [25]. Cette expérience, réalisée avec une pression hydrostatique, fournit la moyenne de plusieurs composantes du tenseur β.

5'1.4. Conclusion. En conclusion, l'ordre de grandeur du coefficient r défini dans la formule (13) peut maintenant être évalué. Avec $E_2 = 10^4 \, \mathrm{V/m}$, $c = 5 \cdot 10^{10} \, \mathrm{N \, m^{-2}}$, $\beta = 10 \, \mathrm{NV^{-1} \, m^{-1}}$, il vaut $r = \beta E_2/2c = 10^{-6}$.

On en déduit, pour cette valeur de champ électrique et pour une impulsion à fréquence $\omega_1 = 2\pi \cdot 10^{10}$ et de durée $\Delta t = 1 \, \mathrm{ms}$, que l'efficacité du renversement du temps définie par

$$(14') \qquad \varepsilon_{\mathrm{réfl}} = \varepsilon_{\mathrm{inc}} \sinh (r\omega_1 \Delta t) \simeq (r\omega_1 \Delta t) \varepsilon_{\mathrm{inc}}$$

vaut

$$\varepsilon_{\mathrm{réfl}} = 6 \cdot 10^{-2} \varepsilon_{\mathrm{inc}} \, .$$

Cet ordre de grandeur qui correspond à une situation de champ assez intense, est suffisamment faible pour justifier l'approximation introduite dans la formule (8) dont découlent toutes les propriétés des fonctions de forme du signal; en même temps, il est assez grand pour que l'on puisse espérer de voir l'écho avec une dynamique raisonnable.

5'2. Processus acoustoélectrique: β'. – Le mécanisme d'interaction acoustoélectrique, qui implique la présence de charges électriques plus ou moins libres dans le cristal, est encore mal clarifié. Il n'existe pas, par exemple, de mesure élémentaire de β'. Mais son importance oblige à avancer quelques idées simples et à indiquer quels sont les facteurs qui semblent gouverner son ordre de grandeur. Ce mécanisme a été mis en évidence [13] d'abord dans les composés II-IV qui sont des semi-conducteurs dont il est aisé de peupler la bande de conduction et les donneurs peu profonds en les éclairant avec une lumière convenable. Le CdS, bien connu pour ses propriétés photoélectriques et pour la grandeur de ses coefficients piézoélectriques, a été le plus utilisé; il existe d'ailleurs une énorme littérature sur les interactions des phonons et des porteurs dans ce cristal.

Les quelques observations suivantes aident à la compréhension du mécanisme:

Une expérience d'échos de phonons nécessite la préparation de l'échantillon par peuplement des niveaux peu profonds ou pièges. Ce remplissage n'est pas permanent; il existe un temps de relaxation de retombée T_T qui dépend de la température. Dans CdS, ce temps peut atteindre une journée à la température de l'hélium liquide.

Un champ électrique de fréquence ω module la densité de charges électriques du cristal à une fréquence 2ω; il libère, en effet, les porteurs piégés, deux fois par période, lorsque son intensité dépasse un certain seuil. Il existe aussi une composante continue. Il y a donc là une non-linéarité qui fait qu'une partie de la modulation de la densité de charges $\delta\varrho$ est proportionnelle au carré du champ électrique.

Dans une expérience d'échos de phonons, la première impulsion engendre une onde élastique qui traîne avec elle, à cause de la piézoélectricité du milieu, un champ électrique de fréquence ω, de vecteur d'onde k et d'intensité $E_{\text{piézo}}$ proportionnelle à ε_{inc} (et donc à E_1). Pendant la seconde impulsion, le champ électrique total E_T est la somme de $E_2(\omega, 0)$ et de $E_{\text{piézo}}(\omega, k)$. Son carré a donc, entre autres, des composantes $(2\omega, 0)$ et $(0, k)$.

Les échos à 2τ s'expliquent alors facilement. La composante de densité de charge $\delta\varrho(2\omega, 0)$, associée au champ piézoélectrique $E_{\text{piézo}}(\omega, k)$, produit un courant de charge

$$j(\omega, -k) \sim \delta\varrho(2\omega, 0) E_{\text{piézo}}^*(\omega, k) \sim [E_2(\omega, 0)]^2 \varepsilon_{\text{inc}}^*(\omega, k) .$$

Les équations de Maxwell et les équations de matière étant linéaires, ce courant est proportionnel à une tension élastique $\sigma(\omega, -k)$ qui engendre une onde réfléchie, détectée sous forme d'écho à 2τ. Pour faire le lien avec la description de la Sect. **3**, on peut écrire que $\delta\varrho(2\omega, 0)$ est produit par un champ effectif $E(2\omega, 0) \sim [E_2(\omega, 0)]^2$. Le deuxième membre de l'éq. (8) est de la forme $E(2\omega, 0)\varepsilon_{\text{inc}}^*(\omega, k)$; le coefficient de proportionnalité définit β'.

Les échos à $T + \tau$ s'expliquent dans le même cadre. La composante de densité de charges $\delta\varrho(0, k)$, proportionnelle à $E_1 E_2$, du fait de son caractère statique, va permettre un repiégeage partiel. C'est un processus d'inscription de l'information au sens où on l'entendait à la Sect. **3** (état figé). Le champ électrique E_3, appliqué à l'instant T, engendre un courant

$$ j(\omega, -k) \sim \delta\varrho^*(0, k) E_3(\omega, 0) , $$

qui permet la production d'un écho à $T + \tau$. Le temps de vie T_1 apparaît, dans ces conditions, comme le temps de vie d'une densité inhomogène de pièges remplis. Il dépend de la température, de l'éclairement; il peut être aussi long que plusieurs heures à 4 °K [13]. C'est ce qui fait tout l'intérêt de ce mécanisme.

Pour achever ces remarques sur le processus acoustoélectrique, on retiendra que tous les échos qui lui sont dus, y compris celui à $T + \tau$, sont de nature dipolaire élastique.

6. – Applications des échos de phonons.

Bien que d'origine récente, le phénomène d'échos de phonons a déjà donné lieu à quelques applications qui intéressent le physicien. Mais, tandis que se dégageaient les concepts nécessaires à leur compréhension, parallèlement et même antérieurement, étaient conçus et réalisés un certain nombre de systèmes intéressant l'ingénieur et ayant pour fonction le traitement des signaux. Utilisant le plus souvent des ondes acoustiques de surface, ces dispositifs, qui réalisent des opérations telles que les produits de convolution ou de corrélation, ont en commun avec les échos de phonons l'utilisation de l'interaction paramètrique $\beta E\varepsilon\varepsilon$ ou $\beta' E\varepsilon\varepsilon$.

Nous allons passer en revue quelques applications, en faisant la distinction (parfois arbitraire), entre celles concernant le physicien du solide et celles intéressant l'ingénieur.

6˙1. *Applications pour physiciens.*

6˙1.1. M e s u r e d e l' a t t é n u a t i o n u l t r a s o n o r e. La mesure de l'atténuation ultrasonore est importante pour le physicien du solide. C'est une mesure propre au sens où elle concerne un seul mode de phonons, de fréquence, de polarisation et de vecteur d'onde déterminés. Elle se différencie par là des mesures de conductibilité thermique ou par impulsions de chaleur. C'est une mesure indispensable car elle renseigne sur les interactions des phonons avec les impuretés, les défauts, les phonons thermiques, éventuellement les électrons libres, présents dans le cristal. En élevant la température, on peut ainsi observer, dans un isolant, le passage d'un régime d'atténuation à un autre: atténuation par les impuretés, régime quantique et régime visqueux. Les contributions de ces différents régimes dépendent de la fréquence des ondes acoustiques.

A haute fréquence, on se heurte à une difficulté technique, bien connue des acousticiens, et qui est illustrée sur la Fig. 4. Sur cette Figure, les rectangles

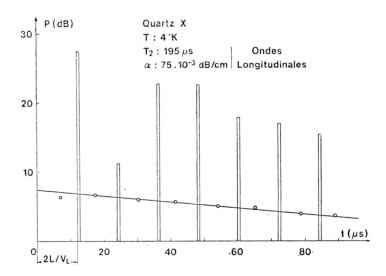

Fig. 4. – Décroissance des échos ultrasonores (représentés par des bâtons) dans un quartz dont le parallèlisme des faces laisse à désirer. La décroissance des échos de phonons est représentée par les points o; le premier n'est pas sur la droite expérimentale à cause d'un défaut d'appareillage aux courts intervalles de temps. La pente de cette droite conduit à une atténuation intrinsèque de $75 \cdot 10^{-3}$ db/cm.

représentent l'intensité des différents échos *ultrasonores* ayant fait des allers et retours dans le cristal; il faudrait, en principe, déduire de cette décroissance, quelque peu anarchique, un décrément logarithmique! L'origine du mal est connue: c'est le défaut de parallélisme des faces terminales de l'échantillon. En effet, la grandeur mesurée est, schématiquement, le flux du champ électrique associé à l'onde à travers la face de réception du cristal. Si celle-ci n'est pas parallèle à la face d'émission et aux plans d'ondes, différentes régions sont excitées en opposition de phase (Fig. 5*a*)), et leurs contributions se retranchent. Un écho ultrasonore peut même être nul alors que les suivants ne le sont pas. La condition pour avoir une mesure correcte est d'autant plus exigeante que la fréquence est plus élevée. La thérapeutique consiste évidemment à améliorer le parallélisme.

Nous allons montrer que la méthode des échos de phonons constitue un remède original [11]. En effet (Fig. 5*b*)), si la propagation de l'onde acoustique

est inversée, l'angle produit par la réflexion sur la face inclinée se trouve rigoureusement compensé au cours du trajet de retour. Ainsi, au moment de la détection sur la face qui a servi pour l'émission, les plans d'onde sont exactement
parallèles à cette face. L'argument est évidemment valable pour un nombre
quelconque de réflexions. Le mode opératoire est donc le suivant. On engendre
et on détecte, par piézoélectricité, les ondes ultrasonores au moyen d'une

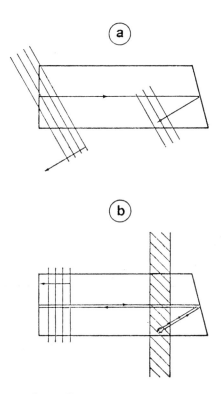

Fig. 5. – Schéma indiquant les méfaits d'un manque de parallèlisme des faces sur les
échos ultrasonores (a). Le retour inverse, dans le cas des échos de phonons (b), élimine
ces inconvénients.

même cavité (dispositif en réflexion) à la fréquence ω. On introduit une région
de l'échantillon dans une cavité résonnant à la fréquence 2ω (Fig. 6); on évite
ainsi l'utilisation des anharmonicités diélectriques du cristal et on dispose d'un
champ $E(2\omega)$ intense. On applique donc une impulsion à la fréquence 2ω au
moment où le train d'ondes ultrasonores passe dans la cavité, à l'aller ou au
retour (instants a_i ou r_i de la Fig. 7). On recueille aux instants $2a_i$ ou $2r_i$ un
écho (au sens des échos de phonons) qui est insensible au défaut de parallèlisme.
Le résultat est visible sur la Fig. 4: le coefficient d'atténuation peut être déduit
avec précision même sur un échantillon dont les faces ne sont pas parallèles.

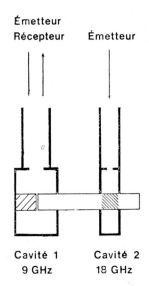

Fig. 6. – Schéma du dispositif utilisé lors des expériences sur le quartz. La zone hachurée représente la région du quartz où le renversement du vecteur d'onde a lieu.

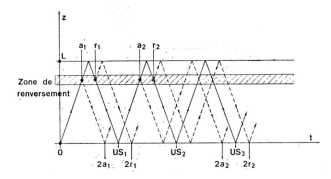

Fig. 7. – Diagramme (z, t) donnant la position des échos ultrasonores et des échos de phonons détectés dans la cavité à 9 GHz. La bande hachurée représente la région du cristal irradiée à 18 GHz.

6ʹ1.2. Les auto-échos. On pourrait définir un acousticien comme un expérimentateur qui envoie une seule impulsion électromagnétique sur un cristal piézoélectrique, et un spécialiste des échos de phonons comme quelqu'un qui envoie au moins deux impulsions. Dans cette optique, il existe un domaine où ils se rejoignent: c'est celui des auto-échos.

Nous avons vu que le phénomène d'échos était lié à l'action de la seconde impulsion par l'intermédiaire d'une non-linéarité. Or, la première impulsion est, elle aussi, soumise aux mêmes non-linéarités; son rôle devient double: par sa fréquence fondamentale, elle engendre des phonons; par son harmonique,

elle renverse le temps. Elle donne ainsi, à elle seule, un phénomène d'écho:
l'auto-écho [10].

Le phénomène d'auto-écho présente évidemment de grandes analogies avec
celui d'écho. Il y a toutefois une petite différence, assez évidente d'ailleurs:
au cours de l'impulsion, il ne peut y avoir renversement que des phonons déjà
émis. Il en résulte que, pour une impulsion de durée Δt, la largeur de l'auto-
écho est $2\,\Delta t$ (Fig. 8). Donc, pour l'acousticien, l'auto-écho se manifeste par
un doublement apparent de la largeur de l'impulsion émise.

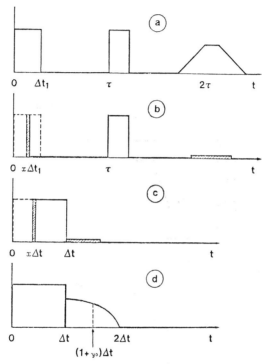

Fig. 8. – Schéma de quelques types d'échos: *a*) Écho à 2τ correspondant à deux
impulsions rectangulaires de durée Δt_1 et Δt_2. *b*) Écho élémentaire relatif à une première
impulsion infinitésimale et à une seconde finie; son amplitude est proportionelle à
$dx\,\Delta t$ et sa durée est $2\,\Delta t$. *c*) Contribution élémentaire à l'auto-écho, construite d'après
le schéma *b*). *d*) Auto-écho, somme des contributions élémentaires.

6˙1.3. Mesures du temps de relaxation T_2. Deux types d'études
systématiques ont été entreprises.

6˙1.3.1. Etude de T_2 en fonction de la température. Une remarque pré-
liminaire s'impose. Lorsque la fréquence s'élève, T_2 diminue. Il s'en suit
qu'une étude en fonction de la température ne peut être menée dans une large
gamme (par exemple, de 4 °K à 300 °K) que si la fréquence n'est pas trop haute,
pratiquement pour $\nu \leqslant 100$ MHz.

Les résultats peuvent se résumer ainsi:

Il y a une certaine analogie entre le comportement de T_2 et celui du temps de vie des ultrasons: un plateau s'étend de la température ambiante aux températures moyennes $((100 \div 150)\ °\mathrm{K})$; en dessous, T_2 s'allonge [3].

Pour les ferroélectriques, au voisinage des transitions de phase, T_2 augmente notablement, ce qui est tout à l'opposé des résultats des mesures ultrasonores où l'on note une forte atténuation critique.

6'1.3.2. Etude de T_2 dans les poudres. On pourrait dire qu'une poudre est un milieu où une onde progressive se heurte fréquemment aux surfaces. Le rôle que ces dernières jouent se trouve donc grossi.

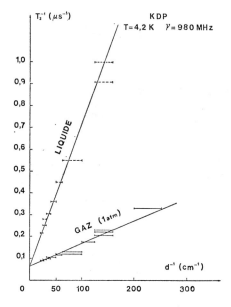

Fig. 9. – Résultats expérimentaux mis sous la forme $T_2^{-1} = f(d^{-1})$. La loi obtenue est $T_2^{-1} = A + Bd^{-1}$. Le coefficient B dépend du milieu baignant la poudre.

Les résultats obtenus sur des poudres de KDP [12] de diamètre compris entre 50 et 500 μm ont permis de mettre en évidence l'existence de plusieurs types de relaxation (Fig. 9):

une relaxation interne, indépendante des dimensions des grains et de la nature du milieu les baignant;

des relaxations externes, dépendant de la taille des grains $(T_{2\mathrm{ext}}^{-1} \sim d^{-1})$. On distingue deux processus:

1) La transmission de phonons vers le milieu extérieur; elle dépend de l'impédance acoustique qui varie selon qu'il est liquide ou gazeux et selon la pression (gaz).

2) La conversion de modes acoustiques qui implique une diminution de l'efficacité du renversement du temps. En effet, lorsqu'une onde se réfléchit, elle donne naissance, en général, à trois nouvelles ondes ayant trois polarisations différentes. Si le coefficient β ou β' de l'interaction paramétrique était le même pour les trois types d'ondes, après l'impulsion de renversement les ondes dont le sens de propagation a été inversé se recombineraient parfaitement et l'effet de la conversion en trois modes distincts serait effacé. Mais les coefficients relatifs à ces ondes sont différents; la reconstruction à la surface n'est que partielle et il y a perte d'amplitude; les effets destructifs de chaque réflexion sont cumulatifs. L'importance de ce processus de relaxation repose sur le disparité des composantes du tenseur β, mais aussi sur les différences entre les vitesses des trois types d'ondes, différences qui conditionnent les proportions d'énergie élastique réfléchie dans chacune d'elles.

6̇1.3.3. Conclusion. En conclusion, on retiendra que la petitesse des cristaux et les réflexions multiples ne sont pas un obstacle à l'observation des échos dans les poudres ou toute expérience ultrasonore serait vouée à l'échec.

6̇1.4. Mesures du temps de relaxation T_1. Peu d'études ont été faites. La signification physique de T_1 dépend du mécanisme invoqué pour expliquer l'écho à $T + \tau$. Le cristal qui semble avoir été le mieux étudié est le sulfure de cadmium CdS où le rôle des porteurs piègés y paraît bien établi.
Les principaux résultats obtenus sont les suivants:

Le temps de relaxation T_1 est très long: selon les échantillons, de 1 s [9] à quelques heures [13] pour une fréquence $\nu = 9$ GHz et une température de 4.2 °K.

La décroissance de l'écho en fonction du temps T n'est pas rigoureusement exponentielle, ce qui traduit la présence de plusieurs temps de relaxation.

La valeur de T_1 dépend de l'illumination subie par l'échantillon.

T_1 diminue très vite lorsque la température s'élève.

T_1 mesuré sur des poudres est indépendant de la pression du gaz qui baigne les grains; ceci montre qu'il n'y a pas émission de phonons vers l'extérieur.

D'autre part, quelques résultats on été obtenus sur le phosphate di-acide de potassium (KDP). Dans les mêmes conditions ($T = 4.2$ °K, $\nu = 9$ GHz), le temps de relaxation y est beaucoup plus court: $T_1 \simeq 1$ ms. De plus, l'amplitude des échos est indépendante de l'illumination.

6˙1.5. Etude des phénomènes de surface. Dans une poudre dont les grains ont une dimension moyenne de 100 μm, une onde élastique donnant lieu à un écho de phonons observé à l'instant $2\tau = 20$ μs a subi approximativement 10^3 chocs sur les surfaces avant d'être détectée. C'est dire combien l'amplitude de l'écho est sensible à tout ce qui se passe à la surface: adsorption d'atomes, dépôt d'un film mince, etc.

Quelques expériences ont été réalisées en déposant sur les surfaces des grains un film mince d'hélium dont l'épaisseur pouvait être changée pour obtenir la résonance mécanique avec des phonons de fréquence $\omega = 2\pi \cdot 10^{10}$. C'est une expérience analogue dans son but à celles réalisées à l'aide d'un spectromètre à phonons [26] ou par une méthode acoustique [27]: le couplage du cristal avec le film résonnant provoque un maximum de transmission de l'énergie dans l'hélium. Du point de vue des échos de phonons, cela se traduit par une vigoureuse diminution du temps de relaxation T_2 que l'on sait relier avec les pertes acoustiques dans l'hélium [28].

6˙1.6. Résumé en forme de conclusion. La méthode des échos de phonons et, en particulier, les mesures de T_2 présentent des analogies avec les méthodes ultrasonores et avec les mesures de conductibilité. Des ultrasons, elles possèdent la monochromaticité; de la conductibilité thermique, elles ont la multiplicité des ondes excitées et l'absence de résolution dans le temps. Au total, la méthode semble assez mal adaptée à la mesure du temps de vie des phonons dans les cristaux. Un avantage, le seul peut-être, est l'absence de préparation préalable des échantillons et, en conséquence, la possibilité d'opérer sur des poudres. L'utilisation de petits grains renforçant l'influence des frontières, il y a là une possibilité d'étude des phénomènes de surface.

6˙2. *Applications pour ingénieurs.* – Les ingénieurs eux-mêmes, peuvent trouver de l'intérêt à cette technique des échos de phonons. Deux orientations sont possibles: la première s'inscrit dans l'effort général fait pour utiliser les ondes acoustiques dans les systèmes de traitement de signaux; la deuxième ouvre quelques possibilités nouvelles pour la mémorisation de l'information.

6˙2.1. Traitement de signaux. On connait le développement qu'a connu depuis quelques années l'utilisation des phénomènes non linéaires auxquels participent les ondes acoustiques. Leur étude a permis la réalisation de lignes à retard, de convoluteurs ou de corrélateurs. On retrouve dans ces dispositifs fonctionnant aussi bien en ondes de surface qu'en ondes de volume les mécanismes dont on a parlé plus haut; tout ce que nous avons essayé d'expliquer quant aux phénomènes non linéaires peut s'appliquer (fonction de forme, largeur en temps, efficacité du processus paramétrique). Plusieurs articles généraux ont fait le point de la question [29].

On ajoutera seulement, pour mémoire, que ces dispositifs sont souvent associés à des amplificateurs acoustoélectriques dont le seul but est de diminuer les pertes d'insertion [30]. Diverses subterfuges tentent, par ailleurs, d'augmenter artificiellement le rendement, par exemple, en concentrant l'énergie acoustique d'une onde de surface sous un ruban métallique; on peut également songer à travailler près d'un point de changement de phase.

Mais tous ces dispositifs ont en commun d'exiger la préparation de l'échantillon avec les contraintes bien connues des acousticiens: poli, orientation, ... Pour des ondes de volume, ils nécessitent l'emploi de transducteurs résonnants; pour les ondes de surface, le dépôt des peignes interdigités, sans être d'une grande difficulté, est tout de même un handicap.

C'est un des gros avantages des échos de phonons de s'affranchir de toutes ces servitudes; on peut ajouter que, par principe, le niveau des signaux parasites, par rapport à l'écho reçu, est extrêmement bas car la convergence en phase de la population de phonons n'est restituée qu'aux instants 2τ ou $T + \tau$.

Sous un dernier angle, enfin, l'existence d'échos à $T + \tau$ peut avoir un intérêt: la fonction de forme de cet écho (formule (20)) montre qu'on peut obtenir à volonté la convolution ou la corrélation en temps réel de deux signaux pourvu que la deuxième ou la troisième impulsion soit une distribution de Dirac $\delta(t)$.

6·2.2. Mémorisation. D'autres applications ne semblent pas s'être encore concrètisées. Plus qu'à la compression d'impulsion par vobulation, on pense ici à la mémorisation d'une information.

L'état figé que nous avons décrit a un temps de vie qui peut presque être allongé à volonté: quelques heures pour une porteuse à 9 GHz. La largeur de bande des signaux étant limitée seulement par les appareils électroniques, la capacité de la mémoire devient extrêmement grande. Par ailleurs, la lecture de l'information est aisée; elle se fait avec un signal de même fréquence porteuse que celle de l'inscription. Si l'on ajoute que le stockage peut se faire en volume dans des poudres cristallines, on dispose là d'un moyen qui, après quelques améliorations techniques, pourrait devenir compétitif.

BIBLIOGRAPHIE

[1]   S. N. Popov et N. N. Krainik: *Sov. Phys. Sol. State*, **12**, 2440 (1971).
[2]   N. N. Krainik, S. N. Popov et I. Mylnikova: *J. Physique*, **33**, C 2, 179 (1972).
[3]   S. N. Popov et N. N. Krainik: *Sov. Phys. Sol. State*, **14**, 2408 (1973).
[4]   A. Kessel, I. Safin et A. Goldman: *Sov. Phys. Sol. State*, **12**, 2488 (1971).
[5]   U. Kopvillem, B. Smolyakov et R. Shapirov: *JETP Lett.*, **13**, 398 (1971).
[6]   Y. Asadullin, U. Kopvillem, V. Osipov, B. Smolyakov et R. Shapirov: *Sov. Phys. Sol. State*, **13**, 2330 (1972).
[7]   Ch. Frenois, J. Joffrin, A. Levelut et S. Ziolkiewicz: *Sol. State Comm.*, **11**, 327 (1972).
[8]   J. Joffrin et A. Levelut: *Phys. Rev. Lett.*, **29**, 1325 (1972).

[9]   A. Billmann, Ch. Frenois, J. Joffrin, A. Levelut et S. Ziolkiewicz: *J. Physique*, **34**, 453 (1973).

[10] Ch. Frenois, J. Joffrin et A. Levelut: *Compt. Rend.*, **278** B, 57 (1974).

[11] Ch. Frenois, J. Joffrin et A. Levelut: *J. Physique*, **34**, 747 (1973).

[12] Ch. Frenois, J. Joffrin et A. Levelut: *J. Physique*, **35**, L 221 (1974).

[13] N. S. Shiren, R. L. Melcher, P. K. Garrod et T. G. Kazyaka: *Phys. Rev. Lett.*, **31**, 819 (1973).

[14] E. Y. Lu et P. A. Fedders: *Appl. Phys. Lett.*, **23**, 502 (1973).

[15] D. F. Nelson et M. Lax: *Phys. Rev. Lett.*, **24**, 379 (1970).

[16] D. F. Nelson et P. D. Lazay: *Phys. Rev. Lett.*, **25**, 1187 (1970).

[17] R. A. Graham: *Sol. State Comm.*, **12**, 503 (1973).

[18] K. Hkuška et V. Kazda: *Czech. Journ. Phys.*, **18** B, 500 (1968).

[19] R. B. Graham: *Phys. Rev. B*, **6**, 4779 (1972).

[20] R. B. Thompson et C. F. Quate: *Journ. Appl. Phys.*, **42**, 907 (1971).

[21] L. O. Svaasand: *Appl. Phys. Lett.*, **15**, 300 (1969).

[22] J. M. Courdille et J. Dumas: *Phys. Rev. B*, **8**, 1129 (1973).

[23] J. M. Courdille et J. Dumas: *J. Physique Lett.*, **36**, L 5 (1975).

[24] C. Pierre, J. P. Dufour et M. Remoissenet: *Sol. State Comm.*, **9**, 1493 (1971).

[25] R. A. Graham: *Sol. State Comm.*, **13**, 1965 (1973).

[26] C. H. Anderson et E. S. Sabisky: *Phys. Rev. Lett.*, **24**, 1049 (1970).

[27] Ch. Frenois, J. Joffrin, P. Legros et A. Levelut: *Phys. Rev. Lett.*, **32**, 1295 (1974).

[28] Ch. Frenois, J. Joffrin et A. Levelut: communication privée.

[29] G. S. Kino, S. Ludvik, H. Shaw, W. Shreve, J. White et D. Winslow: *IEEE Trans. MIT*, **21**, 244 (1973).

[30] L. P. Solie: *Appl. Phys. Lett.*, **25**, 7 (1974).

Wave Propagation in Random Media.

J. J. McCoy

The Catholic University of America - Washington, D. C.

1. – Background discussions.

1˙1. *The medium and its description.* – A random medium is one with material properties that vary in space-time in a manner about which we possess probabilistic information. That is, the fields that describe the variations in material properties are stochastic processes. Thus, an integral part of the concept of a random medium is the idea of an assemblage, or collection, of media. One class of random media is typified by a solid mixture (*e.g.*, a polycrystalline solid, a composite material, etc.). The ensemble in this case would be the collection of solids that emerge from the same manufacturing process. Intuitively, all elements of the ensemble are expected to be identical in some macroscopic sense although they are distinct when viewed on a finer observation scale. A second class of random media is typified by the ocean, or the atmosphere. For this class the dimension of time enters the problem. We can define an ensemble of oceans, required by our statistical formalism, by taking as an element of the ensemble the ocean that exists during a given time interval. The ocean that exists during a subsequent time interval is a second element of the ensemble (*i.e.* a second manifestation of the random ocean). The appropriate time interval for defining a single manifestation is determined by the time interval of the physical experiment in which the ocean is to play a role. For an acoustic signal traversing the ocean medium, the appropriate time interval is determined by the duration of the signal, the sound speed, and characteristic length scales of the fluctuating sound speed field. We note that if the ocean changes little over this time interval one might validly neglect these changes altogether resulting in the idea of a frozen ocean. During the time of a subsequent experiment—a second acoustic pulse—the ocean may again be treated as frozen only now it is a different frozen ocean. The time dimension does not enter the physical process in this situation and plays only the role of identifying different manifestations of the random ocean (*). Of

---

(*) In presenting detailed calculations of a scattering theory I shall concentrate on a specific acoustic experiment involving an ensemble of frozen oceans.

course, if the frequency shifting of energy in the acoustic signal as it propagates through the ocean is a quantity of interest one cannot ignore changes in the ocean that occur during the passage of a single signal. In this type of a problem the time dimension enters in two roles. The rapid temporal variations of the ocean (*i.e.* those measurable on the same time scale as is the acoustic experiment) are an integral part of the physical process; the slower temporal variations identify different manifestations of the random ocean.

Mathematically, a random medium is completely defined by prescribing joint probability functionals on the randomly varying property fields. The enormous amount of information contained in a complete description, as well as its complexity, renders a complete description to be a mathematical ideal to be approached in some limiting sense. In the world of physics we shall have to be content with a very limited portion of this information required for a complete description. One way of collecting partial information of a randomly varying property field that proves useful in many problems of physical interest is via the statistical moments. If we let $T(\boldsymbol{x}, t)$ denote a property field, we can speak of a one-point moment, *e.g.* $\{T(\boldsymbol{x}, t)\}$, a two-point moment, *e.g.* $\{T(\boldsymbol{x}_1, t_1) T(\boldsymbol{x}_2, t_2)\}$, a three-point moment, *e.g.* $\{T(\boldsymbol{x}_1, t_1) T(\boldsymbol{x}_2, t_2) T(\boldsymbol{x}_3, t_3)\}$, etc. The curly brackets are used to denote a statistical, or ensemble average. Thus for example

$$N_1(\boldsymbol{x}, t) \equiv \{T(\boldsymbol{x}, t)\} = \frac{1}{N} \sum_i^N T_i(\boldsymbol{x}, t)$$

and

$$N_2(\boldsymbol{x}_1, \boldsymbol{x}_2, t_1, t_2) \equiv \{T(\boldsymbol{x}_1, t_1) T(\boldsymbol{x}_2, t_2)\} = \frac{1}{N} \sum_i^N T_i(\boldsymbol{x}_1, t_1) T_i(\boldsymbol{x}_2, t_2),$$

where the subscript i identifies a single element of the ensemble of media. Specification of the statistical moments of all orders can be shown to correspond to a complete statistical description of $T(\boldsymbol{x}, t)$.

The definition of the statistical moments given above is unambiguous and is valid for all problems. In most problems of physical interest, however, alternate interpretations can be given them, which prove to be better suited to actually collecting data. To discuss this we introduce two definitions—the stationary medium and the statistically homogeneous medium. The stationary medium is one for which the statistics of $T(\boldsymbol{x}, t)$ are independent of absolute time; the statistically homogeneous medium is one for which the statistics of $T(\boldsymbol{x}, t)$ are independent of absolute position. Thus, for a stationary medium

$$N_1(\boldsymbol{x}, t) = N_1(\boldsymbol{x}),$$

$$N_2(\boldsymbol{x}_1, \boldsymbol{x}_2, t_1, t_2) = N_2(\boldsymbol{x}_1, \boldsymbol{x}_2, \tau = t_1 - t_2),$$

$$N_3(\boldsymbol{x}_1, \boldsymbol{x}_2, \boldsymbol{x}_3, t_1, t_2, t_3) = N_3(\boldsymbol{x}_1, \boldsymbol{x}_2, \boldsymbol{x}_3, \tau_2 = t_2 - t_1, \tau_3 = t_3 - t_1),$$

etc., and for a statistically homogeneous medium

$$N_1(\boldsymbol{x}, t) = N_1(t) \,,$$

$$N_2(\boldsymbol{x}_1, \boldsymbol{x}_2, t_1, t_2) = N_2(\boldsymbol{\rho} = \boldsymbol{x}_2 - \boldsymbol{x}_1, t_1, t_2) \,,$$

$$N_3(\boldsymbol{x}_1, \boldsymbol{x}_2, \boldsymbol{x}_3, t_1, t_2, t_3) = N_3(\boldsymbol{\rho}_2 = \boldsymbol{x}_2 - \boldsymbol{x}_1, \boldsymbol{\rho}_3 = \boldsymbol{x}_3 - \boldsymbol{x}_1, t_1, t_2, t_3) \,,$$

etc. The concepts of stationarity and of statistical homogeneity are to be distinguished from those of a frozen medium or a homogeneous medium. The latter concepts refer to the independence of the field quantity, itself, on time or on space, respectively. A number of physically important stationary or statistically homogeneous media can be said to be ergodic. An ergodic medium is one for which it is valid to interpret an ensemble average in terms of either a temporal average (time-ergodic process) or a spatial average (space-ergodic process). Thus, for example, if $T(\boldsymbol{x}, t)$ denotes the temperature field in a stationary ocean (*) $\{T(\boldsymbol{x})\}$ can be interpreted as the average of the time history of the output signal of a single thermistor positioned at a single point $\boldsymbol{x}$. For a statistically homogeneous ocean (*) $\{T(t)\}$ could be interpreted as the average of outputs, all measured at the time t, of a collection of thermistors uniformly distributed throughout the ocean, or over a plane, or along a line. If, in addition to being statistically homogeneous, the ocean can be assumed to be frozen for the experiments of interest, then the data obtained by the collection of thermistors, in the above example, can be obtained by a single thermistor, successively positioned as would be the collection. In this way the statistics of the spatial variations of the temperature field in a frozen statistically homogeneous ocean are obtainable by towing a thermistor through the ocean, at uniform velocity, and time averaging the output signal.

The above discussion motivates the fact that the mean field, or the one-point moment, is an easily understood and, in principle, an easily measured quantity if the medium is either stationary or statistically homogeneous. Similar arguments can be provided for the higher-order moments, the difference being that the product of the outputs of a suitably positioned thermistor chain replaces the single thermistor required for the one-point moment. Thus, we can see that for a class of stationary or of statistically homogeneous media the statistical moments collect the statistical information in $T(\boldsymbol{x}, t)$ in an easily understood and, in principle, directly measurable hierarchy. What remains to be demonstrated is that the information in the lowest-order moments is important.

(*) Strictly speaking, stationarity and statistical homogeneity requires infinite observation times and unbounded oceans. This is usually interpreted as large relative to all correlation times and lengths of $T(\boldsymbol{x}, t)$.

Intuitively, the nature of the information in the one-point information, or the mean property field, seems to be clear. The two-point moment, or the correlation function, requires more thought but we have encountered it often enough in a multiplicity of problems to feel more or less comfortable with it. It provides information as to the statistical dependence of the values of the property field measured at two different points in the medium at two different times. Further, a collapsed two-point moment for a stationary process, *i.e.* $N_2(x, x, \tau)$, is a quantity of long-standing interest in communication theory ([*]). It is the autocorrelation of the time history of the property field measured at the single point x, and we can identify its Fourier transform, denoted by $\hat{N}_2(x, \omega)$, with the decomposition of the strength of the randomly varying field, *i.e.* of $\{T^2(x)\}$, into characteristic time scales. This decomposition is the temporal power spectrum. For a statistically homogeneous process, the collapsed two-point moment $N_2(\rho, t, t)$ can be interpreted in the same way. Its spatial transform, denoted by $\hat{N}_2(k, t)$, can be identified with the decomposition of the strength of the randomly varying field, *i.e.* of $\{T^2(t)\}$, into characteristic size scales, eddy sizes. The vector dependence of $\hat{N}_2(k, t)$ on k demonstrates a possible anisotropy in the spatial variation of $T(x, t)$ at the time t. For property fields that are both stationary and statistically homogeneous we can transform both with respect to space and time, *i.e.* obtain $\hat{N}_2(k, \omega)$. This last-mentioned transform can be identified with the decomposition of the strength of the randomly varying property field, *i.e.* of $\{T^2\}$, into propagating plane waves. For the fluctuating temperature field in the ocean, the two-point information is very important since the cause of the fluctuations is ultimately due to mixing either by turbulence or by a collection of randomly phased internal waves. To the extent that the temperature can be treated as a passive additive, therefore, its fluctuations are manifestations of fluctuations in the fluid particle velocity field. Thus, the decompositions discussed above can be interpreted in terms of the decomposition of the kinetic energy of the fluid velocity field into characteristic time scales, eddy sizes and propagating plane waves.

The information in the three-point, four-point and higher-order moments requires still further thought. In addition, since we have not encountered them very often in the past and have had little experience with measuring them, we have little intuition as to the nature of the information contained therein. Further, the scattering experiments that I shall discuss do not depend on information of the property field of higher order than two. For these reasons I shall not dwell on these moments during this talk. I would like to make a few comments relating to them, however, since I think they will turn out to be prime targets of study and of exploitation in the not-too-distant future. Briefly,

---

([*]) Professor MONTROSE has devoted several seminars of the series being presented here to the temporal correlation function.

let me consider a random medium that can be classified as a solid mixture or, more precisely, a solid suspension. That is, we have a material of one phase—an inclusion phase—dispersed throughout a second phase—a matrix phase. A description of the geometry of this suspension is contained in the terms inclusion volume fraction, inclusion size, inclusion shape and relative positioning of inclusions. The lowest-order statistical moment, *i.e.* the one-point moment, of a property field contains only volume fraction information. The two-point moment adds, principally, inclusion size information, although, if the size distribution of inclusions is limited, one could interpret the form of the correlation function in terms of shape information. More detailed information of inclusion shape and of inclusion clustering will appear in the higher-order statistical moments and some work has been done on defining shape factors and packing parameters in terms of these moments [2, 3]. I believe that the definition of such measures and the development of measurement procedures for determining values for these measures has enormous potential in a number of areas. Perhaps, an acoustic-scattering experiment could both provide a definition and a measurement procedure.

1`2. *The acoustic field and its description.* – Mathematically, predicting the acoustic field in a random medium requires the solution of a linear partial differential equation—the wave equation—with a coefficient that is a stochastic function of space and time—the sound speed. We shall take the boundary conditions and source terms to be deterministic. A complete solution to the problem, like a complete description of the sound speed field, is given in terms of a joint probability functional defined on the acoustic field and the sound speed field. Again, the enormity and the complexity of the information required for a complete description is such that we do not attempt to formulate the general problem (*) but look instead to formalisms that involve limited portions of the information required for a complete description. This leads us again to statistical moments of the acoustic field. To motivate the choice of appropriate statistical moments let us consider the special case of time-harmonic signals (**) propagating through a time-invariant random media. Mathematically it is convenient to formulate such a case in terms of a complex field—*e.g.* $\hat{p}(x)$, the complex acoustic-pressure field. The actual, or real pressure space-time history, is given by $\mathrm{Re}\,[\hat{p}(x)\exp[ivt]]$, where v is the frequency. The randomly varying property fields manifest themselves in this problem by a randomly varying refractive-index field. The randomly varying refractive-index field results in $\hat{p}$ being a stochastic function of position. There are three

---

(*) Some work has been carried out on developing general formalisms for mathematically similar problems but, to the author's knowledge, no solutions have resulted.
(**) One can generalize the definitions to be given for the mutual coherence to consider stationary signals. See for example [3].

statistical moments of $\hat{p}$ that I shall briefly discuss:

$$\{\hat{p}(\boldsymbol{x})\}, \qquad \{\hat{p}(\boldsymbol{x}_1)\hat{p}^*(\boldsymbol{x}_2)\} \qquad \text{and} \qquad \{\hat{p}(\boldsymbol{x}_1)\hat{p}^*(\boldsymbol{x}_2)\hat{p}(\boldsymbol{x}_3)\hat{p}^*(\boldsymbol{x}_4)\}.$$

Actually, except for the case in which all characteristic lengths of $\{\hat{p}(\boldsymbol{x})\}$ are very large relative to all lengths defined by the refractive-index field, the information in $\{\hat{p}(\boldsymbol{x})\}$ is of limited interest in most physical applications. For the case of a plane-wave signal to be incident upon the random medium, the decay of $\{\hat{p}(\boldsymbol{x})\}$ gives a measure of the rate at which energy is being scattered by the randomly varying refractive-index field. The information in the two-point moment $\{\hat{p}(\boldsymbol{x}_1)\hat{p}^*(\boldsymbol{x}_2)\}$ termed the mutual coherence function proves to be much more useful in the majority of applications. We note that the reduced coherence function $\{|\hat{p}(\boldsymbol{x})|^2\}$ can be related to the acoustic « intensity » at the point $\boldsymbol{x}$. Thus, it is this quantity that provides information on the average location of the center of a beamed signal, as well as information on the averaged beam spread. Further, for $\hat{p}(\boldsymbol{x})$ statistically homogeneous across a plane (an incident plane-wave signal and a statistically homogeneous medium will result in a $\hat{p}(\boldsymbol{x})$ that is statistically homogeneous across a plane normal to incident-signal direction), the coherence function measured at two points in the plane provides—via its Fourier transform—the angular distribution of acoustic intensity across the plane. Thus, it provides the strength of the output signal of a phased array of receivers located in the plane and steered to look in any direction. It is in terms of the mutual coherence function that one can discuss the problem of loss of resolution of an acoustic signal due to scattering by a randomly varying refractive-index field. As we shall also see, measurement of the mutual coherence function provides, in conjunction with an appropriate acoustic model, important information on the statistics of the refractive-index field. The detailed calculations that I shall discuss in the latter part of the lecture will be on the mutual coherence function. The four-point moment $\{\hat{p}(\boldsymbol{x}_1)\hat{p}^*(\boldsymbol{x}_2)\hat{p}(\boldsymbol{x}_3)\hat{p}^*(\boldsymbol{x}_4)\}$ is obviously a good deal more complex than is the two-point moment and we have much less experience in dealing with it. We note that the reduced four-point moment, *i.e.* $\{|\hat{p}(\boldsymbol{x}_1)|^2|\hat{p}(\boldsymbol{x}_2)|^2\}$, correlates the acoustic intensity at two separated points. It contains more refined information on the spreading of a beamed signal due to scattering. The further collapsed four-point moment, *i.e.* $\{|\hat{p}(\boldsymbol{x})|^2\} = \{\hat{I}^2(\boldsymbol{x})\}$, is recognized as the variance of acoustic intensity measured at the point $\boldsymbol{x}$. It is in terms of this quantity that one can discuss the « fluctuation » problem.

I shall not review the literature of the development of models to predict the above-described quantities, or their analogues for other types of radiation fields. A very brief assessment of these results is, however, warranted. Techniques for obtaining prediction models for the mean-field response are well understood and have been applied to a number of different media [5-9]. The two- and four-point coherence functions have received a good deal of attention in the optics literature. For radiation fields with wavelengths that are much smaller than all correlation lengths of the refractive-index field, differential equations have been derived that govern both the two-point [10, 18] and the four-point [18-21] moments for fields emanating from finite sources. A number of solutions of the equation that governs the two-point coherence function have been presented. To my knowledge, only a numerical solution [22] that was obtained for a two-dimensional medium has been accomplished for the equation on the four-point moment. All of the above described work, however, is inapplicable for sound propagation in the ocean of the type discussed by GOODMAN. This is because the acoustic wavelengths required are too large relative to the dimensions of the randomly varying refractive-index field measured in the depth direction. Recently BERAN and myself have derived a differential equation that governs the two-point coherence function for conditions that are applicable in ocean acoustics studies [1, 23]. Further, some solutions to this equation have been obtained. It is this work that I shall discuss in some detail.

In concluding this discussion on the acoustic field and its description I should comment on amplitude and phase fluctuations. This, of course, refers to representing the complex acoustic field in terms of two real fields, its modulus and its argument, and defining some statistical measures on these. The earlier works, which were largely concerned with « single-scatter » (*) solutions of a variety of problems, were often formulated in terms of amplitude and phase statistics [24, 25]. Also, much of the experimental data that have been reported have been in terms of amplitude and phase. In recent years, however, the trend has definitely been toward the coherence functions discussed above, the reason for this being that a general acceptance appears to be developing that these measures contain the information that is of greatest physical interest. Further, the only formulations that purport to be valid in the multiscatter region, which appear to have received a degree of acceptance here, have been given in terms of the mutual coherence function.

2. – Scattering of narrow-band acoustic signals in frozen random media.

GOODMAN has shown that the propagation of a pressure signal in water with a variable index of refraction is

$$(1) \qquad (\nabla^2 - C^{-2} \partial_t^2)\, p = 0\,,$$

where $p(\boldsymbol{x}, t)$ is the excess pressure and $C^2(\boldsymbol{x})$ is the variable sound speed. Although C^2 is shown to be independent of time, we can use eq. (1) to study prop-

---

(*) This term will be considered in detail.

agation of a sequence of acoustic pulses through a time-varying ocean provided the sound speed does not change during the time interval for a single
pulse to traverse a single point in the ocean. In this case the sequence of pulses
traverses a sequence of frozen oceans. It is convenient to rewrite eq. (1) as

$$(2) \qquad \left(\nabla^2 - \frac{\{n^2\}}{\{c\}^2}[1 + \varepsilon\mu(\boldsymbol{x})]\partial_t^2\right)p = 0\,,$$

where the braces indicate an ensemble average, $\{c\}$ is the mean sound speed,
$\mu(\boldsymbol{x})$ denotes a centered, i.e. $\{\mu(\boldsymbol{x})\} = 0$, stochastic process of unit variance,
i.e. $\{\mu^2\} = 1$, $\{n^2\} = \{C\}^2/\{C^2\}$, and ε is a measure of the strength of the index-
of-refraction fluctuation field. We are interested in problems in which $\varepsilon \ll 1$.
To first order therefore $\{n^2\} = 1$.

For narrow-band signals, with central frequency $\bar{\nu}$, it is convenient to introduce the approximation

$$(3) \qquad p(\boldsymbol{x}, t) = \mathrm{Re}\left[\hat{p}(\boldsymbol{x}, \bar{\nu})\exp[2\pi i\bar{\nu}t]\right],$$

where $\hat{p}(\boldsymbol{x}, \bar{\nu})$ is the complex excess pressure field. Substitution into eq. (2) yields

$$(4) \qquad (\nabla^2 + \bar{k}^2[1 + \varepsilon\mu(\boldsymbol{x})])\,\hat{p} = 0\,,$$

where $\bar{k} = 2\pi\bar{\nu}/\{C\}$ is an averaged wave number. Equation (4), together with
boundary conditions that shall be subsequently specified, constitutes a mathematical description of the problems we wish to study.

2`1. Single-scatter solutions. – The smallness of ε suggests that a perturbation
solution of eq. (4) be constructed. Following [1] we write

$$(5) \qquad \hat{p}(\boldsymbol{x}) = \hat{p}_0(\boldsymbol{x}) + \varepsilon\hat{p}_1(\boldsymbol{x}) + \varepsilon^2\hat{p}_2(\boldsymbol{x}) + \ldots\,,$$

where $\hat{p}_i$ are to be independent of ε. Substituting in eq. (4) and equating like
coefficients of ε, we have

$$(6) \qquad \begin{cases} (\nabla^2 + \bar{k}^2)\hat{p}_0 = 0\,, \\ (\nabla^2 + \bar{k}^2)\hat{p}_1 = -\bar{k}^2\mu(\boldsymbol{x})\hat{p}_0\,, \end{cases}$$

etc. Inhomogeneous boundary terms are to be satisfied by $\hat{p}_0$; each of the $\hat{p}_i$
must satisfy homogeneous boundary conditions. An approximation to $\hat{p}$ obtained by truncating eq. (5) after a single term is referred to as a single-scatter
solution, or a Born approximation. We note that its validity is limited to
cases for which $|\hat{p} - \hat{p}_0| \ll |\hat{p}_0|$, i.e. for cases in which the scattered acoustic
field is weak. Intuitively, therefore, the validity of a single-scatter solution

is expected to have a range limitation, no matter how weak the refractive-index fluctuations are, since the effects of scattering are cumulative. Ranges over which a single-scatter solution is valid are referred to as the single-scatter region.

It is instructive to use a single-scatter solution to illustrate the dependence of the angular spectrum of scattered acoustic intensity and the statistics of the refractive-index field. This is readily accomplished by considering an experiment that is schematically represented by Fig. 1.

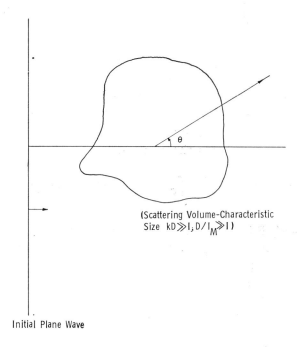

(Scattering Volume-Characteristic Size $kD \gg 1, D/l_M \gg 1$)

Initial Plane Wave

Fig. 1. – Geometry for single-scatter problem.

A plane wave of amplitude of $\hat{I}^{\frac{1}{2}}$, wave number $\bar{k}$ and directed along the z-axis is incident upon a finite « scattering » volume of characteristic size D, where $\bar{k}D \gg 1$ and $D/l_M \gg 1$. Here l_M denotes the largest characteristic length of the fluctuating refractive-index field. Applying the single-scatter formalism, we write for this problem

$$(7) \quad \begin{cases} \hat{p}_0(\boldsymbol{x}) = \hat{I}^{\frac{1}{2}} \exp\left[i\bar{k}z\right] \\ \text{and} \\ \hat{p}_1(\boldsymbol{x}) = -\dfrac{\bar{k}^2 \hat{I}^{\frac{1}{2}}}{4\pi} \int\limits_V \dfrac{\exp\left[i\bar{k}[r(\boldsymbol{x}, \boldsymbol{x}') + z']\right]}{r(\boldsymbol{x}, \boldsymbol{x}')} \mu(\boldsymbol{x}') \, \mathrm{d}\boldsymbol{x}', \end{cases}$$

where $r(x, x')$ is the distance from the scattering point x' to the generic field point x. We intend to calculate the value of the average intensity of the scattered acoustic signal at points far removed from the scattering value. The variation of this scattered intensity with the angular position of the observation point *vis-à-vis* a point in the scattering value gives the desired angular spectrum. Thus we write

$$(8) \qquad \{\hat{I}_s(x)\} \equiv \varepsilon^2 \{\hat{p}_1(x) p_1^*(x)\} =$$
$$= \frac{\bar{k}^4 \hat{I}}{(4\pi)^2} \iint \frac{\exp\left[i\bar{k}[r(x, x') - r(x, x'') + (z' - z'')]\right]}{r(x, x') r(x, x'')} \sigma(x', x'') \, \mathrm{d}x' \, \mathrm{d}x'',$$

where

$$(9) \qquad\qquad\qquad \sigma(x', x'') \equiv \varepsilon^2 \{\mu(x') \mu(x'')\}$$

is the spatial correlation function of the index-of-refraction fluctuations.

Introducing homogeneous statistics, the far-field approximation and the restriction that $D/l_M \gg 1$ enable us to partially accomplish the integration required by eq. (8). We write the result as

$$(10) \qquad\qquad \{\hat{I}_s(x)\} = \frac{\bar{k}^4 \hat{I} V}{(4\pi)^2 R^2} \int \sigma(u) \exp\left[-i\hat{I}(n_s - e_z) \cdot u\right] \mathrm{d}u,$$

where $u = x' - x''$, R is the distance from a point in the scattering volume to the observation point, and n_s is a unit vector directed along the line between these two points. Equation (10) demonstrates the well-known result that in the single-scatter region the angular spectrum of the scattered acoustic intensity and the correlation function of the refractive-index field are related as Fourier-transform pairs. It is a very important result since it enables us to understand the nature of the local scattering. It also serves as a basis for the measurement of $\sigma(u)$ in a scattering experiment.

We proceed in our investigation of eq. (10) by considering a specific correlation function. We choose

$$(11) \qquad\qquad\qquad \sigma(u) = \sigma_0 \exp\left[-\left(\frac{u_x^2}{l_x^2} + \frac{u_y^2}{l_y^2} + \frac{u_z^2}{l_z^2}\right)\right].$$

This functional form is not intended to represent the correlation function of anything as complex as the ocean temperature microstructure. It cannot. Its usefulness rests in the fact that it allows us to evaluate eq. (10) and draw some conclusions that are independent of the details of the functional form. Substituting and integrating we write

$$(12) \qquad \{\hat{I}_s(x)\} = \frac{\sigma_0 \bar{k}^4 \hat{I} V}{(4\pi)^2 R^2} \, \pi^{\frac{3}{2}} l_x l_y l_z \exp\left[-\frac{\bar{k}^2}{4}\left[\frac{l_x^2 x^2}{R^2} + \frac{l_y^2 y^2}{R^2} + l_z^2\left(1 - \frac{z'}{R}\right)^2\right]\right].$$

If we introduce spherical co-ordinates, *i.e.*

(13)
$$
\begin{cases}
x = R \sin \theta \cos \varphi \,, \\
y = R \cos \theta \cos \varphi \,, \\
z = R \cos \theta \,,
\end{cases}
$$

eq. (12) becomes

(14) $\quad \{\hat{I}_s(\boldsymbol{x})\} = \dfrac{\sigma_0 \bar{k}^4 \hat{I} V}{(4\pi)^2 R^2} \pi^{\frac{3}{2}} l_x l_y l_z \cdot$

$$
\cdot \exp\left[-\frac{\bar{k}^2}{4}\left[(l_x^2 \cos^2 \varphi + l_y^2 \sin^2 \varphi) \sin^2 \theta + l_z^2 (1 - \cos \theta)^2\right]\right].
$$

Consider now a few possibilities:

A) $l_x = l_y = l_z = l$, where $\bar{k}l \gg 1$.

This is the usual case discussed in atmospheric-propagation studies. Here

(15) $\qquad \{\hat{I}_s^A(\boldsymbol{x})\} = \dfrac{\sigma_0 \bar{k}^4 \hat{I} V}{(4\pi)^2 R^2} \pi^{\frac{3}{2}} l^3 \exp\left[-\tfrac{1}{2}\bar{k}^2 l^2 (1 - \cos \theta)\right].$

Thus, $\{\hat{I}_s^A(\boldsymbol{x})\}$ is only appreciable if

$$
1 - \cos \theta \ll 1 \,.
$$

Therefore

$$
1 - \cos \theta \approx \theta^2/2
$$

and

(16) $\qquad\qquad\qquad\qquad \theta = O(1/\bar{k}l) \,.$

B) $l_x = l_y = l_z = l$, where $\bar{k}l \ll 1$.

In this case all correlation lengths are small compared to the radiation wavelength. Here

(17) $\qquad\qquad\qquad \{\hat{I}_s^B(\boldsymbol{x})\} = \dfrac{\sigma_0 \bar{k}^4 \hat{I} V}{(4\pi)^2 R^2} \pi^{\frac{3}{2}} l^3 \,,$

and the scattering is isotropic.

C) $l_x = l_z$, $\bar{k}l_x \gg 1$, $\bar{k}l_y \ll 1$.

This is the case of interest in this paper. The vertical direction is the y-direction. Here

(18) $\quad \{\hat{I}_s^C(\boldsymbol{x})\} = \dfrac{\sigma_0 \bar{k}^4 \hat{I} V}{(4\pi)^2 R^2} \pi^{\frac{3}{2}} l_x^2 l_y \exp\left[-\frac{\bar{k}^2}{4} l_x^2 [\cos^2 \varphi \sin^2 \theta + (1 - \cos \theta)^2]\right].$

Scattering in the vertical direction is investigated by choosing

$$x = R \sin \theta \cos \varphi = 0 \, .$$

We find

(19) $$\{\hat{I}_s^c(\boldsymbol{x})\} = \frac{\sigma_0 \bar{k}^4 \hat{I} V}{(4\pi)^2 R^2} \pi^{\frac{3}{2}} l_x^2 l_y \exp\left[-\frac{\bar{k}^2}{4} l_x^2 (1 - \cos\theta)^2 \right].$$

For the exponential term to be appreciable we must have

$$(1 - \cos\theta)^2 \ll 1 \, .$$

Therefore

$$\frac{\bar{k}^2 l_x^2}{4} \frac{\theta_v^2}{4} = O(1)$$

and

(20) $$\theta_v = O\big(1/(\bar{k}l_x)^{\frac{1}{4}}\big) \, .$$

Scattering in the transverse horizontal direction is investigated by choosing $y = R \sin\theta \sin\varphi = 0$ or $\cos\varphi = 1$. Equation (18) is then

(21) $$\{\hat{I}_s^c(\boldsymbol{x})\} = \frac{\sigma_0 \bar{k}^4 \hat{I} V}{(4\pi)^2 R^2} l_x^2 l_y \pi^{\frac{3}{2}} \exp\left[-\frac{\bar{k}^2}{2} l_x^2 (1 - \cos\theta) \right].$$

This is similar to case A) and yields

(22) $$\theta_H = O(1/\bar{k}l_x) \, .$$

Examination of the results of case C) shows that for single scattering we may expect that the horizontal angular spread will be of order $1/\bar{k}l_x$, while the vertical spread will be of order $(1/\bar{k}l_x)^{\frac{1}{2}}$. The same type of analysis holds in general for an arbitrary function $\sigma(\boldsymbol{u})$. We only require that $\bar{k}l_{yM} \ll 1$ and $\bar{k}l_{xm} \gg 1$, where the subscript M denotes the maximum and m denotes the minimum. The statement that

$$\theta = O(1/\bar{k}l_{xm})$$

means that scattering from the smallest scale fluctuation gives $\theta = O(1/\bar{k}l_{xm})$. The largest scales give $\theta = O(1/\bar{k}l_{xM})$. Since, however, $\bar{k}l_{xM} > \bar{k}l_{xm}$ we usually use only the order of magnitude associated with $\bar{k}l_{xm}$.

For the case $\bar{k}l_{yM} \ll (\bar{k}l_{xm})^{\frac{1}{2}}(l_{xm}/l_{xM})^{\frac{1}{2}}$ similar analysis shows that θ is $O\big(1/(\bar{k}l_{xm})^{\frac{1}{2}}\big)$. In general we find that, when $\bar{k}l_{xm} \gg 1$, θ is no greater than $O\big(1/(\bar{k}l_{xm})^{\frac{1}{2}}\big)$ for all values of $\bar{k}l_{yM}$.

2˙2. *Multiscatter solutions.* – As indicated previously, single-scatter solutions have a range limitation due to the cumulative effects of multiscatter. For many important problems, including the application suggested by GOODMAN of using long-range scatter experiments to study internal wave spectra, the need exists to remove this range limitation. This problem received a great deal of study during the last decade and now appears to be solved—at least as far as the lower-order statistics are concerned. There are two derivation procedures that are particularly satisfying to this investigator. One is to use an extended perturbation scheme that results in a partial summation of the series of eq. (5). The method of smoothing, which appears to be the linear counterpart of the Bogoliubov-Krylov-Mitropolski method of averaging non-linear differential equations and the method of diagrams, or renormalization method, can be used to construct very general equations on the any order moment of any linear differential operation [18, 21]. These equations can be specialized to the random Helmholtz operator of eq. (4). The resulting equations for the two-point and the four-point coherence functions are extremely complex and are of little computational value. However, by making use of an assumption that the radiation field crossing any plane z has an angular spectral representation that is confined to a shallow-angled cone centered about the z-direction, considerable simplifications can be introduced into these equations. The second derivation procedure is also limited to radiation fields that are essentially « forward directed ». In this procedure the random medium is viewed as a series of slabs transverse to the principal propagation direction. The slab thickness is taken to be Δz, which is taken to be large relative to the largest correlation length of the refractive-index field measured in this direction but small enough that a single-scatter solution is valid for predicting the desired coherence function at the $z + \Delta z$ plane if one knows the same function at the z-plane. In the multiscatter region, this requires the solution of a difference-differential equation, which may be approximated by a differential equation. Both derivation procedures have led to identical results for all problems to which they have been applied. It is clear, from the above discussion, that these results are only applicable when the radiation field emanating from the $z = 0$ plane has a narrow angular spectrum and when the scattering is essentially a forward scattering.

For problems in which the forward-directed approximation is valid the following equation is to be satisfied by the coherence function measured at two points in the same z-plane, where z is the principal propagation direction [17, 23]:

$$(23) \qquad \frac{\partial\{\widehat{\varGamma}(\boldsymbol{x}_{\perp 1},\, \boldsymbol{x}_{\perp 2},\, z)\}}{\partial z} = - F(\boldsymbol{x}_{\perp 1},\, \boldsymbol{x}_{\perp 2})\{\widehat{\varGamma}(\boldsymbol{x}_{\perp 1},\, \boldsymbol{x}_{\perp 2},\, z)\} +$$

$$+ \frac{i}{2\langle \bar{k}\rangle} [\nabla^2_{\boldsymbol{x}_{\perp 1}} - \nabla^2_{\boldsymbol{x}_{\perp 2}}]\{\widehat{\varGamma}\boldsymbol{x}_{\perp 1},\, \boldsymbol{x}_{\perp 2},\, z)\} + i[\bar{k}(\boldsymbol{x}_{\perp 1}) - \bar{k}(\boldsymbol{x}_{\perp 2})]\{\widehat{\varGamma}(\boldsymbol{x}_{\perp 1},\, \boldsymbol{x}_{\perp 2},\, z)\}.$$

Here $\{\hat{\Gamma}\}$ denotes the two-point coherence function, i.e.

(24) $$\{\hat{\Gamma}(\boldsymbol{x}_{\perp 1}, \boldsymbol{x}_{\perp 2}, z)\} = \{p(\boldsymbol{x}_{\perp 1} + z\boldsymbol{e}_z) p^*(\boldsymbol{x}_{\perp 2} + z\boldsymbol{e}_z)\}\,.$$

The equation can be applied to finite sources and incorporates the possibility of both a spatially varying (in a direction transverse to the principal propagation direction) mean refractive-index field and inhomogeneous statistics. The mean refractive-index field is given by $\bar{k}(\boldsymbol{x}_{\perp})$ and $\langle \bar{k} \rangle$ is an average of $\bar{k}$ taken over a transverse region. The two-dimensional Laplacian is denoted by $\nabla_{\perp}^2$. The scattering term in eq. (23) is the first term on the r.h.s. The function $F(\boldsymbol{x}_{\perp 1}, \boldsymbol{x}_{\perp 2})$ depends on the nature of the local scattering. For a statistically homogeneous and isotropic medium and a radiation field where $\bar{k}l_m \gg 1$, the wavelength is very small relative to the minimum correlation length:

(25) $$F(\boldsymbol{x}_{\perp 1}, \boldsymbol{x}_{\perp 2}) \equiv \bar{\sigma}(0) - \bar{\sigma}(|\boldsymbol{x}_{\perp 1} - \boldsymbol{x}_{\perp 2}|)\,,$$

where

$$\bar{\sigma}(|\boldsymbol{x}_{\perp 1} - \boldsymbol{x}_{\perp 2}|) \equiv \tfrac{1}{4} \int_{-\infty}^{\infty} \sigma(|\boldsymbol{x}_{\perp 1} - \boldsymbol{x}_{\perp 2}|, s_z)\, \mathrm{d}s_z\,.$$

We recall that σ denotes the correlation function of the refractive-index field. With this F the formulation is applicable to the optics problem. For ocean acoustic studies and the highly anisotropic temperature microstructure field, however, we need to use another form for F. The appropriate form for a statistically homogeneous ocean is

$$F(\boldsymbol{x}_{\perp 1} - \boldsymbol{x}_{\perp 2}) \equiv \bar{\sigma}_2(0, 0) - \bar{\sigma}_2(x_{12}, y_{12})\, \delta(y_{12})\,,$$

where

(26) $$\bar{\sigma}_2(x_{12}, y_{12}) = \left(\frac{2}{\pi}\right)^{\frac{1}{2}} \frac{\bar{k}^3}{4} \int_0^{\infty} \cos\left[\frac{\bar{k}y_{12}^2}{2s_z} - \frac{\pi}{4}\right] (\bar{k}s_z)^{-\frac{1}{2}} \left[\int_{-\infty}^{\infty} \sigma(x_{12}, s_y, s_z)\, \mathrm{d}s_y\right] \mathrm{d}s_z\,,$$

and $\delta(y_{12})$ is a projection operator that sets $y_1 = y_2$. We have used x_{12} to denote the horizontal difference co-ordinate and y_{12} to denote the vertical difference co-ordinate.

The second term on the r.h.s. is a diffraction term that must be incorporated at low frequency and/or long ranges. The last term is a refraction term. It is of interest to note that eq. (23) without the first term on the r.h.s. follows directly from the parabolic approximation [26], which has recently received considerable attention in the ocean acoustics community.

We shall consider one solution of eq. (23), with $F(\boldsymbol{x}_{\perp 1}, \boldsymbol{x}_{\perp 2})$ given by eq. (26), and apply this to a potential scattering experiment. We consider a statistically homogeneous ocean and neglect any sound speed profile. Further we assume a plane-wave signal at $z = 0$, *i.e.*

$$(27) \qquad \{\hat{\Gamma}(\boldsymbol{x}_{\perp 1}, \boldsymbol{x}_{\perp 2}, 0)\} = I_0 \, .$$

For the conditions stated, the diffraction and scattering terms do not contribute and we can solve eq. (23) and write

$$(28) \quad \{\hat{\Gamma}(x_{12}, y_{12}, z)\} = I_0 \left(\frac{\bar{\sigma}_2(x_{12}, y_{12})}{\bar{\sigma}_2(x_{12}, 0)} \exp\left[-(\bar{\sigma}_2(0,0) - \bar{\sigma}_2(x_{12}, 0))z \right] + \right.$$
$$\left. + \left(1 - \frac{\bar{\sigma}_2(x_{12}, y_{12})}{\bar{\sigma}_2(x_{12}, 0)} \right) \exp\left[-\bar{\sigma}_2(0,0)z \right] \right) .$$

For points separated by a distance measured along a horizontal line eq. (28) reduces to

$$(29) \qquad \{\hat{\Gamma}(x_{12}, 0, z)\} = I_0 \exp\left[-[\bar{\sigma}_2(0,0) - \bar{\sigma}_2(x_{12}, 0)]z \right] .$$

For points separated by a distance measured along a vertical line eq. (28) reduces to

$$(30) \qquad \{\hat{\Gamma}(0, y_{12}, z)\} = I_0 \left(\frac{\bar{\sigma}_2(0, y_{12})}{\bar{\sigma}_2(0,0)} + \left(1 - \frac{\bar{\sigma}_2(0, y_{12})}{\bar{\sigma}_2(0,0)} \right) \exp\left[-\bar{\sigma}_2(0,0)z \right] \right) .$$

For the two points coincident, *i.e.* $x_{12} = y_{12} = 0$, eq. (28) reduces to I_0. That is, the acoustic intensity is independent of position. This fact is in agreement with energy conservation requirements and with the forward nature of the local scattering that was demonstrated in the last Subsection. Further, noting that $\bar{\sigma}_2(x_{12}, y_{12})$ approaches zero for large enough separation distances for all physically realistic temperature microstructures, we can conclude that

$$(31) \qquad \{\hat{\Gamma}(x_{12}, y_{12}, z)\} \approx I_0 \exp\left[-\bar{\sigma}_2(0,0)z \right]$$

for large x_{12}, y_{12}. The limiting form of the mutual coherence function is that of a plane-wave signal of intensity $I_0 \exp\left[-\bar{\sigma}_2(0,0)z \right]$. This suggests our interpreting this quantity in terms of the relative amount of energy remaining in the completely coherent portion of the signal. Thus, $1/\bar{\sigma}_2(0,0)$ may be interpreted as a characteristic range at which a significant amount of energy has been scattered; it, therefore, provides a measure of the limit of the validity of results that are based on a single-scatter theory.

For values of x_{12}, y_{12} between the two extremes of zero and very large, eqs. (29) and (30) demonstrate that the fall-off of coherence with separation distance is anisotropic. Upon further study one can conclude that the fall-off is much more rapid for distances measured along the vertical. This is in agreement with our single-scatter results. In our subsequent discussion we shall limit ourselves to considering the case of horizontal separation distances, *i.e.* eq. (29).

The environmental data enter eq. (29) via the term $\bar{\sigma}_2(x_{12}, y_{12})$, which is given in terms of the refractive-index correlation function by eq. (26). The amount of information required in this correlation function is extensive and not easily obtained (see Sect. **1**). Fortunately the acoustic model does not require detailed information of the variations of σ measured in the vertical direction. It only requires the integral of this correlation function taken over all separation distances. If we assume that we can approximate

$$(32) \qquad \int_{-\infty}^{\infty} \sigma(x_{12}, s_z, s_y)\, \mathrm{d}s_y = l_y\, \sigma(x_{12}, s_z, 0),$$

where l_y is independent of x_{12}, s_z, then we can obtain a greatly simplified formalism that requires environmental data that can be obtained from horizontal tow runs. We make this assumption, together with the assumption that the statistics of the refractive-index field are isotropic in a horizontal plane, *i.e.*

$$(33) \qquad \sigma(x_{12}, s_z, 0) = \sigma\big(\sqrt{x_{12}^2 + s_z^2},\, 0\big).$$

Finally, we find it convenient to express the environmental data in wave number space, *i.e.* in terms of the one-dimensional spatial power spectrum $\Phi_1(p)$ given by

$$\Phi_1(p) = \frac{1}{2\pi} \int_{-\infty}^{\infty} \sigma(q, 0)\, \exp[ipq]\, \mathrm{d}q,$$

where

$$(34) \qquad q = \sqrt{x_{12}^2 + s_z^2}.$$

As noted in Sect. **1**, Φ_1 has the intuitively satisfying significance of subdividing the power of the fluctuations field into characteristic size intervals, eddy sizes in the turbulence terminology.

By substitution and some degree of manipulation, we can express $\bar{\sigma}_2(x_{12}, 0)$ in terms of $\Phi_1(p)$ and write

$$(35) \qquad \bar{\sigma}_2(0, 0) - \bar{\sigma}_2(x_{12}, 0) = \frac{\bar{k}^{\frac{5}{2}} l_y x_{12}^{\frac{1}{2}}}{2^{\frac{3}{2}}} \int_0^{\infty} F(px_{12})\, \Phi_1(p)\, \mathrm{d}p,$$

where

$$(36) \qquad F(px_{12}) = \frac{1}{(px_{12})^{\frac{1}{2}}} - \frac{\Gamma(\frac{1}{4})}{2^{\frac{3}{2}}}\, (px_{12})^{\frac{1}{4}} J_{-\frac{3}{4}}(px_{12}).$$

Here $\Gamma(\frac{1}{4})$ denotes a gamma-function and $J_{-\frac{3}{4}}$ denotes a Bessel function.

Equation (35) indicates that the entire one-dimensional power spectrum contributes to a degree in determining the coherence at any given separation distance. While this is true, one can show that not all portions contribute to the same degree. Rather, for a given separation distance, the coherence will be determined by the index-of-refraction fluctuations that are measurable on length scales that encompass a limited range of sizes. This is fortunate for several reasons. For the problem of monitoring the refractive index field by means of a scattering experiment it is crucial since it provides us with a selection rule. That is, only the power in the fluctuations in a limited range of sizes will determine the measured coherence for a given separation distance. To determine just which range of size scales is predominant in a given situation will require a detailed analysis of the particular problem. It is not possible to give generally valid rules of thumb. Each case must be treated separately. The important observation from the point of view of the inverse problem, however, is that a selection rule still exists for the multiple-scatter region although it is not as simple, or precise, as in the single-scatter region.

To carry the analysis one step further, we might assume that over the range of size scales that are important for a specific application the power spectrum obeys a power law decay, that is, that

$$(37) \qquad \Phi_1(p) = \begin{cases} \dfrac{A_n^2}{(p^2 + p_M^2)^{n/2}}, & p \leqslant p_m, \\ \\ 0, & p > p_m. \end{cases}$$

Equation (37) gives a power law dependence for $p_M \ll p \ll p_m$. The presence of low and high wave number cut-offs is to be expected from physical considerations. The exact form of the behavior in the vicinity of the cut-off values is arbitrary, but this exact form should not enter our results for separation distances x_{12} that satisfy the inequality $p_m^{-1} \ll x_{12} \ll p_M^{-1}$ if the reasoning leading to the existence of a selection rule is valid. Substitution of eq. (37) into (35) allows us to complete the integration (see ref. [1]). Introducing the restriction that $x_{12} \ll p_M^{-1}$ then enables our simplifying the resulting analytic expression by a power series expansion and truncation. We present here the result obtained for $n = 2$, since data taken at sea and theoretical arguments support a minus two power law for the larger size scale structures that are of interest in ocean acoustic studies:

$$(38) \qquad \{\hat{\Gamma}(x_{12}, 0, z)\} = I_0 \exp\left[-1.1 A_2^2 \, l_y \, \bar{k}^{\frac{3}{2}} \, x_{12}^{\frac{3}{2}} z\right].$$

Equation (38) relates the power in the fluctuations field and the averaged vertical correlation length scale (see eq. (32)) to the rate of decay of acoustic-signal coherence with separation distance measured along a horizontal line. We note that the lack of presence of either p_M or p_m in eq. (38) supports the contention that the form of the power spectrum at either the very large or the very small size scales is not important so long as the separation distance falls within the indicated range. We further note that the power of x_{12} in eq. (38) is directly related to the power of p in eq. (37). Hence, by monitoring the dependence of $\{\hat{\Gamma}\}$ on x_{12}, the theory provides a measurement procedure for n.

REFERENCES

[1]   M. J. BERAN and J. J. McCOY: *Journ. Math. Phys.*, **15**, 1901 (1974).
[2]   M. MILLER: *Journ. Math. Phys.*, **10**, 1968 (1969).
[3]   M. A. ELSAYED: *Journ. Math. Phys.*, to appear.
[4]   M. J. BERAN and G. B. PARRENT jr.: *The Theory of Partial Coherence* (New York, N. Y., 1964).
[5]   J. B. KELLER: *Proc. Symp. Appl. Math.*, **13**, 227 (1962).
[6]   F. KARAL and J. B. KELLER: *Journ. Math. Phys.*, **5**, 537 (1964).
[7]   L. KNOPOFF and J. A. HUDSON: *Journ. Acoust. Soc. Amer.*, **42**, 18 (1969).
[8]   J. J. McCOY: *Inter. Journ. Sol. Struct.*, **8**, 877 (1972).
[9]   J. J. McCOY: *Journ. Appl. Mech.*, **40** E, 511 (1973).
[10] V. I. TATARSKII: *The effects of the turbulent atmosphere on wave propagation*, National Technical Information Service (Springfield, Va., 1971), translated from the Russian of *Wave Propagation in the Turbulent Atmosphere* (Moscow, 1967).
[11] G. KELLER: *Astronom. Journ.*, **58**, 113 (1953).
[12] R. E. HUFNAGEL and N. R. STANLEY: *Journ. Opt. Soc. Amer.*, **54**, 56 (1964).
[13] W. P. BROWN: *IEEE*, AP-**15**, 81 (1967).
[14] D. DE WOLF: *Radio Sci.*, **2**, 1379 (1967).
[15] M. J. BERAN: *Journ. Opt. Soc. Amer.*, **56**, 1475 (1966).
[16] M. J. BERAN: *Journ. Opt. Soc. Amer.*, **60**, 518 (1970).
[17] J. MOLYNEUX: *Journ. Opt. Soc. Amer.*, **61**, 248 (1971).
[18] V. I. SHISHOV: *IVUZ Radiophys.*, **11**, 866 (1968).
[19] M. J. BERAN and T. L. HO: *Journ. Opt. Soc. Amer.*, **59**, 1134 (1969).
[20] J. J. McCOY: *Journ. Opt. Soc. Amer.*, **62**, 30 (1972).
[21] W. P. BROWN: *Journ. Opt. Soc. Amer.*, **62**, 966 (1972).
[22] M. J. BERAN and J. J. McCOY: *Journ. Acous. Soc. Amer.*, **56**, 1667 (1974).
[23] V. TATARSKI: *Wave Propagation in a Turbulent Medium* (New York, N. Y., 1961).
[24] L. CHERNOV: *Wave Propagation in a Random Medium* (New York, N. Y., 1961).
[25] R. H. HARDIN and F. D. TAPPERT: *SIAM Rev.*, **15**, 423 (1973).

Propagation in Fluctuating Media.

R. R. GOODMAN

Naval Research Laboratory - Washington, D. C.

Glossary of symbols.

Symbol	Meaning
c, c_0	sound speed
C	phase velocity of surface waves
d	depth of the water column
$D_n(\varrho)$	structure function of the index of refraction
E	wave number spectral density
$\boldsymbol{g}$	the Earth's average gravity vector at the surface
k, K	wave number
n	frequency of internal waves and surface waves
N	Brunt-Väisälä frequency
p, P	pressure
$P(\eta)$	probability density for surface displacement
R	acoustic range
s	salinity
S	entropy
S, A	wave front and phase of an acoustic wave
t	time
T	temperature
$\mathscr{T}$	surface tension
$\boldsymbol{u}(u, v, w),\ \boldsymbol{U}$	fluid velocity
V	coefficient of variation
x, y, z	Cartesian co-ordinates at the surface with z being positive in the upward vertical direction
α	absorption constant
η	coefficient of viscosity
η	surface displacement
θ, θ_0	angle of intersection of an acoustic ray and the vertical axis
λ	coefficient of expansive friction
ϱ, ϱ_0, R	density
φ	velocity potential
Φ	power spectrum density of internal waves
ω	acoustic frequency
Ω	the Earth's rotation vector

1. – Introduction.

The work that we shall review under the topic of « Propagation in Fluctuating Media » will deal with physical phenomena of macroscopic dimensions many magnitudes larger than what is treated in the other lectures in this course. The topic is not normally considered as part of physical acoustics, but perhaps by the end of this series enough analogies of the physical processes that affect acoustic waves will be identified by you that the field may be considered at least an « honorary member ».

The media of interest in macroscopic phenomena are the atmosphere, the oceans (hydrosphere), and what I believe to be the most interesting and important boundary of all—the sea surface.

The atmosphere, which is known to all of us, needs little introduction. It is known for its stillness, its winds and its turbulence. These all have profound effects on the propagation of sound. The sea, on the other hand, is not known to many of you. It is, without any question, the most remarkable macroscopic medium we have for acoustic propagation. Indeed it is, as we shall see, a worldwide acoustic wave guide. We have some intuitive feelings about the sea surface. We know it to be at different times both remarkably flat and truly dynamic. It is, I believe, the largest dynamic reflecting acoustic surface in the world. But generally we have only a small amount of descriptive information about the water column below it.

Now what are the reasons for an acoustician to concern himself with the oceans? To date almost all interest has been in application-oriented work in communications and detection, where propagation plays only a partial role —reverberation and ambient noise being just as important. However, we now have some experiments and theories that clearly tell us something about oceanographic and atmospheric processes. We have a few which imply something about the boundary as well.

I wish to concentrate in this series of lectures on the propagation in the sea rather than in the atmosphere. The reason for this is twofold. First, it is the least known and therefore offers something potentially new and exciting. Second, because of the nature of the sea, many of the remote measurements must be made by acoustic methods, whereas the atmosphere is amenable to other sensors, such as radar, optical systems, or just simply sensors on aircraft. Obviously the reason we know so much about the overall properties of the atmosphere is because we can see through it, and we readily have access to it. Thus the most important applications of acoustic sensing would appear to be in the sea. The techniques discussed in these talks can be applied to the atmosphere with suitable modification of the sensors and adaptation of the dynamic phenomena to a fluid that is air rather than water.

We will begin by a cursory description of the dynamics of the medium, stressing those features that we know affect acoustic propagation. Then acoustic fields will be introduced in these media to show first the large effects that dominate propagation and then to show the detailed effects due to the dynamic processes of the medium.

After we have some feeling for the general « order of magnitude » and general descriptions, several important experimental results will be discussed to demonstrate the reality of the effects of environment and to bring us to the state of the art. Then some of the interesting possibilities utilizing current knowledge will be introduced in which acoustic fields might be utilized to observe the dynamic processes of the sea and its upper boundary.

2. – Physical processes that produce fluctuations.

It is clear to begin with that those oceanographic quantities that influence the propagation of an acoustic wave are those that change the sound speed, the density, or the boundaries. One would expect, then, that the main contributors in the water column would be temperature, salinity and overall pressure. Indeed this is the case. The oceanographic quantities of interest to us, therefore, are these. We now wish to investigate briefly the oceanographic processes that cause the acoustic variations. Let us begin with a large view of the oceans, one that we shall call « static », and that will be our reference point.

The oceans, in general, exceed depths of 3000 meters over most of their area which, for most of our discussions here, may be considered to be infinite. They are generally of a salinity near 35 parts per thousand with a 10 % excursion from that being rare. Temperatures, of course, vary considerably, both geographically and seasonally and show important changes locally over times from parts of a day to days. Almost all of the change in pressure is due to the weight of the water column, although changes due to motion will also be important. The upper part of the sea, as one would guess, is generally the warmest due to solar heating and interaction with the atmosphere. It clearly is of major significance in energy exchange and transport. This upper layer is known as the *mixed layer*. The deep water below usually has a value near 4 °C in most oceans, and it is believed to be much more tranquil than the upper ocean. The region in between is known as the thermocline.

The model is « static » because of the constancy of the above observations over long times and also because ocean currents are small. A more rigorous meaning to this statement will be given later on.

All of these features, as we shall see later, will be very important to the nature of overall acoustic propagation on which the fluctuations are to be added. They also have a very important role in the development of the equations governing the motion of the sea that give rise to many of the important dynamical processes.

The technique that is used to obtain tractable equations from the nonlinear equations of fluid motion is to obtain a reference state from which variations can be made, and it is hoped that Nature will permit the variations to be small enough to justify a linear expression of terms.

We naturally begin with the equations of motion of a fluid. The first is the conservation of mass

(1)
$$\frac{d\varrho}{dt} + \varrho \nabla \cdot \boldsymbol{u} = 0 ,$$

where

$\varrho =$ density,

$\boldsymbol{u} =$ fluid velocity vector.

The differential operator d/dt refers to the total time derivative that can be written in the Eulerian reference as

$$\frac{\partial}{\partial t} + \boldsymbol{u} \cdot \nabla .$$

The momentum equation or the Navier-Stokes equation is in its general form [1]

(2)
$$\varrho \frac{\partial \boldsymbol{u}}{\partial t} + \varrho (\boldsymbol{u}\nabla) \boldsymbol{u} + \nabla p = \boldsymbol{F} + \left(\frac{4}{3}\eta + \lambda\right) \nabla^2 \boldsymbol{u} - \eta \nabla \times \nabla \times \boldsymbol{u},$$

where

$p =$ pressure,

$\boldsymbol{F} =$ external forces per unit volume,

$\lambda =$ coefficient of expansive friction,

$\eta =$ coefficient of viscosity.

Now we wish to simplify eq. (2) by making the following restricting statements:

1) typical dimensions allow us to use Cartesian co-ordinates with the z-axis being the upward vertical,

2) the only two « Earth » terms will be its rotation, which will be taken as constant, and gravity $\boldsymbol{g}$ in the vertical and constant.

Now we choose a co-ordinate system which rotates with the Earth. Then we have a Coriolis force added to what appears to be the same equation. The well-known transform [2] from a space co-ordinate to an Earth co-ordinate is given by

$$\left(\frac{d}{dt}\right)_{\text{space}} = \left(\frac{d}{dt}\right)_{\text{Earth}} + 2\boldsymbol{\Omega} \times \boldsymbol{u} ,$$

where $\Omega =$ the Earth's rotation. This gives, for a total time derivative

$$(3) \qquad \left(\frac{\partial u}{\partial t} + (u \cdot \nabla) u\right)_{\text{space}} = \left(\frac{\partial u}{\partial t} + (u \cdot \nabla) u + 2\Omega \times u\right)_{\text{Earth}}.$$

The factor of two in the last term is a consequence of du/dt being a second derivative of time.

With these restrictions on eq. (2) we have

$$(4) \qquad \varrho \left(\frac{\partial u}{\partial t} + (u \cdot \nabla) u + 2\Omega \times u\right) + \nabla p - \varrho g = f,$$

where f represents the remaining external forces and, if necessary, any viscous terms. At this point we still do not have a tractable set of equations. Nor do we have enough relating p, ϱ and u. Just as in the development of fundamental equations of acoustics in most textbooks an additional relationship needs to be found between pressure, density and velocity.

The usual technique is to choose some «reference» state and look for excursions from it with the objective of linearizing the equations. In oceanography this state from which linearization begins is called the Boussinesq approximation [3]. It assumes a reference state in which the entropy and salinity are constant, and the fluid is at rest.

By assuming that the pressure is determined to a high degree only by the depth dependence due to the weight of the water column, we can draw some important conclusions. We say for the reference pressure

$$(5) \qquad \frac{dp}{dz} = - g\varrho ,$$

and, if the change in ϱ is also determined by a change in p through an adiabatic process, we then have a relationship well known among acousticians and oceanographers

$$(6) \qquad \begin{cases} d\varrho = \left(\frac{\partial \varrho}{\partial p}\right)_s dp , \\[2mm] \quad = \frac{1}{c^2} dp , \end{cases}$$

where c is the speed of sound and S is entropy. Substituting eq. (6) into eq. (5) gives

$$\frac{d\varrho}{dz} = -\frac{g}{c^2} \varrho ,$$

or

$$(7) \qquad \varrho = \varrho_0 \exp\left[-g \int \frac{dz}{c^2}\right],$$

where ϱ_0 is the density at the surface. Then

$$\frac{\mathrm{d}p}{\mathrm{d}z} = -g\varrho_0 \exp\left[-g\int\limits_0^z \frac{\mathrm{d}z}{c^2}\right],$$

where the right-hand side can be approximated rather well by $\varrho_0 g$ (*). Thus the reference pressure depends on depth as

(8) $p = \varrho_0 g z$.

If we use the adiabatic relationship between pressure and density, the continuity equation may be written as

$$\mathbf{\nabla}\cdot\mathbf{u} - \frac{g}{c^2}w = 0,$$

where w is the vertical component of $\mathbf{u}$. Clearly oceanic processes do not give rise to vertical currents of sufficient magnitude to compensate for the small coefficient g/c^2, thus incompressibility can be implied. (This will not be the case for acoustic fields, obviously.)

The reference pressure can be considered to be hydrostatic and a deviation from it that we shall define as p'. Thus we may then write

(9) $\varrho_0\left[\dfrac{\partial\mathbf{u}}{\partial t} + (\mathbf{u}\cdot\mathbf{\nabla})\mathbf{u} + 2\mathbf{\Omega}\times\mathbf{u}\right] + \nabla p' - (\varrho-\varrho_0)\mathbf{g} = \nu\nabla^2\mathbf{u}$.

Along with the incompressibility condition

(10) $\mathbf{\nabla}\cdot\mathbf{u} = 0$

we have the Boussinesq approximation. These will form the basis of most of what we will discuss. We now must look at processes that will be important to us in the propagation of acoustic waves.

The first one to consider, and by reference to today's oceanographic literature, perhaps the most important, is internal waves. Clearly we cannot go into detail, but we can understand their physical origin and the terms by which they may interact. To keep the algebraic steps to as few as necessary consider a two-dimensional case (x, z) with velocity components (u, w). Write $\varrho - \varrho_0$ as $\Delta\varrho$ and let it have one large term that is the mean value for a given depth $\overline{\Delta\varrho(z)}$ and another term ϱ' which contributes a negligible amount to the

---

(*) Note that in the exponent $g/C^2 \approx (1/225)$ km^{-1}. The oceans being in their deepest parts about 10 km and C varying only a few percent at most [4] we can see that the change in density is small.

continuity equation. That is

$$(11) \qquad \frac{\partial \Delta \varrho}{\partial t} \simeq - w \frac{\partial \overline{\Delta \varrho(z)}}{\partial z} .$$

The two equations from the Boussinesq approximation for momentum, neglecting Earth's rotation, viscosity and higher-order terms, are

$$(12) \qquad \frac{\partial w}{\partial t} + \frac{1}{\varrho_0} \frac{\partial p}{\partial z} + \frac{g}{\varrho_0} \Delta \varrho = 0 ,$$

$$(13) \qquad \frac{\partial u}{\partial t} + \frac{1}{\varrho_0} \frac{\partial p}{\partial x} = 0 .$$

Eliminating the terms in p by taking $\partial/\partial x$ of (12) and $\partial/\partial z$ of (13) and substituting into (12) gives

$$\frac{\partial}{\partial t} \left(\frac{\partial w}{\partial x} - \frac{\partial u}{\partial z} \right) + \frac{g}{\varrho_0} \frac{\partial \Delta \varrho}{\partial x} = 0 .$$

Taking the partial derivative with respect to t and using (1) gives

$$\frac{\partial^2}{\partial t^2} \left(\frac{\partial w}{\partial x} - \frac{\partial u}{\partial z} \right) - \frac{g}{\varrho_0} \frac{\partial w}{\partial x} \frac{\partial \overline{\Delta \varrho(z)}}{\partial z} = 0 .$$

A derivative with respect to x and the use of the irrotational approximation, $\nabla \cdot \boldsymbol{u} = 0$, gives

$$(14) \qquad \frac{\partial^2}{\partial t^2} (\nabla^2 w) - \frac{g}{\varrho_0} \frac{\partial \Delta \varrho}{\partial z} \frac{\partial^2 w}{\partial x^2} = 0 .$$

In general $\partial \overline{\Delta \varrho}/\partial z < 0$ so we have something that is amenable to a wave type of solution under this condition.

The quantity N, defined by

$$(15) \qquad N^2(z) = \frac{g}{\varrho_0} \frac{\partial \overline{\Delta \varrho}}{\partial z} ,$$

is known as the Brunt-Väisälä (B-V) frequency and is of fundamental importance in oceanography. If we had not made all of the approximations, we would have found

$$(16) \qquad N^2 = - \frac{g}{\varrho_0} \frac{\partial \overline{\Delta \varrho}}{\partial z} - \frac{g^2}{c^2} .$$

With the equation

$$\frac{\partial^2}{\partial t^2}(\nabla^2 w) + N^2(z)\frac{\partial^2 w}{\partial^2} = 0$$

we try a wave solution

(17) $$w = W(z)\exp[i(kx - nt)],$$

which leads to

(18) $$n^2\left(\frac{\mathrm{d}^2 W}{\mathrm{d}z^2} - k^2 W\right) + N^2 k^2 W = 0,$$

or

(18') $$\frac{\mathrm{d}^2 W}{\mathrm{d}z^2} + \frac{N^2(z) - n^2}{n^2}k^2 W = 0.$$

Thus we have a wavelike character to the equation in the z-direction as well, if the time period is less than the B-V frequency. If it is greater, the wave is « evanescent or exponentially attenuated in the vertical direction ». The form should look familiar to students of physics. It is much like the well-known one-dimensional potential well problems of quantum mechanics. It is also much like the acoustic-oceanographic wave guide problems that we shall briefly discuss later on. Two typical types of B-V period for the sea are shown in Fig. 1. Two simple solutions are given in Appendix A.

Now we have an equation to work with, if we want to know more about the physics of internal waves. The solutions are straightforward. But of interest, of course, is 1) how are the waves generated? and 2) how do they interact? The latter point is obvious when we return to the procedure we used to obtain eq. (14). All higher-order terms were ignored. If these are restored and collected into a term $F((\boldsymbol{u}\cdot\boldsymbol{\nabla})\boldsymbol{u})$, one sees immediately a source for the interaction of waves of different frequency and wave number. Thus we have an internal wave phenomenon that gives a cascading effect over sufficiently long time that can generate a broad distribution to the power spectrum.

There are serious questions about the stability of internal waves as well [4]. Clearly we have a gradient in the local flow, and this is, of course, the origin of instabilities in fluid mechanics. There are also some extremely interesting results due to the presence of weak shear currents which predict a turning of horizontally propagating waves and a power spectrum distribution of the form

(19) $$\varPhi(z, n) \propto [N^2(z) - n^2]^{\frac{1}{2}}n^{-3}.$$

There is some observational evidence of this type of spectrum in the open ocean [4, 5].

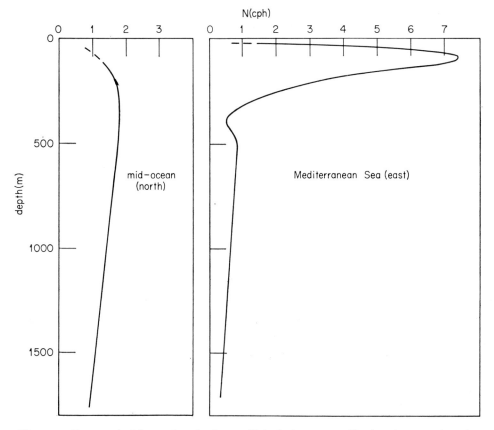

Fig. 1. – Two typical forms for the Brunt-Väisälä frequency distribution as a function of depth.

The sources of waves are also important to consider, but the topic is too broad to discuss in the time we have. Long-period waves can have tidal origins or atmospheric origins as well as complex interactions of current over large regions. Short waves can be excited by currents, surface effects, etc. These are all complex interactions and experimental evidence is too scarce to allow definitive statements to be made. For our purposes it is sufficient to say they exist and can cause fluctuations in acoustic propagation.

In any fluid dynamic system, turbulence should be expected to play a significant role, and indeed it does in the sea. The upper layer or mixed layer that was mentioned earlier is dominated by the turbulent processes initiated by heating, surface waves and winds. Turbulence, of course, is also observed near boundaries where currents exist. The processes are numerous, and many of them are not understood. However, turbulence is a well-known phenomenon, and numerous outstanding texts exist in the field. It is not our intention here to go into the same detail that we did for internal waves but only to outline

the important results that apply in the regime that will affect acoustic propagation significantly. In oceanographic measurements, due to the inherent difficulty of making them, it is not usually possible to be sure that a dynamic
observation is due to turbulence. Indeed, many of the measurements or
observations that are not easily interpreted are considered as due to turbulence even though they may not be.

The gross features of the sea are not isotropic and, therefore, one would
not expect symmetry in the overall turbulent field. Local statistical homogeneity does exist, however, on a small scale, a fact that simplifies considerably
parts of the theory and allows a wave number spectrum to be deduced by
similarity arguments. KOLMOGOROV [6] has obtained the now famous « five-
thirds law », which holds for the part of the spectrum where inertial processes
account for the transfer of energy from one wave number to the next rather
than the viscous range or on the order of the size of the turbulent region. The
equation is given by

$$(20) \qquad\qquad E(k) = A\varepsilon^{\frac{2}{3}}k^{-\frac{5}{3}} ,$$

where A is a universal constant and ε is the value of the net energy transfer
rate. This does not hold in the wave number range in which viscous dissipation
is important and also in the range at which the wave number is of the order of
the inverse dimension of the size of the flow being considered. This, of course,
is a wide range. There is evidence in certain oceanographic experiments of
a « five-thirds law » [4]. Experimental error, especially at sea, however, makes it
difficult to obtain the accuracy necessary to tell five-thirds from something
approximating it. We shall see later on that the upper region of the scale of
turbulence sizes will be important in short-range propagation, and it may be
that some of the distribution spectrum can be measured by acoustical means.

Next, some things should be said about the sea surface and its distribution.
The excellent books by KINSMAN [7] and PHILLIPS [3] are recommended. There
are some important points that should be stated about the sea surface, because
it is an important scattering process for acoustic energy in the sea. Clearly
the physics of the sea surface is readily understandable on first principles. If one
begins with a flat air-sea boundary and distorts it in some way, the restoring
forces to return it to its flat state are gravity and surface tension. Naturally
the role of gravity is overwhelming at large wavelengths and amplitudes. It will
turn out in the brief derivation below that surface tension is the principal
restoring force for small distortions, as one would expect. This range of wavelengths is referred to as capillary waves. The others are called gravity waves.
Both of these play an important role in the dynamic processes on the sea surface and below. The development that follows will demonstrate these regimes
and the relationship between wavelength and wave speed and the dynamic
properties of the sea below the surface due to surface waves. A few things

will then be said about the power spectrum or the probability distribution on the surface.

For our purposes we can consider a two-dimensional sea which has no rotation, is not affected by Coriolis forces, and is inviscid, incompressible and irrotational. Then we have eq. (10)

$$(10) \qquad\qquad \mathbf{\nabla} \cdot \boldsymbol{u} = 0 \,,$$

and

$$(21) \qquad\qquad \nabla^2 \varphi = 0 \,,$$

where

$$(22) \qquad\qquad \boldsymbol{u} = \mathbf{\nabla}\varphi$$

and $\boldsymbol{u} = (u, w)$.

Assume a wave on the surface of the form

$$(23) \qquad\qquad z = A \cos (kx - nt) \,.$$

Then

$$(24) \qquad\qquad \frac{\mathrm{d}z}{\mathrm{d}t} = An \sin (kx - nt) \,.$$

The boundary conditions are therefore

$$w = Ak \sin (kx - nt)$$

at the surface and

$$(25) \qquad\qquad w = 0$$

at the sea floor $(z = -d)$.

Clearly a solution of eq. (21) is

$$(26) \qquad\qquad \varphi = \varphi_0 \cos (kx - nt) F(z) \,,$$

where

$$(27) \qquad\qquad \frac{\mathrm{d}^2 F(z)}{\mathrm{d}z^2} - k^2 F(z) = 0 \,.$$

The boundary conditions are satisfied by the obvious solutions of eq. (27) if

$$(28) \qquad\qquad \varphi = \frac{An}{|k} \frac{\cosh k(z + d)}{\sinh kd} \sin (kx - nt) \,.$$

Note that this is the solution for φ if a wave of the type assumed is possible.

As in most fields of physics, we must look at the boundary to determine what
the requirements are on the wave. We therefore must look at the equations
of motion at the surface.

For our purposes the Boussinesq approximation is to the lowest order

(29)
$$\begin{cases} \varrho\,\dfrac{\partial u}{\partial t} + \dfrac{\partial p}{\partial x} = 0\,, \\[2mm] \varrho\,\dfrac{\partial w}{\partial t} + \dfrac{\partial p}{\partial z} + \varrho g = 0\,. \end{cases}$$

If we assume irrotational flow, a potential φ can be used, giving

(29')
$$\begin{cases} \varrho\,\dfrac{\partial^2 \varphi}{\partial t\,\partial x} + \dfrac{\partial p}{\partial x} = 0\,, \\[2mm] \varrho\,\dfrac{\partial^2 \varphi}{\partial t\,\partial z} + \dfrac{\partial p}{\partial z} + \varrho g = 0\,. \end{cases}$$

These can be combined to give

(30)
$$\nabla\left\{ \varrho\,\frac{\partial \varphi}{\partial t} + p + \varrho g z \right\} = 0\,,$$

which yields an approximation form of the Bernoulli equation

(31)
$$\varrho\,\frac{\partial \varphi}{\partial t} + p + \varrho g z = \text{const}\,.$$

There is now a need to obtain an equation with only φ. Clearly a time derivative
will do it for the last term. Thus at the surface

(32)
$$\varrho\,\frac{\partial^2 \varphi}{\partial t^2} + \frac{\partial p}{\partial t} + \varrho g\,\frac{\partial \varphi}{\partial z} = 0\,.$$

The only remaining term, then, is p. It is, at the boundary, related to the
atmospheric pressure (which we will assume to be constant) by

(33)
$$p = p_a + \mathscr{T}\,\frac{\partial^2 z/\partial x^2}{(1 + (\partial z/\partial x)^2)^{\frac{3}{2}}}\,,$$

where $\mathscr{T}$ is the surface tension. This is easily derived from simple force equilibrium. For small waves the last term may be approximated by $\partial^2 z/\partial x^2$. Thus

(34)
$$\frac{\partial p}{\partial t} = -\mathscr{T}\,\frac{\partial^2}{\partial x^2}\,\frac{\partial z}{\partial t} = -\mathscr{T}\,\frac{\partial^3 \varphi}{\partial x^2\,\partial z}\,.$$

Inserting this into (32) gives the final boundary condition

$$(35) \qquad \varrho \frac{\partial^2 \varphi}{\partial x^2} + \varrho g \frac{\partial \varphi}{\partial z} - \mathscr{T} \frac{\partial^3 \varphi}{\partial x^2 \partial z} = 0 \,.$$

Substituting the solution for φ into this yields the requirement that n and k satisfy

$$(36) \qquad n^2 = gk \left(1 + \frac{\mathscr{T} k^2}{\varrho g}\right) \mathrm{tgh}\, kd \,.$$

(For a rigorous development beginning with general conditions there are several excellent references, [3, 7]).

For our purposes we only wish to consider deep water which, for this solution, only means many wavelengths deep ($kd \gg 1$). Thus

$$(37) \qquad \varphi = \frac{An}{k} \exp\left[kz\right] \sin\left(kx - nt\right)$$

and

$$(38) \qquad n^2 = gk \left(1 + \frac{\mathscr{T} k^2}{\varrho g}\right) \,.$$

The last term, or viscous contribution (capillary waves), is dominant at large wave numbers. The first or gravity wave term dominates for small wave numbers. The phase velocity C is n/k. This gives

$$(39) \qquad C^2 = \frac{g}{k} \left(1 + \frac{\mathscr{T} k^2}{\varrho g}\right) ,$$

which for sea water ($\mathscr{T} \approx 74$ dyn cm^{-1}) is shown in Fig. 2.

We now see what we obtain for sea waves in the lowest approximation: waves that propagate with gravitational restoring forces and those with surface tension restoring forces. It is clear which have most of the energy. It is important to notice that neither has much direct effect on the dynamic properties of the water column due to the exponential attenuation.

The surface itself is an extraordinarily complex surface even though one can show that wave-wave interactions are of low order in the sea. One only needs to observe the dominant waves on the lake near this room to conclude that single waves lose their integrity (and therefore their correlation) over a few wavelengths and, therefore, a stochastic description of the surface is in order. Let the vertical displacement of the sea surface from its mean be given by $\eta(\boldsymbol{x}, t)$, where $\boldsymbol{x}$ is the horizontal position. The observed statistics of the wave height indicate that over a wide range of heights the distribution function that best fits the data is Gaussian. Clearly there has to be a variation from a

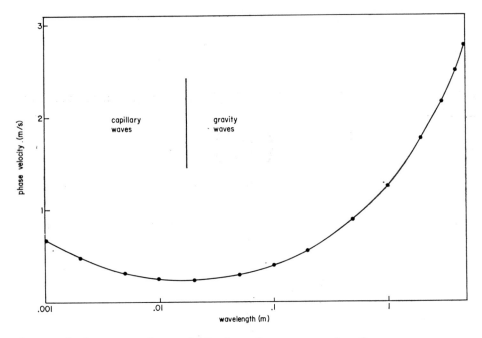

Fig. 2. – Surface wave phase velocity dependence on wavelength.

true Gaussian since waves of very large height do not (and cannot) exist. This is the distribution we would expect for the summation of independent uncorrelated waves. Thus the probability density is approximately

$$(40) \qquad\qquad P(\eta) = \frac{1}{(2\pi\overline{\eta^2})^{\frac{1}{2}}} \exp[-\eta^2/2\overline{\eta^2}] \,.$$

If we take into account the more detailed interactions other forms of distributions can be obtained [3, 7].

The wave number and frequency spectra are also clearly of considerable importance. These, of course, are derived from second-moment statistics or correlation properties $\overline{\eta(\pmb{x}, t)\eta(\pmb{x} + \pmb{X}, t+T)}$. We shall not go into the details of the relationships of this to spectra, since this topic is related to what McCoy will discuss in his lecture. Suffice it to say that theoretical results and empiricism have led to the following spectra for frequency and wave number respectively [3, 8]:

$$(41) \qquad\qquad E(n)\,\mathrm{d}n = B \exp[-\alpha n^4]n^{-5}\,\mathrm{d}n$$

and

$$(42) \qquad\qquad E(k)\,\mathrm{d}k = B'k^{-4} \,.$$

3. – The derivation of the acoustic equations.

It is of interest to develop the acoustic equations for the sea and compare them with what we did in Sect. **2**. We have seen that the Navier-Stokes equation and the continuity equation can be specialized for oceanographic applications in terms of a «reference sea», and cast in a more tractable form, suitable for linearizing as the occasion requires. Here we shall do the same, except our reference will begin with the final equations that were developed for the hydro-dynamics of the sea. Then, by changing the notation to

$$p \to P + p \,, \quad \varrho \to R + \varrho \,, \quad \boldsymbol{u} \to \boldsymbol{U} + u \,,$$

where P, R and $\boldsymbol{U}$ are the reference state of the sea and p, ϱ and $\boldsymbol{u}$ are the acoustically induced changes, we can begin to derive the important equations of acoustics. Substitution of the above quantities into the equations of the sea gives us

$$
(43) \quad
\begin{cases}
\dfrac{\partial \boldsymbol{u}}{\partial t} + (\boldsymbol{u} \cdot \boldsymbol{\nabla})\boldsymbol{U} + (\boldsymbol{U} \cdot \boldsymbol{\nabla})\boldsymbol{u} + \dfrac{1}{R}\boldsymbol{\nabla}p - \nu\nabla^2\boldsymbol{u} + \boldsymbol{\Omega}\times\boldsymbol{u} + \dfrac{\varrho}{R}\boldsymbol{\Omega}\times\boldsymbol{U} - \dfrac{\varrho}{R}\boldsymbol{g} = 0 \,, \\[2mm]
\dfrac{\partial \varrho}{\partial t} + \boldsymbol{\nabla}\cdot(R\boldsymbol{u}) + \boldsymbol{\nabla}\cdot(\varrho\boldsymbol{U}) = 0 \,.
\end{cases}
$$

Clearly the contribution of Coriolis forces is negligible for our purposes. Consider all coupling terms in $\boldsymbol{U}$ to have a small effect. Then (43) become

$$
(44) \quad
\begin{cases}
\dfrac{\partial \boldsymbol{u}}{\partial t} + \dfrac{1}{R}\boldsymbol{\nabla}p - \nu\nabla^2\boldsymbol{u} = 0 \,, \\[2mm]
\dfrac{\partial \varrho}{\partial t} + R\boldsymbol{\nabla}\cdot\boldsymbol{u} = 0 \,.
\end{cases}
$$

Again we use the adiabatic expansion

$$(45) \qquad \delta\varrho = \left(\frac{\partial \varrho}{\partial p}\right)_s \delta p \,,$$

which allows us to write for the continuity equation

$$(46) \qquad \frac{\partial p}{\partial t} + Rc^2\boldsymbol{\nabla}\cdot\boldsymbol{u} = 0 \,,$$

where both R and c can be functions of space and time. By taking the partial time derivative of the second equation and the divergence of the first, simple substitution leads to an equation in terms of only the acoustic pressure p.

It must be remembered that R, v and c are all dependent on position and, in the case of R, also on time. This gives

$$(47) \qquad \frac{1}{c^2}\frac{\partial^2 p}{\partial t^2} - \nabla^2 p + \frac{1}{c^2 R}\frac{\partial R}{\partial t}\frac{\partial p}{\partial t} - \frac{1}{R}\left(\nabla - \frac{1}{R}\right)\cdot(\nabla p) + \frac{v}{c^2}\nabla^2\left(\frac{\partial p}{\partial t}\frac{1}{R}\right) = 0.$$

For the present ignore the viscous term. We see that it adds an absorption term that will be important in later discussions, but it can be treated separately. Now we need to look at the rest of the terms. $\partial R/\partial t$ is from the continuity equation related to U, thus the third term is of the order that we agreed to drop. Thus we have to lowest order

$$(48) \qquad \frac{1}{c^2}\frac{\partial p^2}{\partial t^2} - \nabla^2 p - (\nabla \ln R)\cdot\nabla p = 0.$$

Now in oceanography it is considered to be a good approximation to let R be approximated in the inertial equations by the reference density. This seems clear, since changes in density do not change the inertial term appreciably, and that is where the last term in eq. (48) came from. Thus R is effectively the hydrostatic density which is given by

$$(49) \qquad \left|\begin{array}{l} R = R_0 \exp\left[-g\displaystyle\int_0^z \frac{dz}{c^2}\right], \\[3mm] \ln R = \ln R_0 - g\displaystyle\int_0^z \frac{dz}{c^2}. \end{array}\right.$$

Therefore

$$(\nabla \ln R)\cdot(\nabla p) = -\frac{g}{c^2}\frac{\partial p}{\partial z},$$

where z is the vertical unit vector. This term is generally very small and will be dropped henceforth. This should be expected, of course, since the inverse length coefficient

$$\frac{g}{c^2} \approx \frac{1}{225}\ \text{km}^{-1}$$

is much smaller than those encountered in acoustic fields in the sea.

Thus we have the wave equation for lowest-order term

$$(50) \qquad \left(\nabla^2 - \frac{1}{c^2}\frac{\partial^2}{\partial t^2}\right) p = 0.$$

It appears that we have eliminated all of the effects due to the sea! That is not the case, however, since c is a function of position. Sound speed, of course, being related to the adiabatic bulk modulus $(\partial \varrho / \partial p)_s$ clearly is related to the oceanographic observables, temperature, salinity and pressure (depth). Thus, even if we have stripped the equations of all higher-order interactions with the local flow field, we still have important oceanographic effects appearing in our wave equation. First consider the « static terms ». These will be important in our final discussions about what kind of experiments are feasible.

The value of c^2 is clearly of interest to the oceanographer as well, since it relates to the properties of the equation of state [9, 10].

The equation has been determined empirically to considerable accuracy by several authors [11-13], although there are some differences in the reported higher-order terms. However, for our purposes it is sufficient to use

$$(51) \qquad c = 1.450 + 4.6T + 0.160P + 1.40(s - 35) \,,$$

where T is temperature in °C, P is pressure in kg/cm², and s is salinity in parts per thousand.

MUNK [14] has recently pointed out that the stability of the water column as measured by the Väisälä-Brunt frequency is related to the vertical gradient of $\ln c$ which then may be used to develop a plausible form for sound speed below the mixed layer as a function of depth that has an exponential form. This may result in a better starting point for some of the theories of the fluctuations due to internal wave interactions.

We now must look at some general properties of the medium which are best and quickly described by considering the high-frequency limit which, of course, is the geometrical or ray limit.

This is obtained [15] by the straightforward limit of taking a continuous wave of radial frequency ω and expressing the acoustic pressure in the form

$$(52) \qquad p = A(\boldsymbol{x}) \exp \left[i \big(S(\boldsymbol{x}) - \omega t \big) \right] \,.$$

The approximation to rays is obtained by assuming that the change in amplitude $A(\boldsymbol{x})$ is very slow over many « wavelengths » of an oscillating wave. This substituted into the wave equation leads to two equations, one each for the

$$(53) \qquad \begin{cases} (\boldsymbol{\nabla}S)^2 + \dfrac{\nabla^2 A}{A} = k^2 \,, \\[2ex] 2\boldsymbol{\nabla}A \cdot \boldsymbol{\nabla}S + A\nabla^2 S = 0 \,. \end{cases}$$

If A varies slowly over many wavelengths $2\pi k$, then its « wave numbers » must be very much smaller than k. Thus $\nabla^2 A / A \ll k^2$, so to a good approximation $(\nabla S)^2 = k^2$. Consider the wave front $S(\boldsymbol{x}) = \omega t$. At a time $t + \Delta t$ $S(\boldsymbol{x})$ has progressed to a position $\boldsymbol{x} + \Delta \boldsymbol{x}$ such that

$$S(\boldsymbol{x} + \Delta \boldsymbol{x}) = \omega(t + \Delta t),$$

or

$$(\nabla S) \cdot \Delta \boldsymbol{x} = \omega \, \Delta t,$$
$$= kc \, \Delta t,$$
$$= k|\mathrm{d}\boldsymbol{x}|.$$

Then we can see that ∇S must be parallel to $\Delta \boldsymbol{x}$ since it has a magnitude k. By a simple transformation $\mathrm{d}\boldsymbol{x} = k \, \mathrm{d}\boldsymbol{\sigma}$ it can be shown that

$$(\nabla_\sigma S)^2 = 1,$$

which is the requirement for straight-line rays in σ-space. Thus the path is the minimum distance, such that any other path is longer. Thus the first-order variation in the length gives

$$\delta \int \mathrm{d}\sigma = 0.$$

Thus in the regular x–co-ordinates

(54)
$$\delta \int k \, \mathrm{d}l = \delta \int \frac{\mathrm{d}l}{c} = 0,$$

that is the time is an extremum. This, of course, is Fermat's principle.

For our purposes here it is sufficient to consider an ocean with only vertical structure such that c is only a function of depth. Then it is easy to derive Snell's law

(55)
$$c^{-1} \sin\theta = \mathrm{const},$$

where θ is the angle of the ray with the vertical axis. Thus if we have a ray e-mitted from a source at an angle θ_0 and a point where the sound speed is c_0, we can follow the ray everywhere by a straightforward but rather tedious process. Intensities can also be obtained, of course, but for our present purposes we only wish to gain insight into what the ocean does to acoustic rays. Consider-ing the static ocean for this purpose is satisfactory. Applying the sound speed equation to the ocean model shown in Fig. 3 leads to the sound speed profile shown. Figure 4 shows what happens to rays emanating from sources at various depths. The limiting rays are typically about 12° from the horizontal. The important point to notice is that only on the channel axis is it possible to have

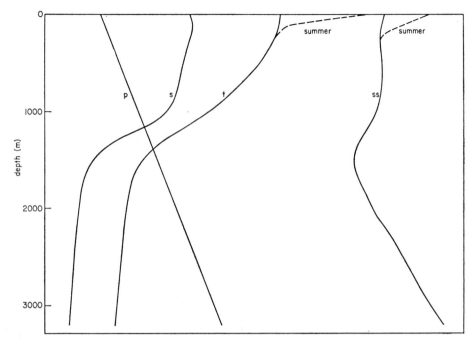

Fig. 3. – Typical temperature, pressure, salinity and sound velocity profiles for the open ocean.

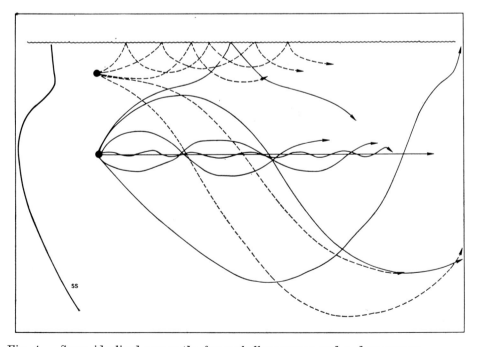

Fig. 4. – Some idealized ray paths for a shallow source and a deep source.

a horizontal ray. Although the curvatures are slight over ranges of a few hundred meters, cumulative changes over longer trajectories (>25 km) cause some tremendous excursions. This will be important later on in the discussion of experimental results.

What can be observed from Fig. 4 is that, for sound emanating from a source on or near the channel axis, much of the acoustic energy stays trapped in it. Thus, unless the sound velocity structure changes or some geologic feature intrudes into it, one can expect extraordinary propagation in the sea. This is indeed the case. There are numerous experiments reporting propagation over thousands of miles, where modest source strengths were used.

Earlier absorption was omitted from consideration in our development of the acoustic equations by dropping all viscous terms. It is important to consider it briefly, because it does have serious consequences on experimental design.

Absorption can be added to the solution of the acoustic equation by an exponentially decreasing term. This is sufficient for most applications. Thus the sound pressure has a solution of the form

$$p(\boldsymbol{x}, t) \exp\left[-\frac{\alpha}{2} R(\boldsymbol{x})\right],$$

where R is the distance traveled and α has been found empirically above about one kilohertz to be a function of the frequency given by [16]

(56)
$$
\begin{cases}
\alpha = 2.34 \cdot 10^{-6} \dfrac{s f_T f^2}{f_T^2 + f^2} + 3.38 \cdot 10^{-6} \dfrac{f^2}{f_T} \text{ neper/m}, \\[3mm]
f_T = 21.9 \cdot 10^{(6-1520/(T+273))} \text{ kHz},
\end{cases}
$$

f is in kilohertz and s is again the salinity. The relaxation term is due to $MgSO_4$ [17, 18]. It should be observed that at frequencies near 100 kHz and ranges in the order of 1000 m pressure attenuation due to absorption is on the order of $\exp[-5]$! Thus high-frequency experiments are considerably limited in range. On the other hand at 1 kHz the absorption over 1000 m is only $\exp[-0.03]$. Thus long-range low frequency causes no particular problem.

Other interesting solutions other than rays are the normal-mode application to acoustics where the form of $c(z)$ shown in Fig. 3 and 4 can be shown to lead to the equivalent of one-dimensional quantum-mechanical potential well [19] problems for the wave form in the z-direction. Ultimately these forms may be useful in discussions of fluctuations, but currently the state of the art is not that advanced.

With this preparation we can now begin to examine some of the experimental evidence that ocean dynamical fluctuations can be observed acoustically. Historically underwater acoustic experiments have gone from high-frequency (tens of kilohertz) short range (hundreds of meters) performed at small depths $((10 \div 500)\text{ m})$ to deeper, low-frequency sources and receivers at ranges of hundreds to thousands of kilometers. This scientific progress has been limited by the technology of instrumentation rather than by experimental design. It is probably best to follow, more or less, history in the presentation of experimental evidence, because by the very nature of geometries and frequencies it addresses different dynamical processes. The earlier results also demonstrate the need for careful experimental design in observing these processes, so that the theoretical interpretations are straightforward. Clearly no overall report on the field can be made in the time allowed, so we shall only consider a few examples.

The first three experiments that we shall discuss concern propagation in the mixed boundary in which we expect turbulence dominates the phenomenon of fluctuations. These experiments are due to SHEEHY [20] and SHVA-CHKO [21, 22].

In these experiments two ships were used, one with a source below it, the other with a receiving hydrophone suspended from it. The two ships moved apart by drifting or slow speed to vary the range. The signals were high-frequency (for underwater acoustics) short pulses of tens of kilohertz with high pulse repetition rates in order to obtain large samples of acoustic fluctuations at each range. Short pulses are required in order to separate the surface-reflected path from the direct path. A resumé of the important features of each experiment is given in Table I.

TABLE I.

Experiment, time and place	Frequency		Range	Depth
SHEEHY, Pacific, 1945 Summer	24	kHz	$(100 \div 2500)$ m	$(50 \div 200)$ m
SHVACHKO, N. Atlantic May 1961	10, 25	kHz	$(180 \div 1100)$ m	$(200 \div 400)$ m
SHVACHKO, N. Atlantic Sea of Norway, May-July 1962	25	kHz	$(200 \div 1500)$ m	$(150 \div 250)$ m
SHVACHKO, Atlantic Sea of Norway, 1964	3, 4	kHz	$(250 \div 9000)$ m	$(150 \div 250)$ m

In Sheehy's 1945 experiment samples of fifty consecutive 24 kHz signals were used to calculate the coefficient of variation (V) of the maximum amplitudes. The scatter of the data is great but a range dependence $V \propto R^{\frac{1}{2}}$ seems appropriate. The data are shown for one experiment in Fig. 5. No significant changes in the coefficient of variation were observed as the depths of sources and receivers were varied from 100 m to 300 m, but a higher variation was observed at 50 m. Unfortunately, no correlative temperature structure data were obtained, undoubtedly due to the lack of sufficient instrumentation in 1945.

Shvachko's first experiment was performed in 1961 in the North Atlantic. Two frequencies, 10 and 25 kHz, were used. Mean square values of the fluctuation were processed. At a frequency of 10 kHz no range dependence was observed. In the 25 kHz data the range dependence on the mean was observed

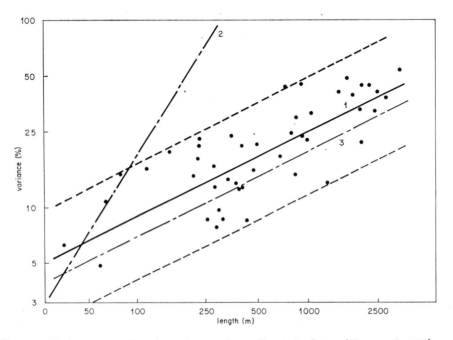

Fig. 5. – Variance as a function of range from Sheehy's data. (From ref. [20].)

to be $R^{0.6}$. During the course of this experiment variations in sound speed occurred along a track at an unspecified depth, presumably at the approximate depth of the acoustic sensors and a « structure function » $D_n(x)$ was obtained for the variations in the index of refraction (μ). A curve of $D_n(x)/2\bar{\mu}^2$ is shown in Fig. 6 which closely follows the value expected from the Kolmogorov spectrum discussed earlier out to about 1.5 m.

This leads to a correlation function with a spatial scale of about the same order. In earlier work by LIEBERMANN [18] a scale of 0.6 m was obtained.

In Shvachko's second experimental report [22] of results from 1962 to 1964, additional similar work is given, also from the Atlantic. Again the index of refraction data gave structure functions very close to the Kolmogorov result. Pressure fluctuations gave varying range dependence, the higher frequencies near 0.5, but the last two lower-frequency results giving $R^{0.77}$ for the Norwegian Sea and $R^{1.21}$ for the Atlantic.

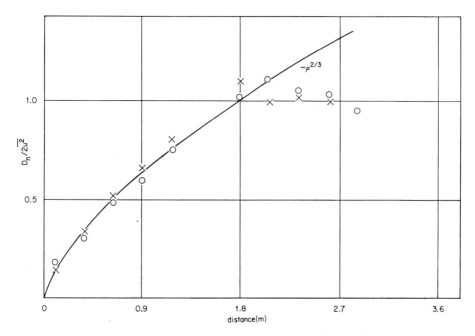

Fig. 6. – Structure function for the index-of-refraction distribution as a function of range. The solid line is for a theoretical Kolmogorov distribution. (From ref. [21].)

These apparent discrepancies are due to the magnitude of the important processes affecting scattering [22]. The theory of propagation through fluctuating media in the Kolmogorov range has been extensively developed by TA-TARSKII in his splendid book [23]. The correct asymptotic form for the variance depends on several parameters: the smallest and largest correlative distances in inertial turbulence, and the product of wavelength times the length of the acoustic path $\sqrt{\lambda R}$. SHVACHKO plausibily explained his results in terms of the Tatarskii theory [22]. MINTZER [24] showed that the results of SHEEHY are plausible in terms of fluctuations, but did not specify the dynamic nature of the fluctuation process.

Although all of these experiments had time as the directly observed variable, there seems to be no question that it is a space-generated fluctuation that they are measuring. Ships are not fixed platforms in the sense of being geographically fixed or drifting with the mean flow at the depths of the experiments. Small components of drift rates of the order of a knot (~ 0.5 m/s) lead to movement in a minute over several significant spatial correlation distances. Temporal changes clearly do not change so rapidly. In Shvachko's experiment the ships were moving in the order of one or two knots.

There seems to be no question that the acoustical fluctuating phenomenon in the mixed layer has the ability to sense certain oceanographically dynamic characteristics. The experimental design, however, must be very carefully considered in order to obtain the appropriate theoretical representation of the experiment.

Experiments that involve propagation through the sea below the water column are generally long range and low frequency. From the ray picture shown in Fig. 4 it is clear that these involve wide vertical excursions of the acoustic energy. These are by their very nature difficult to design for general spatial and temporal fluctuations. However, several recent experiments have demonstrated certain temporal effects [25, 26] which indicate that more carefully constructed experimental systems could yield significant oceanographic information. I shall only report on one of these experiments [26].

The Stanford experiment was between a fixed source at a depth of 527 m located near Eleuthera in the Bahama Islands to two fixed receivers near Bermuda and at depths of approximately 1700 m. The distance between the source and the receivers was 1300 km. The sea has an average depth of 4500 m, which is in excess of the bottom of the sound channel. Thus over most of the range the sea floor played no important role. Experiments were performed during March, April, June, July and September of 1968, giving a wide variation in the sound speed structure throughout the year. This experiment was a « multipath » experiment in that no attempt was made to resolve paths. Instead a continuous-wave, very stable 367 Hz source was used. Amplitude and phase were recorded and processed for statistically significant trends. Examples of amplitude fluctuations are shown in Fig. 7. Power spectral densities

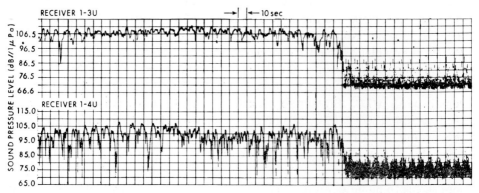

Fig. 7. – 367 Hz amplitude fluctuations observed in the Stanford experiment. (From ref. [26].)

were computed, a typical one being shown in Fig. 8. The frequency fall-off appears to correlate with the internal wave dependence predicted by PHIL-LIPS [4, 26] and given by eq. (19). At higher frequencies the spectrum, according to STANFORD, is affected by near surface phenomena—either surface waves or turbulence. The experiment leaves much to be desired from the standpoint of correlative environmental data. However, the prohibitive cost and elaborate instrumentation needed for obtaining it must be considered. The acoustic energy that arrives at the receiver has, in the summer, probably not been nearer the surface than about 300 m as can be seen from Stanford's ray tracing

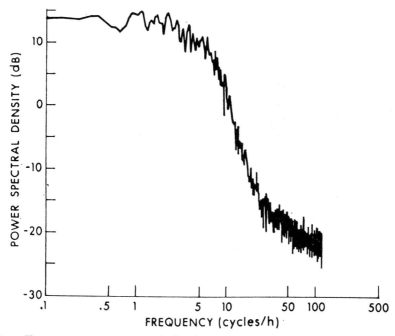

Fig. 8. – Frequency-power spectral density of amplitude fluctuations of a 367 Hz signal from the Stanford experiment.

shown in Fig. 9. As the surface waters cool, the paths come closer to the surface and finally some will reach it. The statistics are now even more complex because it is only over a fraction of the total range that some of the rays are in the upper, more dynamic regions. Phase fluctuations show similar results. It seems clear that even though the data are sparse, this type of experiment, carefully managed and monitored, has promise for remotely sensing oceanographically interesting phenomena.

The last topic that we shall discuss is the reflection from the sea surface. A tremendous number of papers have been written on the subject, most of which, up to 1969, have been listed in a survey paper by FORTUIN [27]. The problem is still with us, however, as important and incomplete. When considering this problem one should also be aware of the large amount of work that has been done on the scattering of electromagnetic waves from rough surfaces.

The topic will only be discussed briefly since it is well covered elsewhere and its usefulness for studying the sea is questionable compared to radar techniques.

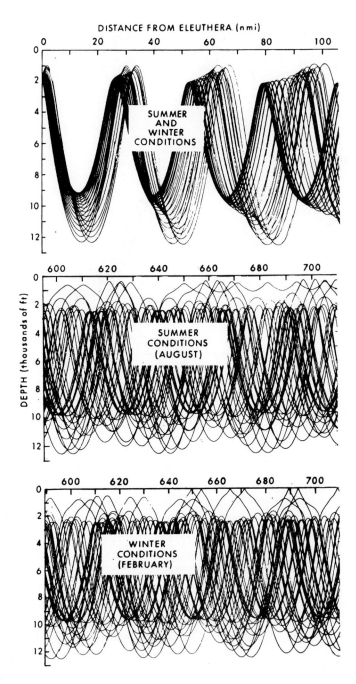

Fig. 9. – Rays of summer and winter conditions over the acoustic range of the Stanford experiment. (From ref. [26].)

There is an important difference between the nature of the sea surface reflection and what we have discussed in transmission. It is a boundary-value phenomenon, but something else that is important is that its motion is significant and frequency smearing takes on two aspects. One, the frequency is smeared out due to the many surface speeds present on the ocean surface, and, two, there turns out to be a special effect on reflection if there is a surface wavelength that is « tuned » to the acoustic wavelength. The surface speed of this wave gives the effect of a Doppler shift. Let us speak of this first by considering a very clever derivation by HOLFORD [28] that gives good insight to the frequency shift. Consider an incoming plane wave in a two-dimensional case (x, z)

$$p_i = \exp\left[i(\mathbf{k} \cdot \mathbf{r} - \omega t)\right] \tag{57}$$

incident on a moving sinusoidal boundary

$$z = \eta(x - Ct), \tag{58}$$

where the wavelength is sufficient to be a gravity wave such that simple dispersion is given from eq. (39) by

$$C^2 = g/K. \tag{59}$$

On the boundary the acoustic pressure vanishes, so the incident and scattered pressures obey

$$p = p_i + p_s = 0 \tag{60}$$

in $z = \eta$. A transformation of the form

$$
\left|
\begin{aligned}
x' &= \frac{x - Ct}{(1 - C^2/c^2)^{\frac{1}{2}}}, \\[2mm]
t' &= \frac{t - Cx/c^2}{(1 - C^2/c^2)^{\frac{1}{2}}}, \\[2mm]
z' &= z
\end{aligned}
\right. \tag{61}
$$

is recognized as a Lorentz transformation. The new equations become

$$
\left\{
\begin{aligned}
\nabla'^2 p' &= \frac{1}{c^2}\frac{\partial^2 p'}{\partial t'^2}, \\[2mm]
p' &= 0 \text{ on } z' = \eta'(x').
\end{aligned}
\right. \tag{62}
$$

The incident wave becomes

(63) $$p'_i = \exp\left[i(\mathbf{k}' \cdot \mathbf{r}' - \omega' t')\right],$$

where

(64)
$$\begin{cases} k'_x = \dfrac{k_x - k(C/c)}{(1 - C^2/c^2)^{\frac{1}{2}}}, \\[2mm] k'_z = k_z, \\[2mm] \omega' = \dfrac{\omega - k_x C}{(1 - C^2/c^2)^{\frac{1}{2}}} = k'c. \end{cases}$$

Now we have a simpler case, namely the fixed sinusoid. This is a well-known problem where there is still controversy over a rigorous solution, but that is not important for our purposes. The solution, due to symmetry, is known to have a form

(65) $$p' = \exp\left[i\mathbf{k}' \cdot \mathbf{x}'\right] + \sum_{n=-\infty}^{+\infty} R'_n \exp\left[i(k'_x + nK')x' + ik'_{nz}z'\right],$$

where

(66) $$K' = \frac{K}{(1 - C^2/c^2)^{\frac{1}{2}}} \quad \text{and} \quad k'^2 = (k'_x + nK')^2 + k'^2_{nz}.$$

Transforming back to the original co-ordinates gives for the second, or scattered term

(67) $$p_s = \sum_{n=-\infty}^{+\infty} R_n \exp\left[i[(k_x + nK)x + k'_{nz}z - i(w + n\Omega)t]\right],$$

where $\Omega = KC$ is the ocean wave frequency. Then

(68) $$(k_x + nK)^2 + k'^2_{nz} = \left(k + \frac{n\Omega}{c}\right)^2.$$

Thus the return is clearly frequency shifted, depending on the direction of the scattered plane wave.

The type of return given by the sea surface has been shown to have these predicted side bands [29, 30]. Even though the ocean surface is composed of many wavelengths each with its own velocity, only the portion in « resonance » with the acoustic wavelength will have a large effect. Thus the side bands are clear and informative, and frequency sweeping does give information on the spectral content of the sea. There remain, naturally, significant data in the remaining portions of the frequency spreading and spatial coherence of the signal that relates to surface dynamics. Unfortunately time does not permit further discussion of this phenomenon.

Appendix A

Two limiting examples of internal waves.

I) Consider a thin layer with a constant density gradient where

$$N_0^2 = \frac{g \, \delta \varrho}{2 \Delta \varrho_0} \, .$$

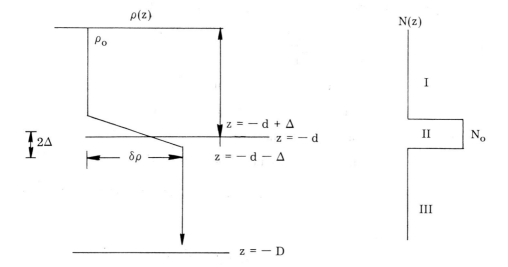

Then if $w = W(z) \exp[i(kx - nt)]$ we have

$$\frac{d^2 W}{dz^2} + \frac{N_0^2 - n^2}{n^2} \, k^2 W = 0 \quad \text{in region II}$$

and

$$\frac{d^2 W}{dz^2} - k \, W = 0 \quad \text{in regions I and III} \, .$$

This gives rise to solutions

$$W_{\mathrm{I}} \ = A_{\mathrm{I}} \ \sinh kz \, ,$$

$$W_{\mathrm{III}} = A_{\mathrm{III}} \sinh k(z + D) \, ,$$

if we require $W = 0$ at $z = 0$ and at $z = -D$.

The solution of interest to us in region II is

$$W = A_{\text{II}} \cos \sqrt{\frac{N_0^2 - n^2}{n^2}}\, k(z + d)\,.$$

The boundary conditions to be satisfied are

$$W_{\text{I}} = W_{\text{II}}\,, \qquad W_{\text{I}}' = W_{\text{II}}' \qquad\qquad \text{at } z = -d + \varDelta\,,$$

$$W_{\text{II}} = W_{\text{III}}\,, \qquad W_{\text{II}}' = W_{\text{III}}' \qquad\qquad \text{at } z = -d - \varDelta\,,$$

where $'$ indicates a derivative with respect to z. Calling

$$k\sqrt{\frac{N_0^2 - n^2}{n^2}} \equiv K\,,$$

$$-A_{\text{I}} \sinh k(d - \varDelta) = A_{\text{II}} \cos K\varDelta\,,$$

$$kA_{\text{I}} \cosh k(d - \varDelta) = -A_{\text{II}} K \sin K\varDelta\,,$$

$$A_{\text{II}} \cos K\varDelta = A_{\text{III}} \sinh k(D - d - \varDelta)\,,$$

$$A_{\text{II}} K \sin k\varDelta = kA_{\text{III}} \cosh k(D - d - \varDelta)\,,$$

which leads to the requirement that, for $K\varDelta \ll 1$, or long wavelength compared to the thickness

$$n^2 = \frac{kg\,\delta\varrho}{\varrho_0}\, \frac{1}{2\varDelta k + \operatorname{ctgh} k(d - \varDelta) + \operatorname{ctgh} k(D - d - \varDelta)}\,,$$

$$n^2 \simeq \frac{kg\,\delta\varrho/\varrho_0}{1 + \operatorname{ctgh} k(d - \varDelta)}\,,$$

for $\varDelta \ll d$

$$n^2 \simeq \frac{kg\,\delta\varrho/\varrho_0}{1 + \operatorname{ctgh} kd}\,,$$

for typical oceans

$$\frac{\delta\varrho}{\varrho_0} \sim 10^{-3}\,.$$

Surface waves in the long-wavelength gravity regime satisfy

$$n^2 \simeq gk\,.$$

Thus internal waves in this approximation show frequencies about 1/30th of that of surface waves.

2) Consider a thick layer as shown below

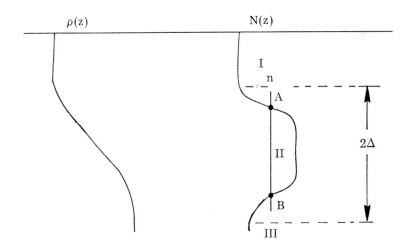

Then the solutions of interest for us are small scale:

$$W(z) \exp [i(k_1 x - nt)] ,$$

where now

$$k_1 \Delta \gg 1 .$$

Away from the regions where $N^2(z) - n^2 \approx 0$ one expects a solution

$$W(z) \sim \exp [ik_2(z)Z]$$

if $N(z)$ changes slowly over a wave number. Then

$$k_2 = k_1 \sqrt{\frac{N^2 - n^2}{n^2}} .$$

This indicates that for higher n the projection of the total wave number $(\exp [i\boldsymbol{k} \cdot \boldsymbol{r}])\boldsymbol{k}$ is smaller in the vertical, and, for $n = N$, the wave must be traveling entirely in the horizontal direction. For a given value of n these waves can be seen to be trapped within the limits A and B on the $N(z)$ curve. At these points the wave is traveling in the horizontal, as it moves away from A and B it turns more and more downward.

REFERENCES

[1]   P. M. MORSE and H. FESHBACH: *Methods of Theoretical Physics* (New York, N. Y., 1953), p. 151.
[2]   H. GOLDSTEIN: *Classical Mechanics* (Cambridge, Mass., 1950), p. 132.
[3]   O. M. PHILLIPS: *The Dynamics of the Upper Ocean* (Cambridge, 1969), p. 14.

[4] O. M. PHILLIPS: *The Dynamics of the Upper Ocean*, Chap. 5 (Cambridge, 1969).
[5] A. HECHT and R. A. WHITE: *Deep Sea Res.*, **15**, 339 (1968).
[6] A. N. KOLMOGOROV: *Dokl. Akad. Nauk SSSR*, **32**, 16 (1941).
[7] B. KINSMAN: *Wind Waves* (Engelwood Cliffs, N. J., 1965).
[8] W. J. PIERSON jr. and L. MOSKOWITZ: *Journ. Geophys. Res.*, **69**, 5181 (1964).
[9] N. P. FOFONOFF: *The Sea*, edited by M. N. HILL (New York, N. Y., 1962), p. 3.
[10] C. ECKART: *Hydrodynamics of Oceans and Atmospheres* (Oxford, London, New York, N. Y., and Paris, 1960).
[11] V. A. DEL GROSSO: *Journ. Acoust. Soc. Amer.*, **53**, 1384 (1973).
[12] W. D. WILSON: *Journ. Acoust. Soc. Amer.*, **23**, 10, 1357 (1960).
[13] M. GREENSPAN and C. E. TSCHIEGG: *Journ. Acoust. Soc. Amer.*, **31**, 75 (1959).
[14] W. H. MUNK: *Journ. Acoust. Soc. Amer.*, **55**, 220 (1974).
[15] I. TOLSTOY and C. S. CLAY: *Ocean Acoustics* (New York, N. Y., 1966).
[16] M. SCHULKIN and H. W. MARSH: *Journ. Acoust. Soc. Amer.*, **34**, 864 (1962).
[17] R. W. LEONARD, P. C. COMBS and L. R. SKIDMORE: *Journ. Acoust. Soc. Amer.*, **21**, 63 (1949).
[18] L. N. LIEBERMANN: *Journ. Acoust. Soc. Amer.*, **20**, 868 (1948).
[19] A. O. WILLIAMS jr.: *Underwater Acoustics*, edited by R. N. B. STEPHANS (London, 1970), p. 33.
[20] M. J. SHEEHY: *Journ. Acoust. Soc. Amer.*, **22**, 24 (1950).
[21] R. V. SHVACHKO: *Sov. Phys. Acoust.*, **9**, 280 (1964).
[22] R. V. SHVACHKO: *Sov. Phys. Acoust.*, **13**, 93 (1967).
[23] V. I. TATARSKII: *The Effects of the Turbulent Atmosphere on Wave Propagation*, translated from the Russian (Springfield, Va., 1971).
[24] D. MINTZER: *Journ. Acoust. Soc. Amer.*, **25**, 922, 1107 (1953).
[25] J. C. STEINBERG and T. G. BIRDSALL: *Journ. Acoust. Soc. Amer.*, **39**, 301 (1966).
[26] G. E. STANFORD: *Journ. Acoust. Soc. Amer.*, **55**, 968 (1974).
[27] L. FORTUIN: *Journ. Acoust. Soc. Amer.*, **47**, 1209 (1970).
[28] R. HOLFORD: unpublished, private communication (1971).
[29] W. I. RODERICK and B. F. CRON: *Journ. Acoust. Soc. Amer.*, **48**, 759 (1970).
[30] M. V. BROWN and G. V. FRISK: *Journ. Acoust. Soc. Amer.*, **55**, 744 (1974).

Finite-Amplitude Effects in Acoustic Propagation in Fluids.

H. O. BERKTAY

Electronic and Electrical Engineering Department, University of Birmingham - Birmingham

1. – Introduction.

The nonlinear nature of the differential equations describing acoustic-wave propagation in fluids has been known for a long time. BLACKSTOCK, in his historical review of nonlinear acoustics [1], suggests that Euler's work in the eighteenth century was the begining of the « classical era » in the theory of finite-amplitude acoustics.

Although the historical development of the subject is instructive and interesting in its own right, in this contribution it is proposed to present the present state of the art, with special reference to the development of interest in underwater acoustics. However, the basic treatment can be extended to finite-amplitude sound in other fluids.

The reader can find interesting treatments of the earlier work in ref. [1-3], while the more detailed analyses used can be found in the various references listed. This contribution is meant to be no more than an introduction to the topic.

More recently, the exploitation of interaction between acoustic waves to produce directional transmission and reception of sonar signals has engendered a great deal of interest. An introductory development of the basic theory of parametric transmitters is given in Sect. **4**.

2. – Waves in a lossless medium.

2‵1. *Basic equations.* – The exact differential equations applicable to the propagation of acoustic waves in a nondissipative fluid are reproduced below, for the Eulerian co-ordinate system.

Equation of continuity:

$$(2.1.1) \qquad \frac{\partial \varrho_T}{\partial t} + \nabla \cdot (\varrho_T \boldsymbol{u}) = 0 .$$

Equation of momentum:

(2.1.2)
$$\varrho_T \left[\frac{\partial \boldsymbol{u}}{\partial t} + (\boldsymbol{u} \cdot \nabla) \boldsymbol{u} \right] + \nabla p = 0 .$$

Equation of state:

For an isoentropic event, the equation of state for a perfect gas is of the form

(2.1.3)
$$p_T = p_0 (\varrho_T / \varrho_0)^\gamma .$$

For a liquid, for constant entropy, the pressure-density relationship can be expanded in the form of a Taylor series around the equilibrium values

(2.1.4)
$$p_T = p_0 + \left(\frac{\mathrm{d} p_T}{\mathrm{d} \varrho_T} \right)_0 (\varrho_T - \varrho_0) + \frac{1}{2} \left(\frac{\mathrm{d}^2 p_T}{\mathrm{d} \varrho_T^2} \right)_0 (\varrho_T - \varrho_0)^2 + \cdots .$$

For realistic intensities it is sufficiently accurate to terminate this series at the square-law term. Then, the equation of state can be written in the form [4]

(2.1.5)
$$p = A (\varrho / \varrho_0) + (B/2)(\varrho / \varrho_0)^2 .$$

Comparison of equations (2.1.4) and (2.1.5) yields

(2.1.6)
$$A = \varrho_0 c_0^2$$

and

(2.1.7)
$$B/A = (\varrho_0 / c_0^2)(\mathrm{d}^2 p / \mathrm{d} \varrho^2) .$$

For a perfect gas, from equations (2.1.3) and (2.1.7),

(2.1.8)
$$B/A = \gamma - 1 .$$

The parameter B/A is a measure of the nonlinearity of a fluid. Its values for various fluids have been measured [4, 5]; for practical purposes, for water, a value of 5.0 has been found to give acceptable results.

In the following Sections, a parameter of nonlinearity β is also used, which is related to B/A by the equation

(2.1.9)
$$\beta = 1 + B/2A .$$

2`2. *Plane waves in a lossless fluid*. – For a plane wave travelling in the x-direction, the exact equations can be reduced to the form [5]

$$(2.2.1) \qquad \frac{\partial u}{\partial t} + (c_0 + \beta u)\frac{\partial u}{\partial x} = 0 .$$

The velocity with which a disturbance u travels is given by

$$(2.2.2) \qquad \left| \begin{array}{l} \dfrac{\mathrm{d}x}{\mathrm{d}t} = -\dfrac{\partial u}{\partial t} \Big/ \dfrac{\partial u}{\partial x} , \\[2mm] i.e. \\[2mm] \qquad = c_0 + \beta u . \end{array} \right.$$

If we consider a wave which is of the form

$$(2.2.3) \qquad u = u_0 f(t)$$

at $x = 0$, this wave becomes progressively distorted as it travels, as the velocity of propagation is a function of the particle velocity.

As an example, if $f(t)$ is a sinusoidal function, the positive half-cycles will travel faster than c_0, while the negative half-cycles will be slower, resulting in the distortion of the wave form.

It is worth noting that the nonlinearity indicated in eq. (2.2.2) is small. But the cumulative effect, as the wave progresses, can be considerable.

From eqs. (2.2.2) and (2.2.3), the plane wave at a distance x may be represented simply, by allowing for retardation at the appropriate velocity, in an implicit form as

$$(2.2.4) \qquad u(x, t) = u_0 f\left(t - \frac{x}{c_0 + \beta u}\right) .$$

At moderate intensities $|u/c_0|$ is very small. Then, as an approximation

$$(2.2.5) \qquad u(x, t) = u_0 f(\tau + \beta u x / c_0^2) ,$$

where τ is the retarded time.

Special case—sinusoidal boundary condition. We assume that

$$(2.2.6) \qquad u(0, t) = u_0 \sin \omega t .$$

Then, if we use symbols representing normalized quantities discussed below, eq. (2.2.5) becomes [6]

$$(2.2.7) \qquad\qquad V = \sin\left(y + \sigma V\right).$$

Here, y is a measure of the phase of the disturbance, in which the « linear » retardation is taken into account; σ is a measure of distance travelled x, normalized with respect to a length $(\beta \varepsilon k)^{-1}$; V is a measure of the instantaneous particle velocity u, normalized with respect to the peak value at $x = 0$.

Rewriting eq. (2.2.7) to isolate the variables V and y, one obtains

$$(2.2.8) \qquad\qquad y = \arcsin V - \sigma V.$$

This equation is in a form suitable for a simple graphical construction of the distorted-wave form at any range σ, by plotting y against V. The range-dependent term is a straight line, the slope of which is $-\sigma$. The difference between the two terms on the right-hand side of eq. (2.2.8) gives the resultant distorted-wave form.

For $\sigma < 1$ the wave is distorted, but single valued. At $\sigma = 1$

$$\left(\frac{\mathrm{d}V}{\mathrm{d}y}\right)_{y=0} \to \infty,$$

i.e. a shock front is formed. For $\sigma > 1$ eq. (2.2.8) indicates a multivalued wave form. This anomalous result (which is not acceptable on physical grounds) arises because thermoviscous effects have been neglected. In a real fluid, as a shock front develops, absorption at this front increases, with the result that the wave form is not permitted to become multivalued.

The propagation of the distorted wave for $\sigma > 1$ can be studied by using the weak-shock theory. If we use Rankine-Hugoniot relations, it can be concluded that in a periodic wave a symmetrical shock front must travel at velocity c_0, which is the average of the velocities immediately before and immediately after the shock. Thus, the wave form for $\sigma > 1$ can be constructed on the basis of eq. (2.2.8), except for the multivalued part—which must be replaced by the shock front. This « weak shock » solution, applied in conjunction with eqs. (2.2.7) and (2.2.8), provides means of determining the distortion of a plane, sinusoidal wave in a lossless fluid.

On this basis the following comments and conclusions appear to be in order:

    i) For $\sigma < \pi/2$ the peak value of the distorted wave does not change.

    ii) For $\sigma > \pi/2$ the peak value is progressively reduced.

    iii) For $\sigma > 1$ the height of the shock ($2V_1$) can be found from eq. (2.2.7) by substituting $y = 0$, *i.e.*

(2.2.9) $V_1 = \sin \sigma V_1$.

iv) For $\sigma \gg 1$ eq. (2.2.9) can be reduced to an approximate form

(2.2.10) $\sigma V_1 \simeq \pi - V_1$,

i.e.

(2.2.11) $V_1 \simeq \pi/(1 + \sigma)$.

The error in using this approximation is less than 2% for $\sigma > 3.6$ and about 10% for $\sigma = 3.0$.

The approximation involved in deriving equation (2.2.10) is equivalent to stating that the distorted wave is of the saw-tooth form.

v) Equation (2.2.11) indicates that the amplitude of the saw-tooth wave is attenuated in a geometrical fashion. Further, for $\sigma \gg 1$

$$V_1 = u_1/u_0 \simeq \pi/\beta \varepsilon k x .$$

As $\varepsilon = u_0/c_0$, one obtains

(2.2.12) $u_1 \simeq \pi c_0/\beta k x$.

Thus, the amplitude of the particle velocity becomes independent of the amplitude of the initial sinusoidal wave.

Frequency spectrum of the distorted wave [7]. As the distorted wave is periodic (with a fundamental « frequency » y), it can be represented in the form of a Fourier series

(2.2.13) $V = \sum B_n \sin ny$,

where

(2.2.14) $B_n = \dfrac{2}{\pi} \displaystyle\int_0^\pi V \sin ny \, \mathrm{d}y$.

a) For $\sigma < 1$ the value of V from eq. (2.2.7) can be used. Then, after some manipulation one obtains

(2.2.15) $B_n = (2/n\sigma) J_n(n\sigma)$,

where J_n is a Bessel function of order n. This is the so-called Fubini solution.

b) For $\sigma > 1$ the expression for B_n can be shown to reduce to [7]

(2.2.16) $B_n = (2V_1/n\pi) + (2/n\pi\sigma) \displaystyle\int_{\text{arcsin } V_1}^{\pi} \cos(n\varphi - n\sigma \sin \varphi) \cdot d\varphi$.

As σ is increased, arcsin V_1 tends to π, thus causing the second term to become small.

Hence, for large σ

(2.2.17) $B_n \simeq 2V_1/n\pi$,

which corresponds to the spectrum of a saw-tooth wave of amplitude V_1.

If we now consider the amplitude of the fundamental component of the distorted wave, $u_0 B_1$, for $\sigma < 1$

(2.2.18) $u_0 B_1 = (2J_1(\sigma)/\sigma) u_0$.

The expression inside the parentheses reduces gradually from 1.0 (for $\sigma = 0$) to 0.88 (for $\sigma = 1$). So, the fundamental component can be expected to be attenuated by 1.1 dB within the pre-shock region.

In the saw-tooth region, from eqs. (2.2.17) and (2.2.11), we obtain

(2.2.19) $u_0 B_1 = 2u_0/(1 + \sigma)$.

For the intermediate region (i.e. $1 < \sigma < 4$) a polynomial approximation has been obtained by BLACKSTOCK and his co-workers in the form

(2.2.20) $B_1 = 1.219\,4 - 0.393\,6\,\sigma + 0.060\,98\,\sigma^2 - 0.003\,47\,\sigma^3$.

2˙3. *Cylindrical and spherical waves in lossless fluids* [8]. – In ref. [8] BLACK-STOCK develops a general differential equation for all one-dimensional waves in the form

(2.3.1) $\dfrac{\partial W}{\partial \sigma} - W \dfrac{\partial W}{\partial y} = 0$,

where

(2.3.2) $W = (r/r_0)^a V$,

r_0 is the source radius, and, for plane waves, $a = 0$,

(2.3.3) $\sigma = \beta\varepsilon kr$,

for cylindrical waves, $a = \frac{1}{2}$,

(2.3.4) $\sigma = 2\beta\varepsilon k \, r_0[\sqrt{r/r_0} - 1]$, for spherical waves, $a = 1$,

(2.3.5)
$$\sigma = \beta \varepsilon k\, r_0 \ln (r/r_0)\,.$$

(The expressions for σ are given for a sinusoidal wave for simplicity.)
    Two points are worth noting at this stage:

    i) for a plane wave eq. (2.3.1) reduces to

(2.3.6)
$$\frac{\partial V}{\partial \sigma} - V \frac{\partial V}{\partial y} = 0\,,$$

which can be obtained directly from eq. (2.2.1) after some manipulation and transformation of variables [5];

    ii) the value of σ at the boundary is zero.

If the wave is sinusoidal at the boundary, eq. (2.3.1) has a solution of the form

(2.3.7)
$$W = \sin (y + \sigma W)\,.$$

This expression is exactly the same as the eq. (2.2.7), albeit in the new variables W and σ. Hence, the discussion in, and the conclusions of, the previous Section can be applied to all one-dimensional waves, with appropriate transformation of the variables.
    Thus, shock formation occurs at $\sigma = 1$, and the wave form (as represented by W) becomes triangular for $\sigma > 3.6$. σ is a transformed range variable, normalized with respect to the shock distance for a plane wave $(\beta \varepsilon k)^{-1}$.
    The spreading of a cylindrical or a spherical wave causes its attenuation, thus reducing the nonlinear effects at a particular range. This is brought out in the variable transformations indicated in eqs. (2.3.2)-(2.3.5).
    If, now, eq. (2.2.13) is rewritten for W in the form

(2.3.8)
$$W = \sum B_n \sin ny\,,$$

the spectral components of W can be calculated in the manner discussed in the previous Section. For example, for large σ the amplitude of the fundamental frequency component of W will be given, approximately, by

(2.3.9)
$$B_1 = 2/\sigma\,.$$

Using eqs. (2.3.2)-(2.3.5), we conclude that the amplitude of the fundamental component of particle velocity for large σ can be expressed, approximately, by $c_0/\beta kr$ for cylindrical waves, and by $2c_0/[\beta kr \ln (r/r_0)]$ for spherical waves.

3. – Waves in lossy media.

3˙1. *Basic concepts and general remarks.* – For the purposes of these notes, the medium is assumed to introduce thermoviscous absorption in acoustic-wave propagation, but effects of relaxation (and of dispersion) are not considered. Further, for simplicity, the treatment is concentrated on an initially sinusoidal wave.

In describing the nonlinear effects in a lossy medium, an acoustic equivalent of Reynolds' number, representing the ratio of the forces producing nonlinearity (*i.e.* the amplitude of the acoustic wave at the boundary) to the absorptive forces, is found to be very useful. The parameter used is also referred to as Gol'dberg number, and is given by [9]

$$(3.1.1) \qquad\qquad\qquad \Gamma = \beta \varepsilon k / \alpha \,,$$

which is the inverse of the small-signal absorption of the sinusoidal wave within the plane-wave shock distance.

For one-dimensional waves the wave equation can be reduced to the general form [10]

$$(3.1.2) \qquad\qquad \frac{\partial W}{\partial \sigma} - W \frac{\partial W}{\partial y} = \left[\left(\frac{r}{r_0}\right)^a \Big/ \Gamma \right] \frac{\partial^2 W}{\partial y^2} \,.$$

For a lossless medium, Γ is infinite, and this equation reduces to eq. (3.1.1) discussed in the previous Section.

As can be seen from eq. (3.1.1), Γ is proportional to the acoustic « Mach » number ε at the boundary, and it can be quite large for even moderate intensities, particularly at low frequencies (*i.e.* for small α). Then, the discussion and the conclusions of the previous Section provide a good insight into wave propagation, with some modification at long ranges.

On this basis, and from physical considerations, the propagation of a finite-amplitude one-dimensional wave can be broken down into three regions.

Near the boundary the nonlinear effects are predominant. The wave is progressively distorted until a shock front is formed. As the wave front steepens (towards a shock) absorption increases, and prevents the formation of an infinite pressure gradient. However, the amplitude of the distorted wave is not greatly attenuated at this stage, and nonlinear effects are still very much in evidence.

As the distortion process continues, the wave form eventually resembles a saw-tooth with finite « shock » thickness. As the amplitude of the saw-tooth wave is attenuated, the nonlinear forces are reduced. Absorption causes the shock thickness to increase. Eventually, the wave reverts to a sinusoidal form and further nonlinear effects can be neglected.

CARY [11] studied solutions of eqs. (3.1.2) on a numerical basis, and concluded that if Γ is less than a certain critical value for a particular geometry, the wave will not reach the stage where its amplitude is severely attenuated. These critical values are given as

(3.1.3)
$$\begin{cases} \Gamma = 4.5 & \text{for plane waves}, \\ \Gamma = 4.5(1 + 0.79/\beta\varepsilon k r_0) & \text{for cylindrical waves}, \\ \Gamma = 4.5 \exp[1.57/\beta\varepsilon k r_0] & \text{for spherical waves}. \end{cases}$$

On this basis one could conclude that, for Γ less than the appropriate critical value, shock formation could be ignored.

Equation (3.1.2) is in the form of Burgers' equation, which has been used in the study of problems of diffusion. For a plane wave (*i.e.* for $a = 0$) it has an exact and complete solution [9]. For $a \neq 0$ an exact solution exists for the saw-tooth region [10]. These solutions are discussed below.

3˙2. Plane waves. – The complete solution of eq. (3.1.2) for plane waves is given [9] in the form of a convolution integral, which also includes a logarithmic transformation of variables. BLACKSTOCK reduced the results into a form which is useful for computing the coefficients B_n (see eq. (2.2.13)).

Without extra-attenuation due to finite-amplitude effects, the amplitude of the fundamental at a range x would be

(3.2.1)
$$B_1 = \exp[-\alpha x].$$

The ratio of this value of B_1 to that calculated from Burgers' equation is a measure of the extra-attenuation. Curves are provided in ref. [9] of the extra-attenuation of the fundamental component of a sinusoidal plane wave, for various values of the parameters Γ and σ.

Before discussing these results in any detail, it may be of interest to mention a more recent development [12] which is useful in determining the value of B_1.

The rate at which energy is lost by the fundamental component of a plane wave is primarily determined by viscous losses and by the interaction of the fundamental with the first-harmonic component. On the basis WESTER-VELT [13] obtained the expression

(3.2.2)
$$\frac{dI_1}{dx} = -2\alpha I_1 - \langle pq \rangle,$$

where I_1 is the intensity of the fundamental frequency component, $\langle\ \rangle$ indicates time averaging over one complete period of the wave, and q is a source function given by [14]

(3.2.3)
$$q = (\beta/\varrho_0^2 c_0^4)\frac{\partial}{\partial t}p^2 .$$

If we use an approximate expression for the first-harmonic component of pressure for p in eq. (3.2.2), the expression can be reduced to a form of Bernoulli's equation which has an exact solution of the form

(3.2.4)
$$I_1 = I_0 \exp[-2\sigma/\Gamma]\{1 + \Gamma^2[1 - \exp[-2\sigma/\Gamma]]^2/16\}^{-1},$$

where I_0 is the acoustic intensity at $x = 0$.

As σ/Γ is $\propto x$, extra-attenuation of the fundamental component can now be calculated in a closed form

(3.2.5) $\Big\{$
$$\text{EX dB} = 10 \log[I_0 \exp[-2\sigma/\Gamma]/I_1],$$

i.e.

$$= 10 \log\{1 + (\Gamma/4)^2[1 - \exp[-2\sigma/\Gamma]]^2\}.$$

The extra-attenuation values obtained from eq. (3.2.5) are said to agree well with those given in ref. [9]. The biggest difference found was 1.0 dB, which occurred for $\sigma = 3$.

The simplicity of this result makes it very useful for estimating the behaviour of the fundamental frequency component. However, for more general results we revert to ref. [9].

For $\sigma < 1$ the effect of absorption is to reduce the extra-attenuation of the fundamental component. As Γ is reduced, absorption becomes more effective.

For $\sigma > 1$ various approximate expressions have been proposed. A simple form (which provides the values of B_1 with an error of less than 2% for $\sigma > 3.0$) is

(3.2.6)
$$V = \frac{2}{\Gamma}\sum\frac{\sin ny}{\sinh(n\pi\Delta/2)},$$

where

(3.2.7)
$$\pi\Delta/2 = (1 + \sigma)/\Gamma.$$

Equation (3.2.6) is a Fourier-series representation of a periodic wave form, a single cycle of which can be represented by

(3.2.8)
$$V = (2/\pi\Gamma\Delta)[-y + \pi \tgh(y/\Delta)]$$

for y in the range $-\pi$ to π.

Equation (3.2.8) was obtained by SOLUYAN and KHOKHLOV [16], and is an exact (but not complete) solution of eq. (3.1.2) for plane waves.

In this expression Δ is a measure of the thickness of the shock. The hyperbolic tangent changes sign with y for values near $y = 0$, and reaches ± 0.964 for $y/\Delta = \pm 2$. Hence, this transition from -1 to $+1$ will be very rapid for small values of Δ. As Δ increases, the transition becomes more gradual.

The spacing, along y, of the peak and the trough of the wave T can be found from eq. (3.2.8), in the form

$$(3.2.9) \qquad T = 2\Delta \operatorname{arccosh} \sqrt{\pi/\Delta} \ .$$

In physical terms, for a given value of Γ, as σ increases, the shock thickness is increased as a result of the absorption at the shock front. Eventually, for large σ, absorption predominates, and further nonlinear effects become negligible.

This effect can be illustrated from eqs. (3.2.6) and (3.2.7). For $\sigma \gg \Gamma$, $\pi\Delta/2 \gg 1$ and (3.2.6) becomes

$$(3.2.10) \qquad \left|
\begin{aligned}
V &\simeq \frac{1}{\Gamma} \sum \exp\left[-n\sigma/\Gamma\right] \sin ny \,, \\
&i.e. \\
&\simeq \frac{1}{\Gamma} \sum \exp\left[-n\alpha x\right] \sin ny \ .
\end{aligned}
\right.$$

Thus, each frequency component is exponentially attenuated.

The amplitude of the fundamental component of the particle velocity becomes

$$(3.2.11) \qquad u_1 = (\alpha c_0/\beta k) \exp\left[-\alpha x\right] \ .$$

This expression is independent of the boundary value u_0. One can conclude that the fundamental component of the particle velocity at a point some distance away from the source cannot be increased beyond the value indicated in eq. (3.2.11); any further increase in the transmitted intensity is absorbed in increased extra-attenuation. From the curves in ref. [9] this conclusion would appear to be valid for $\sigma/\Gamma > 2$ (see Fig. 1).

In the shock region, on the other hand, $\sigma \gg 1$ and $\sigma/\Gamma \ll 1$. Then, $\pi\Delta/2 \ll 1$, and from eqs. (3.2.6) and (3.2.7)

$$(3.2.12) \qquad V \simeq [2/(1+\sigma)] \sum \frac{1}{n} \sin ny \ .$$

This is the same result as that obtained for a plane wave in a lossless medium. The amplitude of the fundamental component of the particle velocity becomes

(3.2.13)
$$u_1 = \frac{2c_0/\beta k}{(\beta \varepsilon k)^{-1} + x} .$$

As ε is increased, u_1 reaches a maximum value of

(3.2.14)
$$u_1 = 2c_0/\beta kx ,$$

a result which is again independent of u_0.

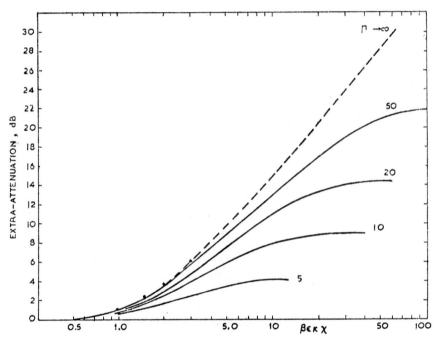

Fig. 1. – Extra-attenuation of the fundamental frequency component of a single-frequency plane wave.

The saw-tooth region may be taken to start at $\sigma = \pi/2$, and to continue until the shock thickness becomes comparable with π and the wave profile loses its triangular nature. BLACKSTOCK [9] suggests an upper limit for T of $2\pi/5$, which corresponds to

(3.2.15)
$$\pi \varDelta/2 \leqslant 0.6 .$$

From eqs. (3.2.7) and (3.2.15), the saw-tooth region can be said to exist for the range

$$\pi/2 < \sigma \leqslant 0.6 \, \varGamma - 1 .$$

For $\Gamma \gg 1$ the upper limit corresponds to a range x_M where

(3.2.16) $$\alpha x_M \simeq 0.6 \text{ neper} .$$

From eqs. (3.2.1), (3.2.6) and (3.2.16) the extra-attenuation of the fundamental component to this point can be calculated:

(3.2.17)
$$\left| \begin{array}{l} \text{Extra-attenuation} = \dfrac{\Gamma}{2} \sinh \dfrac{\pi \varDelta}{2} \exp\left[-\sigma/\Gamma\right] , \\[2mm] \qquad\qquad\qquad = 0.349(\Gamma/2) . \end{array} \right.$$

The asymptotic value of the extra-attenuation (as $\sigma \to \infty$) is clearly $\Gamma/4$. Hence, the fundamental component of the wave suffers an additional attenuation of about 3.1 dB before the nonlinear effects on the fundamental component become negligible. (A more detailed analysis of ref. [9] shows that this asymptotic value of extra-attenuation is valid only for large Γ; the error for $\Gamma \geqslant 20$ is less than 0.5 dB.)

In ref. [17] the upper limit of the saw-tooth region is calculated by balancing the absorption and the nonlinear attenuation of the fundamental component of the particle velocity. Then, the upper limit is given as

(3.2.18) $$\pi \varDelta/2 < 1.0 .$$

This would suggest that a further nonlinear attenuation of 1.3 dB would occur beyond the saw-tooth region.

3˙3. Spherical and cylindrical waves. – An exact, closed-form solution of eq. (3.1.2) is not available for $a \neq 0$. Methods have been developed for numerical solution for particular boundary conditions [11].

Equation (3.2.8) is a solution of eq. (3.1.2) (valid in the saw-tooth region only) for cylindrical and spherical waves. Only the parameter $\varDelta$ needs to be related to the primary variables. A general expression for $\varDelta$, applicable for all three one-dimensional waves, is

(3.3.1) $$\varDelta = (r/r_0)^a (1 + \sigma)(2/\pi \Gamma) .$$

Thus, eq. (3.2.6) can be used (as a Fourier-series representation of the distorted-wave form) also for cylindrical and spherical waves, valid in the saw-tooth region only.

$\varDelta$ is, again, a measure of the shock thickness. But, for a divergent wave, the spreading loss, as well as absorption, causes the shock thickness to increase. Eventually the wave profile loses its saw-tooth nature, and eq. (3.2.8) ceases to be applicable for divergent waves.

The region of validity of eq. (3.2.8) for divergent waves is of some significance. BLACKSTOCK suggests [17] that the upper limit of the saw-tooth region
should correspond to the range at which $\pi\Delta/2 = 1$. MUIR [19] assumed that
for a spherical wave the saw-tooth region extends to the range where $\pi\Delta/2 = 1$,
and that there is no extra-attenuation of the fundamental component beyond
this point. On this basis the range dependence of the fundamental component
of a spherical wave can be calculated. Results predicted on this basis agreed
well with experimental data [19].

In the absence of finite-amplitude effects, the normalized amplitude of
the particle velocity in a spherical wave at range r would be

$$(3.3.2) \qquad\qquad B_1' = (r_0/r) \exp\left[-\alpha(r - r_0)\right].$$

From eq. (3.2.8), the corresponding term is

$$(3.3.3) \qquad\qquad B_1 = (2/\Gamma)[\sinh(\pi\Delta/2)]^{-1}.$$

The ratio B_1'/B_1 is the extra-attenuation due to finite-amplitude effects for a
spherical wave. Hence

$$(3.3.4) \quad \text{extra-attenuation} = (\Gamma r_0/2r)\sinh(\pi\Delta/2)\exp\left[-\alpha r_0[(r/r_0) - 1]\right].$$

A similar expression can be developed for cylindrical waves.

On plotting the extra-attenuation for spherical and cylindrical waves against
the range variable σ (for various values of $\beta\varepsilon k r_0$ and of Γ') from eq. (3.3.4), it
was noted that the rate of change of extra-attenuation with σ *increased* for
$\pi\Delta/2 > 0.6$ [18]. On physical grounds this result appears unacceptable. Therefore, one must conclude that the solution given in eq. (3.2.8) ceases to be valid
(for spherical and cylindrical waves) when $\pi\Delta/2 > 0.6$. A rough estimate made
(by using a perturbation technique) suggests that the fundamental component
may be further attenuated beyond this point by about 1 dB for a spherical
wave and $(1.5 \div 2)$ dB for a cylindrical wave.

The extra-attenuation of a spherical wave (calculated along the lines indi-
cated above) for $\beta\varepsilon k r_0 = 1$ is plotted in Fig. 2, for various values of Γ. The
points corresponding to $\pi\Delta/2 = 0.6$ and $\pi\Delta/2 = 1$ are shown together with
the asymptotic values of extra-attenuation predicted in the two different ways
discussed above. In practice, the two predictions of the asymptotic value
for long ranges agree to within 1 dB.

If we indicate by r_M the range at which $\pi\Delta/2 = 0.6$, then the amplitude
of the fundamental component of the particle velocity at that point (calculated
from eq. (3.3.3)) is

$$(3.3.5) \qquad\qquad u_1(r_M) = 3.14\,\alpha c_0/\beta k,$$

and is independent of the initial particle velocity u_0.

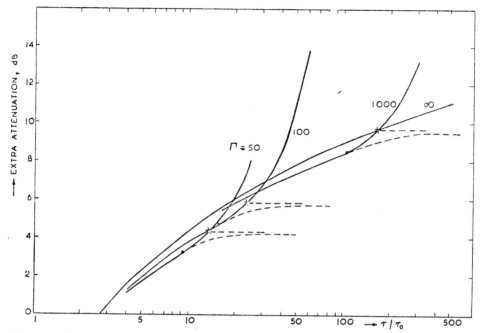

Fig. 2. – Extra-attenuation of the fundamental frequency component of a single-frequency spherical wave for $\beta \varepsilon k r_0 = 1$: ● the range where $\pi \Delta/2 = 0.6$, × the range where $\pi \Delta/2 = 1.0$, —— from equation (3.3.4), ———— extrapolated for $\pi/2 > 0.6$.

If the nonlinear effects are neglected, the amplitude of the particle velocity at the same range would be

$$(3.3.6) \qquad u_1'(r_M) = u_0(r_0/r_M) \exp\left[-\alpha(r_M - r_0)\right].$$

The asymptotic value of the extra-attenuation (for $r > r_M$) is expected to be about 1 dB more than that at r_M; the latter can be calculated as the ratio

$$u_1'(r_M)/u_1(r_M).$$

The value of r_M can be calculated from (3.3.1) by putting $\pi \Delta/2 = 0.6$. That is, for a spherical wave,

$$(3.3.7) \qquad (r_M/r_0)[1 + \beta \varepsilon k r_0 \ln (r_M/r_0)] = 0.6(\beta \varepsilon k r_0/\alpha r_0).$$

The variation of r_M/r_0 with αr_0, calculated for various values of $\beta \varepsilon k r_0$, is shown in Fig. 3. Various values of Γ are also indicated in the Figure. Using these results, the extra-attenuation to r_M was calculated, and is shown in Fig. 4, again for various values of $\beta \varepsilon k r_0$ and of Γ. The solid line in Fig. 4 indicates the range of applicability of this analysis. In other words for points below this line a « shock » does not form. This limiting line agrees with Cary's results (see eq. (3.1.3)).

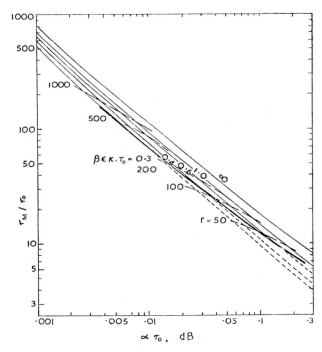

Fig. 3. – Variation of r_M/r_0 with αr_0 for a spherical wave (calculated from equation (3.3.7)).

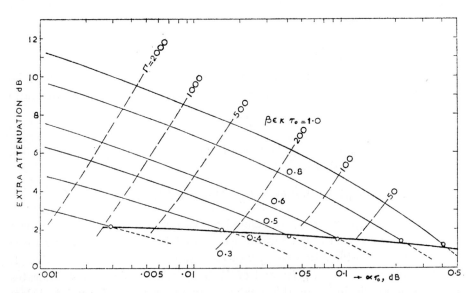

Fig. 4. – Extra-attenuation of the fundamental component of a spherical wave to the range r_M where $\pi/2 = 0.6$. (The thick line indicates the region of the applicability of the results.)

3˙4. *Discussion and conclusion.* – In the foregoing Subsections the behaviour of one-dimensional finite-amplitude acoustic waves in a homogeneous, thermoviscous medium has been discussed, with particular reference to sinusoidal boundary conditions.

The following summary of the results may be useful:

i) If the acoustic Reynolds number Γ is sufficiently high at the boundary, a « weak shock » wave form will develop as a result of the progressive distortion of the wave.

ii) The « saw-tooth » wave retains its form until losses at the shock front reduce its amplitude. Eventually, absorption predominates over the nonlinear forces, the shock thickness becomes comparable with π, and the wave loses its triangular form. This occurs at a range

$x = x_M, \ \alpha x_M \simeq 0.6$ for a plane wave,

$r = r_M, \ \alpha r_M \simeq 0.3$ for a cylindrical wave and

$(r_M/r_0) \ln (r_M/r_0) \simeq 0.6/\alpha r_0$ for a spherical wave.

iii) The fundamental component of particle velocity at a given range increases with Γ until a « saturation » value is reached. Any further increase in Γ increases the extra-attenuation of the fundamental component so that the measured particle velocity (or pressure) remains unaltered.

The « saturation » value of the particle velocity reached depends upon whether the range is smaller or greater than x_M or r_M, as appropriate.

If $x < x_M$ (or $r < r_M$), the weak-shock solution of Sect. **2** is applicable.

If $x > x_M$ (or $r > r_M$), the results presented in Sect. **3** give the saturation values for the 3 one-dimensional waves.

Radiation from a real transducer. It is of interest to apply the analysis presented here to a monochromatic radiation from a real transducer.

Even at infinitesimal intensities, the acoustic field in the vicinity of a transducer is rather complex. However for a square or circular piston in an infinite rigid baffle, an approximate model can be formulated.

The acoustic energy within the near field of such a transducer is concentrated within a column, the cross-section of which is roughly the same as the dimensions of the transducer. The near-field distance can be taken to be

(3.4.1) $$R_0 = S/\lambda \,,$$

where S is the area of the transducer.

For ranges greater than R_0 the field gradually tends to a spherically spreading wave, with a directivity pattern depending upon the size of the transducer and the wavelength λ.

If $\alpha R_0 > 1$ neper, the finite-amplitude effects can be expected to occur mostly within the near field of the transducer. In this case one may expect the plane-wave theory to provide useful results. Some comparisons made on this basis [20] with experimental data show reasonable agreement.

If $\alpha R_0 \ll 1$, the bulk of the nonlinear effects can be said to occur in the far field of the transducer. Then, the spherical-wave theory should be applicable. However, an effective source radius r_0 needs to be defined. Empirically, it has been found that a source radius of the order of half the near-field distance provides acceptable agreement with experimental results. The results discussed in ref. [19] are very encouraging.

Because of the directionality of the spherical waves, finite-amplitude effects are most noticeable on or near the acoustic axis of the transducer. This results in the flattening of the beam pattern and in the enhancement of the relative amplitude of the side lobes, which would not be expected to suffer to the same extent from extra-attenuation [19, 20].

If the transmitted acoustic intensity is such that

$$(\beta \varepsilon k)^{-1} < R_0 \,,$$

shock formation occurs within the collimated-wave region of the field, and extra-attenuation can be expected to increase rapidly.

4. – Parametric acoustic arrays.

So far, we have mainly concentrated upon the propagation of one-dimensional acoustic waves, by looking for solutions for the exact hydrodynamic equations for the given geometry and for prescribed boundary conditions.

WESTERVELT [14] proposed the use of nonlinear interactions of sound waves for directional transmission or reception of relatively low-frequency acoustic waves.

In its simplest form, in a parametric transmitter, two sinusoidal waves, both of relatively high frequencies, are transmitted from the same transducer assembly. Nonlinear interaction between these two monochromatic waves results in the production of various harmonic frequencies. In practice, these harmonic waves also interact with the primary waves, and with one another, producing additional interaction components. A complete analysis becomes very difficult and, for many practicable systems, the simple quasi-linear approach developed in ref. [14] suffices. In this contribution the higher-order interactions will be ignored, and the basis of the quasi-linear treatment of such parametric devices will be discussed.

We assume that the primary waves are centred at frequencies ω_1 and ω_2, such that the average frequency ω_0 is much greater than the difference frequency $\omega_- = \omega_1 - \omega_2$.

The primary waves will be considered to be either monochromatic or narrow band. In the latter case it is assumed that the frequency spectrum associated with the difference frequency waves does not overlap with the primary spectra.

4'1. *The source function.* – WESTERVELT [21] extended earlier work by LIGHTHILL, and developed an effective volumetric source (of volume density q) which could be used to account for the scattering of sound waves from a point in the primary field. Using this approach, and neglecting the absorption of the scattered waves, we can write the velocity potential of the scattered sound Φ_s in the form

(4.1.1) $$\Box^2 \Phi_s = -q \,,$$

where

(4.1.2) $$q = (\beta/\varrho_0^2 c_0^4) \frac{\partial}{\partial t} (p_i^2) \,.$$

Here p_i is total acoustic pressure at the particular point.

The expression for q is quite general, and can be shown to be accurate for all practicable intensities. If, however, higher-order interactions are not negligible, the complete solution of the problem may become rather difficult, if at all possible.

4'2. *One-dimensional waves.* – In the absence of higher-order interactions, if both the primary and the scattered waves are constrained to one-dimensional mode of propagation, eqs. (4.1.1) and (4.1.2) can be solved directly.

It is of more general use, however, to develop integral solutions in these cases. In the following, for simplicity, the two primary waves are assumed to be monochromatic (with absorption coefficients α_1, α_2 and wave numbers k_1, k_2) and the scattered wave is assumed to consist of the difference frequency component (α_- and k_- refer). Further, it is assumed that $k_- r_0 \gg 1$ for cylindrical and spherical waves. At the end of this Section, a time-domain solution is discussed, in a slightly different form.

The primary waves may be represented in the form

(4.2.1) $$p_i = P_1 \exp\left[-\alpha_1 x\right] \cos \omega_1 \tau + P_2 \exp\left[-\alpha_2 x\right] \cos \omega_2 \tau$$

for plane waves and

(4.2.2) $$p_i = \left(\frac{r_0}{r}\right)^a \{P_1 \exp\left[-\alpha_1(r - r_0)\right] \cos \omega_1 \tau + P_2 \exp\left[-\alpha_2(r - r_0)\right] \cos \omega_2 \tau\}$$

for cylindrical and spherical waves.

The difference frequency component of the source function can then be calculated from (4.1.2) for plane and for spreading waves as

(4.2.3) $\qquad q_- = - (\omega_- \beta / \varrho_0^2 c_0^4) P_1 P_2 \exp\left[- (\alpha_1 + \alpha_2) x\right] \sin \omega_- \tau$

and

(4.2.4) $\quad q_- = - (\omega_- \beta / \varrho_0^2 c_0^4)(r_0/r)^{2a} P_1 P_2 \exp\left[- (\alpha_1 + \alpha_2)(r - r_0)\right] \sin \omega_- \tau \, .$

A co-phasal source surface will radiate one-dimensional waves as well. A thin shell of sources of thickness δx (or δr, as applicable) can be considered to radiate with a surface velocity $\delta u(x, t)$ where, allowing for radiation in the $- x$-direction as well,

(4.2.5) $\qquad\qquad\qquad\qquad \delta u(x, t) = q_-(x, t) \cdot \delta x / 2$

and

(4.2.6) $\qquad\qquad\qquad\qquad \delta u(r, t) = q_-(r, t) \cdot \delta r / 2 \, .$

If we allow for the absorption, the retardation and the attenuation of these radiated waves, the difference frequency particle velocity at a range R can be calculated in the form of an integral. If we use also the impedance relation $p = \varrho_0 c_0 u$, the results can be put in the form

(4.2.7) $\qquad p_-(R, t) = (k_- \beta / 2 \varrho_0 c_0^2 \alpha_T) P_1 P_2 \exp\left[- \alpha_- R\right] \cdot$

$$\cdot \left[1 - \exp\left[- \alpha_T R\right]\right] \cos\left(\omega_- \tau + \pi/2\right)$$

for plane waves and

(4.2.8) $\qquad p_-(R, t) = (k_- \beta / 2 \varrho_0 c_0^2) P_1 P_2 (r_0/R)^a \exp\left[- \alpha_-(R - r_0)\right] \cdot$

$$\cdot \cos\left(\omega_- \tau + \pi/2\right) \int_{r_0}^{R} (r_0/r)^a \exp\left[- \alpha_T (r - r_0)\right] dr$$

for spreading waves. The integral in eq. (4.2.8) yields

$$r_0 (\pi / \alpha_T r_0)^{\frac{1}{2}} \exp\left[\alpha_T r_0\right]\left[\operatorname{erf} \sqrt{\alpha_T R} - \operatorname{erf} \sqrt{\alpha_T r_0}\right]$$

for a cylindrical wave and

$$r_0 \exp\left[\alpha_T r_0\right]\left[E_1(\alpha_T r_0) - E_1(\alpha_T R)\right]$$

for a spherical wave.

An interesting feature for the plane waves is that for $\alpha_T R \ll 1$ the amplitude of the difference frequency pressure increases linearly with range.

Also, in all these expressions, the influence of the primary frequencies is only through the control of the absorption parameter α_T.

Time domain solution. For simplicity we neglect the absorption at the difference frequency and, further, we assume that the primary waves may be represented in the form

(4.2.9) $$p_i(r, t) = P_0 h(r) f(\tau) \cos \omega_0 \tau ,$$

where $h(r)$ has the following form:

(4.2.10) $$h(r) = \exp\left[-\alpha_0 r\right]$$

for a plane wave and

(4.2.11) $$h(r) = (r_0/r)^a \exp\left[-\alpha_0(r - r_0)\right]$$

for a spreading wave. (Equation (4.2.9) can be reduced to eqs. (4.2.1) and (4.2.2) by using suitable expressions for α_0 and $f(\tau)$.)

Then, the low-frequency content of the source function can be reduced to

(4.2.12) $$q_- = (P_0^2 \beta / 2\varrho_0^2 c_0^4) \, h^2(r) \frac{\partial}{\partial t} f^2(\tau) .$$

This expression can be used to calculate the pressure at range R, in the manner discussed above. An interesting conclusion emerges, however, when it is remembered that integration with respect to the spatial variable r is effected to produce the result. Thus, each result is proportional to

$$\frac{\partial}{\partial t} f^2(\tau) .$$

At the boundary (*i.e.* for $x = 0$ or for $r = r_0$) $f(\tau)$ becomes $f(t)$, the envelope function of the transmitted signal. Hence, provided that the spectrum of $f^2(t)$ does not overlap with that of the transmitted band, a low-frequency component proportional to the time derivative of the square of the transmitted envelope function will be received at distance R from the source.

For a plane wave geometry the result can be shown to be

(4.2.13) $$p_- = (P_0^2 \beta / 8\varrho_0 c_0^3 \alpha_0)[1 - \exp\left[-2\alpha_0 R\right]] \frac{\partial}{\partial t} f^2(\tau) .$$

To compare this result with eq. (4.2.7) one may consider the special case of

$P_1 = P_2$. Then, if we substitute

$$P_0 = P_1 + P_2 ,$$

$$f(t) = \cos(\omega_- t/2) ,$$

$$\omega_0 = (\omega_1 + \omega_2)/2$$

and

$$2\alpha_0 = \alpha_T ,$$

the two results can be shown to agree.

4'3. Westervelt's model. – In the previous Section the primary and the « scattered » waves were assumed to propagate in a one-dimensional form. This is only possible if the extent of a planar transducer or the length of a cylindrical transducer is infinite, or a large spherical transducer is used.

WESTERVELT [14] considered the primary waves to be planar, and confined within a column of finite cross-sectional area S. Then, the two primary waves can be represented in a complex form as

$$\underline{p_1} = P_1 \exp[-(\alpha_1 + jk_1)x]$$

and

$$\underline{p_2} = P_2 \exp[-(\alpha_2 + jk_2)x] .$$

The difference frequency component of the source function can be calculated from

(4.3.1) $$q_- = j(\beta k_-/\varrho_0^2 c_0^3)\, \underline{p_1} \cdot \underline{p_2}^* ,$$

where the asterisk indicates the complex conjugate of the associated function. Hence

(4.3.2) $$q_- = j(\beta k_-/\varrho_0^2 c_0^3) P_1 P_2 \exp[-(\alpha_1 + \alpha_2 + jk_-)x] .$$

If $\mathbf{s}$ is the vector joining the source point to the field point, the pressure element produced by a source element $q_- \cdot \delta v$ is given by

(4.3.3) $$\delta \underline{p} = (\varrho_0/4\pi s) \exp[-(\alpha_- + jk_-)s]\, j\omega_- q_-\, \delta v .$$

Hence, the scattered difference frequency pressure at a point (R, θ) relative to the transducer can be written in the form of a volume integral

$$(4.3.4) \qquad \underline{p}_- = - P_1 P_2 (\beta k^2_- / 4\pi \varrho_0 c_0^2) \iiint (1/s) \cdot$$
$$\cdot \exp\left[-(\alpha_1 + \alpha_2 + jk_-) x - (\alpha_- + jk_-) s \right] dv .$$

The limits of integration must be chosen to cover the whole of the « source » volume.

If the dimensions of the cross-sectional area of the column of primary waves are small compared to the wavelength at the difference frequency, integration for constant x yields the area S as a factor. Integration with respect to x can also be simplified by assuming the field point (R, θ) to be well in the far field of the source region. Then

$$(4.3.5) \qquad\qquad s \simeq R - x \cos \theta$$

and

$$(4.3.6) \qquad\qquad 1/s \simeq 1/R .$$

Equation (4.3.4) becomes

$$(4.3.7) \qquad
\begin{cases}
\underline{p}_-(R, \theta) \simeq - P_1 P_2 (\beta k^2_- / 4\pi \varrho_0 c_0^2)(S/R) \exp\left[-(\alpha_- + jk_-) R \right] \cdot \\
\qquad\qquad\qquad \cdot \displaystyle\int_0^\infty \exp\left\{ -[\alpha_T + jk_-(1 - \cos \theta)] x \right\} dx , \\
i.e. \\
\qquad\qquad \simeq - P_W(R, 0) D(\theta) ,
\end{cases}$$

where

$$(4.3.8) \qquad P_W(R, 0) \simeq (P_1 P_2 S \beta k^2_- / 4\pi \varrho_0 c_0^2 \alpha_T R) \exp\left[-(\alpha_- + jk_-) R \right]$$

and

$$(4.3.9) \qquad\qquad D(\theta) = [1 + j(2k_-/\alpha_T) \sin^2 (\theta/2)]^{-1} .$$

The last expression is a complex directivity function. On the axis of the primary waves $(\theta = 0)$ its value is unity. Its magnitude is reduced to $1/\sqrt{2}$ (i.e. the half-power point is obtained) when $\theta = \theta_d$, where

$$(4.3.10) \qquad\qquad \sin (\theta_d/2) = \sqrt{\alpha_T/2k_-} .$$

In most instances $\sqrt{\alpha_T/2k_-} \ll 1$. Then

$$(4.3.11) \qquad\qquad \theta_d \simeq \sqrt{2\alpha_T/k_-} .$$

It is worth noting that $D(\theta)$ is the directivity pattern of a continuous, semi-infinite end-fire array with an exponential taper, $|D(\theta)|$ is a monotomic function of θ; it demonstrates no sidelobes.

The beam width of the difference frequency radiation $2\theta_d$ is controlled primarily by the absorption at the transmitted frequencies, and the wave number at the difference frequency.

In practice, the high-frequency waves would be transmitted from a finite transducer. Then, within the near field of the transducer the waves may be assumed to be collimated. If the absorption of the primary waves is such that the interaction volume is effectively confined to the near-field region, then the results discussed in this Section must apply. It must be remembered, however, that the transmitted intensities must be such that the primary waves do not form shocks—otherwise, higher-order interactions cannot be neglected.

If a substantial part of the interaction volume is in the far field of the transducer, then the results of this Section are not applicable.

Time domain solution. If the transmitted plane waves are assumed to be given as in eq. (4.2.9), the low-frequency source function of eq. (4.2.12) would be obtained.

Using the Fourier transform of the square of the envelope function, it is possible to estimate the directivity of the low-frequency signal [22]. However, the pressure along the acoustic axis (*i.e.* for $\theta = 0$) can be estimated quite simply.

For $\theta = 0$, $s = R - x$. Therefore, using eq. (4.2.12) for $q_-(t, x)$, eq. (4.3.3) can be written in the time domain. For $R \gg 1/\alpha_0$, and neglecting $\alpha_- R$, we obtain

(4.3.12)
$$\delta p_-(R, t) \simeq (\varrho_0/4\pi R) \frac{\partial}{\partial t} q_-(t - R/c_0) \delta v ,$$

i.e.

$$\simeq (P_0^2 \beta / 8\pi \varrho_0 c_0^4 R) \exp\left[-2\alpha_0 x\right] \frac{\partial^2}{\partial t^2} f^2(\tau) \delta v .$$

If we integrate, we obtain

(4.3.13)
$$p_-(R, t) \simeq (P_0^2 \beta S / 16\pi \varrho_0 c_0^4 \alpha_0 R) \frac{\partial^2}{\partial t^2} f^2(\tau) .$$

Thus unlike the time domain solution for one-dimensional waves, the low-frequency pressure wave form along the axis of the parametric array is proportional to the second derivative of the square of the envelope function $f(t)$.

As discussed previously, eq. (4.2.9) may be used to represent the transmission of two monochromatic primary waves. But, equally, it can represent the transmission of a single-frequency, pulsed-carrier type of wave form. If eqs. (4.2.13) and (4.3.13) are interpreted on the basis of a pulsed-carrier type transmission, the low-frequency component must be in the form of a single pulse, the shape of which is either the first or the second derivative of the square of the pulse envelope, depending on the geometry of the primary waves.

4˙4. *Interactions in the far field of a transducer.* – The primary fields in the far field of a transducer can be represented in the form

$$p_1 = (P_1'/r) D_1(\gamma, \varphi) \exp\left[-(\alpha_1 + jk_1)r\right]$$

and

(4.4.1) $$p_2 = (P_2'/r) D_2(\gamma, \varphi) \exp\left[-(\alpha_2 + jk_2)r\right],$$

where $P_{1,2}'$ are the pressure amplitudes referred back to 1 m range, and $D_{1,2}$ are two-dimensional directivity functions at the two frequencies.

Then, the difference frequency pressure at a field point (R, θ, η) can be written in the form

(4.4.2) $$p_-(R, \theta, \eta) = -(P_1' P_2' k_-^2 \beta / 4\pi \varrho_0 c_0^2) \cdot$$

$$\cdot \iiint \frac{D_1(\gamma, \varphi) \cdot D_2(\gamma, \varphi)}{8r^2} \exp\left[-(\alpha_1 + \alpha_2 + jk_-)r - (\alpha_- + jk_-)s\right] dv .$$

The volume integral can be evaluated numerically for any set of parameters. Although, strictly, the integral over the variable r needs to be taken out to ∞, in practice it is sufficient to take the upper limit to be R, as « backscattering » from sources for $r > R$ should be small. This technique for evaluating the characteristics of parametric arrays has been used with great success [23].

For an observer in the far field of the interaction region, some useful, normalized results can be obtained [24]. The geometry used is shown in Fig. 5.

If we make the far-field substitutions for s (eqs. (4.3.5) and (4.3.6)) and integrate with respect to r from zero to ∞, eq. (4.4.2) yields

(4.4.3) $$p_-(R, \theta, \eta) \simeq -(P_1' P_2' k_-^2 \beta / 4\pi \varrho_0 c_0^2 \alpha_T) \exp\left[-(\alpha_- + jk_-)R\right] \cdot$$

$$\cdot \int\limits_{-\pi/2}^{\pi/2}\!\!\int \frac{D_0^2(\gamma, \varphi) \cos \gamma \, d\gamma \, d\varphi}{1 + j(k_-/\alpha_T)(1 - u)} ,$$

where $u = \cos \gamma \cos \theta \cos (\varphi - \eta) + \sin \gamma \sin \theta$.

Here, the directivity functions at the two frequencies are assumed to be the same.

If the significant values of all the angles are assumed to be small compared with 1 radian, then

$$(k_-/\alpha_T)(1 - u) \simeq \left(\frac{\gamma - \theta}{\theta d}\right)^2 + \left(\frac{\varphi - \eta}{\theta d}\right)^2 .$$

Further, for a transducer of area S radiating at a frequency $\omega_0 = (\omega_1 + \omega_2)/2$, the equivalent plane-wave pressure amplitude near the surface (P_1, say) can be related to the far-field value referred to 1 m range (P_1') through the expression

$$(4.4.4) \qquad\qquad P_1' = P_1 R_0 \, ,$$

where R_0 is the near-field distance of the transducer given by

$$(3.4.1) \qquad\qquad R_0 = S/\lambda_0 \, .$$

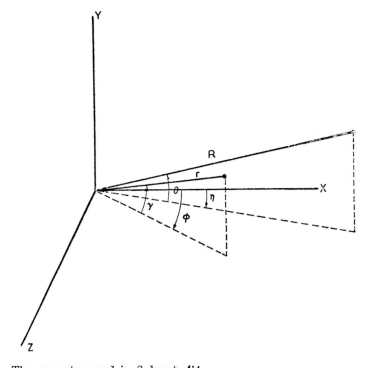

Fig. 5. – The geometry used in Subsect. 4'4.

By comparing the expression before the double integral in eq. (4.4.3) with eq. (4.3.4), it can be shown that the former expression reduces to

$$- P_w(R, 0)(S/\lambda_0^2) \, .$$

Using this result and eq. (4.4.4), one can write eq. (4.4.3) in the form

$$(4.4.5) \qquad\qquad p_-(R, \theta, \eta) \simeq - P_w(R, 0) \cdot V(\gamma_1, \varphi_1, \theta, \eta) \, ,$$

where, for a highly directional beam,

(4.4.6) $$V(\gamma_1, \varphi_1, \theta, \eta) \simeq (\theta_d^2 S/\lambda_0^2) \iint \frac{D_0^2(\gamma, \varphi)\, d\gamma'\, d\varphi'}{1 + j[(\gamma' - \theta')^2 + (\varphi' - \eta')^2]}.$$

Here, the primed symbols represent the various angles, normalized with respect to θ_d, e.g. $\gamma' = \gamma/\theta_d$.

For a rectangular transducer with sides l and m (where $l = L\lambda_0$ and $m = M\lambda_0$)

(4.4.7) $$D_0(\gamma, \varphi) \simeq \frac{\sin(\pi L\gamma)}{\pi L\gamma} \frac{\sin(\pi M\varphi)}{\pi M\varphi}.$$

This directivity pattern will have half-power beam widths of $2\gamma_1$ and $2\varphi_1$ in the two planes of symmetry, where

(4.4.8) $$\pi L\gamma_1 = \pi M\varphi_1 = \sqrt{2}.$$

The directivity function can now be expressed in the form of normalized angles γ' and φ' by substituting

(4.4.9) $$\begin{vmatrix} \pi L\gamma = \sqrt{2}\,\gamma/\gamma_1, \\ i.e. \\ \quad = \sqrt{2}\,\gamma'/\psi_y \end{vmatrix}$$

and

(4.4.10) $$\pi M\varphi = \sqrt{2}\,\varphi'/\psi_z,$$

where

$$\psi_y = \gamma_1/\theta_d$$

and

(4.4.11) $$\psi_z = \varphi_1/\theta_d.$$

Thus

$$D_0(\gamma, \varphi) = D_0'(\gamma', \varphi') \quad \text{say},$$

where the function $D_0'(\ )$ depends on the normalized parameters ψ_y and ψ_z, which are the ratios of the half-power beam widths of the primary waves in the two planes of symmetry to $2\theta_d$.

Further

(4.4.12)
$$\begin{cases} \theta_d^2 S/\lambda_0^2 = LM\theta_d^2\,, \\[2mm] \text{i.e.} \\[2mm] = 2/\pi^2\,\psi_y\,\psi_z\,. \end{cases}$$

With these substitutions the integral in eq. (4.4.6) can be evaluated numerically for any given values of ψ_y and ψ_z. The resulting expression $V(\cdot)$ is a function of ψ_y, ψ_z, θ' and η'. The magnitude of $V(\cdot)$ for $\theta' = \eta' = 0$ can be substituted in eq. (4.4.5) to obtain the difference frequency sound pressure level along the acoustic axis of the tranducer, while the ratio $|V(\theta', \eta')|/|V(0, 0)|$ gives the normalized two-dimensional directivity pattern of the difference frequency sound waves. An interesting feature of this result is that both $V(0, 0)$ and the normalized directivity function are functions of ψ_y and ψ_z only, and hence can be represented in the form of normalized curves. Further, the angles θ and η appear in the results in their normalized forms (θ' and η', respectively) and the half-power beam widths in $\eta = 0$ and $\theta = 0$ planes ($2\theta_0$ and $2\eta_0$) are obtained in normalized forms θ_0/θ_d and η_0/θ_d.

The curves of $|V(0, 0)|$, θ_0/θ_d and η_0/θ_d for various values of ψ_y and ψ_z are provided in ref. [24]. For the purposes of these notes it is proposed to concentrate upon a discussion of parametric sources using a square transducer to launch the primary waves. The use of these results will be illustrated later on by considering an example.

If the primary waves have two-dimensional symmetry, $\psi_y = \psi_z$, and $V(\cdot)$ need only be evaluated in the plane $\eta' = 0$, say. The normalized values of the difference frequency beam width $2\theta_0/2\theta_d$ are shown in Fig. 6 as functions of $\psi_y = 2\gamma_1/2\theta_d$.

It is also instructive to plot $2\theta_0/2\gamma_1$ against ψ_y, which provides an indication of the relationship between the beam width of the primary waves and that at the difference frequency. ($2\theta_0/2\gamma_1$ can be obtained simply by calculating the ratio $(2\theta_0/2\theta_d)/\psi_y$.) These results are also shown in the Figure.

A study of these curves indicates clearly that $2\theta_0$ is of the order of $2\theta_d$ for $\psi_y < 1$, and of the order of $2\gamma_1$ for $\psi_y > 1$. In these two regions the beam width at the difference frequency can be said to be controlled mainly by the end-fire effect and by the geometry of the primary waves, respectively.

Using a value of $\psi_y > 1$ provides a means of producing a low-frequency wave the beam width of which varies only marginally as the frequency is varied over a wide range. This point can be illustrated by considering radiation from a given transducer of primary waves, the frequencies of which are much higher than the difference frequency. Then, the difference frequency can be varied over a range of some octaves by small fractional changes of the primary frequencies. This change in the primary frequencies will result in a small variation in the primary beam width which may be ignored as a first approximation. From Fig. 6 it can be noted that the ratio $2\theta_0/2\gamma_1$ varies only by about 10% over the range $1.5 < \psi_y < 4.0$. If γ_1 is assumed to be fixed, this range of ψ_y would correspond to a 7-to-1 variation in the difference frequency (as $\theta_d \propto k_-^{-1/2}$). By utilizing this characteristic of difference frequency waves, wide-band parametric transmitters (with constant-beam-width property) can be produced.

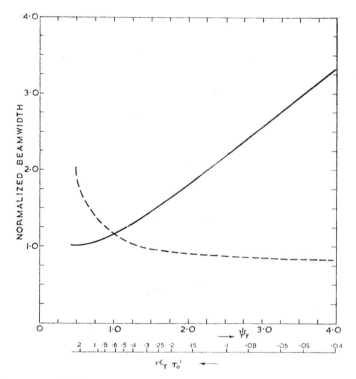

Fig. 6. – The normalized beam width of the difference frequency waves $(2\theta_0)$: —— θ_0/θ_d, $----\; \theta_0/\gamma_1$ $(\psi_y = \gamma_1/\theta_d)$.

The effective source level of a parametric transmitter utilizing a square transducer can be calculated from the expression

$$(4.4.13) \qquad p_-(R, 0) = - P_W(R, 0) \cdot V(0, 0),$$

where, in terms of the parameters used in this Subsection,

$$(4.4.14) \quad \left|\begin{array}{l} P_W(R, 0) = (P_1' P_2' k_-^2 \beta/4\pi\varrho_0 c_0^2 \alpha_T R)(\lambda_0^2/S)\exp\left[-(\alpha_- + jk_-)R\right], \\[2mm] i.e. \\[2mm] \qquad = [(P_1' P_2' k_- \beta/2\varrho_0 c_0^2 R)\alpha_T r_0']\exp\left[-(\alpha_- + jk_-)R\right], \end{array}\right.$$

where

(4.4.15) $$r'_0 = (k_0/k_-)\,R_0\,,$$

and R_0 is the near-field limit of the transducer given by eq. (3.4.1).

Thus, the amplitude of the difference frequency pressure component along the acoustic axis, referred to 1 m, becomes

(4.4.16) $$p_-(1, 0) = (P'_1 P'_2 \, k_- \beta / 2 \varrho_0 \, c_0^2) |V(0, 0)|/\alpha_T r'_0\,.$$

As mentioned above, $|V(0, 0)|$ is given in ref. [24] for various values of ψ_y. Also, $\alpha_T r'_0$ can be shown to be related to ψ_y.

From eqs. (4.4.15) and (3.4.1)

$$\alpha_T r'_0 = (2\alpha_T/k_-)(\pi S/\lambda_0^2)\,,$$

i.e.

$$= \pi \theta_a^2 \, LM\,.$$

Hence, if we use eq. (4.4.12),

(4.4.17) $$\alpha_T r'_0 = 2/\pi \psi_y \, \psi_z\,.$$

For the particular case of $\psi_y = \psi_z$

(4.4.18) $$\alpha_T r'_0 = 2/\pi \psi_y^2\,.$$

Thus, the parameter

(4.4.19) $$N = |V(0, 0)|/\alpha_T r'_0$$

can be evaluated as a function of ψ_y (or of $\alpha_T r'_0$) by using the results given in ref. [24]. The parameter N is shown plotted in Fig. 7.

The expression given in (4.4.16) leads to a very simple formula for computing the effective source level of a parametric source. For water, if we use

$$\beta = 3.5\,,$$

$$\varrho_0 = 10^3\ \text{kg/m}^3$$

and

$$c_0 = 1500\ \text{m/s}\,,$$

the r.m.s. source level at the difference frequency can be obtained from

(4.4.20) $$\text{(SL)}_- = \text{(SL)}_1 + \text{(SL)}_2 + 20 \log F_- + 20 \log N(\psi_y) -$$

$$- 286.7\ \text{dB rel. } \mu\ \text{Pa m}\,.$$

Here, $(SL)_1$ and $(SL)_2$ are the primary source levels (r.m.s.) in dB relative μ Pa m, F_- is the difference frequency in kHz and the parameter N can be obtained from Fig. 7.

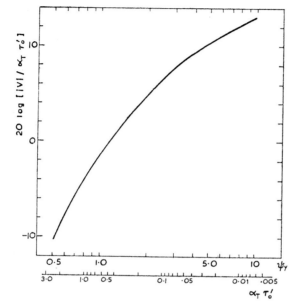

Fig. 7. – The normalized source level parameter $N = |V(\gamma_y)|/\alpha_T r_0'$.

Thus by using the information provided in Fig. 6 and 7, together with the simple formulae presented here, the source level and the beam width of a parametric source can be estimated.

The treatment presented has been confined mainly to parametric transmitters using a square transducer. If a rectangular transducer is being considered, the values of ψ_y and ψ_z can be calculated and the curves provided in ref. [24] can be used to evaluate the difference frequency beam widths in the planes $\eta' = 0$ and $\theta' = 0$ and the value of $|V(0, 0)|$. The latter result can be used in eq. (4.4.20) to obtain the difference frequency source level.

It is shown in ref. [24] that the results for square transducers can also be used for circular transducers. In this case the half-power beam width of the primary waves $2\gamma_1$ is related to the transducer radius a through the expression

$$k_0 a \gamma_1 = 1.616 \, ,$$

which leads to the results

$$\psi_d = \gamma_1/\theta_d = 1.616/k_0 a \theta_d$$

and

(4.4.21)
$$\alpha_T r_0' = (0.808/\psi_d)^2 \, .$$

Asymptotic expressions [24]. The discussion presented in the previous Subsection may have served to indicate the significance of the various parameters used in the analysis of parametric transmitters using rectangular or circular transducers. In this Subsection it is proposed to discuss briefly the asymptotic behaviour of the integral in eq. (4.4.6) for $(\psi_y, \psi_z) \ll 1$ and for $(\psi_y, \psi_z) \gg 1$.

The integral mentioned is a two-dimensional convolution integral involving two directivity functions:

i) $D_0^2(\theta, \eta)$, the product of the directivity functions of the primary waves;

ii) $D(\theta, \eta) = [1 + j(\theta'^2 + \eta'^2)]^{-1}$, which is the directivity function associated with Westervelt's model (see eq. (4.3.9)), represented in the co-ordinate system used in Fig. 5, and hence leads to a beam width of $2\theta_d$ in the $\eta = 0$ plane.

For $(\psi_y, \psi_z) \ll 1$ the primary beam widths are much smaller than $2\theta_d$. In other words, $D_0^2(\theta, \eta)$ varies much more rapidly than $D(\theta, \eta)$. Then, the integral can be evaluated in the form of an asymptotic series, and the most significant term of this series gives

$$V(\theta, \eta) \simeq D(\theta, \eta)(S/\lambda_0^2) \int\int D_0^2(\gamma, \varphi) \, d\gamma \, d\varphi .$$

The double integral in this expression is well known in studies of radiation from an aperture (note: for better accuracy, the form given in eq. (4.4.3) may be used without affecting the present argument) and yields λ_0^2/S. Hence, for γ_1 and φ_1 much smaller than θ_d

$$V(\theta, \eta) \simeq D(\theta, \eta) .$$

In other words, the parametric transmitter reverts to Westervelt's model discussed in Subsect. 4˙3. In these circumstances

$$V(0, 0) \simeq D(0, 0) = 1 .$$

For a square or circular transducer, for example, when $\psi_y \ll 1$ (*i.e.* $\alpha_T r_0' \gg 1$) the parameter $N(\psi_y)$ approaches $-20 \log (\alpha_T r_0')$.

If, on the other hand, $(\psi_y, \psi_z) \gg 1$, $|D(\theta, \eta)|$ varies much more rapidly than $D_0^2(\theta, \eta)$, and the most significant term in the asymptotic expansion becomes

$$V(\theta, \eta) \simeq (\theta_d^2 S/\lambda_0^2) D_0^2(\theta, \eta) \int\int \frac{d\gamma' \, d\varphi'}{1 + j(\gamma'^2 + \varphi'^2)} .$$

Clearly, in this case, the difference frequency beam pattern approaches the product of the primary directivity functions.

These asymptotic results indicate quite clearly that directivity of the difference frequency waves is largely controlled by the product of the directivity functions of the primary waves when $(\psi_y, \psi_z) \gg 1$, and approaches Westervelt's results when $(\psi_y, \psi_z) \ll 1$.

Limitations of analysis. The analysis presented above, leading to the formulation of the scattering integral (eq. (4.4.2)), is based upon Westerverlt's single-scattering approach. Therefore, extra-attenuation of the primary waves and the resulting effects on the difference frequency source distribution are ignored. Experience has shown, however, that, if the primary source levels are such that the «shock formation distance» $1/\beta\varepsilon k_0$ is not less than the near-field limit R_0, the results presented agree well with experimental data. When $\beta\varepsilon k_0 R_0 > 1$ the extra-attenuation effects become prominent, the difference frequency beam width increases, and the source level is reduced below the value predicted from eq. (4.4.20). A model proposed in ref. [25] is said to provide acceptable results, predicting the characteristics of parametric transmitters at high intensities as well as at lower intensities. However, for the purposes of the present treatment, it is intended to establish the conditions imposed on the design of a parametric source by the use of the criterion $\beta\varepsilon k_0 R_0 < 1$ rather than discussing the effects of extra-attenuation at length. This approach is particularly useful in considering a narrow-beam parametric transmitter for use in a long-range system (*e.g.*, an echo sounder). It could, however, lead to a suboptimum design for a shorter-range application (*e.g.*, a subbottom profiler for shallow-water use) where the beam broadening arising from finite-amplitude effects may be a worth-while sacrifice in return for higher difference frequency source levels.

If, as before, a primary wave is assumed to be planar in the near field of the transducer, with the pressure amplitude P_1, then

$$\beta\varepsilon k_0 R_0 = (\beta k_0/\varrho_0 c_0^2) P_1 R_0 ,$$

i.e.

$$= (\beta k_0/\varrho_0 c_0^2) P_1' ,$$

where P_1' is the far-field pressure amplitude, referred to a range of 1 m, and is related to the r.m.s. primary source level $(\mathrm{SL})_1$ through the expression

$$(\mathrm{SL})_1 = 20 \log P_1' + 117 \ \mathrm{dB \ rel.} \ \mu \ \mathrm{Pa \ m} .$$

Using the appropriate values of β, ϱ_0 and c_0, for water $\beta\varepsilon k_0 R_0 = 1$ corresponds to

$$(4.4.22) \qquad (\mathrm{SL})_1 + 20 \log F_0 = 280.5 \ \mathrm{dB \ rel.} \ \mu \ \mathrm{Pa \ m} ,$$

where F_0 is the arithmetic mean of primary frequencies in kHz.

Hence, for the results presented above to be applicable, the primary source levels must be maintained at values below that indicated by eq. (4.4.22) for the primary frequencies used. It is of interest to note that although eq. (4.4.20) does not *directly* depend upon the primary frequencies, consideration of finite-amplitude effects introduces f_0 in the expression for $(SL)_-$ in an explicit form.

For a given set of parameters, the maximum obtainable difference frequency source level (without serious finite-amplitude effects) can be calculated by using the limiting values of $(SL)_1$ and $(SL)_2$ as indicated by eq. (4.4.22) in eq. (4.4.20). This results in a new expression for $(SL)_-$, $(\widehat{SL})_-$ say, given by

$$(4.4.23) \qquad (\widehat{SL})_- \simeq 274 + 20 \log F_- - 40 \log F_0 + 20 \log N \text{ dB rel. } \mu \text{ Pa m}.$$

In many instances, especially when attempting to minimize the size of the transducer for a particular application, the transmitted intensities required to obtain the primary source level given by eq. (4.4.22) may be prohibitively high. In these cases the characteristics of the parametric device would be controlled by the intensity limit imposed by cavitation considerations, or by the power-handling capacity of the transducer. Then, the difference frequency source level can be calculated from eq. (4.4.20).

Example. To illustrate how the analysis and the data presented above can be used in the design of parametric sources an example will be considered in some detail.

The parametric transmitter to be designed is to produce a wave at 8 kHz, with a beam width of the order of $(3 \div 4)°$, using a circular transducer of diameter $D = 30$ cm. The characteristics of the parametric source for a difference frequency of 3.5 kHz is also of interest.

The main variable in a problem of this kind is the mean primary frequency f_0. If f_0 is too high, the primary source levels which may be used without incurring in extra-attenuation effects, and hence the difference frequency source level, are reduced. If f_0 is too low (as the transducer size is fixed), the primary beam width $2\gamma_1$ becomes too large for the difference frequency beam width $2\theta_0$ to have values between 3° and 4°. However, within the constraints discussed here, there is no unique « optimum » value of f_0. To help in choosing a suitable value of f_0, the values of various relevant parameters of the device have been calculated as a function of f_0. These results are presented in Fig. 8. The calculations were made for each value of f_0 in the following manner.

The value of the absorption coefficient in sea water α_0 was used (α_0 data were for 35 % salinity and 10 °C temperature) to calculate $2\theta_d$ (see eq. (4.3.11)). At 8 kHz k_- is 33.5. Then the expression for $2\theta_d$ reduces to

$$2\theta_d = 39.6 \sqrt{\alpha_0} \text{ degrees},$$

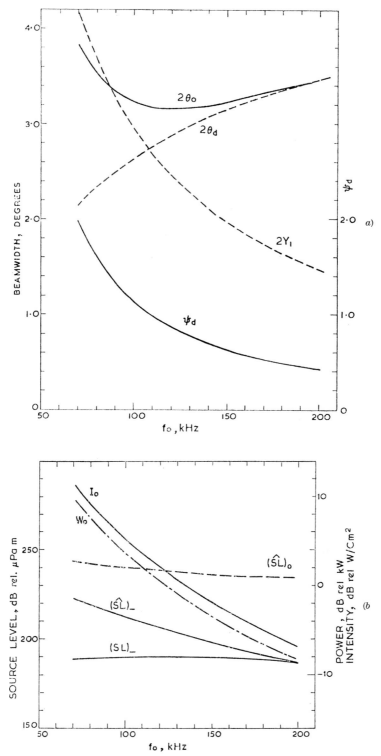

Fig. 8. – The values of the various parameters of the device considered in the design example for a difference frequency of 8 kHz.

where α_0 is in neper per metre (or, $2\theta_d = 13.4 \sqrt{\alpha_0}$ degrees, where α_0 is in decibel per metre).

The primary beam width $2\gamma_1$ was calculated using the expression

$$2\gamma_1 = 294.5/F_0 \text{ degrees} ,$$

where F_0 is the mean primary frequency in kHz.

The value of ψ_d (calculated as the ratio $2\gamma_1/2\theta_d$) was used to determine the ratio $2\theta_0/2\theta_d$ from Fig. 6 and N (in dB) from Fig. 7. Thus, the difference frequency beam width ($2\theta_0$) and, from eq. (4.4.23), the maximum difference frequency source level $\widehat{(SL)}_-$ were calculated for each value of F_0.

The corresponding value of the primary source level $\widehat{(SL)}_0$ was calculated from eq. (4.4.22), and the transmitted acoustic power per frequency W_0 from the expression

$$(SL)_0 = 10 \log W_0 + (DI)_0 + 170.8 \text{ dB rel. } \mu \text{ Pa m} ,$$

where W_0 is in watt and $(DI)_0$ is the directivity index of the transducer at frequency f_0.

As can be seen from Fig. 8a), any frequency in the range $(80 \div 200)$ kHz may be used as the mean primary frequency to produce difference frequency waves at 8 kHz with a beam width in the range $(3 \div 3.5)°$. For any value of f_0 in the range $(100 \div 150)$ kHz, $2\theta_0$ is approximately $3.2°$. Thus, the choice of primary frequencies must depend on other considerations.

From Fig. 8b) the difference frequency source level obtainable $\widehat{(SL)}_-$ increases considerably as f_0 is reduced. However, the transmitted acoustic power per frequency (and hence the acoustic intensity) also increases at lower values of f_0. The intensity (I_0) varies from 13.8 W/cm² at 70 kHz to 0.21 W/cm² at 200 kHz. It is worth-while to note that, when two frequencies are transmitted simultaneously, the total acoustic intensity at the peak of the envelope will be four times that per frequency. Clearly, in a practical situation, some constraints will be introduced as regards the maximum acoustic intensity which may be launched. These constraints may arise from the need to avoid cavitation or, ultimately, the stress or power-handling capacity of the transducer.

In the present example, for the higher values of f_0, the acoustic intensities which may be used (and hence the difference frequency source level obtainable) are still likely to be limited by the need to avoid extra-attenuation effects, whereas at lower primary frequencies cavitation or transducer characteristics will be the more restrictive consideration.

For example, if the transducer is to be operated near the sea surface, and the total intensity is not to exceed 1 W/cm² in order to avoid cavitation,

I_0 must not be greater than $\frac{1}{4}$ W/cm² (*i.e.* $I_0 < -6$ dB rel. W/cm²). Then, at $f_0 = 70$ kHz for example, each primary source level will be 17.4 dB below $(\widehat{SL})_0$, causing a reduction of 34.8 dB in the difference frequency source level $(SL)_-$. The values of $(SL)_-$ calculated on this basis are also shown in Fig. 8*b*). Clearly, with this additional constraint, the source level obtainable at 8 kHz varies only marginally with the primary frequencies for 85 kHz $< f_0 <$ 180 kHz.

On the basis of these results the following observations appear to be in order:

a) if the only constraint on the acoustic intensities which can be used is the extra-attenuation effects, then the primary frequencies could be selected at about 90 kHz, resulting in a source level of about 215 dB relative μ Pa m, and a beam width of about 3.3°, at 8 kHz;

b) if I_0 is not to exceed $\frac{1}{4}$ W/cm², $f_0 \simeq 150$ kHz $\left(\text{with } (SL)_- \simeq 190\text{ dB}\right.$ relative μ Pa m and $2\theta_0 \simeq 3.2°\left.\right)$ might be more advantageous.

Clearly, there is no uniquely optimum design, and the final choice of parameters might well be influenced by such factors as the difficulties in the design of a transducer to permit the launching of the two primary frequencies simultaneously.

To evaluate the parameters at a difference frequency of 3.5 kHz, we note that $2\theta_d$ is increased by $(8.0/3.5)^{\frac{1}{2}}$ (and hence ψ_d is reduced by the same factor) for a given value of f_0, as compared with the corresponding values for a difference frequency of 8 kHz. The beam width and the source level at 3.5 kHz were calculated in the manner explained above. The results are shown in Fig. 9. As can be seen, if $f_0 \simeq 90$ kHz, one obtains $2\theta_0 \simeq 4.2°$ and $(\widehat{SL})_- \simeq$ $\simeq 204$ dB relative μ Pa m.

For $f_0 \simeq 150$ kHz, $2\theta_0 \simeq 4.8°$ and $(SL)_- \simeq 176$ dB relative μ Pa m.

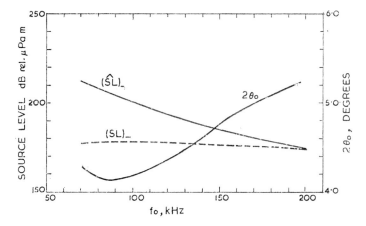

Fig. 9. – For the same device, for a difference frequency of 3.5 kHz.

Comparing these results with those obtained for a difference frequency of 8 kHz, we note that in the two examples considered the beam is broader (by 35% and 50%, respectively) and the source level is reduced by 11 dB and 14 dB.

If the transducer can be operated at some depth, the permissible value of I_0 is increased. For example, if $I_0 = 1$ W/cm², the difference frequency source level can be increased by 12 dB for all primary frequencies below 135 kHz; this would make the device more attractive.

The effects of increasing the transducer diameter are somewhat complex, and depend very much upon the constraints on the system. However, a new set of curves can be produced for the new transducer size, and a choice of the primary frequencies and power made along the lines outlined above.

As an example we consider the effects of increasing the diameter of the transducer to 45 cm. At a particular frequency f_0 the value of $\widehat{(\mathrm{SL})_0}$ is not changed. But, as $(\mathrm{DI})_0$ has increased by 1.75 dB the corresponding value of W_0 is reduced by that amount, and the intensity by 3.5 dB. If the intensity limit (imposed by cavitation considerations) is $\frac{1}{4}$ W/cm², each primary-frequency sourve level will be increased by 3.5 dB for values of $f_0 < 150$ kHz. As $2\gamma_1$ (and hence ψ_d) is reduced by a factor of 1.5, however, N will be reduced. For example, at $f_0 = 150$ kHz the new value of ψ_d becomes 0.41, causing a reduction in N of about 3.5 dB. Thus, an overall increase in $(\mathrm{SL})_-$ of about 3.5 dB can be obtained.

4`5. *Discussion and conclusions*. – The advantages of parametric transmitters can be summarized as follows:

   i) narrow beams can be obtained at low frequencies, using relatively small transducers;

   ii) the difference frequency beams do not exhibit any sidelobes;

   iii) the difference frequency can be varied over a wide range with small variation in the beam width.

These properties of parametric devices make them very useful in long-range, narrow-beam echo sounding, in subbottom profiling and in shallow-water sonar and communication applications where absence of sidelobes reduces unwanted forward- and/or back-scattering effects.

In the example discussed in Subsect. 4`4 above, the transducer diameter (30 cm) was about 1.6 wavelengths at the difference frequency of 8 kHz. If a wave at 8 kHz were transmitted from this transducer directly, the beam width would be about 37° and, for an intensity of 1 W/cm², a source level of 213 dB relative μ Pa m would be obtained for a radiated power of about 700 W.

In the discussions on the parametric device it was pointed out that an intensity of 1 W/cm² at the peak of the transmitted envelope corresponded to $\frac{1}{4}$ W/cm² at each frequency, and the expected source level at 8 kHz (for 350 W total acoustic power) was about 190 dB relative μ Pa m. The beam width of the difference frequency waves was estimated to be about 3.2°.

Thus, for a given transducer size, a very high increase in directivity can be achieved by using a parametric device. In this instance the transmitted power is halved, and the source level is reduced by 23 dB. Two possible ways of improving the effective source level may be considered.

Either by placing the transducer at some depth or by using short pulses at a relatively high frequency the cavitation threshold may be increased. In this case the difference frequency source level increases as $20 \log I_0$.

If longer, band-spread signals (*e.g.* « chirp » pulses) are transmitted and a form of matched-filter reception is used, the resulting signal-processing gain would improve the effective source level.

In this introductory treatment of parametric transmitters the aim has been to provide an insight into the behaviour of such devices for targets in the far field of virtual sources. More detailed treatment of other aspects of the devices is still the subject of research. The theoretical results presented here agree well with computer analyses of specific devices [23], with other generalized analyses [25, 26] and with a wealth of experimental data.

The author's personal feeling is that parametric devices lead to more economical solutions for some engineering and scientific instrumentation problems in underwater acoustics.

REFERENCES

[1]   D. T. BLACKSTOCK: *History of nonlinear acoustics and a survey of Burgers' and related equations*, in *Nonlinear Acoustics*, edited by T. G. MUIR (Austin, Tex., 1970).
[2]   Lord RAYLEIGH: *Theory of Sound*, 1877, second edition (New York, N. Y., 1945), re-issue.
[3]   H. LAMB: *The Dynamical Theory of Sound*, 1931, second edition (New York, N. Y., 1960), re-issue.
[4]   See, for example, R. T. BEYER and S. V. LETCHER: *Physical Ultrasonics*, Chap. 7 (New York, N. Y., 1969).
[5]   D. T. BLACKSTOCK: *Journ. Acoust. Soc. Amer.*, **34**, 9 (1962).
[6]   S. I. SOLUYAN and R. V. KHOKHLOV: *Vestn. Mosk. Univ. Ser. III, Fiz. Astron.*, **3**, 52 (1961).
[7]   D. T. BLACKSTOCK: *Journ. Acoust. Soc. Amer.*, **39**, 1019 (1966).
[8]   D. T. BLACKSTOCK: *Journ. Acoust. Soc. Amer.*, **36**, 217 (1964). Also K. A. NAUGOL'NYKH, S. I. SOLUYAN and R. V. KHOKHLOV: *Sov. Phys. Acoust.*, **9**, 42 (1963).
[9]   D. T. BLACKSTOCK: *Journ. Acoust. Soc. Amer.*, **36**, 534 (1964).
[10] For example, L. D. ROZENBERG, Editor: *High Intensity Ultrasonic Fields*, Chap. 1 and 2 (New York, N. Y., 1971).
[11] B. B. CARY: *Journ. Acoust. Soc. Amer.*, **43**, 1364 (1968).
[12] H. M. MERKLINGER: *Journ. Acoust. Soc. Amer.*, **54**, 1760 (1973).

[13] P. J. WESTERVELT: *Proceedings of the 3rd I.C.A., Stuttgart, 1959*, Vol. **1** (Amsterdam, 1961), p. 316.

[14] P. J. WESTERVELT: *Journ. Acoust. Soc. Amer.*, **35**, 535 (1963).

[15] B. B. CARY: *Journ. Acoust. Soc. Amer.*, **42**, 88 (1967).

[16] See ref. [9, 10] for details.

[17] D. T. BLACKSTOCK: *Journ. Acoust. Soc. Amer.*, **39**, 1019 (1966).

[18] H. O. BERKTAY and B. V. SMITH: unpublished work.

[19] T. G. MUIR: *An analysis of the parametric acoustic array for spherical wave fields*, Applied Research Laboratories (University of Texas at Austin), Technical Report ARL-TR-71-1 (AD 723 241). See also J. A. SHOOTER, T. G. MUIR and D. T. BLACKSTOCK: *Journ. Acoust. Soc. Amer.*, **55**, 54 (1974).

[20] H. O. BERKTAY: *Some finite-amplitude effects in underwater acoustics*, in *Nonlinear Acoustics*, edited by T. G. MUIR (Austin, Tex., 1970).

[21] P. J. WESTERVELT: *Journ. Acoust. Soc. Amer.*, **29**, 199 (1957).

[22] H. O. BERKTAY: *Journ. Sound Vib.*, **2**, 435 (1965).

[23] T. G. MUIR and J. G. WILLETTE: *Journ. Acoust. Soc. Amer.*, **52**, 1481 (1972).

[24] H. O. BERKTAY and D. J. LEAHY: *Journ. Acoust. Soc. Amer.*, **55**, 539 (1974).

[25] R. H. MELLEN and M. B. MOFFETT: *A model for parametric sonar radiator design*, Naval Underwater Systems Center Technical Memorandum No. PA 41-229-71 (1971).

[26] F. H. FENLON: *Journ. Acoust. Soc. Amer.*, **55**, 35 (1974).

List of symbols.

$a = 0$	for a plane wave
$a = \frac{1}{2}$	for a cylindrical wave
$a = 1$	for a spherical wave
c_0	velocity of propagation for infinitesimal waves
k	wave number (the subscripts denote the wave numbers for various frequencies)
p_0	pressure in the absence of disturbance
p_T	total pressure
r_0	source radius for a cylindrical or spherical wave
t	time variable
T	shock thickness, defined as the spacing between the peak and the trough along the x-axis
u	particle velocity
$V = u/u_0$	normal measure of particle velocity
$W = (r/r_0)^a V$	
x, r	range variable
$y = \omega\tau$	phase angle, or nondimensionalized measure of retarded time
α	absorption coefficient (subscripts used for various frequencies)
$\beta = 1 + B/2A$	a parameter of nonlinearity of the fluid ($\beta \simeq 3.5$ for water)
$\Gamma = \beta\varepsilon k/\alpha$	acoustic Reynolds number
Δ	a measure of shock thickness
$\varepsilon = u_0/c_0$	acoustic « Mach » number
ϱ_0	density of the fluid when at rest
ϱ_T	total density of the fluid
σ	nondimensionalized range parameter
τ	retarded time

Finite-Amplitude Propagation in Solids.

R. W. B. STEPHENS

Chelsea College, University of London - Wandsworth, London

1. – Introduction.

The subject of nonlinearity has received a great deal of attention in the case of intense ultrasonic waves propagated in fluid media [1]. The nonlinearity of the equation of motion and of the equation of state of the fluid gives rise to such phenomena as shock wave formation, nonlinear interaction between waves, generation of combination frequencies etc. A basic knowledge of nonlinearity is also most important in such industrial processes as ultrasonic cleaning, welding, emulsification, etc.

The interest in finite wave propagation in solids however has only heightened during the last decade or so, but the phenomena and the basic theory do have a close parallel to those of the fluid state. Since the nonlinearity involves finite displacements, the stress in a solid is no longer linearly related to the strain, i.e. it deviates from Hooke's law. A brief resumé will therefore be made concerning the notation involved in higher-order elasticity.

2. – Higher-order elasticity theory.

The elastic energy stored in a deformed isotropic or anisotropic material may be written in tensor notation as [2]

$$\varphi = \tfrac{1}{2} c_{ijkl} S_{ij} S_{kl} + c_{ijklmn} S_{ij} S_{kl} S_{mnt} \dots ,$$

where

$$i, j, k, l, m, n = 1, 2 \text{ or } 3 ,$$

and where S_{ij} etc. are the elastic strains.

If only the first term in the expression for φ is included, we have « first order » or « linear » elasticity theory. The coefficients are known as « second order » elastic constants (or stiffnesses) because the first term contains strain products of the second degree.

The coefficients c_{ijklmn} are called 3rd-order elastic constants or stiffnesses.

For a crystal of lowest symmetry (triclinic) there are 21 independent 2nd-order and 56 independent 3rd-order elastic constants.

In 1st-order elasticity theory only the 2 Lamé constants λ and μ are required for an isotropic body. Once an isotropic body is appreciably strained however, the linear theory is no longer adequate and three additional 3rd-order elastic constants are necessary.

HUGHES and KELLY [3] obtained 7 equations relating the velocities of specific longitudinal and transverse waves to the magnitude of the prevailing hydrostatic pressure or simple uniaxial stresses in isotropic materials.

TOUPIN and BERNSTEIN [4] also derived a relation between these quantities and determined conditions such that data on ultrasonic propagation in elastic materials be compatible with 2nd-order elastic theory. They define a linear coefficient of acoustic birefringence A which directly relates the difference in velocities between the two shear waves polarized normal and parallel to a uniaxial stress and propagating normal to the direction of this stress to the magnitude of the stress itself.

Thus

$$v_{\parallel} - v_{\perp} = AT \, ,$$

where T is the magnitude of the stress and A is given by

$$A = \frac{(1 + \sigma)(v_3 + \mu)}{E v_0 \varrho_0} \, ,$$

where σ = Poisson's ratio,

E = Young's modulus,

v_3 is one of the 3rd-order elastic constants for an isotropic material,

μ = shear modulus

v_0 = velocity of either shear mode

and

ϱ_0 = density of medium.

TRUESDELL [5] has shown that in any isotropic elastic material the waves travelling down a principal axis of stress are either pure longitudinal or pure transverse modes. He derived exact general formulae for the velocities of the 6 pure transverse and the 3 pure longitudinal waves which are possible in a 3-dimensional stress system. He has evaluated unique values of 2nd- and 3rd-order constants from wave velocities, providing certain compatibility conditions are satisfied.

3rd-order constants are often negative and an order of magnitude larger than 2nd order, but 4th-order constants are smaller than those of the 3rd order.

Estimates of 4th-order constants for copper have been obtained from shock wave measurements [6].

3. – Nonlinear elastic waves.

Nonlinear effects in elastic-wave propagation may result from a number of different causes such as i) a wave amplitude which is so large that finite strains arise, ii) a normally linear material showing nonlinear behaviour in infinitesimal wave propagation when it is subjected to an external uniaxial or hydrostatic stress, iii) a medium behaving locally in a nonlinear manner, due to the presence of various energy-absorbing mechanisms.

Nonlinear propagation differs from linear elastic waves in that the initially sinusoidal longitudinal stress wave of a given frequency becomes distorted as propagation proceeds and energy is transferred from the fundamental to the harmonics that develop. The degree of the distortion and the strength of the harmonic generation will be directly dependent on the initial wave amplitude. Furthermore although a pure-mode longitudinal nonlinear wave may propagate as such, a pure transverse nonlinear wave will necessarily be accompanied by a longitudinal wave during propagation. However a nonlinear transverse wave does not distort when propagated through a defect-free solid. Nonlinear elastic waves can interact with other waves in a solid and in the region of the interaction of two ultrasonic beams a third ultrasonic beam may be generated.

When large displacements are considered in a solid, the stress is no longer linearly related to the strain and, if we follow the treatment of THURSTON and SHAPIRO [6] and take the special case of one-dimensional motion in an isotropic solid, or along certain directions in anisotropic media, the equation of motion becomes

$$\varrho_0 \ddot{u} = \frac{\partial^2 u}{\partial x^2}\left[M_2 + M_3\frac{\partial u}{\partial x} + M_4\left(\frac{\partial u}{\partial x}\right)^2 + \ldots\right],$$

where x is the Lagrangian co-ordinate in the direction of motion of a particle whose displacement is u.

M_2 is a linear combination of second-order elastic coefficients, M_3 is a linear combination of second- and third-order coefficients, while M_4 is a linear combination of second-, third- and fourth-order coefficients.

M_2, M_3 etc. may be associated with the nonlinear parameters A and B for liquids and with γ, the ratio of principal specific heats for perfect gases, viz.

$$-\frac{M_3}{M_2} = 2 + \frac{B}{A} = \gamma + 1$$

and

$$\frac{M_4}{M_2} = 3 + \frac{3B}{A} + \frac{C}{2A} = \frac{1}{2}(\gamma + 1)(\gamma + 2).$$

Also $M_2 = K_2$ and $M_3 = K_3 + 2K_2$, where K_2 and K_3 are related to the second- and third-order elastic constants.

If the fourth and higher orders are omitted, then the wave equation may be written in the approximate form

$$\ddot{u} = \frac{M_2}{\varrho_0} \frac{\partial^2 u}{\partial x^2} \left[1 + \frac{M_3}{M_2} \frac{\partial u}{\partial x} \right].$$

If c_0 is the speed of propagation of an infinitesimal amplitude wave, then $c_0^2 = M_2/\varrho_0$.

The modification to the « lossless » wave equation to allow for energy dissipation is obtained by adding a term to include the frequency-dependent attenuation coefficient, $\alpha = \alpha(\omega)$, to the right-hand side of equation, viz.

$$\ddot{u} = c_0^2 \frac{\partial^2 u}{\partial x^2} + \left(3 + \frac{K_3}{K_2} \right) c_0^2 \frac{\partial^2 u}{\partial x^2} \frac{\partial u}{\partial x} + \frac{2\alpha}{\omega^2} c_0^3 \frac{\partial^3 \dot{u}}{\partial x^2 \partial t},$$

which is the same approximate nonlinear wave equation as for plane-wave propagation in a viscous fluid, and hence similar theoretical techniques may be used for its solution in the case of a solid. For example the modified wave equation may be reduced to the form of Burgers' equation

$$\frac{\partial V}{\partial t} + V \frac{\partial V}{\partial X'} = \nu \frac{\partial^2 V}{\partial X'^2},$$

where V is a particle velocity, X' is a moving co-ordinate, t is time and ν is a viscosity parameter.

In general the dissipative mechanisms increase with frequency and the wavefront will attain a maximum steepness when the transfer of energy to higher harmonics due to nonlinearity is just equalized by the increase of absorption at the higher frequencies. However it is only a relative stabilization of the wave profile as, due to damping, the wave gradually returns towards its initial sinusoidal shape. The transfer of energy to the higher harmonics appears as an attenuation of the fundamental frequency (f_0) component of the wave in excess of the attenuation due to the absorption of a monofrequency wave of f_0.

4. – Monitoring of finite-amplitude waves in solids.

The arrangement used by RICHARDSON et al. [7] is shown in Fig. 1. In order to detect the n-th harmonic, the light source and the detector are set at the appropriate Bragg angle, viz. $\theta_B(n\omega) = \sin^{-1}[\lambda/\Lambda(n\omega)]$, where λ is the light wavelength, Λ is the wavelength of the n-th acoustic harmonic and ω

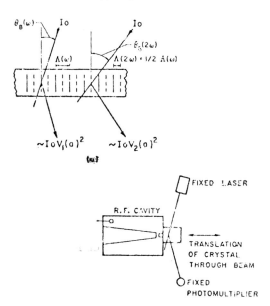

Fig. 1.

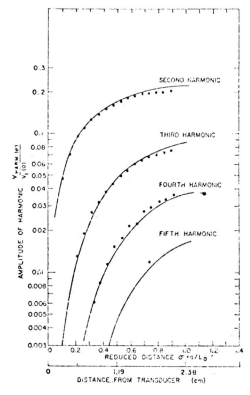

Fig. 2. – Harmonic content of a longitudinal wave propagating along the z-axis of quartz at a fundamental frequency of 562 MHz.

is the fundamental acoustic frequency. The intensity of the light diffracted is proportional to the intensity of the laser and of the acoustic harmonic $V_n(A)^2$. The spatial growth of the harmonic is measured by moving the crystal through the laser beam, keeping the light source and the detector in a fixed position. Figure 2 shows the experimental observations along the z-axis of quartz at 562 MHz, together with the theoretical results (solid lines), taking $L = 2.38$ cm and $\varGamma = 4.75$. The latter quantity is equal to $(\alpha_0 L)^{-1}$, where α_0 is the attenuation at the fundamental frequency for infinitesimally small amplitudes. $\varGamma$ is taken as a measure of the importance of nonlinearity relative to that of dissipation.

5. – Reflection of finite-amplitude ultrasonic waves.

With the need for correct interpretation of harmonic generation in solids and in liquids, the latter case including the special area of underwater acoustics, the question of the reflection of finite-amplitude ultrasonic waves at interfaces has become a significant problem. For a nonsinusoidal wave the phase shift upon reflection at an interface may cause a relative spatial shifting of the Fourier harmonic components. Assuming independent reflection of the Fourier harmonics VAN BUREN and BREAZEALE [8] were able to use linear theory to calculate this change in phase angle between the fundamental and the second-harmonic components. In the liquid-solid interfaces considered there was good agreement between theory and experiment, but when the energy transfer between harmonics takes place in a less restricted region than that of a boundary it is possible to utilize a nonlinear stress-strain relationship [9]; Fig. 3 and 4 indicate the velocity and displacement profiles for incident and reflected distorted waves within a solid. It is seen that when the incident velocity pro-

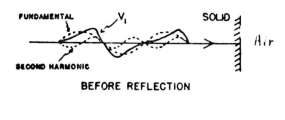

BEFORE REFLECTION

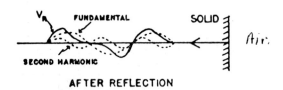

AFTER REFLECTION

Fig. 3. – Internal reflection in a solid.

file is saw-tooth then the reflected wave is a reverse saw-tooth with respect to the initial direction of propagation.

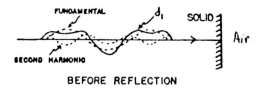

BEFORE REFLECTION

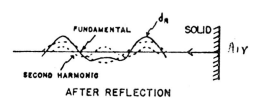

AFTER REFLECTION

Fig. 4. – Internal reflection in a solid.

If we consider now that the incident wave originated from one of finite amplitude and has distorted during propagation, with each of the individual Fourier harmonics undergoing the same phase shift. Since, however, the acoustic wavelengths of the harmonics differ, the phase relationship between the harmonics will change leading to an unstable wave form. In solving the problem VAN BUREN and BREAZEALE used a computer, basing their calculations on the phenomenological model used by FOX and WALLACE. In their technique the propagation distance is divided into small intervals and the wave is allowed first to distort over an interval and is then corrected for absorption over the interval distance; the process is repeated over each successive interval. VAN BUREN and BREAZEALE concluded that the behaviour of a finite-amplitude wave which has been phase modified can be explained if the return to its natural or stable saw-tooth profile can be considered to take place with a modification of the harmonics that possess an unnatural phase relationship, this modification taking place by the generation of new harmonics which possess the proper phase relationship with the existing next lower harmonic.

BREAZEALE [10] has used an optical probe to investigate experimentally the incidence of a distorted ultrasonic wave travelling in water when it is reflected at a solid interface. The latter is situated at the discontinuity distance for the 5 MHz wave which is reflected at an angle of approximately 15°. The light beam from a point source transverses in water the region of interaction of the incident and reflected waves and suffers a double diffraction. The optical-diffraction pattern obtained will be asymmetrical in consequence of the distorted ultrasonic waveform and can be interpreted as the resultant effect when each of the diffraction orders produced by the incident wave is diffracted by the reflected wave. The wave reflected from a resilient boundary is distorted in an opposite sense to that from a rigid boundary and moreover is in an unstable condition since the nonlinear effects cause the higher harmonics to decrease. For normal incidence the asymmetry of the diffraction pattern indicates that standing waves are symmetrical for a rigid and asymmetrical for a resilient boundary. The latter case is analogous to the stress-free boundary condition obtaining when an ultrasonic wave travelling within a solid strikes an end face. The propagation of a finite-amplitude wave in a nonlinear solid can be expressed as

$$\varrho_0 U_{tt} = K_2(U_{xx} + 3U_x U_{xx}) + K_3 U_x U_{xx} \, ,$$

where U_{tt} is the second time derivative of the displacement. For a cubic crystal the coefficients K_2 and K_3 are combinations of the ordinary elastic constants and the third-order elastic constants respectively and will depend on the propagation direction with respect to the crystal axes (see Table I).

TABLE I. – K_2 and K_3 for a cubic crystal in terms of elastic constants of 2nd and 3rd order respectively.

Direction	K_2	K_3
(100)	C_{11}	C_{111}
(110)	$(C_{11} + C_{12} + 2C_{44})/2$	$(C_{111} + 3C_{112} + 12C_{166})/4$
(111)	$(C_{11} + 2C_{12} + 4C_{44})/3$	$(C_{111} + 6C_{112} + 12C_{144} + 24C_{166} + 2C_{123} + 16C_{456})/9$

If nonlinearity is neglected, pure longitudinal waves may be propagated in cubic crystals in the directions [100], [110] and [111]. Pure transverse waves may also propagate in the directions defined by a) $k_1 = k_2$, k_3 arbitrary and b) $k_3 = 0$, k_1 and k_2 arbitrary. k_1, k_2 and k_3 are the three components of the propagation vector. If nonlinear terms are included, it is found that pure transverse modes are nonexistent, but pure longitudinal modes can continue to be propagated in the three directions stated.

It should be noted that, if the equation is solved for the particle velocity V, the expression is

$$V = U_t = 2V_0 \sum_{n=1}^{\infty} \frac{J_n(nx/L)}{nx/L} \sin n(\omega t - kx) \, ,$$

which is identical with the expression obtained by KECK and BEYER [11] for a sinusoidal driver in a fluid medium.

For a liquid the combination $-(3K_2 + K_3)/K_2$ is replaced by $B/A + 2$, where A and B are the coefficients of the Taylor expansion of the pressure in terms of the condensation. The distance travelled by a wave in a medium for an initially sinusoidal wave to develop a discontinuity in particle velocity and known as the discontinuity distance L is given by

$$L = (K_2/\varrho_0) \bigg/ \left(-\left(\frac{3K_2 + K_3}{K_2}\right) 2\pi^2 f^2 U_0 \right),$$

where U_0 is the particle displacement amplitude at the source and f is the fundamental frequency of excitation.

As an interesting comparison of the values of L for a liquid and solid taking $U_0 = 10^{-8}$ cm and $f = 30$ MHz, then L for water $= 15$ cm, whereas for germanium it would be between 120 and 500 cm, depending on the direction of propagation in the crystal (table II).

TABLE II. – *Discontinuity distance in germanium (after* BREAZEALE).

Direction	K_2 (dyn·cm^{-2})	K_3 (dyn·cm^{-2})	L (for U_0 expressed in cm) (cm)
(110)	$1.288 \cdot 10^{12}$	$-2.20 \cdot 10^{13}$	$5.04 \cdot 10^{-6}/U_0$
(110)	$1.053 \cdot 10^{12}$	$-3.93 \cdot 10^{13}$	$1.18 \cdot 10^{-6}/U_0$
(111)	$0.975 \cdot 10^{12}$	$-3.36 \cdot 10^{13}$	$1.2 \ \cdot 10^{-6}/U_0$

BREAZEALE points out that for propagation distances small compared with L it is sufficiently accurate to use a perturbation solution of the nonlinear differential equation to calculate the third-order elastic constants of solids, and this condition is easily satisfied with the ordinary size of specimen used in laboratory experiments.

A perturbation solution of the equation above gives

$$U = U_0 \sin (kx - \omega t) + \frac{U_0 x}{4L} \cos 2(kx - \omega t) + \dots,$$

which indicates that the second harmonic in the distorted wave is proportional to the distance from the sinusoidal source and to the square of the fundamental amplitude (L is proportional to $1/U_0$). These relations provide a useful test on experimental procedure, and it is to be noted that $L \to \infty$, *i.e.* giving a strict linear behaviour for the solid if $K_3 = -3K_2$.

6. – Harmonic generation in crystals.

The creation of harmonics in an ultrasonic stress wave will take place even in a perfectly ordered lattice since the anharmonicity of interatomic forces is characteristic of all solids. A growing interest in this harmonic generation

arises for example from the increasing power used in ultrasonic applications
and the consequent need to evaluate higher-order elastic constants. The measurement of the amplitude of the second harmonic of a wave provides one method
of determining the third-order elastic constants. In an imperfect lattice there
is also the possible additional source of harmonic generation arising from dislocations in the lattice and this provokes the problem of separating the contributions from the two sources. The presence of an external static stress influences
the harmonic contribution from dislocations, but does not ease appreciably the
difficulty of separation. Turning to the third harmonic in general the lattice
contribution may be negligible compared with that due to dislocations and
moreover this harmonic is also sensitive to an external static stress. This
stress however must only be sufficient to cause changes in dislocation loop
length rather than in the degree of bowing-out.

A number of workers have carried out experiments on harmonic generation
and only a brief mention will be made here of the work of BRITTON and SCORER
on single crystals of sodium-chloride. In this material dislocation pinning points
in the form of colour centres may be produced easily by X-irradiation, so
enabling dislocation-point defect interactions to be studied. Owing however
to difficulties of high purity the dislocation loop lengths were short compared
with those obtainable with metal crystals. Fundamental waves were observed
which contained typically 1% of the second harmonic and 0.33% of the third
harmonic. The apparatus employed was a modification of the usual ultrasonic
pulse-echo technique (see Fig. 5) using quartz transducers although the receiving

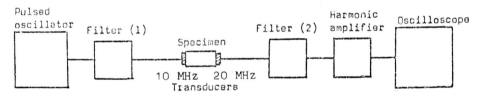

Fig. 5. – Simplified block diagram of a second-harmonic system.

element was for some observations substituted with a capacitative transducer.
Static bias stresses were applied to the specimen crystal by suspending it
and attaching weights to the lower end. This arrangement was modified for
compressive stress measurements. Observations were also made on aluminium
single crystals and there was a general agreement with the predictions of
HIKATA and ELBAUM [12] in their theory of dislocation harmonic generation.
An interesting observation on third-harmonic measurements in a sodium-
chloride crystal was the existence of internal stresses remaining after plastic
deformation which had previously been reported by HIKATA, CHICK and EL-
BAUM [13] in second-harmonic studies in aluminium crystals. However in the

present case the likely explanation was different and suggested the requirement of an actual unpinning of dislocations by the internal stresses.

Since theory shows that the amplitude A_3 of the third harmonic is proportional to the cube of the fundamental amplitude A_0, then a plot of $\log A_0$ vs. $\log A_3$ should give a straight line of slope 3 (see fig. 6).

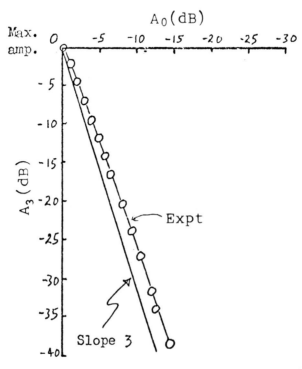

Fig. 6. – Relation between amplitudes of the fundamental and third harmonic (after BRITTON et al. [13a]).

7. – Acoustic second-harmonic generation in piezoelectric crystals.

In extending the acoustical theory for second-harmonic generation to non-piezoelectric crystals additional terms are required to include piezoelectric and higher-order cross-terms in the expression for the crystal free energy. In consequence there must be distinctions between elastic stiffness constants measured under conditions of constant electric field (or constant electric displacement vector) and between dielectric constants measured at constant stress or strain. This piezoelectric « stiffening » leads to an increase in phase velocity over that for the propagation of the unstiffened wave in the same direction. There are two unique directions in quartz for which the two unstiffened transverse phase velocities are equal but the stiffened velocities are unequal. The complicated character of the problem adds to the difficulties in the design of beam steering and focussing devices.

As a direct consequence of piezoelectric coupling oscillating strain fields due to ultrasonic excitation will be accompanied by oscillating electric fields which will contribute to the higher-order terms of the crystal free energy thus causing changes in the higher-order elastic constants. In consequence third-order elastic constants determined by means of second-harmonic generation are not equal to the free-field third-order elastic constants [14].

It should be noted that piezo-active waves directed opposite to each other can interact, but a purely elastic nonlinearity cannot give rise to interaction of elastic waves of the same frequency travelling in opposite directions. Various acousto-electric phenomena can arise in piezoelectric semiconductors due to the interaction of the free electrons with the longitudinal electric field caused by the nonlinear piezoelectric effect and in ferroelectrics due to the field interaction with domain walls.

The nonlinear phenomena are of technical interest in convolution techniques [15, 16].

SINGH and AGARWAL have derived optimum conditions for maximum harmonic generation and find, compared with purely piezoelectric materials, that the second-harmonic flux is a maximum at higher frequencies in nonpiezoelectric semiconductors and a maximum at lower frequencies for semiconductors possessing both piezoelectricity and deformation potential.

8. – Resonance interaction of elastic waves in isotropic solids.

This class of phenomena has attracted some recent interest in relation for example to nonlinear resonance interaction between a longitudinal pump wave and two parametrically amplified transverse waves in an attempt to construct a parametric ultrasonic amplifier.

In considering the resonance interaction in isotropic solids the conditions to be satisfied for the wave vectors are

$$\omega = \omega_1 + \omega_2 \quad \text{and} \quad \boldsymbol{k} = \boldsymbol{k}_1 + \boldsymbol{k}_2 .$$

In this case the amplitudes of certain waves will grow effectively at the expense of the others. Two cases may be considered:

1) if all three wave vectors have the same direction, then resonance interaction between three longitudinal or three transverse waves is possible,

2) if $\boldsymbol{k}_1$ and $\boldsymbol{k}_2$ have different directions, then ω and $\boldsymbol{k}$ must be longitudinal, while

  a) $\boldsymbol{k}_1$ and $\boldsymbol{k}_2$ are both transverse, or

  b) one is longitudinal and one is transverse.

Furthermore it has been shown that, when two transverse waves of frequencies ω_1 and ω_2 intersect at an angle θ and satisfy the resonance condition $\cos \theta = R^2 + (R^2 - 1)(1 + \delta^2)/2\delta$, where $R = C_t/C_l$ and $\delta = \omega_2/\omega_1$, then the scattered wave is longitudinal of frequency $\omega_3 = \omega_1 + \omega_2$. C_t and C_l are transverse and longitudinal velocities respectively.

9. – Optical-acoustical interaction in photoconducting piezoelectrics.

The fundamental nonlinearity of photoconductivity results in two interfering light beams forming an inhomogeneous current-carrier concentration that varies periodically in space. This periodic variation of conductivity in a piezoelectric crystal gives rise to a number of resonance and antiresonant phenomena in the propagation of elastic waves.

CHABAN [17, 18] in a recent paper first considers the generation of sound due to an external alternating electric field so that at a certain resonance frequency the stresses induced by the piezoelectric properties of the crystal will produce a sound wave. By making suitable assumptions CHABAN shows that it is possible to generate a sound wave of considerable intensity in any arbitrarily chosen region of the crystal. He points out the existence of at least two mechanisms that give rise to the generation of a nonpiezoactive sound wave (applicable also to nonpiezoelectric photoconducting crystals). The first is associated with the action of the alternating electric field upon the space charge distribution and the second is related to the electrostriction force. A calculation by CHABAN shows that these mechanisms in a CdS crystal are capable of generating a nonpiezoactive sound wave having a strain amplitude of the order of 10^{-6}.

10. – Nonlinearity in surface waves.

The existence of nonlinearity effects in surface acoustic waves (SAW) is of particular interest since their energy is confined approximately to a thin surface layer of the order of a wavelength deep. Hence pronounced nonlinear effects should be easily observable, as quite low acoustic source powers will produce the requisite high power densities.

LEAN and TSENG [19] propagated SAW on a Y-cut, Z-oriented LiNbO$_3$ slab having two optically polished surfaces and the resulting periodic surface deformation provided a moving phase grating to an optical laser probing beam (Fig. 7). The angular dependence of the diffracted beam is given by

$$\sin \theta_m = \sin \theta_0 + m\lambda/\Lambda , \tag{1}$$

where $m = 0, \pm 1$, etc., Λ is the wavelength of sound and λ is the wavelength of laser light.

The intensity of the first-order diffracted light I_m with respect to the zeroth-order intensity I_0 is given by

$$(2) \qquad I_m/I_0 = \sum_{ln} [1/(l!)^2]\{[A/(nf)^2]P_d(nf)\}^l ,$$

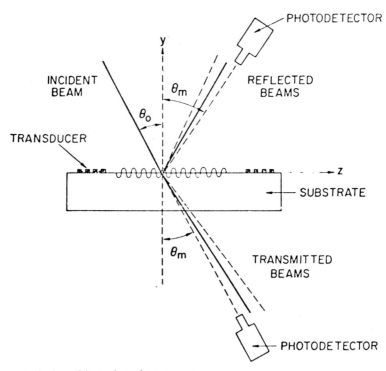

Fig. 7. – Optical probing of surface waves.

where the summation is over all of the positive integers l and n such that $l \times n = m$. A is a proportionality constant which depends on the acoustical-optical properties of the medium. $P_d(nf)$ is the acoustic-power density (APD) of the n-th harmonic, where APD = input acoustic power$/\Lambda \times$ beam width. Each term in expression (2) is the l-th-order diffraction from the n-th–harmonic acoustic wave.

The contributions from the higher-order diffraction terms of f_0 are usually negligible compared with the first-order diffractions of the m-th harmonic, i.e. $l = 1$. Hence eq. (2) may be written as

$$(3) \qquad I_m/I_0 = [A/(mf)^2]P_d(mf) ,$$

where $m = 0, 1, 2, \ldots.$ Unlike optical second-harmonic generation the absence of dispersion of Rayleigh waves ensures that the phase-matching condition is satisfied for all frequencies, so that in principle large numbers of harmonics can be generated in the substrate. Although the problem of accurate theoretical evaluation is very difficult, the growth of each harmonic as a function of interaction length and input pump power level can be directly observed by means of the optical probe.

In parametric mixing, when two waves of different frequencies are injected, the phase-matching condition again becomes significant and the sum and difference signals generated will in turn mix with the fundamentals and with each other to generate more sums and differences and harmonics. For a practical system the harmonic frequencies higher than the pump frequency must be suppressed and one procedure is to create a dispersive delay line by the deposition of a thin film over the beam path.

The presence of nonlinearity is seen both as an asset and a liability in surface wave devices. The harmonic generation limits the power-handling capabilities of SAW delay lines and leads to signal distortion, but on the other hand signal-processing applications such as convolution and correlation rely on the presence of strong acoustic nonlinearity.

It should be added that although a number of SAW signal-processing devices have been designed using first-order theory, yet others have shown a difference between their predicted and actual performances, especially those employing a high-coupling material such as lithium-niobate. A number of second-order effects have been quoted as possibly contributing to this discrepancy such as a constant cycling of energy between the electrical and acoustical fields when a SAW transverses an interdigital transducer, the acoustical reflections occurring at the transducer electrode edges, the diffraction and beam steering of the SAW which depend on crystal cut and transducer geometry etc. A preliminary investigation has revealed that the major cause of distortion arises from the acoustic mismatch at the electrode discontinuities of the metallization on the piezoelectric substrate.

As an alternative to the use of SAW for convolution and correlation it has been found possible to use the nonlinear interaction of Lamb waves on polarized plates of PZT using interdigital transducers [20]. In convolution with bulk or surface waves the modulated acoustic waves are launched from the opposite ends of the specimen and a large-area transducer covers the entire region of interaction and serves as an integrator of the product signal, thus giving the convolution. On the other hand, correlation with bulk waves depends on the differing velocities of longitudinal and transverse modes; this is not of course applicable to the SAW case. The procedure for SAW is to propagate a SAW towards the interaction region below a large-area transducer to which the other signal is applied and produces a standing-wave system below it. The frequency of this signal is chosen so that the nonlinear interaction between the propagating and the standing waves gives a difference frequency signal which can propagate and is received by an interdigital transducer. A similar procedure may be followed using Lamb plane waves in the same or different modes. These waves combine the advantages of both bulk wave and SAW methods for they exist with a broad range of velocities and yet retain the high energy density and accessibility of SAW.

The design of useful SAW devices requires a detailed knowledge of the propagation characteristics such as particle displacements, attenutation, wave velocity, energy profile and the direction of power flow. Optical measurement techniques permit the determination of these various parameters. Because of the high power densities associated with the concentration of the surface acoustic energy within a depth from the surface of approximately one acoustic wavelength, care has to be exercised in interpreting the results of optical probing. LEAN and TSENG have used the linear region to observe the Fresnel near-field diffraction of Rayleigh waves propagated from an interdigital transducer and to ascertain in anisotropic substrates the beam steering resulting from misalignment of transducers. They also operated in the linear region to measure the attenuation and velocity of the acoustic waves and also to determine the reflection and transmission coefficients of a tuned transducer. In the non-linear region the optical probe was utilized to observe the spatial growth of each harmonic generated by the elastic nonlinearity of the substrate.

11. – Surface waves on a liquid.

The propagation of waves over the surface of a pure liquid is controlled by surface tension and gravity forces and provides examples of nonlinearity with strong dispersion (see Fig. 8). For the longer waves the control is that of gravity, while for the shorter waves, known as ripples or capillary waves, surface tension plays the major role in the control. When both controls are equal, the velocity of propagation will have its minimum value given by $c_{\text{min}} = (4\sigma g/\varrho)^{\frac{1}{4}}$, where σ is the surface tension and ϱ the density of the liquid, g being the ac-

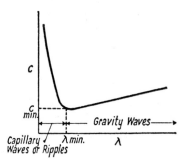

Fig. 8. – Dispersion of surface waves on a liquid.

celeration due to gravity. If the depth of the liquid (h) is greater than $\lambda/2$ (λ being the wavelength), then $c^2 = g\lambda/2\pi + 2\pi\sigma/\varrho\lambda$ gives the velocity of the surface waves in deep water as defined. For water waves of large amplitude the velocity of propagation becomes greater than that of classical small-amplitude theory and the water particles actually have a forward movement, *i.e.* the particle paths are no longer closed. The profile of the waves is now not sinusoidal but may be considered as following the trochoidal wave which would be traced out by a point at a distance a (the wave amplitude) from the centre of a rolling circle.

KRASILNIKOV *et al.* [21] have recently shown that the strong dispersion in capillary waves (*i.e.* those with wavelength $< \lambda_{\text{min}}$ in Fig. 8) prevents the development of nonlinear effects in finite-amplitude capillary waves. For frequencies up to $\sim 10^6$ Hz the capillary-wave velocity may be taken as $c = [2\pi\sigma/\varrho\lambda]^{\frac{1}{2}}$. KRASILNIKOV also shows that for the surface wave the dependence of attenuation on wave amplitude is small compared to the case of volume acoustical waves; this may be explained by the influence of dispersion as well as by the linear dependence of the capillary-wave attenuation coefficient on frequency.

It was FRANKLIN [22] who first made an attempt to give a scientific reason for the phenomenon long known to seamen of calming a rough sea by the pouring of vegetable oils on the surface. He concluded that the oil mainly affected the damping of the capillary waves and furthermore said « there are continually rising on the back of every great wave, a number of smaller ones, which roughen its surface and give the wind hold, as it were, to push it with greater force ». In recent years a keen interest in capillary waves has been shown by surface and colloid chemists with the objective of finding what other physical properties besides surface tension contribute to the resistance of the surface against deformation. This information is of particular importance in the control of various industrial technological processes such as emulsification, foaming, detergency, extraction, etc. A very important factor in surface behaviour is the presence of a monolayer of surface-active material at the surface since its elasticity will provide an additional resistance to the periodic deformation created by the wave motion. Until twenty-five years ago only the case of a pure liquid and that of a liquid with a nearly incompressible monolayer have been considered.

An excellent account is given by LUCASSEN-REYNDERS and LUCASSEN [23] of the development of the subject and the problem of associating with readily measurable parameters the four rheological coefficients: surface dilational elasticity ε_D and surface dilational viscosity η_D which are concerned with resistance against changes in surface area, and surface shear elasticity ε_s and the surface shear viscosity η_s which provide the resistance against changes in the shape of a surface element.

Assuming the energy dissipation is small and the damping coefficient is appreciably less than λ^{-1} (where λ is the wavelength), then KELVIN [27] obtained the following relation, using the normal stress boundary condition: $\sigma k^3 + \varrho g k = \omega^2 \varrho$, where $k = 2\pi/\lambda$. Water has a sufficiently low viscosity to satisfy the assumptions and the experimental results agree well with theoretical values.

The recent theories of surface wave propagation differ from that of KELVIN in that the surface is endowed with elastic properties. This elasticity could be due to interactions between the surface molecules or, more likely, to the fact that the surface tension decrease arising from the presence of adsorbed organic molecules increases with their concentration at the surface. Any expansion of the surface will involve a change in surface concentration of the adsorbed molecules and lead to an increase in surface tension which will resist the expansion and so gives rise to a dilatation surface elasticity ε. This coefficient is sometimes termed the Marangoni elasticity and is defined by $\varepsilon = \mathrm{d}\sigma/\mathrm{d}(\ln A_l)$, where A_l is the area of an element of surface. If the adsorbed molecules are soluble in the substrate and can pass freely into solution to establish a dynamic equilibrium between adsorption and desorption, then this time-dependent component of elasticity is effectively a dilational surface viscosity. The analysis of LUCASSEN and HANSEN is for negligible dilatational surface viscosity and a brief reference will be made to an experimental test of the theory made by SCOTT [25] using the apparatus shown in Fig. 9, in which moiré

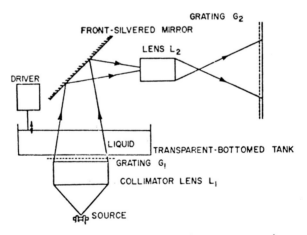

Fig. 9. – Schematic diagram of the wave-measuring apparatus (from ref. [25]).

fringes are used to record the wave system, through the distortion of the fringes. Figure 10 is a specimen of the form of record obtained. This technique produces instantaneous records of the wave field from which the wavelength and damping may be readily obtained and moreover is particularly useful for recording time-variant properties. In Scott's experiment polyethylene oxide

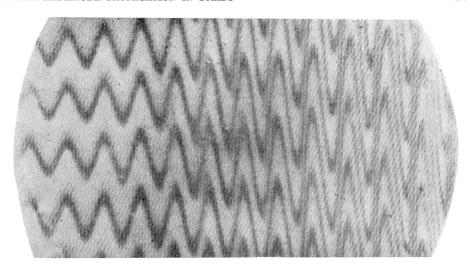

Fig. 10. – Ripples travelling from right to left (after ref. [25]).

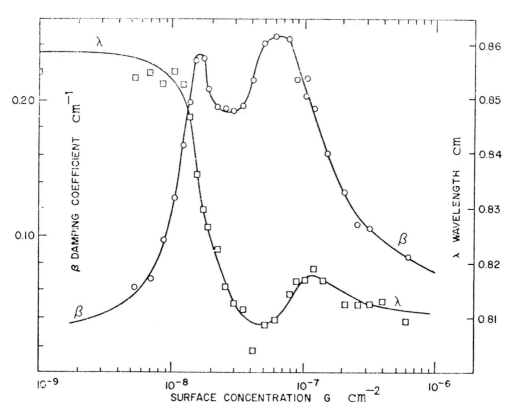

Fig. 11. – Wavelength and damping-coefficient measurements on spread monolayers of poly(ethylene) oxide.

was the polymer used and although highly soluble in water the monolayers showed no tendency to be dissolved in the water substrate.

According to LUCASSEN and HANSEN [26] damping maxima are postulated to occur at certain values of ε for which the transverse ripples lose energy in some form of « resonance interaction » with a highly damped longitudinal wave which is associated with compressions and rarefactions in the « elastic » monolayer. The L-H theory predicts a single maximum for materials simpler in structure than polyethylene oxide, but it has been shown to be valuable in interpreting the damping behaviour of a surface-active material exhibiting two damping peaks (Fig. 11), since both peaks occur at coincidences of σ and ε where the longitudinal-wave « resonance » is expected.

12. – Acoustic streaming etc.

When dealing with acoustic-wave propagation in fluids at appreciable amplitudes the time-independent contributions to the displacement, velocity and fluid stresses become significant and are evident in the measurable phenomena of radiation pressure and acoustic streaming. In order to account for these effects in a linear homogeneous isotropic fluid the dynamical and continuity equations for sound propagation in the x-direction become respectively

$$\varrho[\partial\boldsymbol{u}/\partial t + (\boldsymbol{u}\cdot\nabla)\,\boldsymbol{u}] = -\nabla p + [\mu' + \tfrac{4}{3}\mu]\nabla\cdot\nabla\boldsymbol{u} - \mu\nabla\times\nabla\times\boldsymbol{u}$$

and

$$\partial\varrho/\partial t + \nabla\cdot\varrho\boldsymbol{u} = 0\,,$$

where p, ϱ and u are respectively the independent variables of pressure, density and particle velocity, while μ and μ' are the shear and volume viscosity coefficients respectively. The equation of state for the medium will also in general be nonlinear.

Each dependent variable may be expressed as a series of functions of decreasing order of magnitude, e.g. $p - p_0 = p_1 + p_2 + ...$, where p_0 is the zeroth-order quantity, p_1 is a first-order quantity occurring in the linear equation for very small amplitudes and p_2 is a second-order quantiy arising from nonlinearity.

Dealing only with the time-independent variables NYBORG [27] obtains the expression $\boldsymbol{F} = \nabla p_2 - \mu\nabla x\nabla xu_2$, where $\boldsymbol{F}$ is interpreted as « an effective force per unit volume ». When $\boldsymbol{F}$ is determined, the acoustic problem becomes one in viscous time-independent hydrodynamics of determining the steady-state flow field (u_2) and static pressure field (p_2) for a given steady exciting force. If the force field is irrotational, i.e. when $\nabla\times F$ is zero, streaming is absent and only a radiation pressure field exists. The impingement of the wave on boundaries at oblique or tangential incidence will set up boundary layers so that the force field is no longer irrotational and acoustic streaming will take place. Such a situation was considered by RAYLEIGH when the fluid medium was confined between rigid parallel walls normal to the z-direction, the direction of propagation being along the x-axis. If we assume no slip at the walls, then boundary layers are set up and RAYLEIGH [28] showed that the streaming was in the form of a regular array of eddies a quarter of acoustic wavelength apart. The dust ridges that collect in the Kundt's tube experiment is an example of the above type of motion.

A recent experiment with liquid crystals [29] as the fluid medium is illustrative of the presence of streaming. The investigation was concerned with the effect of the high-frequency perturbation of one containing boundary (the

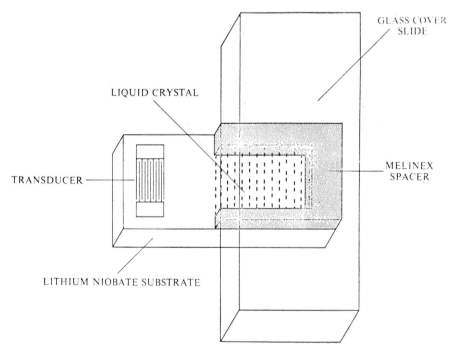

Fig. 12. – Liquid-crystal cell and surface wave generator.

base) of a shallow cell containing the nematic liquid crystal. The perturbation was obtained by the repetitive pulsing of a mechanical surface wave of a high ultrasonic frequency (Fig. 12). Additionally, since liquid crystals react to the presence of electric fields, it was further predicted that, if a piezoelectric medium formed the base of the cell and its surface was disturbed by the passage of the mechanical surface wave, then the rapidly changing electric fields created at the boundary might produce effects similar to dynamic scattering.

By generating surface waves at 67 and at 22 MHz on the free surface of lithium-niobate corresponding wavelengths of 50 and 150 µm respectively were obtained. At these wavelengths with a continuous wave of $5V$ peak to peak and under the assumption of negligible insertion loss, then field gradients of the order of $200 \, kV \, m^{-1}$ and $67 \, kV \, m^{-1}$ respectively should be realized. Such gradients are comparable to those used in liquid-crystal display cells. Observations indicated that the mechanical action of the surface waves predominated over that of the electric field. Measurements of fluid particle velocity by using a cine film indicated a proportionality with the acoustic intensity as might be expected.

13. – Nonlinearities in cochlear hydrodynamics.

One of the fundamental assumptions of the auditory theory of HELMHOLTZ was that the cochlea functioned as a perfectly linear system and any nonlinear phenomena such as the production of aural harmonics were attributed to the middle ear.

Cochlear travelling waves show similar properties [30] to capillary waves, whose motion is basically nonlinear for a number of reasons and in particular from the associated boundary layer effects. Mechanical models of the cochlea show two types of nonlinearity that might explain a number of auditory and/or electrophysiological phenomena. Firstly an amplitude-independent asymmetrical displacement of the partition for appropriate signals leading to demodulation. These experiments were performed at input levels below those needed for eddies to form. The appearance of the beat rate as a new frequency component on applying two signals of closely related frequencies was evidence of the existence of nonlinearity. Owing to a filtering action with distance of travel along the partition a low-frequency component propagated further along the partition (and through the cochlear fluids) than reached by the primary frequencies.

Thus the appearance of the beat frequency involved the process of demodulation, or envelope detection. The new low-frequency component corresponded always to the periodicity of the envelope.

The travelling-wave motion of the cochlear partition tends to motivate the fluid particles of the boundary layer and although the coupling is not very efficient it improves as the travelling wave steepens with distance, leading to an increased acceleration of the fluid stream. As a result, there is a thinning of the boundary layer and since the travelling-wave velocity decreases with distance, a point is reached when it becomes less than the fluid stream which breaks away to form an eddy motion. The harmonic distortion of the wave form arises from the presence of the eddies and does not involve the material properties of the partition.

14. – Scattering of sound by sound.

As already mentioned the wave form of a nonlinear sound wave becomes distorted during its propagation and the phenomenon can be regarded as a nonlinear interaction of the sound wave with itself. If the wave is initially of a single frequency ω_0, then harmonics of this will be generated through self-interaction until a shock front develops. Similarly two sound waves of frequencies ω_1 and ω_2 propagating simultaneously in a nonlinear medium will interact and give rise to sum and difference frequencies. The influence of the interaction may be confined to the interaction region but also may give rise to a secondary field outside, which is known as the scattered radiation field [31].

15. – High-power ultrasonics (macrosonics) and nonlinear effects in crystalline solids.

Although it has been common knowledge for some time that mechanical vibrations of high amplitude may affect the physical properties of materials, it was not until about a decade ago that research and development in macrosonics was really started. In 1955 BLAHA and LANGENECKER [32] reported that the superposition of ultrasonic vibrations on zinc single crystals at the same time as they were subjected to tensile deformation gave rise to a softening effect. Since then, as a result of many workers in the field, it has been possible to divide the observed effects broadly into two categories: a « volume » and a « surface » effect. The former includes all processes which influence the internal friction of the test piece, while under the « surface effect » are classified those processes which affect the external friction between a forming tool and the workpiece.

In their experiments BLAHA and LANGENECKER immersed the tensile specimen in a vessel containing carbon-tetrachloride, the bottom of the vessel being coupled to the transducer of an ultrasonic generator operating at 800 kHz. Figure 13 shows the result of their experiments with zinc crystals. The significant decrease in stress when ultrasound is imposed indicates the appreciable softening. In the upper curve the tensile stress is seen to return to its initial value when intermittent insonation ceases. In later experiments longitudinal vibrations were transferred to the specimen through an acoustic horn thus avoiding the use of a liquid for the transmission of ultrasonic energy. In this case it was found that the softening occurred without the application of external stress. The « Blaha » effect, as the phenomenon is known, was explained by BLAHA and LANGENECKER as arising from the creation and consequent movement of dislocations due to the energy supplied by the ultrasonic field. Later NEVILL and BROTZEN [33] showed that the stress in low-carbon steel

wires was reduced by an amount which was directly proportional to the amplitude of vibration, but, unlike BLAHA and LANGENECKER, they did not find any
effect at stresses below the macroscopic yield point. They discussed mechanisms
of resonance, relaxation and hysteresis by which the dislocations could absorb

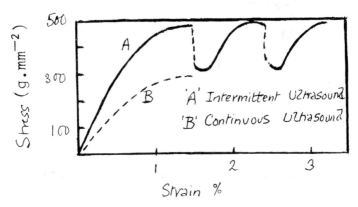

Fig. 13. – Stress-strain curves for Zn crystals. Softening effect of insonation (after
BLAHA and LANGENECKER).

energy but did not find any of them to be satisfactory and their final explanation
was in terms of a macroscopic process arising from the superposition of steady
and alternating stresses. LANGENECKER and his associates also conducted experiments in which a transducer was fixed to the lower end of the zinc crystal,
the transducer itself being attached to the tensile machine. The form of the
stress-strain curves obtained appeared similar for low amplitudes over a frequency range from 15 Hz to 1.5 MHz, but on exceeding a pressure amplitude
of $2.5 \cdot 10^7$ dyn/cm^2 the vibrations produced work-hardening which was directly
proportional to the stress amplitude over a considerable range, but it became
less following repeated applications of ultrasound at a constant amplitude and
finally approached a saturation level corresponding to about $5.5 \cdot 10^7$ dyn/cm^2
pressure amplitude.

It should also be mentioned that LANGENECKER [34, 35] has proposed that
the actual mechanism for the various effects observed could arise from the
localized heating taking place in regions around dislocations and other imperfections when the ultrasonic waves are scattered. However, other investi-
gators have reported that the temperature rise is only of the order of 1 °C per
insonation cycle and that about 10^7 times more thermal energy than ultrasonic
energy is required in order to produce the same stress reduction.

GREEN [36, 37] has considered the relatively simple case of a metal specimen
subjected to a tensile test at a constant rate of elogation and develops an
equation to describe the behaviour of the machine-specimen interaction. Based
on nonlinear elastic considerations alone, he shows that the effective Young's

modulus E^* should decrease in a linear fashion with increased applied tensile stress. Finally he works out for a polycrystalline copper test specimen that $(E - E^*)/E = 0.7\%$, a decrease in the value of Young's modulus from its value E in the unstressed state. In this calculation GREEN assumes a stress equal to the reduction obtained by POHLMAN and LEHFELD [38] in 20 kHz imposed ultrasonic vibrations on polycrystalline copper.

16. – Note on geometrical approach.

In general, nonlinear differential equations are not solvable by analytical means and this note is to draw attention to the fact that many features of the solutions can be derived by a geometrical approach. POINCARÉ and LIAPUNOV developed the idea of representing the state of a quantity for a system with one degree of freedom by a point on a phase plane in which the two variables are respectively the position (x) of the quantity and the first time derivative ($\dot{x}$). The motion of the point in this phase plane will then exhibit the changing state of the system under examination [38].

The behaviour of forced vibrations of systems with nonlinear elasticity characteristics grows in significance as larger forces become involved and is of particular interest for example in the case of piezoelectric oscillators or resonators. If we assume a sinusoidal excitation of constant amplitude, then for small damping a steady-state peak response amplitude of the system will vary with the frequency as shown in Fig. 14 where the dashed line AB corresponds to the nonlinear free-vibration amplitude-frequency curve. It is seen that the location of the peak response will depend upon the direction (upwards or downwards) in which the input frequency is slowly swept. This so-called jump phenomenon, whereby under certain circumstances an infinitesimal change of either the frequency or amplitude of the output signal causes quite a large

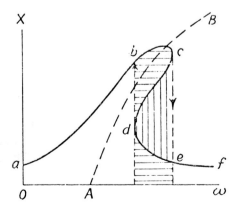

Fig. 14. – Nonlinear free-vibration amplitude v frequency curve and « jump phenomenon ».

and discontinuous jump in the amplitude of the output, is obtained with servo-control systems and has also been shown by SEED to exist when exciting quartz resonators at large amplitudes. The AT « cut » quartz crystal is seen (Fig. 15) to correspond to a hard spring and the BT « cut » crystal to a soft spring. From the experimental observations estimates may be made of the higher-order elastic constants of quartz [39].

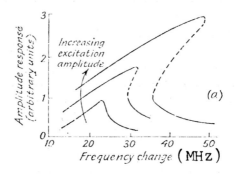

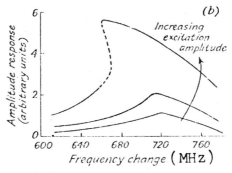

Fig. 15. – a) AT cut quartz crystal resonant frequency 4.894 919 MHz (large-amplitude excitation). b) BT cut quartz crystal resonant frequency 4.989 723 MHz (large-amplitude excitation) (after SEED).

17. – Shock waves in solids.

An account of nonlinearity in solids would not be complete without a brief mention of shock propagation. A great deal of work has been done on the propagation of small-amplitude stress waves in solids and also on such large stress amplitudes that the materials behave somewhat like fluids. Shock waves should be propagated in solids when the stress-strain relation becomes nonlinear, but, if the stress exceeds that which corresponds to the proportionality limit, then plastic flow can occur and so results in plastic-wave propagation [40-42]. As differing from fluids the particle velocity in solids compared with the phase velocity is small, even for quite large stresses. It is in fact the change of elastic modulus with strain which is largely responsible for any distortion of finite-amplitude pulses in solids. However for most solids the elastic modulus decreases beyond the proportionality limit and so plastic rather than shock waves are set up when the deformations are sufficiently large. Some solids, *e.g.* high polymers (perspex, etc.), undergo orientation of long-chain molecules (leading to an increase of modulus) when subjected to large tensile strains which make it possible for the growth of shock waves from large-amplitude waves. In general therefore in order to generate shock waves in a solid an explosive has to be placed in contact with, or a high projectile fired at, the surface of the specimen. By using various strength shocks and making corresponding measurements of the wave and particle velocities in the shocked state a « Hugoniot $(p\text{-}v)$ » curve may be plotted from which the equation of state of the solid may be derived.

YASUMUTO [43] and colleagues have used an electromagnetic-induction type of transducer to generate acoustic shock pulses in an aluminium bar for the purpose of investigating nonlinear properties. Although the stress-strain relation did not show nonlinearity, interesting variations of the form of the pulse were obtained at the higher shock pressure pulses and at the higher temperatures of annealing the specimen. This variation of waveform was attributed to the change of internal structure of the solid.

REFERENCES

[1] L. BJØRNØ, Editor: *Finite amplitude wave effects in fluids*, in *Proceedings of the 1973 Symposium, Copenhagen* (London, 1973).

[2] R. T. SMITH and R. W. B. STEPHENS: *Effects of anisotropy on ultrasonic propagation in solids*, in *Progress in Applied Materials Research*, Vol. **5** (London, 1964).

[3] D. S. HUGHES and J. L. KELLY: *Phys. Rev.*, **92**, 1145 (1953).

[4] R. A. TOUPIN and B. BERNSTEIN: *Journ. Acoust. Soc. Amer.*, **33**, 216 (1961).

[5] (*a*) C. TRUESDELL and (*b*) A. SEEGER: *International Symposium on Second Order Effects in Elasticity, Plasticity and Fluid Mechanics* (Israel, 1962).

[6] R. N. THURSTON and M. J. SHAPIRO: *Journ. Acoust. Soc. Amer.*, **41**, 1112 (1967).

[7] B. A. RICHARDSON, R. B. THOMPSON and C. D. W. WILKINSON: *Journ. Acoust. Soc. Amer.*, **44**, 6 (1968).

[8] A. L. VAN BUREN and M. A. BREAZEALE: *Journ. Acoust. Soc. Amer.*, **44**, 4 (1968).

[9] A. L. VAN BUREN: University of Tennessee Tech. Report No. 4, NONR 4289 (01) (Dec. 1967) for extension of the above work.

[10] M. A. BREAZEALE: *Finite amplitude waves in liquids and solids*, in *V International Congress on Acoustics* (Liège, 1965).

[11] W. KECK and R. J. BEYER: *Phys. Fluids*, **3**, 346 (1960).

[12] A. HIKATA and C. ELBAUM: *Phys. Rev.*, **144**, 469 (1966).

[13] A. HIKATA, B. CHICK and C. ELBAUM: *Journ. Appl. Phys.*, **36**, 229 (1965).

[13a] W. G. B. BRITTON, G. P. PARKES, F. J. BURGUM and S. E. EVANS: *Acustica*, **33**, 4 (1975).

[14] D. H. McMahon: *Journ. Acoust. Soc. Amer.*, **44**, 4 (1968).

[15] V. V. Lemanov and N. K. Yushkin: *Sov. Phys. Solid State*, **15**, 11 (1974).

[16] R. S. Woollett and C. L. Le Blanc: *I.E.E.E. Trans. Sonics Ultrasonics*, SU-**20**, 1 (1973).

[17] A. A. Chaban: *Sov. Phys. Acoust.*, **19**, 3 (1973).

[18] A. A. Chaban: *Sov. Phys. Solid State*, **9**, 2622 (1968).

[19] E. G. Lean and C. C. Teng: *Journ. Appl. Phys.*, **41**, 10 (1970).

[20] E. S. Ferguson and V. L. Newhouse: *I.E.E.E. Trans. Sonics Ultrasonics*, SU-**20**, 4 (1973).

[21] V. A. Krasilnikov, V. I. Pavlov, V. P. Vorinin and L. K. Zarembo: *Nonlinear effects on propagation of capillary waves of finite amplitude*, in *Finite Amplitude Wave Effects in Fluids*, edited by L. Bjørnø (London, 1973).

[22] B. Franklin: *Phil. Trans.*, **64**, 445 (1774).

[23] E. H. Lucassen-Reynders and J. Lucassen: *Properties of capillary wave*, in *Advances in Colloid and Interface Science* (Amsterdam, 1969).

[24] W. Thomson (Lord Kelvin): *Phil. Mag.*, **42**, 368 (1871).

[25] J. C. Scott and R. W. B. Stephens: *Journ. Acoust. Soc. Amer.*, **52**, 3 (1972).

[26] J. Lucassen and R. S. Hansen: *Journ. Colloid-Interface Sci.*, **22**, 32 (1966).

[27] W. L. Nyborg: *Concepts of nonlinear acoustics applied to macrosonics*, in *Proceedings of the I International Symposium on High-Power Ultrasonics, Graz, Austria*, (London, 1970).

[28] Lord Rayleigh: *Theory of Sound*, Sect. **352** (New York, N. Y., 1945).

[29] W. G. B. Britton, T. F. North and R. W. B. Stephens: *Effects on nematic liquid crystals of excitation by acoustic surface waves*, in *V International Liquid Crystal Conference* (Stockholm, 1974).

[30] J. Tonndorf: *Journ. Acoust. Soc. Amer.*, **47**, 2 (1970).

[31] P. J. Westervelt: *Journ. Acoust. Soc. Amer.*, **47**, 2 (1970).

[32] F. Blaha and B. Langenecker: *Naturwiss.*, **42**, 556 (1955).

[33] G. F. Nevill jr. and F. R. Brotzen: *Proc. Amer. Soc. Test. Mat.*, **57**, 751 (1957).

[34] B. Langenecker: *A.T.A.A. Journ.*, **1**, 80 (1963).

[35] B. Langenecker: *Proc. Amer. Soc. Test. Mat.*, **62**, 602 (1962).

[36] R. E. Green jr.: *Nonlinear effects of high-power ultrasonics in crystalline solids*, in *Ultrasonics International* (London, 1973).

[37] R. E. Green jr.: *Ultrasonic Investigation of Mechanical Properties: Treatise on Materials Science and Technology*, Vol. **3** (New York, N. Y., 1973).

[38a] R. Pohlman and F. Lehfeldt: *Ultrasonics*, **4**, 178 (1966).

[38b] A. E. Seed: Ph. D. Thesis, University of London.

[39] L. C. Barcus: *I.E.E.E. Trans. Sonics Ultrasonics*, SU-**22**, 4 (1975).

[40] J. M. Bowsher: *Can. Journ.*, **37**, 1017 (1959).

[41] R. T. Beyer: *Nonlinear acoustics*, Chap. 10, in *Physical Acoustics*, Vol. **2**, Part B (New York, N. Y., 1965).

[42] R. W. B. Stephens and A. E. Bate: *Acoustics and Vibrational Physics* (London, 1966).

[43] Y. Yasumoto, A. Nakamura and R. Takeuchi: *Acustica*, **30**, 5 (1974).

A Physical Approach to Elastic Surface Waves.

J. DE KLERK

Westinghouse Research Laboratories - Pittsburgh, Pa.

Introduction.

In recent years a great deal of interest has been focused on elastic surface waves, mainly for two reasons; firstly, energy can be tapped from these waves at any point along their propagation path, and, secondly, the surface wave velocity is 10^5 times slower than the electromagnetic counterpart. The combination of these two characteristics has enabled engineers to design and fabricate very sophisticated signal-processing devices which are orders of magnitude smaller than electromagnetic devices which perform the same functions. The electrical-engineering approach to elastic surface waves uses equivalent circuits which enable the engineer to ignore elasticity completely and think only in terms of electrical transmission lines. Although this method may have certain advantages, it does not adequately describe the physical phenomena encountered in the generation, propagation and detection of Rayleigh waves. To obtain a more complete concept of the behavior of the elastic waves, it is necessary to use a physical rather than an electrical-engineering approach. In Sects. **1** and **2** the equations of motion are derived for propagation in isotropic and anisotropic media. In Sect. **3** different methods of generation are described and an analytical method of determining the optimum crystal orientations and propagation directions is developed. This method is then applied to the most frequently used surface wave materials. In Sect. **4** the effect of the piezoelectric properties of the material on surface wave reflection from, and regeneration by, interdigital grids on the surface is considered. Finally, in Sect. **5** the basic concepts of two surface wave devices using phase and frequency modulation are described. Surface wave amplification and other filter applications are not dealt with here as these subjects will be treated in detail by other speakers at this series of lectures on new directions in physical acoustics.

1. – Rayleigh waves in isotropic media.

Although there are several different types of waves called « surface » waves, such as those named after LAMB, LOVE, STONELEY, GULYAEV or BLEUSTEIN and Lord RAYLEIGH, only the latter type of waves has so far been extensively exploited in the service of signal processing.

Rayleigh waves, which exist on the free surface of a half-space, have retrograde elliptical particle motion at the surface. These waves, which are dispersionless, rapidly attenuate below the surface. Lamb waves are plate waves, with each surface of the plate supporting a Rayleigh-type wave. When the thickness of the plate approaches the wavelength, these waves become very dispersive. Stoneley waves are similar to Rayleigh waves and propagate at the interface between two half-spaces. Love waves have displacements which are transverse to the propagation direction but parallel to the surface. These waves may extend many wavelengths into the medium. Gulyaev or Bleustein waves are similar to Love waves but are accompanied by piezoelectric fields and hence can only exist in certain special directions on a piezoelectric medium.

Any unbounded isotropic solid can support two and only two types of elastic waves, *viz.* longitudinal and transverse. However, if there is a bounding surface present, a third type of wave may be propagated on this surface. These surface waves were first studied by Lord RAYLEIGH [1], who showed that their velocity of propagation was slower than either bulk wave and that they only extended below the surface to a depth of approximately two wavelengths.

If one considers the propagation of a plane wave through an isotropic elastic medium with a bounding surface, the equations of motion [2] for the waves in the medium are

$$(1.1) \qquad \varrho \frac{\partial^2 u}{\partial t^2} = (\lambda + \mu) \frac{\partial \Delta}{\partial x} + \mu \nabla^2 u \, ,$$

in which

$$(1.2) \qquad \nabla^2 = \frac{\partial^2}{\partial x^2} + \frac{\partial^2}{\partial y^2} + \frac{\partial^2}{\partial z^2} \, ,$$

$$(1.3) \qquad \varrho \frac{\partial^2 v}{\partial t^2} = (\lambda + \mu) \frac{\partial \Delta}{\partial y} + \mu \nabla^2 v$$

and

$$(1.4) \qquad \varrho \frac{\partial^2 w}{\partial t^2} = (\lambda + \mu) \frac{\partial \Delta}{\partial z} + \mu \nabla^2 w \, .$$

In these equations

$$(1.5) \qquad \Delta = \varepsilon_{xx} + \varepsilon_{yy} + \varepsilon_{zz} \, ,$$

$$(1.6) \qquad = \frac{\partial u}{\partial x} + \frac{\partial v}{\partial y} + \frac{\partial w}{\partial z}$$

is the dilatation and represents the change in volume of a unit cube, μ and λ are elastic constants known as Lamé constants, u, v and w are displacements along the x, y and z axes, and ϱ is density.

Let us now consider the boundary to be the xy-plane and the positive z-direction towards the medium interior. For a wave to propagate along this plane, it must satisfy the condition that the boundary must be free from stress. If the plane wave propagates along the x-direction, the displacements will be independent of y and two potential functions, φ and ψ, may be defined such that

(1.7)
$$u = \frac{\partial \varphi}{\partial x} + \frac{\partial \psi}{\partial z}$$

and

(1.8)
$$w = \frac{\partial \varphi}{\partial z} - \frac{\partial \psi}{\partial x} .$$

If we use the eqs. (1.7) and (1.8), the dilatation $\varDelta$ is given by

(1.9)
$$\varDelta = \frac{\partial u}{\partial x} + \frac{\partial w}{\partial z} = \nabla^2 \varphi ,$$

and the rotation in the xz-plane, $\bar{\omega}_y$, by

(1.10)
$$2\bar{\omega}_y = \frac{\partial u}{\partial z} - \frac{\partial w}{\partial x} = \nabla^2 \psi ,$$

which separates the effects of dilatation and rotation in the medium.

By substituting eqs. (1.7) and (1.8) in eqs. (1.1) and (1.4) respectively, we obtain

(1.11)
$$\varrho \frac{\partial}{\partial x} \left(\frac{\partial^2 \varphi}{\partial t^2} \right) + \varrho \frac{\partial}{\partial z} \left(\frac{\partial^2 \psi}{\partial t^2} \right) = (\lambda + 2\mu) \frac{\partial}{\partial x} (\nabla^2 \varphi) + \mu \frac{\partial}{\partial z} (\nabla^2 \psi)$$

and

(1.12)
$$\varrho \frac{\partial}{\partial z} \left(\frac{\partial^2 \varphi}{\partial t^2} \right) - \varrho \frac{\partial}{\partial x} \left(\frac{\partial^2 \psi}{\partial t^2} \right) = (\lambda + 2\mu) \frac{\partial}{\partial z} (\nabla^2 \varphi) - \mu \frac{\partial}{\partial x} (\nabla^2 \psi) .$$

Equations (1.11) and (1.12) will be satisfied if

(1.13)
$$\frac{\partial^2 \varphi}{\partial t^2} = [(\lambda + 2\mu)/\varrho]\nabla^2 \varphi = C_1^2 \nabla^2 \varphi$$

and

(1.14)
$$\frac{\partial^2 \psi}{\partial t^2} = [\mu/\varrho]\nabla^2 \psi = C_2^2 \nabla^2 \psi ,$$

where C_1 is the velocity of the propagation of the bulk dilatation or longitudinal waves, *i.e.*

(1.15) $C_1 = [(\lambda + 2\mu)/\varrho]^{\frac{1}{2}}$,

and C_2 is the velocity of propagation of bulk distortion or transverse waves, *i.e.*

(1.16) $C_2 = (\mu/\varrho)^{\frac{1}{2}}$.

For a sinusoidal wave propagating along the X-direction with velocity C, wavelength $2\pi/f$ and frequency $p/2\pi$, the following solutions to eqs. (1.13) and (1.14) could be tried:

(1.17) $\varphi = F(z) \exp\left[i(pt - fx)\right]$

and

(1.18) $\psi = G(z) \exp\left[i(pt - fx)\right]$,

where $F(z)$ and $G(z)$ are functions which describe the behavior of the amplitude of the waves below the surface.

By substitution of (1.17) in (1.13)

(1.19) $-\dfrac{p^2}{C_1^2} F(z) = -f^2 F(z) + \dfrac{d^2 F(z)}{dz^2}$,

i.e.

$$\dfrac{d^2 F(z)}{dz^2} - \left(f^2 - \dfrac{p^2}{C_1^2}\right) F(z) = 0 ,$$

which can also be written, if $h = p/C_1$, as

(1.20) $\dfrac{d^2 F(z)}{dz^2} - (f^2 - h^2) F(z) = 0 .$

The solution of this equation in general form is

(1.21) $F(z) = A \exp\left[-(f^2 - h^2)z\right] + A' \exp\left[(f^2 - h^2)z\right]$,

which, if $q^2 = f^2 - h^2$, becomes

(1.22) $F(z) = A \exp\left[-qz\right] + A' \exp\left[qz\right]$.

Since the second term in (1.21) increases with increasing z, *i.e.* the amplitude grows with depth, unlike a Rayleigh wave, the constant A' must be zero and (1.22) reduces to

(1.23) $F(z) = A \exp\left[-qz\right]$.

The expression for $G(z)$ which can be similarly derived using eqs. (1.14) and (1.18) is

(1.24) $$G(z) = B \exp[-sz],$$

where

(1.25) $$s^2 = f^2 - K^2 \quad \text{and} \quad K = p/C_2.$$

Substituting (1.23) and (1.24) respectively in (1.17) and (1.18) to obtain values for φ and ψ results in

(1.26) $$\varphi = A \exp[-qz + i(pt - fx)]$$

and

(1.27) $$\psi = B \exp[-sz + i(pt - fx)].$$

The boundary condition for surface waves to propagate is that the surface shall be stress free, *i.e.* the stress components σ_{zz}, σ_{zx}, σ_{zy} become zero when $z = 0$. By substituting for φ and ψ from (1.26) and (1.27) in the expressions for σ_{zz} and σ_{zx} which are

(1.28) $$\sigma_{zz} = \lambda \Delta + 2\mu \frac{\partial w}{\partial z}$$

and

(1.29) $$\sigma_{zx} = \mu \left(\frac{\partial u}{\partial z} + \frac{\partial w}{\partial x} \right),$$

and some mathematical manipulation [2], the following equation can be derived:

(1.30) $$K_1^6 - 8K_1^4 + (24 - 16\alpha_1^2) K_1^2 + (16\alpha_1^2 - 16) = 0,$$

where

(1.31) $$\alpha_1^2 = \frac{\mu}{\lambda + 2\mu} = \frac{1 - 2\nu}{2 - 2\nu},$$

(1.32) $$K_1 = K/f$$

and $\nu =$ Poisson's ratio.

Equation (1.30) may be solved numerically if the value of Poisson's ratio is known for the material. Now, from (1.32) and (1.25)

(1.33) $$K_1 = K/f = p/fC_2,$$

thus K_1 gives the ratio of the propagation velocities for Rayleigh and bulk shear waves in the medium, since p/f is the velocity of surface waves and C_2 is the velocity of bulk distortion or shear waves from eq. (1.16). Hence, the velocity of propagation of Rayleigh waves is independent of frequency, $p/2\pi$, and is solely dependent upon the elastic constants of the material. These waves are therefore dispersionless and propagate without change in form.

The decay rate of Rayleigh waves below the surface depends upon q and s, the attenuation factors. From eqs. (1.21) and (1.25) it is clear that

$$(1.34) \qquad \frac{q^2}{f^2} = 1 - \alpha_1^2 K_1^2$$

and

$$(1.35) \qquad \frac{s^2}{f^2} = 1 - K_1^2 .$$

Thus, if the value of K_1 is known, the values of q/f and s/f can be calculated. Now, using eqs. (1.7), (1.8), (1.17) and (1.18) and taking the real parts we can obtain the following expressions for the displacements u and w:

$$(1.36) \qquad u = Af\left[\exp\left[-qz\right] - \frac{2qs \exp\left[-sz\right]}{s^2 + f^2}\right] \sin\left(pt - fx\right)$$

and

$$(1.37) \qquad w = Aq\left[\exp\left[-qz\right] - \frac{2f^2 \exp\left[-sz\right]}{s^2 + f^2}\right] \cos\left(pt - fx\right) .$$

For an isotropic material which has a Poisson's ratio of 0.25 and for which the square of the ratio of the shear to compressional velocities is 0.33, i.e. $\nu = \frac{1}{4}$ and $\alpha_1^2 = \frac{1}{3}$, eq. (1.30) reduces to

$$(1.38) \qquad 3K_1^6 - 24K_1^4 + 56K_1^2 - 32 = 0 .$$

By factorization this becomes

$$(1.39) \qquad (K_1^2 - 4)(3K_1^4 - 12K_1^2 + 8) = 0 ,$$

which has solutions

$$(1.40) \qquad K_1^2 = 4 , \qquad 2 + \frac{2}{\sqrt{3}} \quad \text{or} \quad 2 - \frac{2}{\sqrt{3}} .$$

The first two solutions yield negative values of the attenuation factors q and s and hence do not describe a Rayleigh wave, but the third solution yields a value of K_1 of approximately 0.9. Hence, in an isotropic material, which has a Poisson's ratio of 0.25, the velocity of propagation of Rayleigh waves will be about 0.9 times the velocity of the distortion or shear waves.

Using eqs. (1.34) and (1.35) and the value of $K_1 = 2 - 2/\sqrt{3}$ from eq. (1.40), we see that the calculated values of q/f and s/f are respectively 0.8475 and 0.3933. The rate at which the dilatational or compressional component of the Rayleigh wave attenuates with depth below the surface for this material is given by the factor

$$(1.41) \qquad \exp[-qz] - \frac{2qs\exp[-sz]}{s^2 + f^2}$$

from eq. (1.36).

This factor reduces to

$$(1.42) \qquad \exp[-0.8475fz] - 0.5773\exp[-0.3933fz]$$

by using the numerical values of q/f and s/f calculated above. A curve of the amplitude of the compressional component in this material as a function of depth below the surface in terms of wavelengths can now be derived from (1.42) by allowing fz to assume suitable values. Equation (1.42) decreases rapidly for increasing values of fz, and reaches zero when $fz = 1.21$. The displacement u is thus zero for all values of x and t, resulting in a plane where there is no com-

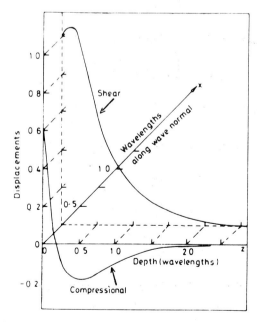

Fig. 1. – Amplitude of displacement of the shear and the compressional component of a Rayleigh wave as a function of depth below the surface in an isotropic medium.

pressional displacement, *i.e.* at a depth 0.193 of a wavelength, since f was defined previously as $2\pi/\Lambda$, where Λ is the wavelength of the Rayleigh wave. At depths greater than 0.193Λ the value of u once again becomes finite but of opposite sign. As indicated in Fig. 1, below the zero-displacement plane u reaches a maximum value at about half a wavelength then decreases exponentially to zero within two wavelengths.

The behavior of w, the displacement amplitude of the distortion or shear component below the surface, can be calculated from eq. (1.37) by the factor

$$(1.43) \qquad \exp\left[-qz\right] - \frac{2f^2 \exp\left[-sz\right]}{s^2 + f^2} .$$

If we use the numerical values for q/f and s/f appropriate for this isotropic material as before, eq. (1.43) becomes

$$(1.44) \qquad \exp\left[-0.8475fz\right] - 1.7321 \exp\left[0.3933fz\right] .$$

As this factor does not change sign, unlike the compressional component, there is no finite depth at which the displacement w becomes zero. Assigning fz different values as before results first in an increase in w with increasing depth, reaching a maximum at a depth of $0.076\,\Lambda$, and then in an exponential decrease, reaching a value of 0.19 of its surface value at a depth of one wavelength. This type of behavior is also shown in Fig. 1. From eqs. (1.36) and (1.37) it is clear that the phases of the compressional and shear displacement components are $\pi/2$ radians or a quarter of a wavelength apart. The variation of displacements of the two components with depth and distance along the wave normal, as expressed in eqs. (1.36) and (1.37), is illustrated in Fig. 2. Another way of

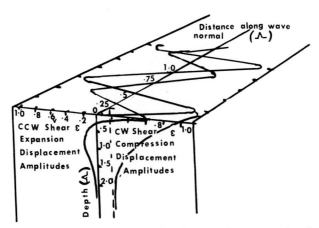

Fig. 2. – Variation of displacements of the shear and compressional components of a Rayleigh wave with depth below the surface and along the direction of propagation.

presenting the particle motion in a Rayleigh wave is shown schematically in Fig. 3. In this presentation the points of maximum shear and compression are indicated at the surface, and it will be seen that they are spatially separated by a quarter of wavelength.

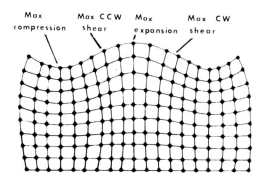

Fig. 3. – Particle motion in a Rayleigh wave shown schematically. (The wave is moving from left to right.)

2. – Rayleigh waves in anisotropic media.

In an unbounded anisotropic medium three types of elastic waves may be propagated, *viz.* one longitudinal and two transverse. If a boundary surface is present in this anisotropic medium, a fourth type of wave may be propagated in certain directions along the surface. This type of surface wave, a Rayleigh wave, is more complicated than its counterpart in isotropic media. The velocity of propagation of bulk waves varies with direction in the crystalline medium, and in addition the direction of energy flow does not in general coincide with the wave normal. Only in certain specific crystallographic directions do pure-mode directions of propagation exist.

2'1. *General relations between acoustic velocities and elastic constants.* – For the propagation of plane elastic waves in a solid it is possible to derive[3] the following three equations relating the displacement vector (A_x, A_y, A_z), the wave normal (l, m, n), the acoustic velocity (v) and the elastic constants (c_{ij}):

$$(2.1) \qquad (A - \varrho v_i^2) A_x + \alpha\beta A_y + \gamma\alpha A_z = 0 \,,$$

$$(2.2) \qquad \alpha\beta A_x + (B - \varrho v_i^2) A_y + \beta\gamma A_z = 0 \,,$$

$$(2.3) \qquad \gamma\alpha A_x + \beta\gamma A_y + (C - \varrho v_i^2) A_z = 0 \,,$$

where

(2.4) $A = l^2 c_{11} + m^2 c_{66} + n^2 c_{55} + 2mn c_{56} + 2nl c_{15} + 2lm c_{16}$,

(2.5) $B = l^2 c_{66} + m^2 c_{22} + n^2 c_{44} + 2mn c_{24} + 2nl c_{46} + 2lm c_{26}$,

(2.6) $C = l^2 c_{55} + m^2 c_{44} + n^2 c_{33} + 2mn c_{34} + 2nl c_{35} + 2lm c_{45}$,

(2.7) $\beta\gamma = l^2 c_{56} + m^2 c_{24} + n^2 c_{34} + mn(c_{23} + c_{44}) + nl(c_{36} + c_{45}) + lm(c_{25} + c_{46})$,

(2.8) $\gamma\alpha = l^2 c_{15} + m^2 c_{46} + n^2 c_{35} + mn(c_{36} + c_{45}) + nl(c_{13} + c_{55}) + lm(c_{14} + c_{56})$,

(2.9) $\alpha\beta = l^2 c_{16} + m^2 c_{26} + n^2 c_{45} + mn(c_{25} + c_{46}) + nl(c_{14} + c_{56}) + lm(c_{12} + c_{66})$.

The conditions for nonzero solutions of (A_x, A_y, A_z) is given by the determinant

(2.10)
$$\begin{vmatrix} A - \varrho v_i^2 & \alpha\beta & \gamma\alpha \\ \alpha\beta & B - \varrho v_i^2 & \beta\gamma \\ \gamma\alpha & \beta\gamma & C - \varrho v_i^2 \end{vmatrix} = 0 .$$

Equation (2.10) is a cubic equation in ϱv^2, indicating that, in general, in an unbounded medium there are three possible velocities of elastic-wave propagation associated with any wave normal (l, m, n). Thus, the velocity surfaces for these three modes of propagation are represented by eq. (2.10). In general, the direction (A_x, A_y, A_z) of the displacement vector of each mode is neither coincident with nor orthogonal to the wave normal (l, m, n). It can also be shown [3] that the displacement vectors of the three possible modes associated with each wave normal form an orthogonal triad.

In order to find those directions along which the displacement vector and the wave normal are parallel or orthogonal, i.e. pure-uncoupled-mode propagation directions, it is necessary to compute the velocity surfaces for each of the three modes. Along these desired directions, the normal to the tangent plane to the velocity surface and the radius vector or wave normal are parallel.

2˙2. Relations between velocity and elastic constants for hexagonal materials. – As an example the general equations for hexagonal materials will be derived using the previous Section. The velocity surfaces for a specific material, e.g. ZnO, will be computed and plotted. From Table I it will be seen that the only nonzero elastic constants are c_{11}, c_{12}, c_{13}, $c_{22} = c_{11}$, $c_{23} = c_{13}$, c_{33}, c_{44}, $c_{55} = c_{44}$, $c_{66} = \frac{1}{2}(c_{11} - c_{12})$. By retaining only these nonzero elastic constants in eqs. (2.4) through (2.9), remembering that

(2.11) $l^2 + m^2 + n^2 = 1$,

TABLE I. – *The c_{ij}-matrices for the different crystal classes* [4].

Key to notation

· zero component,

• nonzero component,

•——• equal components,

○ the numerical equal to the heavy-dot component to which it is joined,

•——○ components numerically equal, but of opposite sign,

× $\frac{1}{2}(c_{11}-c_{12})$.

All the matrices are symmetric about the leading diagonal.

The number in parentheses gives the number of independent moduli.

Triclinic

Classes 1 and $\bar{1}$

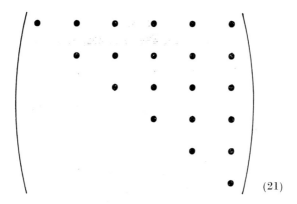

(21)

Monoclinic

Classes 2, m, $2/m$

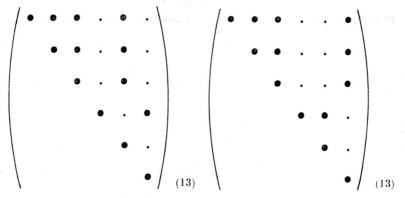

(13)   (13)

dyad $\| x_2$ (standard orientation)   dyad $\| x_3$

T<small>ABLE</small> I (*continued*).

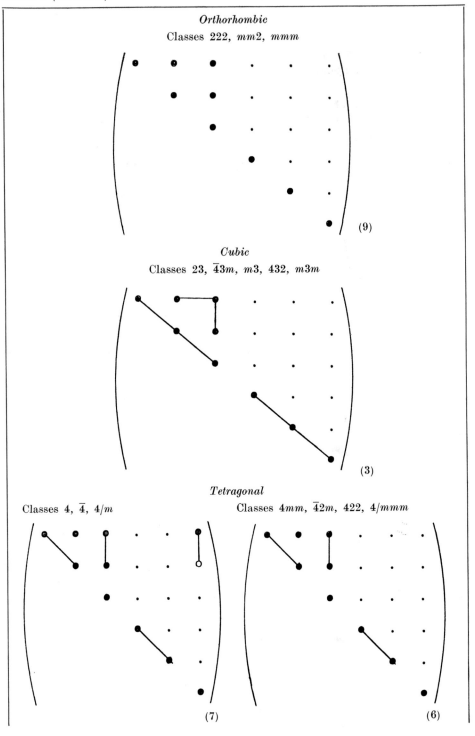

TABLE I (*continued*).

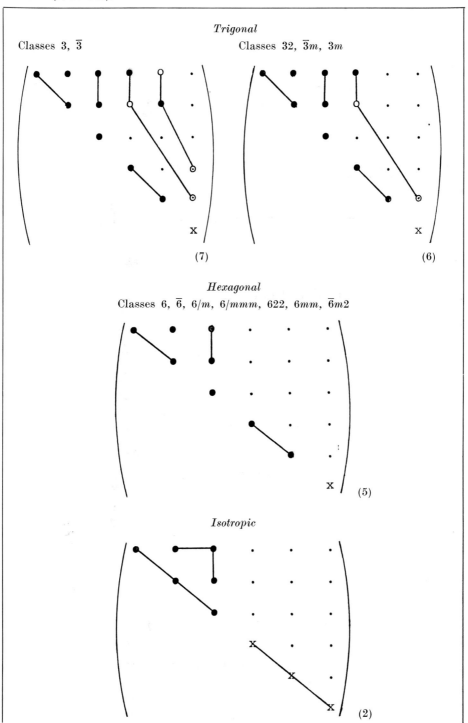

and writing

(2.12)
$$a = c_{11} - c_{44},$$

(2.13)
$$c = c_{11} - c_{12} - 2c_{44},$$

one can see that equation (2.4) reduces to

(2.14)
$$A = l^2 a + m^2 \frac{c}{2} + c_{44}.$$

Similarly eq. (2.5) becomes

(2.15)
$$B = \frac{l^2}{2}(c_{11} - c_{12}) + m^2 c_{11} + n^2 c_{44} = l^2 \frac{c}{2} + m^2 a + c_{44},$$

and from eq. (2.6)

(2.16)
$$C = l^2 c_{44} + m^2 c_{44} + n^2 c_{33} = n^2 h + c_{44},$$

where

(2.17)
$$h = c_{33} - c_{44}.$$

Likewise from eq. (2.7)

(2.18)
$$\beta\gamma = mn(c_{13} + c_{44}) = mnd,$$

where

(2.19)
$$d = c_{13} + c_{44}.$$

From eq. (2.7) one obtains

(2.20)
$$\gamma\alpha = nl(c_{13} + c_{44}) = nld,$$

and from eq. (2.8)

(2.21)
$$\alpha\beta = lm(c_{12} + c_{66}) = lm\left(a - \frac{c}{2}\right).$$

By substituting these values for A, B, C, $\alpha\beta$, $\beta\gamma$, $\gamma\alpha$ in eqs. (2.1), (2.2), (2.3) and (2.10) and putting

(2.22)
$$\varrho v_i^2 - c_{44} = H,$$

the velocity equation becomes

(2.23)
$$\begin{vmatrix} l^2 a + m^2 \dfrac{c}{2} - H & lm\left(a - \dfrac{c}{2}\right) & nld \\[2ex] lm\left(a - \dfrac{c}{2}\right) & l^2 \dfrac{c}{2} + m^2 a - H & mnd \\[2ex] nld & mnd & n^2 h - H \end{vmatrix} = 0 \,,$$

which can be expanded to the form

$$(2.24) \qquad H^3 - \left\{ n^2 h + (1 - n^2)\left(a + \dfrac{c}{2}\right) \right\} H^2 +$$

$$+ (1 - n^2)\left\{ (1 - n^2)\dfrac{ac}{2} + n^2\left[h\left(a + \dfrac{2}{c}\right) - d^2 \right] \right\} H - n^2(1 - n^2)^2 \dfrac{c}{2}(ah - a^2) = 0 \,.$$

Since the only direction cosine which appears in (2.24) is n, the velocity surfaces are circularly symmetric about the X_3- or Z-axis. Thus, the circles of intersection of the three velocity surfaces with the basal plane can be obtained by allowing n to become zero in eq. (2.24), which then reduces to

$$(2.25) \qquad\qquad H^3 - (a + \tfrac{1}{2}c)H^2 + \tfrac{1}{2}acH = 0 \,.$$

This equation can be factorized as follows:

$$(2.26) \qquad\qquad (H - a)\left(H - \dfrac{c}{2}\right)H = 0 \,,$$

which yields three roots

$$(2.27) \qquad\qquad H_L = a \,,$$

$$(2.28) \qquad\qquad H_{T_1} = \dfrac{c}{2} \,,$$

$$(2.29) \qquad\qquad H_{T_2} = 0 \,,$$

where L refers to the compressional mode, T_1 to the shear mode with displacement vector parallel to the basal plane and T_2 to the shear mode with displacement vector normal to the basal plane as shown in Fig. 4.

By substituting eqs. (2.12), (2.13) and (2.17) in eqs. (2.27), (2.28) and (2.29), one obtains

$$(2.30) \qquad\qquad \varrho v_L^2 = c_{11} \,,$$

$$(2.31) \qquad\qquad \varrho v_{T_1}^2 = \tfrac{1}{2}(c_{11} - c_{12}) \,,$$

$$(2.32) \qquad\qquad \varrho v_{T_2}^2 = c_{44} \,.$$

Hence, if the elastic constants c_{11}, c_{12}, c_{44} and the density ϱ are known, the three velocities in the basal plane can be calculated. Alternatively, the elastic constants c_{11}, c_{12} and c_{44} can be calculated if the velocities of propagation in the basal plane and the density can be measured.

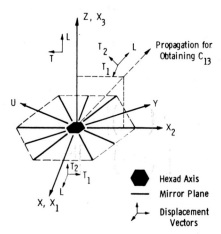

Fig. 4. – Hexagonal ZnO symmetry element.

Equation (2.24) can be expressed as a product of a linear and a quadratic term, thus

$$(2.33) \quad [H - \tfrac{1}{2}c(1 - n^2)][H^2 - \{n^2 h + (1 - n^2) a\} H + n^2 (1 - n^2)(ah - d^2)] = 0 .$$

In order to obtain expressions for the velocities along the X_3- or Z-axis it is only necessary to put $n = 1$ in eq. (2.33). The linear term then becomes

$$(2.34) \quad\quad\quad\quad\quad\quad\quad H_{T_1} = 0 ,$$

and the quadratic term reduces to

$$H^2 - hH = 0 ,$$

i.e.

$$(2.35) \quad\quad\quad\quad\quad\quad\quad H(H - h) = 0 ,$$

from which one obtains the roots

$$(2.36) \quad\quad\quad\quad\quad\quad\quad H_{T_2} = 0$$

and

$$(2.37) \quad\quad\quad\quad\quad\quad\quad H_L = h .$$

Equations (2.34) and (2.36) indicate that the shear modes are degenerate along the Z-axis and by substitution from eqs. (2.17) and (2.22) one obtains for propagation along the Z- or X_3-axis

$$(2.38) \qquad\qquad \varrho v_T^2 = c_{44}$$

and

$$(2.39) \qquad\qquad \varrho v_L^2 = c_{33} \, .$$

Thus, the shear velocity along the X_3- or Z-axis provides a check on the value of c_{44} as derived from eq. (2.32) using the slow shear velocity in the basal plane.

In order to determine the last of the independent elastic constants, *viz.* c_{13}, a value of n must be chosen so that the term involving d in eq. (2.33) does not vanish. If one chooses a direction in the $X_2 X_3$-plane where $l = 0$ such that $n = m = 1/\sqrt{2}$, eq. (2.33) reduces to

$$(2.40) \qquad (H - c/4)\{H^2 - \tfrac{1}{2}(h + a)H + \tfrac{1}{4}(ah - d^2)\} = 0 \, ,$$

which has roots

$$(2.41) \qquad\qquad H_{T_1} = \frac{c}{4}$$

and

$$(2.42) \qquad H_{L, T_2} = \tfrac{1}{2}[\tfrac{1}{2}(a + h) \pm \tfrac{1}{4}(a + h)^2 - (ah - d^2)]^{\frac{1}{2}} \, .$$

Thus, one obtains the velocity equation for propagation in the $X_2 X_3$-plane at $45°$ to the X_3-axis by substitution from the appropriate equations

$$(2.43) \qquad\qquad \varrho(v_{T_1})^2 - c_{44} = \tfrac{1}{4}(c_{11} - c_{12} - 2c_{44})$$

or

$$(2.44) \qquad\qquad \varrho(v_{T_1})^2 = \tfrac{1}{4}(c_{11} - c_{12} + 2c_{44}) \, .$$

From eq. (2.42)

$$(2.45) \qquad \varrho(v_L)^2 = \tfrac{1}{4}(c_{11} + c_{33} + 2c_{44}) + \tfrac{1}{4}\{(c_{11} - c_{33})^2 + 4(c_{13} + c_{44})^2\}^{\frac{1}{2}}$$

and

$$(2.46) \qquad \varrho(v_{T_2})^2 = \tfrac{1}{4}(c_{11} + c_{33} + 2c_{44}) - \{(c_{11} - c_{33})^2 + 4(c_{13} + c_{44})^2\}^{\frac{1}{2}} \, .$$

Thus, propagation in any direction at $45°$ to the X_3-axis yields two equations for determining c_{13} and a check for c_{11}, c_{12} and c_{44} and indirectly for c_{33}.

The curves of intersection of the velocity surfaces of the L, T_1 and T_2 modes with any plane containing the X_3-axis can now be determined by allowing n in eq. (2.33) to assume all values from $+1$ to -1, and substituting the computed values of c_{11}, c_{12}, c_{13}, c_{33} and c_{44} in this equation.

The preceding derivations are correct for nonpiezoelectric hexagonal materials, *i.e.* for all materials belonging to crystal classes $6/m$ and $6/mmm$. For all other hexagonal crystal classes, as they have piezoelectric properties, correction terms must be added to eqs. (2.4) through (2.9) to allow for piezoelectric stiffening.

As an example the correction terms for crystal class $6mm$ are given below. These involve both piezoelectric and dielectric matrices. The dielectric matrix for class $6mm$ is given [4] in eq. (2.47):

$$(2.47) \qquad \xi = \begin{vmatrix} \xi_{11} & 0 & 0 \\ 0 & \xi_{11} & 0 \\ 0 & 0 & \xi_{33} \end{vmatrix},$$

and the piezoelectric matrix [4] in (2.48):

$$(2.48) \qquad e_{ij} = \begin{vmatrix} 0 & 0 & 0 & 0 & e_{15} & 0 \\ 0 & 0 & 0 & e_{15} & 0 & 0 \\ e_{31} & e_{31} & e_{33} & 0 & 0 & 0 \end{vmatrix}.$$

From eq. (2.47) the dielectric correction factor can be written

$$(2.49) \qquad \xi_{rs} k_r k_s = \xi_{11} l^2 + \xi_{11} m^2 + \xi_{33} n^2 = \xi_{11}(l^2 + m^2) + \xi_{33} n^2 = S_\xi.$$

From eq. (2.48) the piezoelectric correction factors can be obtained. Thus, eq. (2.4) becomes

$$(2.50) \qquad A' = A + S_1^2/S_\xi,$$

where

$$(2.51) \qquad S_1^2 = nl(e_{15} + e_{31}),$$

and S_ξ is given by eq. (2.49).

Likewise eq. (2.9) becomes

$$(2.52) \qquad (\alpha\beta)' = \alpha\beta + S_1 S_2/S_\xi,$$

where

$$(2.53) \qquad S_2 = mn(e_{24} + e_{32}) = mn(e_{15} + e_{32}),$$

since from (2.48) $e_{24} = e_{15}$.

Also, eq. (2.8) becomes

(2.54) $$(\gamma\alpha)' = \gamma\alpha + S_1 S_3/S_\xi,$$

where

(2.55) $$S_3 = (l^2 + m^2)\,e_{15} + n^2\,e_{33},$$

and eq. (2.5) becomes

(2.56) $$B' = B + S_2^2/S_\xi.$$

Similarly, eq. (2.7) becomes

(2.57) $$(\beta\gamma)' = \beta\gamma + S_2 S_3/S_\xi,$$

and eq. (2.6) becomes

(2.58) $$C' = C + S_3^2/S_\xi.$$

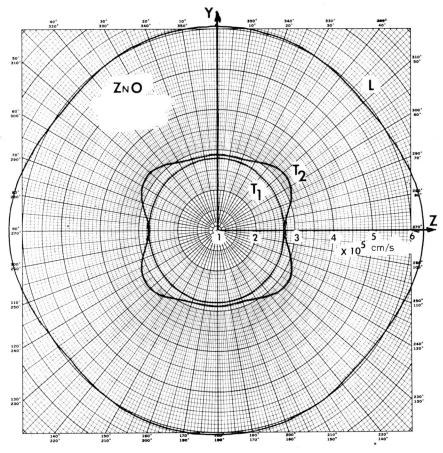

Fig. 5. – Curves of intersection of any plane containing the Z-axis with the velocity surfaces in ZnO.

Hence, the conditions for nonzero solutions of (A_x, A_y, A_z) for class $6mm$ materials are given by the determinant

(2.59)
$$
\begin{vmatrix}
A' - \varrho v_i^2 & (\alpha\beta)' & (\gamma\alpha)' \\
(\alpha\beta)' & B' - \varrho v_i^2 & (\beta\gamma)' \\
(\gamma\alpha)' & (\beta\gamma)' & C' - \varrho v_i^2
\end{vmatrix} = 0 .
$$

The velocity surfaces for ZnO have been calculated by substituting the unique ZnO values of elastic, piezoelectric and dielectric constants in the solutions

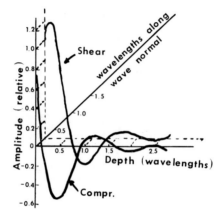

Fig. 5a. – Amplitudes of displacement of the shear and the compressional components of a Rayleigh wave in an anisotropic medium.

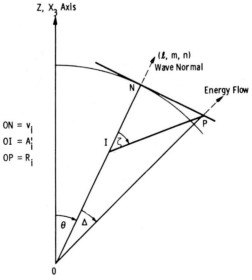

Fig. 6. – Relations between wave normal and energy flow directions for bulk waves in a crystalline material.

of these equations. Figure 5 shows the intersection of any plane continuing the Z-axis with the three velocity surfaces of ZnO. It is clear from Fig. 5, that, apart from the basal plane in which all three modes are pure for all directions, the only direction along which all three bulk modes can propagate as pure modes is along the X_3- or Z-axis.

2'3. *Wave surfaces*. – It is necessary to compute the bulk wave surfaces in order to determine the directions along which pure Rayleigh waves may propagate, since both shear and compressional components must be pure modes. Continuing with the example of hexagonal crystals, the curves of intersection of the wave surfaces with any plane containing the Z-axis are loci of points R such that

$$R_i^2 = \frac{(v_i - D_i)^2}{(\cos \zeta_i)^2} + 2D_i v_i - D_i^2$$

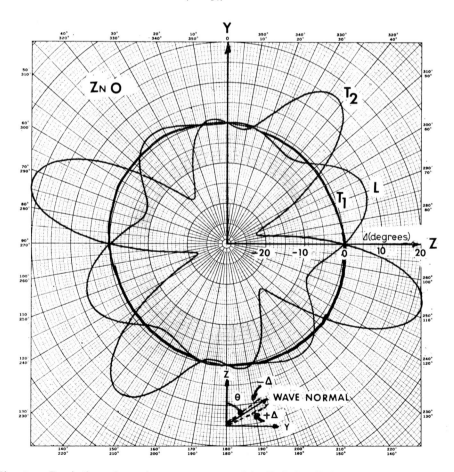

Fig. 7. – Deviation of ray from wave normal in ZnO for L, T_1 and T_2 modes in any plane containing the Z-axis.

or

$$(2.60) \qquad R_i^2 = \left(\frac{H_i}{\varrho v_i \cos \zeta_i}\right)^2 + 2 D_i v_i - D_i^2 ,$$

where $i = L, \ T_1, \ T_2,$

$$(2.61) \qquad \cos \zeta_i = \left\{\frac{m^2 n^4 d^4}{[(H_i - m^2 a)^2 + m^2 n^2 d^2]^2} + \frac{(H_i - m^2 a)^4}{n^2[m^2 n^2 d^2 + (H_i - m^2 a)^2]^2}\right\}^{\frac{1}{2}}$$

$$(2.62) \qquad D_i = \frac{c_{44}}{\varrho v_i} ,$$

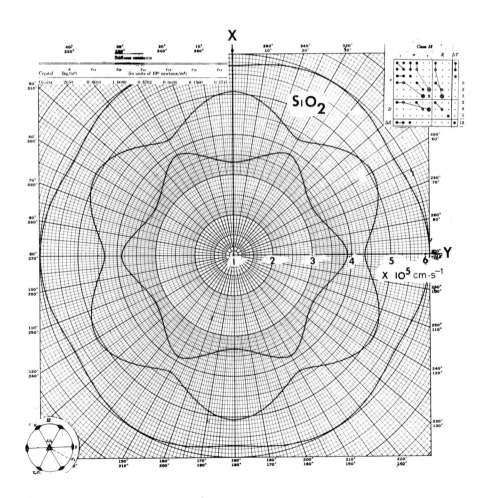

Fig. 8a. − Curves of intersection of the velocity surfaces with the XY plane of quartz.

and H_i is as defined in eq. (2.22). The parameters R_i, ζ_i, D_i and v_i are as defined in Fig. 6. The angle $\varDelta_i$ between the wave normal and the direction of energy flow is defined by

$$(2.63) \qquad\qquad \operatorname{tg} \varDelta_i = \frac{v_i - D_i}{v_i} \operatorname{tg} \zeta_i \ .$$

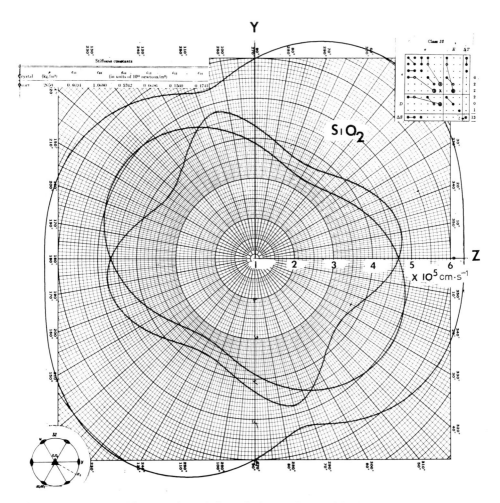

Fig. 8b. – Curves of intersection of the velocity surfaces with the YZ plane of quartz.

Figure 7 shows how the ray direction, or energy flow, deviates from the overall normal for each of the modes L, T_1, T_2 as a function of θ, which is the angle between the Z-axis and the wave normal in any plane containing the Z-axis for ZnO.

In a similar manner the velocity and wave surfaces can be determined for crystalline materials belonging to crystalline symmetries other than hexagonal. From the equations of motion the pure-mode directions for each bulk mode

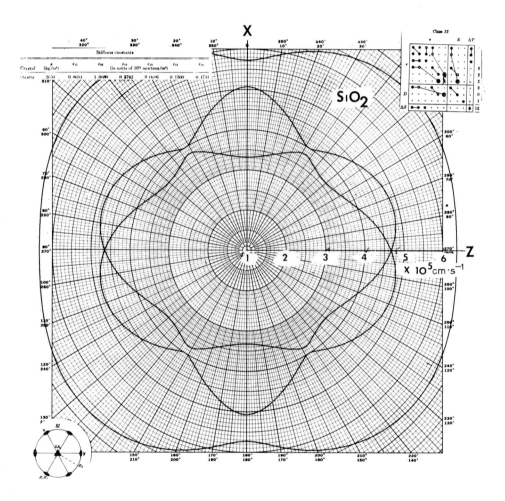

Fig. 8c. – Curves of intersection of the velocity surfaces with the ZX plane of quartz.

can be determined and hence used in the analysis necessary to determine crystallographic planes and directions which will support pure mode Rayleigh waves. The velocity surfaces for quartz, lithium-niobate and bismuth germanium oxide, the three most frequently used materials for surface wave devices, are given in Fig. 8, 9 and 10. These surfaces will be used to determine the most suitable orientations of these materials for surface wave applications.

2'4. *Surface waves in anisotropic media.* – The equations for surface wave propagation [12] in anisotropic and piezoelectric materials can be obtained by substituting the solutions to eqs. (2.10) and (2.59) respectively into the boundary

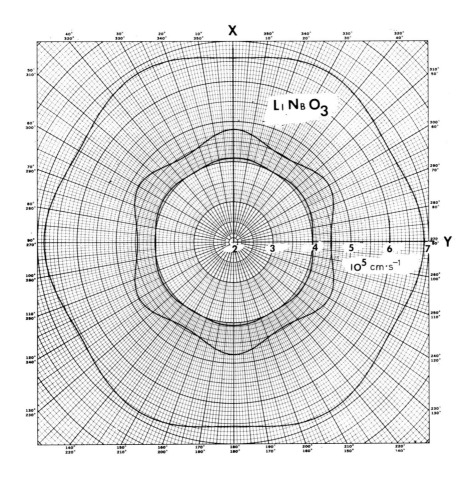

Fig. 9a. – Curves of intersection of the velocity surfaces with the XY plane of lithium-niobate.

conditions for Rayleigh-wave propagation in these media. For a nonpiezo-electric anisotropic medium the boundary conditions are that the boundary surface shall be stress free. For piezoelectric media two added conditions are that the normal component of electric displacement and the tangential electric field shall be continuous. Solutions to the equations of motion for all aniso-tropic media generally must be obtained numerically [15]. There are certain significant differences between Rayleigh-wave propagation in isotropic and anisotropic media, due to the more complex physical nature of crystalline media. Whereas the particle displacements in an isotropic medium are always in the

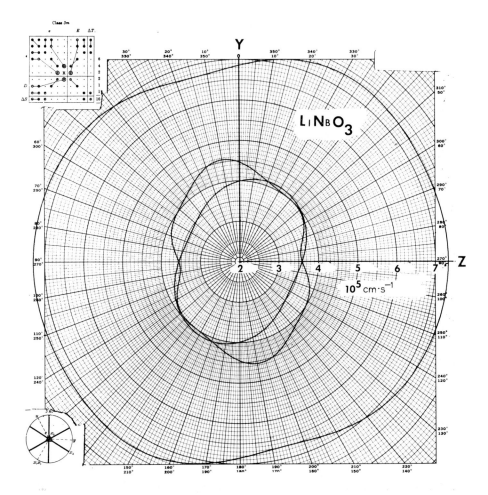

Fig. 9b. – Curves of intersection of the velocity surfaces with the YZ plane of lithium-niobate.

sagittal plane, *i.e.* the plane contained by the surface and wave normals, in anisotropic media, if the sagittal plane is not a symmetry plane, the particle displacement may contain a third orthogonal component. In this case the retrograde elliptical particle motion will be oblique rather than normal to the surface. As Rayleigh waves have two components of displacement, *viz.* compressional and shear, the directions of the wave normal and energy flow are not usually parallel in anisotropic media (see Subsect. 2·3). Thus, with the exception of pure-mode directions, the phase and group velocities are different.

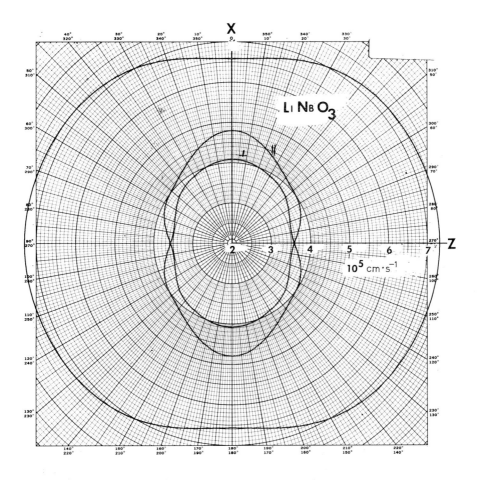

Fig. 9c. – Curves of intersection of the velocity surfaces with the ZX plane of lithium-niobate.

As the basal plane of an hexagonal material has elastic isotropy, Rayleigh waves on such a plane have properties identical to those on isotropic media. The attenuation factors for the two displacement components in anisotropic media may be complex and hence could exhibit an oscillatory behavior with depth below the surface, such as that shown in Fig. 5a. The rates of decay and oscillation are determined by the elastic constants of the medium. Along certain orientations where the direction of energy flow of one or the other component is not in the surface plane, the Rayleigh wave tends to propagate as a « leaky wave », *i.e.* the energy leaks away into the bulk.

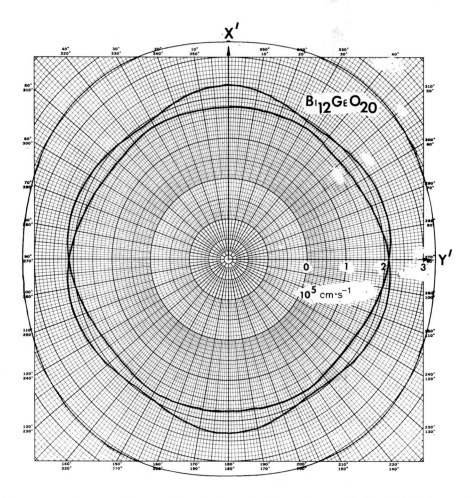

Fig. 10*a*. – Curves of intersection of the velocity surfaces with the $X'Y'$ plane of bismuth-germanium oxide.

Surface waves propagating in a piezoelectric medium are usually accompanied by a traveling electric field (see Sect. **3**), which is generated by the propagating stress waves coupling to the direct piezoelectric effect defined in eq. (3.2). These fields extend above the surface of the piezoelectric medium as well as down into the bulk. An approximate formula [5, 12] for the strength of coupling between the electromagnetic and mechanical energies in a surface

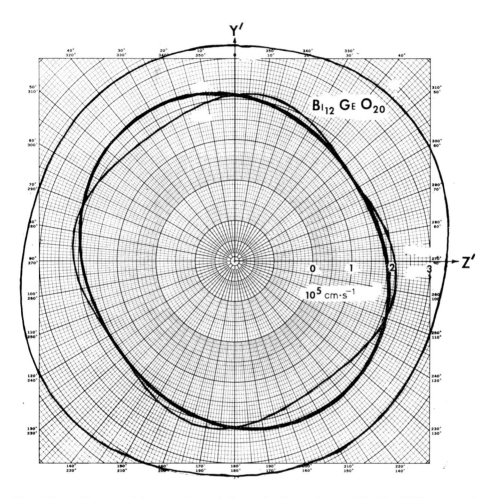

Fig. 10b. – Curves of intersection of the velocity surfaces with the $Y'Z'$ plane of bismuth-germanium oxide.

wave can be written

(2.64) $$k^2 = 2(\Delta v/v_0).$$

In this formula v_0 is the surface wave velocity with a metallized surface, *i.e.* with the external field shorted out, and Δv is the change in velocity when the metallization has been removed.

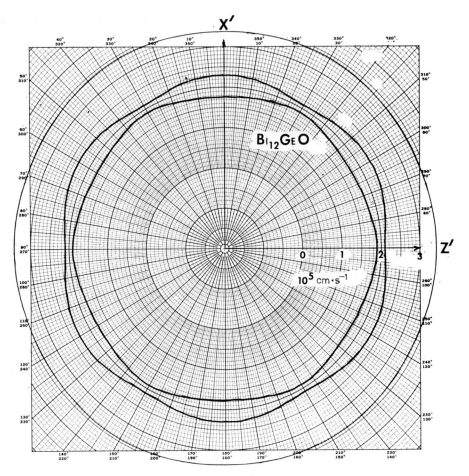

Fig. 10c. – Curves of intersection of the velocity surfaces with the $Z'X'$ plane of bismuth-germanium oxide.

3. – Generation and detection of Rayleigh waves.

Many different methods of launching Rayleigh waves have been described in the literature [5, 6]. One fairly frequently used method employs a wedge which is acoustically bonded to the surface of the material along which the Rayleigh waves are to be propagated. This arrangement, as illustrated in Fig. 11, uses a piezoelectric transducer on the sloping face of the wedge for launching compressional waves into the wedge. The material and angle θ of the wedge, for optimum excitation, are chosen so that

$$(3.1) \qquad \sin \theta_R = \frac{v_L}{v_R}.$$

Thus, it is seen that the compressional velocity in the wedge must be less than the Rayleigh-wave velocity on the surface of the sample. Plastic materials, which have low bulk wave velocities, are usually used for the wedge to satisfy

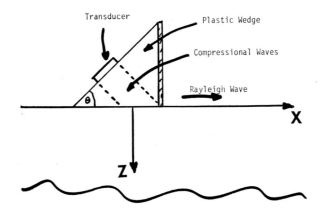

Fig. 11. – Rayleigh-wave generation by means of a wedge.

this requirement. As the angle θ_R is greater than the angle for total internal reflection [2] of both compressional and shear waves, the amplitude of the disturbance transmitted into the sample dies away very rapidly with distance

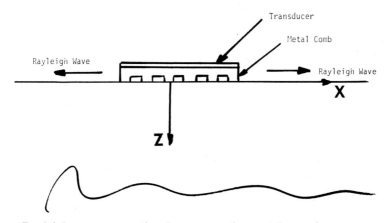

Fig. 12. – Rayleigh-wave generation by means of a metal « comb ».

below the interface of wedge and sample. Hence, a periodic disturbance is created at the wedge sample boundary with a spatial period equal to the Rayleigh wavelength. By this means one Rayleigh wave is launched from the vertical edge of the wedge on the sample surface. Most other methods of generation launch two waves traveling in opposite directions. The level of bulk modes launched by the wedge method of Rayleigh-wave generation is at least 20 to 30 dB below the Rayleigh-wave level.

It is also possible to generate Rayleigh waves by the wedge method using shear waves with displacements in a plane containing the normal to the sample surface and the direction of propagation. In this case the shear velocity in the wedge must be less than the Rayleigh velocity in the sample. Materials such as brass or steel can be used instead of plastics. This would have some advantages over plastics as machining would be easier and better matching of acoustic impedances of wedge and sample would allow more efficient coupling into the sample.

Another method of generating Rayleigh waves employs a metal comblike structure and a bulk compressional wave transducer as shown in Fig. 12. This type of structure launches Rayleigh waves from both edges parallel to the slots in the comb and in addition launches numerous unwanted bulk modes.

All of the above methods of generation only provide one of the two displacements necessary for Rayleigh-wave propagation, *viz.* either shear or compression. Hence, only a small percentage of the acoustic energy generated is converted to Rayleigh waves. Much greater conversion efficiencies are possible when both displacements can be provided with the correct phase relationship. This can readily be achieved by using a piezoelectric sample to which appropriate electric fields can be applied for coupling to suitable piezoelectric moduli of the sample. Figure 13 shows how electric fields can be generated at and below the surface of a piezoelectric material. Part *a*) shows the geometry of the grid, part *b*) the electric field distribution within the material and part *c*) the phase relationship between E_1, the electric-field component along the X_1-axis, and E_2, the electric-field component along the X_2-axis. It is now only necessary to choose a piezoelectric material suitably oriented to allow the two electric-field components to generate the required strains. These are a shear strain with displacement normal to the surface and a compressional strain parallel to the surface with energy flow for both components along the wave normal. These must be 90° out of phase with each other, as indicated in Fig. 13*d*).

In order to do this it is necessary to use the piezoelectric matrix which relates the applied electric field to the generated strain (converse piezoelectric effect) and the generated electric polarization to the applied stress (direct piezoelectric effect). The generalized expression for the « direct » piezoelectric effect can be written

$$(3.2) \qquad\qquad P_i = d_{ij}\sigma_j \, ,$$

and the generalized expression for the « converse » piezoelectric effect can be written

$$(3.3) \qquad\qquad \varepsilon_j = d_{ij}E_i \, ,$$

where P_i is electric polarization, d_{ij} is piezoelectric modulus, ε_j is strain, σ_j is stress, E_i is electric field, $i = 1, 2, 3$ and $j = 1, 2, ..., 6$. The relationships between the applied stresses and the resultant electric polarizations as

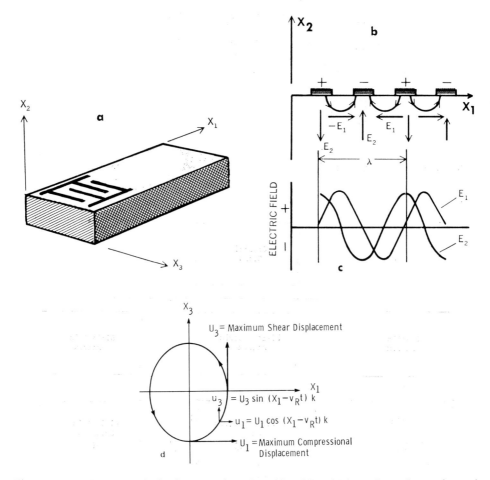

Fig. 13. – a), b), c), method of generating electrid fields within and at the surface of a piezoelectric material by means of a metallic interdigital grid on the surface; d) phase and amplitude relationships between shear and compressional components of a Rayleigh wave.

well those between the applied electric fields and the resultant strains are shown in Fig. 14, the generalized piezoelectric matrix.

An analytical method of determining the most suitable crystallographic planes and directions in those planes for launching and propagating Rayleigh waves is outlined schematically in Fig. 15. The first characteristic to be determined is the crystal class to which the material belongs [7]. Having ascertained this information, the elastic, piezoelectric and dielectric matrices are known [4].

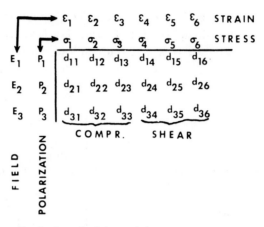

Fig. 14. – The generalized piezoelectric matrix.

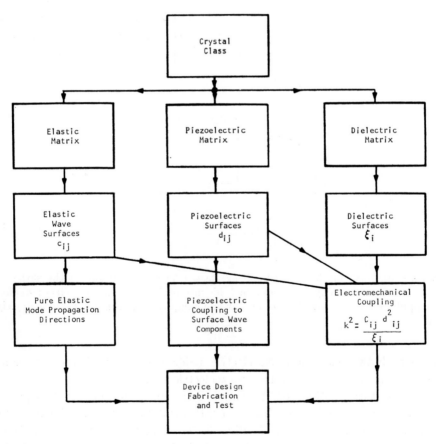

Fig. 15. – Schematic representation of the method of analysis of a crystalline material for surface wave applications.

As shown in Sect. **2**, the elastic matrix yields the appropriate velocity equations, from which the velocity and wave surfaces [3] for bulk waves of the material can be determined, assuming a knowledge of the elastic moduli c_{ij}. It can be seen from these surfaces (see for example Fig. 8 to 10) that in general the three possible bulk modes are not pure modes except along certain unique axes. The wave surfaces can thus be used to determine those directions along which pure bulk modes exist and hence define directions contained in appropriate planes for pure-mode Rayleigh-wave propagation.

The piezoelectric matrix can be used to determine crystallographic planes and directions along which piezoelectrically coupled Rayleigh waves can be launched. When one of these latter directions coincides with an elastic-pure-mode direction for both the compressional and the relevant shear modes, piezo-electrically coupled pure Rayleigh waves can readily be launched, propagated and detected. The piezoelectric matrix also yields piezoelectric surfaces for the material for determining the piezoelectric properties in directions other than pure-mode directions.

In a similar manner the dielectric matrix defines the dielectric surface, which can be used, in conjunction with the elastic and piezoelectric surfaces, to estimate the strength of electromechanical coupling in any desired direction. The relationship between k, the electromechanical coupling coefficient, and the piezoelectric d, elastic c, and dielectric moduli ξ is given by

$$(3.4) \qquad\qquad k_{ij}^2 = c_{ij}\,\frac{d_{ij}^2}{\xi_i}\,.$$

For a chosen crystallographic plane and direction in a piezoelectric material, the piezoelectric matrix can also be used to determine how much of the total converted energy will propagate as Rayleigh-wave energy, as well as the directions and particle displacements of the energy of the concomitantly generated unwanted bulk modes. This procedure facilitates the choice of optimum piezoelectric and elastic directions for Rayleigh-wave propagation on a chosen material.

This analytical method will now be used to determine the best cuts and propagation directions for several materials commonly used in surface wave applications. Figure 8 indicates that pure elastic bulk waves can be propagated along both X_1 and X_3 or Z axes of quartz. Since two strains, *viz.* shear and compression, must be generated and the piezoelectric matrix for quartz (trigonal class 32, see Table II) indicates that no piezoelectric moduli exist along the Z-axis of quartz, the X_1-axis is chosen as the propagation axis and the X_2- or Y-axis as the « cut », or the direction normal to the surface on which Rayleigh waves will be launched. This arrangement is shown in Fig. 13a). The electric fields below the ID grid, shown in Fig. 13b) and c), can be resolved into two components, E_1 parallel to the surface and E_2 perpendicular to the surface, which

TABLE II. – *The d_{ij}-matrices for the piezoelectric crystals of different classes* [4].

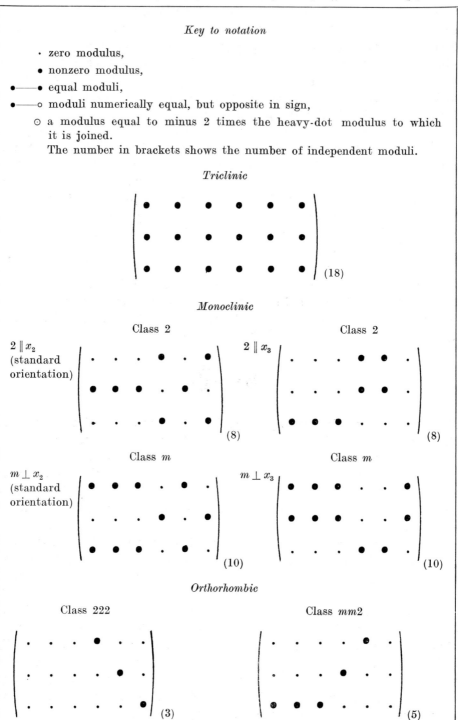

Key to notation

- · zero modulus,
- • nonzero modulus,
- •——• equal moduli,
- •——○ moduli numerically equal, but opposite in sign,
- ⊙ a modulus equal to minus 2 times the heavy-dot modulus to which it is joined.

The number in brackets shows the number of independent moduli.

Triclinic

(18)

Monoclinic

Class 2   Class 2

$2 \parallel x_2$
(standard
orientation)   (8) $2 \parallel x_3$ (8)

Class *m*   Class *m*

$m \perp x_2$
(standard
orientation)   (10) $m \perp x_3$ (10)

Orthorhombic

Class 222   Class *mm2*

(3)   (5)

TABLE II (*continued*).

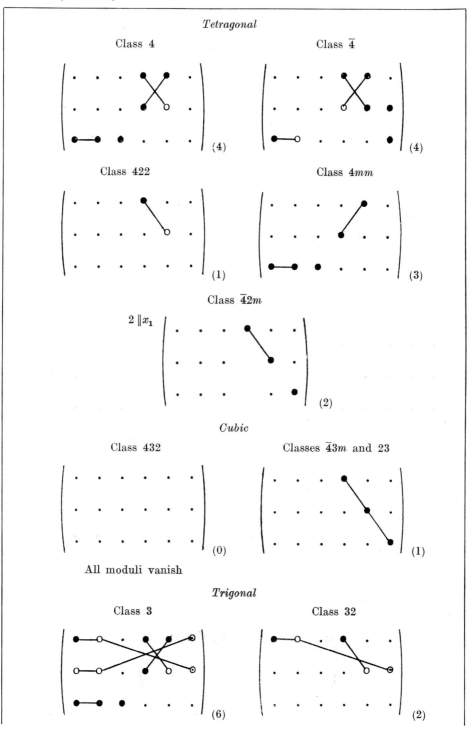

TABLE II (*continued*).

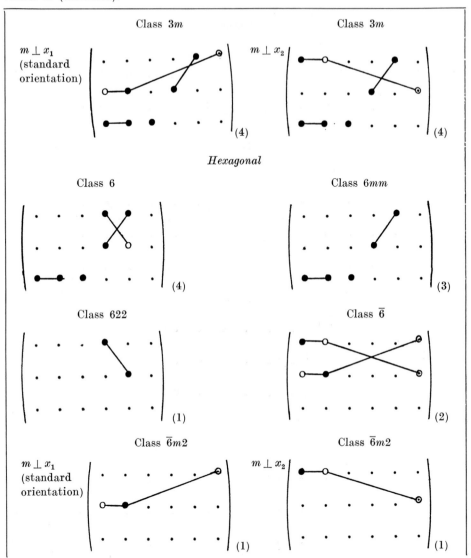

are 90° out of phase with each other. If one writes the piezoelectric matrix for quartz thus

(3.5)
$$\begin{pmatrix} d_{11} & d_{12} & \cdot & d_{14} & \cdot & \cdot \\ \cdot & \cdot & \cdot & \cdot & d_{25} & d_{26} \\ \cdot & \cdot & \cdot & \cdot & \cdot & \cdot \end{pmatrix},$$

it is clear that the E_1 field component will generate the following strains between the ID fingers:

$$(3.6) \qquad \varepsilon_1 = d_{11} E_1, \qquad \varepsilon_2 = d_{12} E_1 \qquad \text{and} \qquad \varepsilon_4 = d_{14} E_1.$$

ε_1 is a compressional strain with displacement and energy flow along the X_1-axis. This strain provides the required compressional component of the Rayleigh wave. The E_2 field component will generate the following strains at the ID fingers:

$$(3.7) \qquad \varepsilon_5 = d_{25} E_2 \qquad \text{and} \qquad \varepsilon_6 = d_{26} E_2.$$

ε_6 provides a shearing strain with particle displacement along the X_2-axis and energy flow along the X_1-axis. This is the required shear component of the Rayleigh wave. Thus, if an alternating potential is applied to the ID grid, alternating elastic strains, ε_1, $-\varepsilon_1$, ε_6, $-\varepsilon_6$, will be generated in the quartz plate. Since the velocity of propagation is the product of the frequency and the wavelength, Rayleigh waves will be most efficiently launched when the distance between adjacent ID fingers of the same electrical polarity is exactly one wavelength.

The strength of coupling for each component of elastic strain can be determined by the expression for the electromechanical coupling coefficient k for bulk waves given [8] in eq. (3.4).

It is clear from eqs. (3.6) and (3.7) that all the energy converted does not

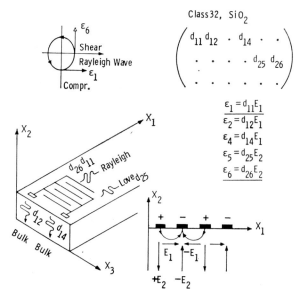

Fig. 16. – Strains generated by an interdigital grid on the X_1X_3-plane of quartz for propagation along X_1.

propagate as a Rayleigh wave, since only ε_1 and ε_6 combine to form the Rayleigh wave. The strains ε_2 and ε_4 respectively propagate as bulk compressional and bulk shear modes as indicated in Fig. 16. The strain ε_5 will propagate as a shear mode at the surface with particle displacement parallel to the X_3-axis and energy flow parallel to the X_1-axis. The coupling to this shear mode, at the Rayleigh-

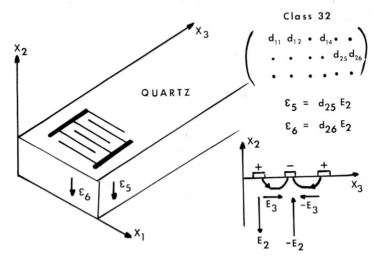

Fig. 17. – Strains generated on the X_1X_3-plane of quartz for propagation along X_3.

wave frequency, will be small as the velocity of this shear mode is much greater than that of the Rayleigh wave (see Fig. 8). Under this condition the strain contributions ε_5 from the N finger pairs of the ID grid will not be in phase, causing partial destructive interference.

Using this method of analysis, one can see that Rayleigh waves cannot be launched along the X_3-axis as neither of the displacement components can be generated for Y-cut Z-propagating quartz. Figure 17 shows which strains can be generated by this orientation of quartz. These are ε_5 and ε_6. Thus

$$(3.8) \qquad \varepsilon_5 = \varepsilon_{31} = \varepsilon_{13} = d_{25}\,E_2$$

and

$$(3.9) \qquad \varepsilon_6 = \varepsilon_{12} = \varepsilon_{21} = d_{26}\,E_2 \,.$$

ε_5 provides shearing about X_2, *i.e.* displacements along X_3 with energy flow along X_1 or displacements along X_1 with energy flow along X_3. ε_6 provides shearing about X_3, *i.e.* displacements along X_2 with energy flow along X_1 or displacements along X_1 with energy flow along X_2. As the required shear displacement for Rayleigh-wave propagation along X_3 on the X_1X_3-plane is a displacement along X_2 with propagation along X_3, *i.e.* $\varepsilon_{23} = \varepsilon_4$, this can only be generated by a field component E_1, which does not exist for the arrangement shown in Fig. 17.

By means of the same analytical method, the most suitable orientations and directions for launching and propagating pure Rayleigh waves in ZnO can be determined. In Sect. **2** it was shown that all three bulk modes in the basal

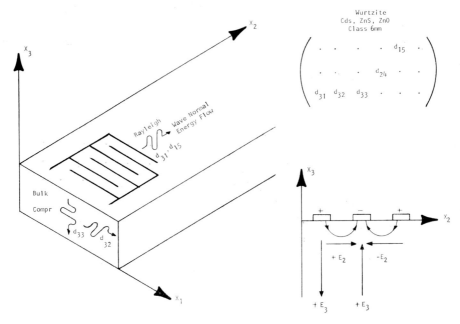

Fig. 18. – Rayleigh-wave generation on the basal plane of ZnO.

plane of ZnO propagate as pure modes. As this material is elastically and piezo-electrically isotropic in the basal plane, it is only necessary to consider one direction of propagation for this plane. From the arrangement shown in Fig. 18, and the piezoelectric matrix for class **6mm** crystals to which ZnO belongs

$$(3.10) \qquad \begin{pmatrix} \cdot & \cdot & \cdot & \cdot & d_{15} & \cdot \\ \cdot & \cdot & \cdot & d_{24} & \cdot & \cdot \\ d_{31} & d_{32} & d_{33} & \cdot & \cdot & \cdot \end{pmatrix},$$

it is clear that a Rayleigh wave can readily be launched by means of an ID grid. The electric-field components E_2 and E_3 couple the piezoelectric moduli active along X_2 and X_3 to generate the following strains:

$$(3.11) \qquad \varepsilon_4 = d_{24} E_2, \qquad \varepsilon_3 = d_{33} E_3, \qquad \varepsilon_2 = d_{32} E_3 \quad \text{and} \quad \varepsilon_1 = d_{31} E_3.$$

The strain ε_4 is a shearing about the X_1-axis with particle displacement along X_3 and energy flow along X_2, which is the appropriate shear mode for propagation of a Rayleigh wave in the basal plane along X_2. The strain ε_2 is a compressional strain with particle displacement and energy flow along the X_2-axis, which is the appropriate compressional mode for the Rayleigh wave. At the synchronous frequency ε_3 and ε_4 combine to launch a Rayleigh wave in both directions normal to the ID grid fingers. The strains ε_1 and ε_3 are bulk compressional strains with displacements and energy flow respectively along the X_1 and X_3 axes.

This orientation is extremely useful for launching Rayleigh waves along a suitable direction on a nonpiezoelectric material. Thin films of ZnO can readily be grown with the orientation shown in Fig. 18.

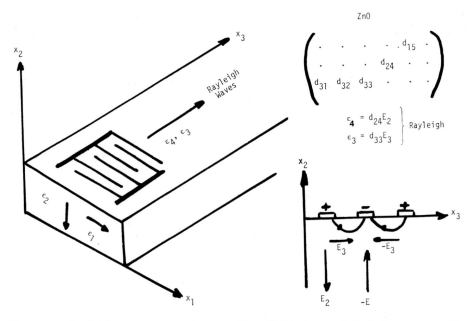

Fig. 19. – Rayleigh-wave generation on the X_1X_3-plane of ZnO for propagation along X_3.

A higher coupling orientation of ZnO would be that shown in Fig. 19. If we use eq. (3.10) it will be seen that the strain generated by the E_2 field component will be

$$(3.12) \qquad \varepsilon_4 = \varepsilon_{23} = \varepsilon_{32} = d_{24}E_2 ,$$

which is a shearing strain about the X_1-axis with particle displacement along X_2 and energy flow along X_3. Thus, ε_4 provides the shear component required for Rayleigh-wave generation.

The strains generated by the E_3 field component will be

(3.13) $\varepsilon_1 = d_{31}E_3\,,\quad \varepsilon_2 = d_{32}E_3\quad$ and $\quad \varepsilon_3 = d_{33}E_3\,;$

ε_3, which is a compressional strain with particle motion and energy flow along X_3, provides the compressional component required for Rayleigh-wave generation. As ε_4 is generated *at* the ID fingers and ε_3 *between* fingers, these strains are $90°$ out of phase with each other as required for Rayleigh-wave generation. Strains ε_1 and ε_2 will propagate as undesired bulk modes along the X_1 and X_2 axes as indicated in Fig. 19.

Lithium-niobate, another material frequently used for surface wave applications, will next be examined. Figure 9 shows that in the XZ and YZ planes only along two directions, *viz.* along the X_3- or Z-axis in the XZ-plane and the X_1-axis in the same plane, can all three bulk modes be propagated as pure modes. The two shear modes are degenerate along the Z-axis in so far as they propagate with the same velocity. Hence, the most suitable plane for propagating pure Rayleigh waves will be the XZ-plane. Generation along the Z-axis on the XZ-plane will first be considered. This orientation is shown in Fig. 20. The piezoelectric matrix for crystal class $3m$, to which LiNbO$_3$ belongs, is given

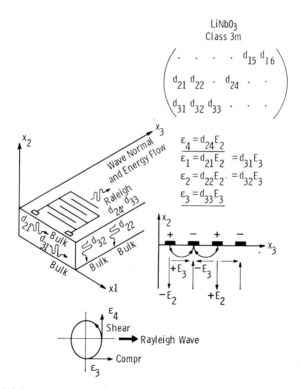

Fig. 20. – Rayleigh-wave generation on Y-cut Z-propagating LiNbO$_3$.

by

$$(3.14) \qquad \begin{pmatrix} \cdot & \cdot & \cdot & \cdot & d_{15} & d_{16} \\ d_{21} & d_{22} & \cdot & d_{24} & \cdot & \cdot \\ d_{31} & d_{32} & d_{33} & \cdot & \cdot & \cdot \end{pmatrix}.$$

The E_2 field component couples to the d_{2j}-moduli resulting in the generation of

$$(3.15) \qquad \varepsilon_1 = d_{21}E_2, \qquad \varepsilon_2 = d_{22}E_2 \quad \text{and} \quad \varepsilon_4 = d_{24}E_2.$$

ε_1 and ε_2 are compressional strains which propagate along the X_1 and X_2 axes and consequently are of no value in generating a Rayleigh wave. ε_4, however, provides a shearing strain about X_1 with particle displacement along X_2 and energy flow along X_3. Hence, the required shear components are generated *at* the ID fingers. The E_3 field component couples to the d_{3j}-moduli resulting in the generation of

$$(3.16) \qquad \varepsilon_1 = d_{31}E_3, \qquad \varepsilon_2 = d_{32}E_3 \quad \text{and} \quad \varepsilon_3 = d_{33}E_3.$$

ε_3 provides compressional strains with energy flow along X_3. These are the required compressional-strain components which are generated *between* ID fingers resulting in a phase shift of $\pi/2$ required for Rayleigh-wave generation. Thus, pure-mode Rayleigh waves can readily be generated, propagated and detected with the arrangement shown in Fig. 20.

If the ID grid of Fig. 20 is rotated through 90°, the arrangement shown in Fig. 21 will result. The E_2 field components will generate strains

$$(3.17) \qquad \varepsilon_1 = d_{21}E_2, \qquad \varepsilon_2 = d_{22}E_2 \quad \text{and} \quad \varepsilon_4 = d_{24}E_2.$$

ε_1 is a compressional strain with displacement and energy flow along X_1. ε_2 is a compressional strain which propagates along X_2 as an undesired bulk mode. ε_4 provides shear strains about X_1 with particle displacement along X_2 and energy flow along X_3 and consequently propagates as an undesired bulk shear mode.

The E_1 field components generate the strains

$$(3.18) \qquad \varepsilon_5 = d_{15}E_1 \quad \text{and} \quad \varepsilon_6 = d_{16}E_1.$$

ε_5 shears about X_2 with particle displacement along X_3 and energy flow along X_1, whereas ε_6 shears about X_3 with particle displacement along X_2 and energy flow along X_1. Hence, ε_6 and ε_1 combine to provide the strain components required for Rayleigh-wave generation. However, the measured values [9] of $d_{31} = 0.1 \cdot 10^{-11} C/N$, and $d_{16} = 4.2$ are much lower than $d_{33} = 0.6$ and $d_{15} = 6.8$

which are involved in generating Rayleigh waves along the X_3-axis. Consequently, although Rayleigh waves can be launched by the arrangement of Fig. 21, they will be almost an order of magnitude lower than those generated

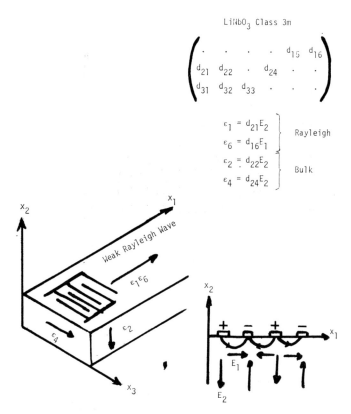

Fig. 21. – Rayleigh-wave generation on Y-cut X-propagating LiNbO$_3$.

by the arrangement in Fig. 20, since

$$d_{31} \times d_{16} = 0.42 \cdot 10^{-11} \, C/N \quad \text{and} \quad d_{33} \times d_{15} = 4.08 \cdot 10^{-11} \, C/N \,.$$

It is also possible to generate Rayleigh waves on the YZ-plane of LiNbO$_3$ with propagation along X_3. This configuration is illustrated in Fig. 22. The E_1 field component generates strains

$$(3.19) \qquad\qquad \varepsilon_5 = d_{15} E_1 \quad \text{and} \quad \varepsilon_6 = d_{16} E_1 \,.$$

ε_5 shears about X_2 with particle displacement along X_1 and energy flow along X_3, which is the required shear strain for Rayleigh-wave propagation along X_3. (ε_6 shears about X_3 and propagates as a bulk shear either along X_1 or X_2.) The

E_3 field component generates strains

$$(3.20) \qquad \varepsilon_1 = d_{31}E_3, \qquad \varepsilon_2 = d_{32}E_3 \qquad \text{and} \qquad \varepsilon_3 = d_{33}E_3.$$

ε_3 provides compressional strains with particle displacement and energy flow along X_3. Thus, ε_5 and ε_3 combine to launch a Rayleigh wave along X_3 on the

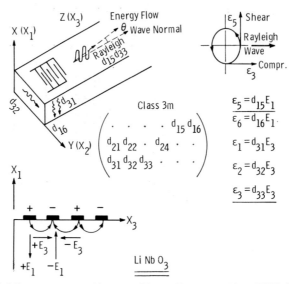

Fig. 22. – Rayleigh-wave generation on X-cut Z-propagating LiNbO$_3$.

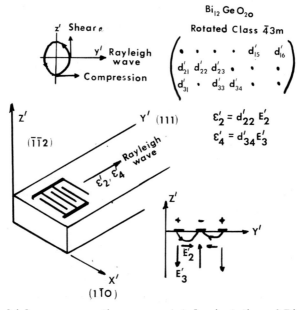

Fig. 22a. – Rayleigh-wave generation on a rotated orientation of Bi$_{12}$GeO$_{20}$.

YZ-plane of LiNbO$_3$, and result from the same piezoelectric moduli $d_{24} = d_{15}$ and d_{33} interacting with the electric-field components beneath the ID grid. However, by examining the velocity surfaces for LiNbO$_3$ in Fig. 9, it will be seen that while the shear modes are pure on the XZ-plane, pure-mode propagation on the YZ-plane does not occur. Consequently, the direction of Rayleigh-wave energy flow on the YZ-plane does not coincide with the wave normal or Z-direction. The experimental measurement of θ (see Fig. 22) was found to be approximately 4°.

Materials belonging to other crystal classes can be analyzed in the same way. Sometimes it becomes necessary to seek orientations which are not along major crystallographic directions such as those on the materials discussed above. Materials which have a very high degree of symmetry such as those which have tetragonal or cubic symmetry sometimes only have one independent piezoelectric modulus, *viz.* shear. An example of this is class $\bar{4}3m$ to which Bi$_{12}$GeO$_{20}$ belongs. In these cases it becomes necessary to search for rotated directions with lower symmetry, to obtain the compressional moduli necessary for generating the appropriate strains. As an example, the standard orientation of bismuth germanium oxide (Bi$_{12}$GeO$_{20}$) has the piezoelectric matrix

$$(3.21) \qquad \begin{pmatrix} \cdot & \cdot & \cdot & d_{14} & \cdot & \cdot \\ \cdot & \cdot & \cdot & \cdot & d_{14} & \cdot \\ \cdot & \cdot & \cdot & \cdot & \cdot & d_{14} \end{pmatrix}.$$

A compressional strain cannot be generated by ID grids on the standard orientation, *i.e.* on any of the the cube faces with ID fingers parallel to any of the X, Y or Z axes. However, by rotating the orientation to that shown in Fig. 22a so that the new axes are $X' = (1\bar{1}0)$, $Y' = (111)$ and $Z' = (\bar{1}\bar{1}2)$, the rotated piezoelectric matrix is shown as

$$(3.22) \qquad \begin{pmatrix} \cdot & \cdot & \cdot & \cdot & d'_{15} & d'_{16} \\ d'_{21} & d'_{22} & d'_{23} & \cdot & \cdot & \cdot \\ d'_{31} & \cdot & d'_{33} & d'_{34} & \cdot & \cdot \end{pmatrix}.$$

If we use a crystal of Bi$_{12}$GeO$_{20}$ oriented to provide a Z'-cut Y'-propagating sample and fabricating an ID grid on the $X'Y'$-surface with fingers parallel to the X'-axis, electric-field components E'_2 and E'_3 can be generated. The E'_2 field component will generate a compressional strain ε'_2 with displacement and energy flow along the Y'-axis. The E'_3 field component will generate a shear strain ε'_4 about the X'-axis, *i.e.* with displacements along Z' and energy flow along Y'. These strains, ε'_2 and ε'_4, combine to generate a Rayleigh wave which will propagate along the Y' or (111)-axis.

4. – Backcoupling and reflection effects.

As both direct and converse piezoelectric effects operate simultaneously, one might expect to observe some differences in the behavior of weak- and strong-coupling materials. Since each pair of fingers in an ID grid makes a contribution to the amplitude of, and hence power in, a launched surface wave, it was suggested [10] that the change in power converted with increasing N,

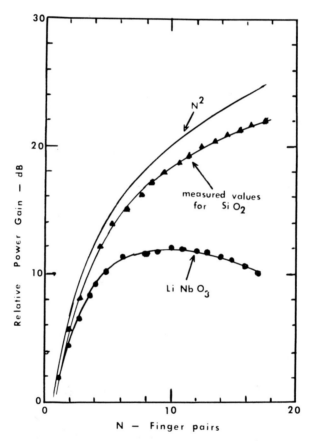

Fig. 23. – Relative power gain with increasing number of fingers in an ID grid for SiO$_2$ and LiNbO$_3$.

the number of finger pairs in the ID grid, would be proportional to N^2 for low-coupling materials, such as quartz. However, the experimental measurements [11] plotted in Fig. 23 show marked departures from the above-mentioned small-coupling theory, even for quartz. The power gain curve for Y-cut Z-propagating LiNbO$_3$ reaches a maximum for $N = 10$ and then declines for $10 < N < 17.5$ by approximately 2 dB. These deviations from the N^2 prediction can be accounted for by the presence of piezoelectrically generated reverse electric fields. If we use the expressions for the direct and converse piezoelectric effects given in eqs. (3.2) and (3.3) and Hooke's law [4], which can be expressed as

$$(4.1) \qquad\qquad \sigma = c\varepsilon \, ,$$

the polarization P' due to the stress σ generated by applying an electric field E to a piezoelectric can be written

$$(4.2) \qquad\qquad P_i' = d_{ij}\sigma_j = d_{ij}(c_{ij}\varepsilon_j) \, ,$$

and by substitution from eq. (3.3)

$$(4.3) \qquad\qquad P_i' = d_{ij}(c_{ij}d_{ij}E_i) = c_{ij}d_{ij}^2 E_i \, .$$

But since $P = \xi E$, where ξ is the dielectric constant, eq. (4.3) can be written as

$$(4.4) \qquad\qquad \xi_i E_i' = c_{ij}d_{ij}^2 E_i$$

or

$$(4.5) \qquad\qquad E_i' = \frac{c_{ij}d_{ij}^2}{\xi_i} E_i \, .$$

By substitution from eq. (3.4)

$$(4.6) \qquad\qquad E_i' = k_{ij}^2 E_i \, .$$

Hence the « operative » electric field can be written

$$E_{\mathrm{op}} = E - E' \, ,$$

i.e.

$$(4.7) \qquad\qquad E_{\mathrm{op}} = (1 - k_{ij}^2) E_i \, .$$

Equation (4.7) was used to compute the curve shown in Fig. 24. The « operative » electric field E_{op} decreases with increasing electromechanical coupling k, and reaches zero for $k = 1$. This decrease in E_{op} with increasing k is responsible for departures of the experimental curves from the small-coupling theoretical curve of Fig. 23, because the effective electromechanical coupling for an ID grid increases with increasing N.

The backcoupling is also responsible for the reflection of surface waves from ID grids. In most surface wave filter applications, reflection of the surface wave from the ID grids presents severe practical problems. Measurements [11] of the amplitude of the reflected wave as functions of the number

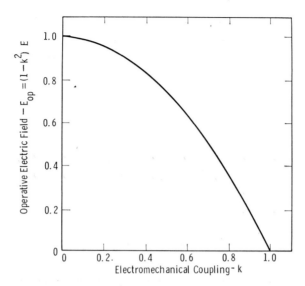

Fig. 24. – Variation of «operative» electric field as a function of electromechanical coupling coefficient for piezoelectric materials.

of fingers in, and the electrical termination of, the reflecting grid are shown in Fig. 25 for quartz and LiNbO$_3$. The arrangement of the ID grids for these measurements is shown inset in Fig. 25. The launching and receiving grids

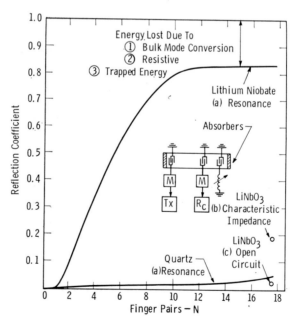

Fig. 25. – Reflection coefficient for surface waves on LiNbO$_3$ and SiO$_2$ as a function of the number of finger pairs in the reflecting ID grid for various electrical terminations.

shown on the left and center were electrically matched in impedance respectively to the transmitter and receiver.

The electrical termination of the reflector grid, on the right, could be switched to assume three chosen values, *viz.* *a*) electrical resonance, *b*) characteristic impendance and *c*) open circuit, when condition *a*) applied the circuit at the reflector ID grid shown in Fig. 25 was used. An adjustable inductor was used to resonate with the ID grid capacity at the synchronous frequency, *i.e.* the center frequency of the propagating surface wave. For condition *b*) a resistor equal to the characteristic impendance was connected between the inductor and ground. The resistor and inductor were disconnected from the ID grid for condition *c*). Reflection of surface waves from both ends of the delay line was prevented by placing wax on the surface at each end.

As Fig. 25 indicates, the reflection coefficient for LiNbO$_3$ is a very sensitive function of N for termination condition *a*). The reflection coefficient rapidly increases as N increases from 1 to 10. No further increase occurs for $N > 10$. At $N = 10$, 82 % of the incident surface wave power is thus reradiated as surface waves by the ID grid due to the backcoupling. Some of the incident power is reradiated as bulk modes (see Sect. **3**), and in addition some of the incident power is dissipated as resistive losses in the ID grids and the external electrical matching circuit.

A single point, for $N = 17.5$, is also plotted for termination conditions *b*) and *c*) for LiNbO$_3$. For the condition *b*), the characteristic impedance termination, only 18 % of the incident power is reflected as surface waves. When the open-circuit termination is used, *i.e.* condition *c*), only about 2 % of the incident power is reflected. Under these conditions acoustic impedance discontinuities exist at the edges of each finger due to the mass loading by the 1000 Å gold fingers, and the slight change in velocity [12] under the metal fingers.

The lower curve in Fig. 25 for quartz under termination condition *a*) shows that only 4 % of the incident surface wave power is reflected by 17.5 finger pairs each of 1000 Å gold. The same percentage of power is reflected on LiNbO$_3$ for the same termination condition when $N = 2$. From this it is clear that the reflection coefficient is directly related to the effective electromechanical coupling coefficient k_{ef} of the material and to the electrical termination condition. Thus, an ID grid with large N will not respond to an alternating electric field at its fundamental frequency except at the two extreme ends because the value of E_{op} will be zero except at the ends. An ID grid with uniform finger width and spacing will thus act as a band stop filter at its center frequency for large values of N.

5. – Applications.

Since the velocity of Rayleigh waves is approximately 10^5 times slower than that of electromagnetic waves, the surface wavelength is reduced by the same factor. This large reduction in wavelength allows similar reductions in device size compared to electromagnetic devices. As the surface wave energy is available for sensing at any point along its path, some very sophisticated filter functions are possible. Phase-coded surface wave trains can be launched and detected by arranging the placement of the fingers in an ID grid such that two adjacent fingers have the same electrical potential and effectively spatially remove half a wavelength from the acoustic wave train as shown in Fig. 26.

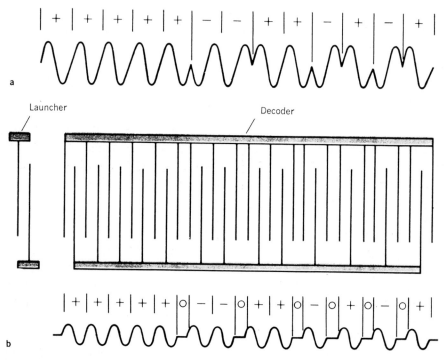

Fig. 26. – A phase-coded ID grid showing the electrical and acoustic signals associated with such an array. The phase-coded sequence in the electromagnetic signal is shown in part *a*) and the acoustic signal launched by the single finger pair in part *b*).

This Figure shows an ID array consisting of two parts, *viz.* a launcher consisting of one finger pair and a decoder in which two adjacent fingers are electrically connected at each point where the electrical phase is abruptly reversed. Between these connected fingers no electrical potential exists and consequently no elastic strains are generated. If an electrical signal like that in Fig. 26*a*) is applied to the launcher, the elastic signal shown in Fig. 26*b*) will result and propagate under the decoder.

As the acoustic signal propagates under the decoder, each finger pair will respond to the elastic strains and cause an electrical signal to be developed between the two rails of the decoder, the electrical polarity of which will be determined by the acoustic polarity and the ID grid bit polarity. When these polarities are the same, a positive signal will be developed, but when they are opposite a negative signal will be developed. The signal output from the decoder will have six unit-amplitude leading sidelobes, the correlation peak of amplitude 13, then six trailing sidelobes of unit amplitude as the surface wave emerges from the decoder, as shown in Fig. 27. The phase sequence for the 13 bit

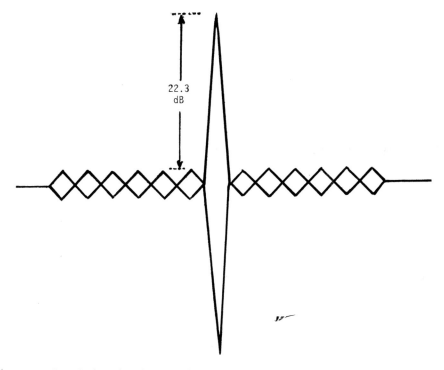

Fig. 27. – Correlation signal output from phase-coded decoder of a 13 bit Barker code.

Barker code [13] shown in Fig. 26 is $+ + + + + - - + + - + - +$. The code can be read forwards or backwards and the polarities of all the bits can be reversed without altering the correlation signal to sidelobe amplitude of 13:1, or 22.3 dB. The above description is true for weak-coupling materials only, such as quartz. When strong-coupling materials, such as $LiNbO_3$ or $Bi_{12}GeO_{20}$, are used, the strong backcoupling causes internal reflections [14] to occur and in addition the acoustic-signal amplitude decreases as it penetrates further under the decoder. These effects will be evident by nonuniformity in the amplitude and shape of the sidelobes, some being greater in amplitude than unity, and in addition a reduction in the correlation peak amplitude. It has been found possible to realize the full theoretical peak-to-sidelobe ratio by using quartz.

Although the phase velocity of propagation is the same for all frequencies, *i.e.* the velocity is nondispersive, artificial dispersion can be designed into a filter if so desired. This is readily achieved by allowing the distance of propagation to vary with frequency, as shown in Fig. 28. The transit or propagation time for the highest frequency is $d_{\text{hi}}/v_{\text{R}}$, while that for the lowest frequency is $d_{\text{lo}}/v_{\text{R}}$. With the arrangement of the two frequency-modulated ID grids shown in Fig. 28, a frequency-modulated radio frequency pulse can readily be

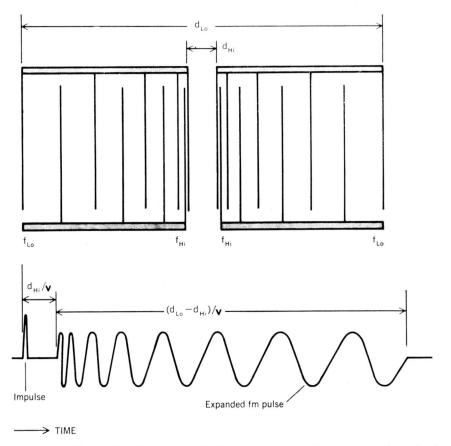

Fig. 28. – A filter with «built-in» velocity dispersion. The transit time for low frequencies is greater than that for high frequencies.

generated by applying an electrical impulse to the left ID grid and monitoring the electrical output of the right-hand grid. The expanded frequency-modulated pulse shown in the lower part of this Figure will be detected at the right-hand grid. The band width of this expanded pulse will be

$$(5.1) \qquad\qquad \Delta f = f_{\text{hi}} - f_{\text{lo}} ,$$

and the pulse width will be

(5.2)
$$\Delta t = \frac{d_{lo} - d_{hi}}{v_R} ,$$

where v_R is the surface wave velocity. A compression filter can readily be made using the same pair of frequency-modulated grids of Fig. 28, by placing the low-frequency ends adjacent to one another and the high-frequency ends remote. By this means the distances of propagation for the high and low frequencies are interchanged. By applying the expanded pulse to such a compression filter all frequencies arrive simultaneously at their respective ID detector fingers, thus generating a peak pulse the amplitude of which will be the sum of the outputs from each finger pair. The compression ratio for this filter will be

(5.3)
$$\Delta f \, \Delta t = (f_{hi} - f_{lo}) \, \frac{d_{lo} - d_{hi}}{v_R} .$$

A complete expansion and compression filter is depicted in Fig. 29. In this case the piezoelectric material was $Bi_{12}GeO_{20}$ with four frequency-modulated

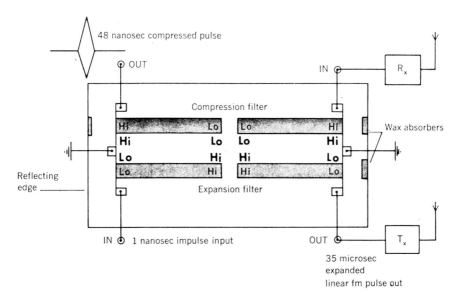

Fig. 29. – Pulse expansion and compression filter. The pulse lengths indicated were achieved by using $Bi_{12}GeO_{20}$ with frequency-modulated grid which operated between 16 and 37 MHz.

ID grids each covering from 16 to 37 MHz. A one-nanosecond impulse applied to the input grid at the lower left was stretched to 35 microseconds at the lower right output grid. When this expanded signal was fed to the upper right re-

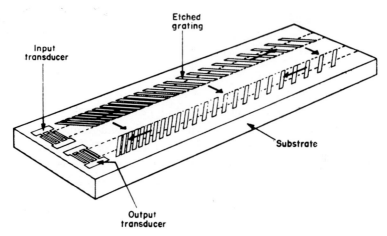

Fig. 30. – Schematic diagram of a reflective array compressor (after WILLIAMSON and SMITH (ref. [16])).

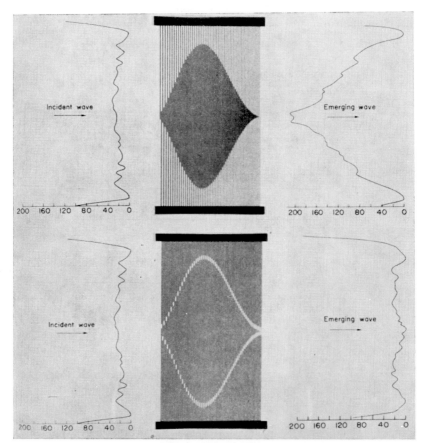

Fig. 31. – Phase distortion in surface waves launched by two types of amplitude and frequency modulated interdigital grids (after TANCRELL and WILLIAMSON (ref. [18])).

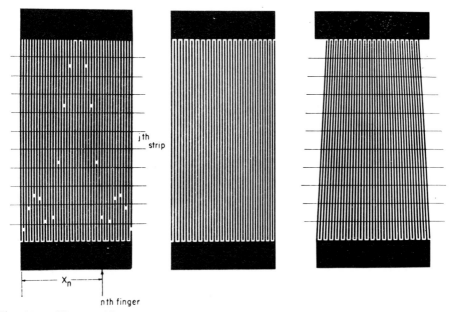

Fig. 32. – Three grid structures used in surface wave filter design. Left to right: $\sin x/x$, AM; linear FM and tapered.

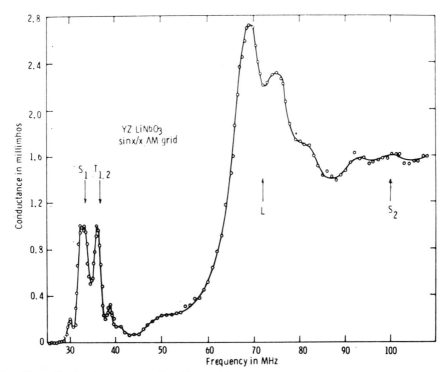

Fig. 33. – Conductance as a function of frequency for the $\sin x/x$ AM grid (after DANIEL and DE KLERK (ref. [19])).

ceiver input grid, it developed a compressed pulse of 48 nanoseconds duration at the upper left output grid. The compression ratio for this filter was thus

$$\Delta f \Delta t = 21 \cdot 10^6 \times 35 \cdot 10^{-6} = 735 \, .$$

Another method [16] of obtaining pulse compression is by reflecting the surface waves from a grating on the surface as shown in Fig. 30. Frequency-

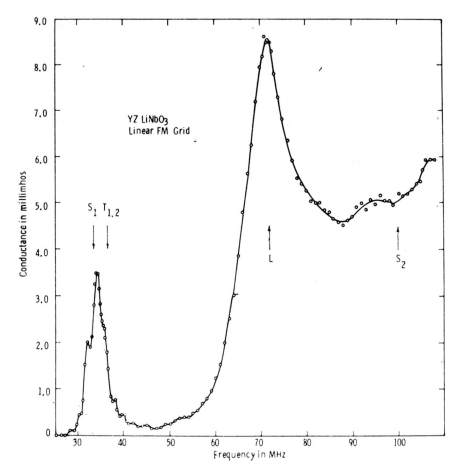

Fig. 34. – Conductance as a function of frequency for the FM grid (after DANIEL and DE KLERK (ref. [19])).

modulated surface waves are launched by a broad-band input transducer reflected by two dispersive gratings and received at the broad-band output transducer. The spacings between grating elements which are etched into the surface and the depths of the elements determine which frequencies will be selectively reflected. For this type of device the spacings between elements are a wavelength, not half a wavelength. In the Figure shown high frequencies are reflected by those parts nearest the transducers, while those furthest away respond most strongly to low frequencies. One such device with center frequency 1 GHz and bandwidth 512 MHz has been reported [17].

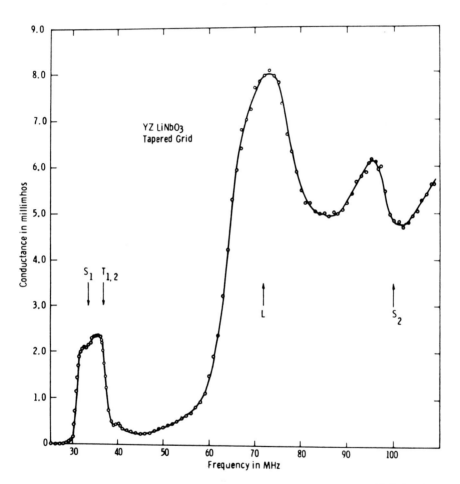

Fig. 35. – Conductance as a function of frequency for the tapered grid (after DANIEL and DE KLERK (ref. [19])).

The presence of a metal film on the surface, as mentioned in Sect. **2**, influences the velocity of the surface wave. This effect is extremely important for devices which use high-coupling materials such as $LiNbO_3$ and $Bi_{12}GeO_{20}$. In dispersive devices, such as pulse compression filters, this change in velocity under metal strips can lead to large phase distortions [18], as shown in Fig. 31. The finger overlap is identical in both interdigital grids. However, in the upper grid, the excess finger lengths have been removed, while in the lower grid the excess finger lengths were retained but tied electrically to the opposite busbar. It is clear that the velocity of propagation under the upper grid is lowest along the axis of the grid and highest nearest the busbars. Hence, the phase distortion is greatest along the axis. This effect is largely eliminated by the use of « dummy » fingers as shown in the lower section of Fig. 31.

Unwanted bulk mode generation by an interdigital surface wave grid often leads to undesirable device characteristics. The strengths of coupling to surface and bulk waves for three types of interdigital grid [19], shown in Fig. 32, is illustrated in Fig. 33, 34 and 35. It can readily be seen from these Figures that maximum frequency of operation of surface wave devices using $LiNbO_3$ is restricted to less than the second harmonic. However, the frequency range can be extended by adjusting the geometry of the delay line or filter to prevent bulk waves from reaching the receiver grid with phase coherence. This can be achieved by sloping all faces of the delay medium with reference to the surface supporting the surface waves.

A comprehensive review and bibliography of surface wave applications is contained in ref. [20].

6. – Conclusions.

It has been shown that a complete description of elastic surface waves in isotropic media can be given in relatively simple mathematical terms. The treatment of Rayleigh waves on anisotropic surfaces, however, is much more complicated. Nevertheless, it is not necessary to resort to complicated computer calculations to determine the most suitable crystal orientations, pure-mode propagation and coupling directions, as this can be done much more readily using an analytical method based on the elastic, piezoelectric and dielectric matrices of the crystal class to which the propagation medium belongs. This method also accounts for all the elastic strains generated by an interdigital grid on the surface of a piezoelectric material. It is shown that, in addition to the desired elastic surface wave, numerous unwanted bulk waves are also usually launched. The use of high-coupling materials leads to serious reflection and regeneration problems in surface wave devices.

* * *

Part of the work reported here was supported by the U.S. Office of Naval Research under Contract No. N00014-69-C-0118.

Symbols.

A_x, A_y, A_z displacement vectors in Cartesian co-ordinates,

c_{ij} elastic moduli ($i, j = 1, 2, ..., 6$),

C_1 velocity of propagation of the dilatation waves in an unbounded isotropic medium $= [(\lambda + 2\mu)/\varrho]^{\frac{1}{2}}$,

C_2 velocity of propagaton of distortion waves in an unbounded isotropic medium $= (\mu/\varrho)^{\frac{1}{2}}$,

d_{ij} piezoelectric moduli (strain coefficients) ($i = 1, 2, 3$; $j = 1, 2, ..., 6$),

e_{ij} piezoelectric constants (stress coefficients) ($i = 1, 2, 3$; $j = 1, 2, ..., 6$),

E_i electric-field component,

h $= p/C_1 = [\varrho p^2/(\lambda + 2\mu)]^{\frac{1}{2}}$ for Rayleigh waves,

i $\sqrt{-1}$ (in complex equations),

ID   interdigital,

k_{ij} electromechanical coupling coefficient $= (c_{ij} d_{ij}^2/\xi_i)^{\frac{1}{2}}$,

K $= p/C_2 = [\varrho p^2/\mu]^{\frac{1}{2}}$ for Rayleigh waves,

K_1 ratio of propagation velocities for Rayleigh and bulk shear waves,

l, m, n direction cosines,

L subscript for compressional mode,

p $= 2\pi$ times frequency of sinusoidal waves,

P_i electric polarization component,

q attenuation factor for compressional displacement component of Rayleigh wave,

S_i correction terms for elastic moduli for piezoelectric stiffening,

s attenuation factor for shear displacement component of Rayleigh wave,

t time,

T_1, T_2 subscripts for shear modes,

u, v, w displacements in x, y, and z directions respectively in Cartesian co-ordinates,

v_i acoustic phase velocity in crystalline media ($i = L, T_1, T_2$),

v_R Rayleigh-wave velocity,

X_i crystallographic axes in Cartesian co-ordinates ($i = 1, 2, 3$),

X, Y, Z crystallographic axes in Cartesian co-ordinates,

α attenuation factor for stress waves in dissipative medium,

Δ $=$ dilatation $= \varepsilon_{xx} + \varepsilon_{yy} + \varepsilon_{zz}$,

ε strain,

$\varepsilon_{xx}, \varepsilon_{yy}, \varepsilon_{zz}$ components of strain in Cartesian co-ordinates,

λ Lamé's constant,

Λ acoustic wavelength $= 2\pi/f$,

μ rigidity modulus,

ν Poisson's ratio,

ξ_i dielectric moduli ($i = 1, 2, 3$),

ϱ density,

σ stress,

$\sigma_{xx}, \sigma_{yy}, \sigma_{zz}$ components of stress in Cartesian co-ordinates,

φ potential function of displacement for irrotational strain,

ψ potential function of displacement for rotational strain,

$\bar{\omega}_x, \bar{\omega}_y, \bar{\omega}_z$ components of rotation in Cartesian co-ordinates,

∇^2 Laplacian operator ($= \partial^2/\partial x^2 + \partial^2/\partial y^2 + \partial^2/\partial z^2$ in Cartesian co-ordinates).

REFERENCES

[1]   Lord RAYLEIGH: *Proc. Lond. Math. Soc.*, **17**, 4 (1885).
[2]   H. KOLSKY: *Stress Waves in Solids* (Oxford, 1953), p. 10.
[3]   M. J. P. MUSGRAVE: *Proc. Roy. Soc.*, A **226**, 339 (1954).
]4]   J. F. NYE: *Physical Properties of Crystals* (Oxford, 1960), p. 124.
[5]   R. M. WHITE: *Proc. IEEE*, **58**, 1238 (1970).
[6]   I. A. VIKTOROV: *Rayleigh and Lamb Waves. Physical Theory and Applications* (New York, N. Y., 1967).
[7]   J. D. H. DONNAY, Editor: *Crystal Data Determinative Tables* (Washington, D. C., 1963).
[8]   W. G. CADY: *Piezoelectricity* (New York, N. Y., 1964), p. 759.
[9]   A. W. WARNER, M. ONOE and G. A. COQUIN: *Journ. Acoust. Soc. Amer.*, **42**, 1223 (1967).
[10] C. C. TSENG: *IEEE Trans. Elect. Dev.*, ED-**15**, 586 (1968).
[11] J. DE KLERK: Invited, *Proceedings of the 1970 Ultrasonics Symposium*, p. 94, IEEE Cat. No. 70C69SU (1971).
[12] J. J. CAMPBELL and W. R. JONES: *IEEE Trans. Son. Ultrason.*, SU-**15**, 209 (1968).
[13] E. C. FARNETT, T. B. HOWARD and G. H. STEPHENS: *Radar Handbook*, edited by M. I. SKOLNIK (New York, N. Y., 1970), Capt. 20, p. 18.
[14] W. S. JONES, C. S. HARTMANN and T. D. STURDIVANT: *IEEE Trans. Son. Ultrason.*, SU-**19**, 368 (1972).
[15] G. W. FARNELL: in *Physical Acoustics*, edited by W. P. MASON, Vol. **6** (New York, N. Y., 1970), p. 109.
[16] R. C. WILLIAMSON and H. I. SMITH: *IEEE Trans. Son. Ultrason.*, SU-**20**, 113 (1973).
[17] R. C. WILLIAMSON, V. S. DALAT and H. I. SMITH: *1973 Ultrasonic Symposium Proceedings*, IEEE Cat. No. 73CH0807-8SU.
[18] R. TANCRELL and R. C. WILLIAMSON: *Appl. Phys. Lett.*, **19**, 456 (1971).
[19] M. R. DANIEL and J. DE KLERK: *1973 Ultrasonic Symposium Proceedings*, IEEE Cat. No. 73CH0807-8SU, p. 449.
[20] M. G. HOLLAND and L. T. CLAIBORNE: *Proc. IEEE*, **62**, 582 (1974).

Surface Acoustic-Wave Devices.

C. ATZENI and L. MASOTTI

Istituto di Ricerca sulle Onde Elettromagnetiche del C.N.R. - Firenze

1. – Introduction.

For many years bulk acoustic waves propagating inside solids have played important roles in electronics. Initially these waves were used in oscillators and filters; later, devices employing the propagation of bulk acoustic waves in solids over paths many wavelengths long were made for the storage, delay or processing of signals. Recently, engineers in many different countries have directed attention to *surface* acoustic waves because of the possibility of constructing signal-processing devices which employ these waves. Currently most of the major electronics-systems firms in the USA and a large number in Europe have some research effort devoted to surface acoustic waves (s.a.w.).

Several features of s.a.w. may be identified that cause so much attention to be drawn to them:

a) they are small planar structures compatible with integrated circuit manufacturing techniques;

b) the s.a.w. device performance is almost exclusively controlled through the transducer design, thus the mask ensures reproducibility of performance;

c) because the information is on the surface it is accessible for processing, for example, by tapping off and recombining different parts of the signal.

In all the surface wave devices we shall be considering, the launching of an acoustic wave from an electrical signal and subsequently reconverting it to an electrical signal occurs [2].

The acoustic surface wave, the medium on which it propagates, the transducer have all been treated in detail by other speakers at this series of lectures on « New Directions in Physical Acoustics ». We shall deal here with the design of transducers for implementing s.a.w. signal processors, the problems associated with the matching circuits of the transducers and a review of the current status of s.a.w. devices and applications.

2. – Interdigital transducers.

There are numerous methods of generating and detecting surface acoustic waves electrically, but of these we shall only consider here the interdigital transducer which has become the pre-eminent one. Because of the nature of the structure and the complexity of the behaviour of the anisotropic piezo-electric substrate, an accurate theory is very laborious and, from an engineering standpoint, unrewarding. It is more profitable to consider an approximate simplified theory and add sophistication when the occasion demands.

The system can be reduced to a one-dimensional configuration by assuming that the electric field is merely parallel to the propagation vector (« in-line » model) or normal (« crossed-field » model). This second model is most popular, since it leads to predictions in good agreement with experimental measurements for materials having a strong electromechanical coupling. This type of theory has been given by SMITH *et al.* [1] who considered the transducer as an array of sources, each analogous to a piezoelectric plate transducer for launching bulk waves. The transducer is considered to have N sections, each of length L as in Fig. 1, so that the total number of electrodes is $2N+1$. To each sec-

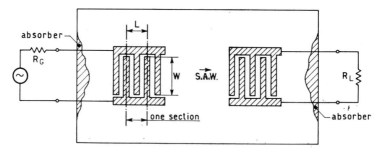

Fig. 1. – S.a.w. delay line showing configuration of interdigital transducers.

tion, by analogy with bulk wave transducer, is assigned an equivalent circuit, originally introduced by MASON for bulk-resonant transducers, in which the electromechanical coupling constant will have a value appropriate for surface waves. The equivalent circuit of the whole transducer is formed by cascading the circuits of individual sections, and considering the N sections acoustically in cascade (to represent generation from fingers) and electrically in parallel. The impedance of the transducer as seen at the electrical port, $Z(f)$, can be represented by either of the series or parallel circuits shown in Fig. 2, which are equivalent.

Here C_T is the total electrical capacity presented by the electrode structure on the solid, given by NC_s, C_s being the interelectrode capacitance of a single section. For frequency near the acoustic synchronism (the synchronism frequency is $f_0 = v/L$, v being the s.a.w. phase velocity) $G_a(f)$ and $B_a(f)$ are approximately given by

$$G_a(f) = G_a(f_0)\left(\frac{\sin x}{x}\right)^2,$$

$$B_a(f) = G_a(f_0)\frac{\sin 2x - 2x}{2x^2},$$

where

$$x = N\pi\frac{f - f_0}{f_0},$$

$$G_a(f_0) = 8K^2 f_0 C_s N^2,$$

and $Z(f)$ can be calculated as

$$Z(f) = R_a(f) + jX_a(f) + \frac{1}{2\pi jfC_T} = \frac{1}{G_a(f) + jB_a(f) + j2\pi fC_T},$$

$G_a(f)$ is maximum at $f = f_0$ and its fractional bandwidth is inversely proportional to the number N of periodic sections ($G_a(f)$ has zeros at values of f resulting from $(f - f_0)/f_0 = \pm 1/N$).

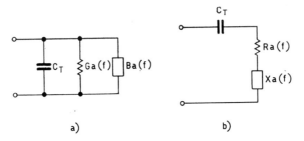

Fig. 2. – Parallel (a)) and series (b)) equivalent circuits for an interdigital transducer.

Assuming that Ohmic losses can be ignored, the acoustic power radiated is given by the power dissipated in the resistive part of the equivalent circuit, thus giving the conversion loss and bandwidth of the transducer.

This frequency response coincides with the fundamental frequency response that can be expected if, as discussed further on, each finger is regarded as a radiating source.

In the passband the loss can be minimized by using an inductor to correct the reactive part of the transducer impedance at frequency f_0, so that the transducer then appears as purely resistive. The common series-tuning configuration is shown in Fig. 3.

At the synchronous frequency f_0 the tuned transducer appears as a resistor $R_a(f_0)$ which is made equal to R_G, the generator internal resistance, in order to minimize the conversion loss.

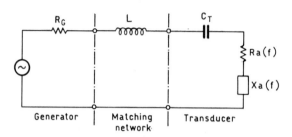

Fig. 3. – Use of a series tuning inductor to minimize conversion loss.

$R_a(f_0)$, for $4K^2N \ll \pi$, is independent of N and proportional to the inverse of the aperture W of the transducer. A radiation resistance of 50 Ω can be achieved at f_0 by choice of aperture in the range $50\,\lambda_0$ to $100\,\lambda_0$ depending on the piezoelectric material. For the series inductance tuning, with source and radiation resistance matched, the electrical Q_e is given by $Q_e = \omega_0 C_T/G_a(f_0) = \pi/4K^2N$. On the other hand, the acoustic Q_a of the periodic structure, as previously seen, is about N. Since the transducer bandwidth is limited by the greatest of these Q's, the largest bandwidth is obtained when they are equal, *i.e.* when $N^2 = \pi/4K^2$ [3]. For LiNbO$_3$ substrate the value of N for optimum bandwidth is thus $\sqrt{5\pi}$ ($\simeq 4$), because $K^2 \simeq 0.05$. Since both $G_a(f_0)$ and C_T are proportional to the finger length, a designer can choose this length to match the mid-band radiation conductance to the source, and, within limits, the number of fingers that are required to obtain a given bandwidth. Figure 4

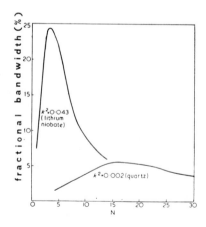

Fig. 4. – Bandwidth of a series-tuned transducer to the 1.5 dB points of the conversion loss (crossed-field model).

shows the bandwidth obtained as a function of N for quartz and lithium-niobate substrates [4]. The bandwidth refers to the 1.5 dB points of the conversion loss curve, *i.e.* the 3 dB points of insertion loss for a delay line such as that in Fig. 1.

The circuit theory also yields the reflection and transmission coefficients for an incident surface wave (the conversion coefficient to electrical energy is, by reciprocity, the same as that for excitation).

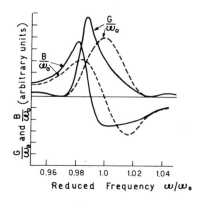

Fig. 5. – Frequency dependence of the admittance of a 30-finger-pair transducer on LiNbO$_3$. Crossed-field model (dashed line) and theory incorporating multiple reflections (continuous line).

For tuned and matched transducers the power reflection coefficient is -6 dB. In delay lines this causes a spurious output known as the triple-transit signal, corresponding to three transits between the transducers, with a level 12 dB below the main output signal. Better triple-transit suppression can be obtained at the expense of increased insertion loss (by electrically mismatching the transducers) or by using a unidirectional transducer made, for example, using multistrip couplers.

When the number N of sections becomes too high (*i.e.* NK^2 is not small) the crossed-field model breaks down, and considerable distortion of the $G_a(f)$

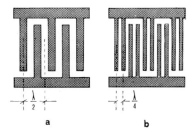

Fig. 6. – Two types of interdigital surface wave transducers: *a*) conventional i.d. transducer, *b*) « split finger » i.d. transducer.

and $B_a(f)$ curves is found. This can be theoretically predicted by an equivalent circuit which includes mismatched sections to take into account the change in wave velocity under the metal fingers with respect to that in the gaps. The effect of the periodically mismatched transmission path introduces stop-band characteristics in the vicinity of normal peak response so that the maximum-frequency response is lower than the synchronous one (Fig. 5). The use of a « split finger » transducer removes this complication (Fig. 6).

2˙1. S.a.w. *delay line performance*. – Under usual conditions the frequency response of the delay line shown schematically in Fig. 1 will be dictated by launching and detecting transducers including their respective matching networks and loads. If we call these responses T_1 and T_2 we have

$$T_1(f) \cdot T_2(f) = T(f) ,$$

$$h_1(t) * h_2(t) = h(t) ,$$

where $h(t)$ is the impulse response. This ignores the frequency response of the propagating medium; strong diffraction and absorption effects may make it necessary to take it into account and treat the medium as other than a frequency-independent delay line. The delay line with identical periodic input and output transducers has a $[(\sin x)/x]^4$ frequency response modified by the electrical circuits.

TABLE I [5]. – *Comparison of lithium-niobate and quartz delay lines.*

	Y-Z	LiNbO$_3$	ST-X	Quartz
Finger pairs	4		21	
	tuned (*)	untuned	tuned (*)	untuned
Bandwidth (3 dB)	25%	25%	4%	4%
Insertion loss (dB)	8	∼26	8	∼50

(*) Series inductive tuning.

In Table I characteristics of delay lines on Y-X lithium-niobate and ST-X quartz are listed. The insertion loss has been increased by 2 dB as a result of the minor diffraction losses, finger resistance losses, and bulk-mode losses, that are normally encountered in a short (10 µs) delay line. The number of finger pairs is chosen to optimize bandwidth and their aperture chosen to match a 50 Ω source and load.

At an operating frequency of 100 MHz their dynamic range is estimated to be about 100 dB. Intrinsically, quartz offers a superior dynamic range than LiNbO$_3$ because of its lower nonlinearity.

2'1.1. Techniques for suppressing spurious signals in s.a.w. delay lines. Two types of spurious signals affect the performance of s.a.w. delay lines: the triple-transit echo and bulk-mode signals. Some techniques for suppressing or reducing the spurious signals are schematically shown in Fig. 7.

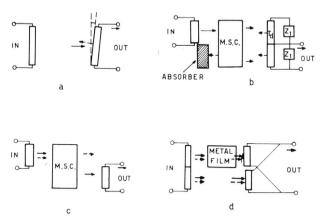

Fig. 7. – Techniques for suppressing triple-transit signals: *a*) tilting the output transducer rotates the phase front of the reflected signal and produces a reduced triple-transit signal, *b*) the multistrip coupler (m.s.c.) « echo trap », power from the input transducer is split by the m.s.c. The reflected signal from the output transducer combines with that from the dummy output transducer T_d and is dissipated in the absorber. Techniques for suppressing bulk waves: *c*) The m.s.c. couples only s.a.w. to the output transducer. *d*) The metal modifies the velocity of the s.a.w. producing constructive interference at the output while unaffected bulk waves destructively interfere.

2'1.2. High-frequency delay lines. An approach to large-bandwidth delay lines is to use periodic transducers with high synchronous frequencies.

Standard photolithographic techniques involving evaporated or sputtered metallic films and optical exposure of positive or negative photo-resist have been used for frequencies up to several hundreds of MHz. For higher frequencies, micron and submicron lines with good edge definition are required and electron beam exposure of special electron-sensitive resists (such as polymethyl-methacrylate) appears to be essential. With this technique a delay line on Y-Z LiNbO$_3$ having a centre frequency of 2.5 GHz, a 3 dB bandwidth of 210 MHz, a 1.5 µs delay and an insertion loss of 29 dB has been made [6].

2'1.3. Delay lines for recirculating storage. If we use one of the techniques for broad-bandwidth transduction, the s.a.w. delay line offers a means of forming a high–bit-density storage quite competitive in cost per bit and power consumption per bit. In its simplest form, the bit circulates as a short-tone pulse which is reshaped electronically on each circulation.

In Fig. 8 the scheme of a low-temperature-coefficient, wide-bandwidth delay line (see Table II) is shown. It is a composite structure which unites the strong coupling of Y-Z lithium-niobate (for a large-bandwidth and low-insertion losses) with the temperature insensitivity of ST-X quartz.

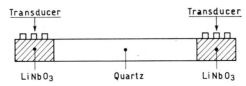

Fig. 8. – Low-temperature coefficient, wide-bandwidth delay line using $LiNbO_3$ and quartz.

TABLE II [5].

Performance of s.a.w.	Recirculating memory
Bit rate	75 Mbit s^{-1}
Bit per channel	5100 bit·s
Power per bit	$(5 \div 10)$ μW
Input voltage	3 V
Output voltage	$5 \cdot 10^{-3}$ V

2'1.4. Long delay lines. The long (> 100 μs) s.a.w. delay line faces several problems: how can *a*) the s.a.w. be « folded » so that the delay exceeds the limits imposed by crystal length/s.a.w. velocity, *b*) excessive diffraction losses be avoided, *c*) cross-talk kept to an acceptable level and *d*) an acceptable bandwidth and dynamic range be maintained. The « folding » schemes which have been tried or proposed are summarized in Fig. 9.

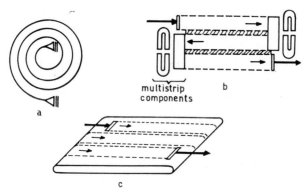

Fig. 9. – Long-delay–line structures: *a*) spiral wave guide, *b*) folded delay line using multistrip component, *c*) helical wrap around delay line.

3. – Frequency filters.

The above description was confined to uniform transducers. Nonuniform transducers in which either the periodicity or the aperture or both are allowed to vary offer a great deal of versatility and are the basis of many s.a.w. devices. Applying a voltage impulse to an i.d. transducer will produce a mechanical deformation of the substrate which is directly related to the electrode structure of the transducer. The s.a.w. component of this disturbance radiates out from the transducer onto the surrounding surface, thereby creating a propagating « image » of the transducer. Conversely, when an impulse, propagating along the line, passes an array of interdigital electrodes, the resulting response is a set of samples reproducing in time the spatial configuration of the array. An array of electrodes graded in spacing and aperture thus provides a set of impulses accordingly delayed and weighted.

3‘1. *Concept of transversal filter.* – In linear transversal filters, first described by KALLMANN [1], the filtering function is synthesized by tapping at appropriate points the signal launched in a lossless nondispersive delay line, and weighting and summing the signal contributions from the taps (Fig. 10).

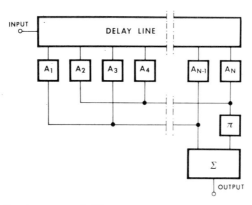

Fig. 10. – Scheme of a transversal filter.

Let $h(t)$ represent the filter's impulse response and τ_n and A_n the occurrence time and the value of the n-th sample, respectively. Then, the corresponding tap must be delayed by τ_n, weighted by $|A_n|$, and its polarity determined according to the sign of A_n. If we assume lightly coupled nonreflecting taps, the transversal-filter impulse response is given by the sample sequence

$$h_s(t) = \sum_n A_n \delta(t - \tau_n) \,,$$

where δ is the Dirac impulse function. As is well known, the corresponding frequency response consists of the desired transfer function, *i.e.* the Fourier transform of $h(t)$, and of a number of harmonic components.

This procedure permits, in principle, the synthesis of any linear filter with impulse response of finite duration.

The practical implementation of such a device, however, requires a means for controlling the spacing and weighting of the tapping elements. S.a.w. delay lines offer this capability, since, as mentioned above, continuous tapping and weighting of surface waves can be achieved by controlling the spacing and aperture, respectively, of surface-electrode transducers.

3˙2. *Phase-sampling procedure.* – As shown previously, the design of the tapping-electrode configuration of an s.a.w. transversal filter requires the determination of a sampled form of the wanted impulse response. The time characteristics of the impulse-response function are given, or may be calculated as the inverse Fourier transform of a transfer function approximating the given design data; thus from the *a priori* knowledge of the impulse-response function we have the possibility of choosing the sampling procedure most suitable for the design of a simple and efficient tapping structure. The sampling procedure that we have proposed can be understood by considering the synthesis of a dispersive filter having the impulse response given by

$$h(t) = \begin{cases} \cos\varphi(t) & \text{for } |t| < \dfrac{T}{2}, \\ 0 & \text{elsewhere}. \end{cases}$$

Since the impulse response has a rectangular envelope, samples all having the same amplitude can be obtained by sampling $h(t)$ in correspondence with its positive and negative peaks, *i.e. where the instantaneous phase $\varphi(t)$ assumes values multiples of π* [7].

Since the position of these sampling points depend on the PM law, the requirement of uniform sample amplitudes leads, in general, to a nonuniform sampling interval. Figure 11 shows schematically a s.a.w. dispersive filter used for phase dispersion (top) or phase equalization (bottom).

When the impulse response is amplitude modulated, *i.e.* has the form

$$h(t) = a(t) \cos \varphi(t),$$

where $a(t)$ is a positive envelope function, the samples determined according to the above criterion are no longer of constant amplitude. In this case different weights are introduced by grading the length of the fingers corresponding to the samples. With reference to Fig. 12 let ΔZ_{n-1}, ΔZ_n and ΔZ_{n+1} be the lengths of three adjacent fingers exceeding the middle line of the two combs.

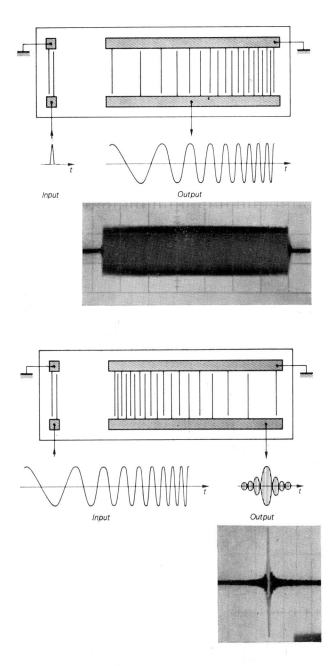

Fig. 11. – Schematic representation of a s.a.w. dispersive filter used for phase dispersion (top) and phase equalization (bottom).

It can be easily recognized that the total facing length of the n-th finger is

$$2 \Delta Z_n + \Delta Z_{n-1} + \Delta Z_{n+1}.$$

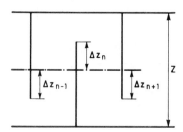

Fig. 12. – Elementary cell of an interdigital array with length-modulated fingers.

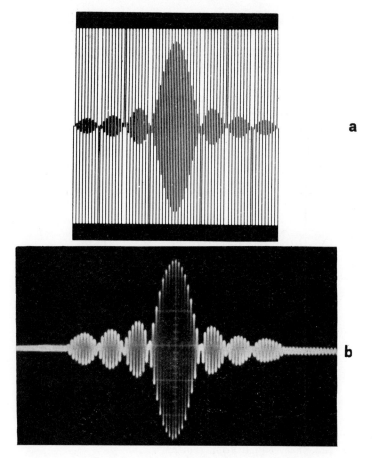

Fig. 13. – Apodized transducer for a band pass filter: *a*) i.d. transducer, *b*) impulse response.

The response of the n-th finger is assumed proportional to this quantity. The variation of the finger lengths must be designed so that the total facing length is proportional to the required weight a_n (the sample at the time τ_n).

However, when the impulse response has low fractional bandwidth, it can be easily shown that the required weight is achieved simply by making ΔZ_n proportional to a_n.

Figure 13 schematically shows an apodized transducer for the synthesis of a band-pass characteristic (top) and the experimental impulse response (bottom).

4. – S.a.w. oscillators.

Since the use of high–Q-factor bulk-wave quartz crystals to control oscillator circuits is widespread, it is natural to consider whether s.a.w.-based

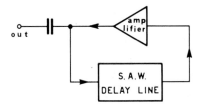

Fig. 14. – Scheme of a s.a.w. oscillator.

devices have relevance in this context. One approach is the s.a.w. delay line oscillator; it is simply a delay line utilized to form an oscillator by feeding the output back to the input with sufficient gain to overcome the loss in the delay line (Fig. 14).

Advantages attributed to the s.a.w. oscillator include

    a) operation up to low gigahertz frequencies without multipliers,

    b) freedom from satellite modes,

    c) ruggedness,

    d) modulability,

    e) the usual advantage of s.a.w. devices.

The allowed oscillation frequencies of a delay line oscillator are those having excess gain and a net phase shift around the loop of $2\pi n$, where n is an integer. The phase condition is

$$\varphi_{\mathrm{el}} + 2\pi f_n \frac{L_0}{V} = 2\pi n \,,$$

where L_0 is the acoustic path length, f_n the frequency of the n-th mode and φ_{el} the phase shift through the feedback loop and that due to the reactive part of the transducers. The allowed modes therefore form a comb of frequencies from which all but one may be suppressed by suitable choice of transducer geometry. The flexibility of s.a.w. delay lines allows such single-mode oscillators to be designed with varying sensitivity to the electrical part of the circuit by a suitable choice of the relative size of $2\pi f_n L_0/V$ and φ_{el}. For high stabilities the acoustic phase shift should dominate, while oscillators requiring modulation capability will demand a lower phase-shift ratio (the oscillator can be tunable using, for example, a varactor phase-shift network).

In contrast to conventional quartz oscillators, the frequency is not determined by the dimensions of the crystal but by the transducer geometry, so that devices can be made as rugged as desired, even at the highest frequency.

The temperature stability of an s.a.w. oscillator using ST-X quartz is adequate for many applications (variation $\simeq 3$ parts in 10^6 per °C). The short-term stability of these devices was 10^{-7} measured over 1 s, a value which is adequate for many applications. Many aspects of the s.a.w. oscillator remain to be fully evaluated, $e.g.$ ageing, short-term stability and reproducibility, but at the present time their future looks bright as oscillators of intermediate stability having a modulation capability.

5. – Multistrip couplers and their applications.

Multistrip couplers were introduced by MARSHALL and PAIGE in England in 1971 [3, 5]. Their basic structure is shown in Fig. 15.

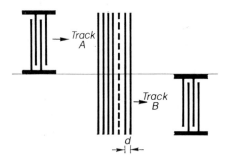

Fig. 15. – Basic structure of a multistrip coupler.

It consists of an array of parallel metallic strips deposited on the piezo-electric substrate lying in and perpendicular to the path of an acoustic beam which is of smaller aperture than the complete width of the coupler. In Fig. 15 the aperture of the incident wave is half the width of the coupler.

The two i.d. transducers define two tracks, A and B, of equal aperture. If an acoustic surface wave is launched into one half of the structure (along track A for example), the incident wave is coupled to track B. The coupling action occurs because the wave in track A induces charges and voltages on the metallic strips, that, in turn, like a transducer, launch a wave in track B. In other words, the alternate-potential difference set up between adjacent metallic strips of the coupler generate a new acoustic wave which also appears in track B.

The fraction of incident acoustic energy coupled to track B depends on the number of the strips in the structure. The behaviour of the coupler can be described in terms of the propagation modes of the structure or in terms of a simplified equivalent circuit. Both approaches lead to the result that, for a given piezoelectric material, there exists a number of strips for which *complete* energy transfer between tracks A and B occurs. This number N_T results inversely proportional to the electromechanical coupling coefficient of the material K^2, and for LiNbO$_3$ is of the order of 100. In order to understand this phenomenon in a simplified manner, let us consider the case in which the number of strips is $N_T/2$. In this case the energy of the incident wave is equally divided between tracks A and B, but the phase of the wave in track B is 90° in advance of that in track A, as illustrated in Fig. 16.

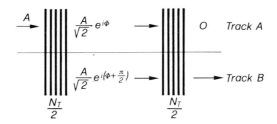

Fig. 16. – Complete energy transfer in a multistrip coupler.

Assume now that the waves emerging from the first coupler impinge on a second coupler having $N_T/2$ strips. Since the waves are in phase quadrature and the second coupler introduces a further phase quadrature between the tracks, it appears that the waves combine so that all the input energy emerges from track B. Thus, the overall effect of a multistrip coupler of N_T strips is to transfer all the incident power from one track to the other.

The multistrip coupler exhibits a very good bandwidth, but has a stop-band in correspondence to the frequency $f_s = v/2d$, where d is the period of the structure. For this frequency, in fact, strong reflections occur, and the simple coupler action breaks down.

A useful rule of the thumb is to set the working frequency at approximately three-quarters of the stop-band frequency, where the coupler bandwidth is near 120 percent and coupler strip reflections due to the proximity of the stop-band are negligible.

A unique feature of the multistrip coupler is that the strips may be bent, thus allowing one to direct the coupled wave at an angle to the incoming signal, as shown in Fig. 17.

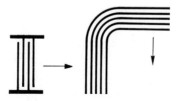

Fig. 17. – Acoustic-wave redirection by bent m.s.c.

This flexibility has led to a family of novel devices, that perform functions not previously available to the s.a.w. system designer. Let us review the main applications of multistrip couplers.

$5\,'1.$ *Bulk-wave suppression.* – Since the strips couple to the surface waves and not to the bulk waves generated by an interdigital transducer, the insertion of a multistrip coupler is an efficient means of reducing $\left(\text{by} \sim (15 \div 20)\,\text{dB}\right)$ the undesired effect of bulk waves.

$5\,'2.$ *Coupling between separate substrates.* – The multistrip coupler allows one to couple two separate substrates, thus overcoming the limitation due to the finite size of piezoelectric substrates in the implementation of long delay lines.

$5\,'3.$ *Surface wave reflector.* – Consider the U-shaped structure of the m.s.c. (multistrip coupler) having $N_T/2$ strips shown in Fig. 18.

It is easy to see that the phase quadrature inserted by the two arms of the U (as discussed above) and the redirection of the coupled waves determined by the m.s.c. geometry are such that all the incident power is reflected.

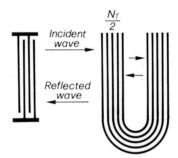

Fig. 18. – M.s.c. reflector.

5'4. *Unidirectional transducers.* – Consider the U-structure discussed above and an interdigital transducer placed within the U, but offset from the centre by $\lambda/8$. The phase difference of the waves launched by the i.d. transducer on the two arms of the U will be 90°. Because of this initial phase difference and of the phase quadrature introduced by the m.s.c., all the energy emerges along only one direction.

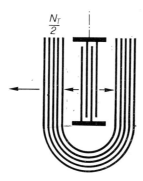

Fig. 19. – M.s.c.-based unidirectional transducer.

5'5. *Beam width compressor.* – Consider the « stepped » structure shown in Fig. 20, where one-half of the m.s.c. having $N_T/2$ strips is offset by $\lambda/4$ from the other half.

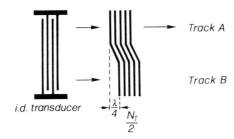

Fig. 20. – M.s.c. beam width compressor.

The i.d. transducer at the left launches a wave along both tracks A and B, but the wave in track A arrives on the m.s.c. 90° in advance. Because of the phase quadrature introduced by the m.s.c., it is easy to verify that the incident power emerges only from track A. In this manner the incident beam, which occupies the full width of the coupler, is compressed in width by a factor of 2 and accordingly raised in intensity. By cascading more of these structures, large acoustic-power densities can be achieved over small beam apertures.

6. – Convolution using parametric interactions of acoustic surface waves.

As demonstrated first by QUATE and THOMPSON, the nonlinear interaction between two acoustic waves passing in opposite directions through an acoustic medium can give rise to a third signal that is the convolution of the two waves. The basic process involved is the parametric interaction between two r.f. signals of frequencies ω_1 and ω_2 and propagation constants k_1 and k_2 respectively. When two such signals are passed in opposite directions through an acoustic medium, any nonlinearity present in the medium gives rise to an acoustic strain component with a frequency

$$\omega_3 = \omega_1 + \omega_2$$

and a propagation constant

$$k_3 = k_1 - k_2 .$$

If the two waves are surface acoustic waves, the power densities are high, because the energy is confined to a surface layer, and hence the nonlinear interaction can take place at relatively low-power densities, as was first demonstrated by SVAASAND. Such an interaction occurs by inserting the signals at frequencies ω_1 and ω_2 on transducers on opposite ends of the substrate (Fig. 21).

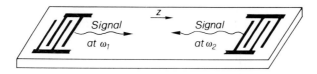

Fig. 21. – Parametric interaction of surface acoustic waves.

Let the oscillating parts of the two propagating signals be $\exp[j(\omega_1 t - k_1 z)]$ and $\exp[j(\omega_2 t - k_2 z)]$ respectively. From their product we obtain

$$\exp[j(\omega_1 t - k_1 z)]\exp[j(\omega_2 t - k_2 z)] = \exp\left[j[(\omega_1 + \omega_2)t - (k_1 + k_2)z]\right] .$$

Consider the special case in which $\omega_1 = \omega_2 = \omega$ and $k_1 = k_2$. In this case

$$\omega_3 = 2\omega , \qquad k_3 = 0 .$$

The interaction of the two signals thus produces a signal of uniform spatial

distribution and frequency 2ω. This can be detected using an output transducer that consists of two continuous metal films on each side of the acoustic
medium (Fig. 22).

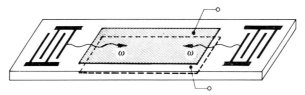

Fig. 22. – Surface acoustic-wave convolver.

If the size of this transducer is such that the signals interact wholly under
it, the output signal is the integral of the product of the two signals launched
by the opposite transducers. If, therefore,

$$s_1(t) = a(t) \exp [j\omega t] , \qquad s_2(t) = b(t) \exp [j\omega t]$$

are the two signals, the output is given by

$$u(t) = v \int_{-\infty}^{\infty} a\left(t - \frac{z}{v}\right) b\left(t + \frac{z}{v}\right) \mathrm{d}\frac{z}{v} .$$

With the substitution

$$t + \frac{z}{v} = \tau ,$$

$$t - \frac{z}{v} = 2t - \tau ,$$

one can write

$$u(t) = v \int_{-\infty}^{\infty} a(2t - \tau) b(\tau) \, \mathrm{d}\tau ,$$

which is recognized as the convolution of the two input signals. If the two
signals are identical, but reversed in time, i.e.

$$s_2(t) = s_1(- t) ,$$

we obtain the *autocorrelation* of the signal given by

$$u(t) = v \int_{-\infty}^{\infty} a(\tau) \, a(\tau - 2t) \, \mathrm{d}\tau .$$

The device acts in this case as a correlator. The significance of the device for signal-processing applications lies in the fact that any signal can be correlated with its stored replica, provided that this replica is previously time reversed.

The device becomes equivalent to a variable, real-time matched filter. It has no constraints on the form of the signal and offers therefore an enormous flexibility, of great interest, for example, in modern radar systems, where change of the transmitted signal as a precaution against jamming and deception is often required.

Note the factor of 2 that multiplies t in the above expression of the output signal. This factor derives from the fact that the waves are contradirected, and, because of their relative motion, everything goes as if the propagation velocity were doubled. This fact generates a time compression by a factor of two.

Finally, it must be noted that in systems where the time of arrival of the received signal is not known *a priori* as in radar, it is continuously necessary to refresh the reference signal entering the convolver. The reference signal must therefore be fed at a sufficiently high rate, and a complex synchronization system is needed that offsets the simplicity of the convolver device.

7. – Interaction of light with surface acoustic waves.

7˙1. *Light diffraction by surface acoustic waves.* – The principle of diffraction of light by surface acoustic waves is illustrated in Fig. 23.

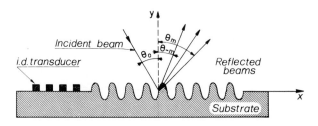

Fig. 23. – Schematic diagram for light diffraction by surface acoustic waves (reflection mode).

The propagating surface wave presents both a surface deformation and a periodic variation of the index of refraction within a surface layer of the order of an acoustic wavelength. The surface deformation and periodic index variation provide a moving phase grating for the incident light beam and thus diffract the incident beam into many side orders. For the reflected beam only the surface deformation contributes to the light diffraction, while for the transmitted beam both the surface deformation and the periodic index of refraction within the surface layer contribute to the diffracted light.

The surface deformation can be written

$$d = d_0 \cos (\omega_s t - k_s z) \,,$$

where ω_s and k_s are the acoustic angular frequency and wave number, respectively, and d_0 is the maximum surface deformation. As the incident light is reflected by this sinusoidal surface deformation, a phase grating type of diffraction results. The angular relation of the m-th–order diffracted light is given by the same relation used for a line grating

$$\sin \theta_m = \sin \theta_0 + m \frac{\lambda}{\Lambda}, \qquad m = 0, \pm 1, \pm 2, \dots \,,$$

where θ_m is the diffraction angle of the m-th order measured with respect to the surface normal, θ_0 is the angle of the incident beam, λ is the optical wavelength and Λ is the acoustic wavelength. From the Raman-Nath theory of acoustic diffraction, it can be shown that the ratio of the m-th–order diffracted light intensity I_m to zeroth-order undiffracted beam intensity I_0 is given by the square of the m-th–order Bessel function

$$\frac{I_m}{I_0} = J_m^2 (\nu) \,,$$

where

$$\nu = \frac{4\pi}{\lambda} d_0 \cos \theta_0 \,.$$

Under normal experimental conditions ν is much smaller than one, so that the only appreciable diffracted order is the first-order one, and

$$\frac{I_1}{I_0} \sim \left(\frac{\nu}{2} \right)^2 .$$

Since ν^2, depending on d_0^2, is related to the surface acoustic power, the first-order diffracted light is a linear function of the input acoustic power (as long as the above expansion of $J_1(\nu)$ is valid). In this linear region diffraction of light by surface acoustic waves was used as an optical probing technique to measure the energy profiles of surface waves launched from an interdigital transducer: a laser beam scanned the surface wave across its beam width at various distances from the i.d. transducer, and a photodetector, located at the appropriate angle, measured the first-order diffracted light.

In the transmission case the incident light is modulated both by the surface deformation and the periodic variation of the index of refraction in the surface layer. Similar results are obtained also in this case, but the two contributions to diffraction must be accounted for.

7'2. *Interaction of surface acoustic waves and surface optical guided waves.* – Optical waves can propagate in a thin-film wave guide deposited on the surface of a substrate. There is a natural parallel between research in optical guided waves and acoustic surface waves. It is not therefore surprising that acoustic surface waves can be a convenient tool for controlling optical guided waves.

There are remarkable similarities in the methodology and technology used. Both deal with waves confined to the surface of a substrate, and the techniques for coupling to the surface are quite similar. Although the frequency ranges of interest are orders of magnitude apart, the wavelength ranges, which determine the actual geometry of the device, are quite close, as shown in the following Tables.

TABLE III. – *Acoustic surface waves.*

Band	VHF	Low micro
frequency	30 MHz	3 GHz
wavelength	100 μm	1 μm

TABLE IV. – *Optical guided waves.*

Band	Infra-red	Visible
frequency	30 THz	600 THz
wavelength	10 μm	0.5 μm

A surface acoustic wave causes Raman-Nath or Bragg diffraction of an optical guided wave, deflecting it in the plane of the film. The acoustic wave, propagating on the substrate surface, forces a strain wave in a thin-film optical wave guide deposited on the substrate. The strain imposed on the film is seen by the optical guided wave as a periodic variation of the optical refractive index, so that diffraction of light occurs in a way similar to that discussed above.

First experiments of guided-light diffraction by acoustic surface waves were performed by KUHN at IBM. The optical wave guide was a glass film, 0.8 μm thick, $n \simeq 1.73$, sputtered onto a quartz crystal substrate. A laser beam was coupled into and out of the film by suitable grating couplers. The surface acoustic wave had a frequency of 191 MHz, and wavelength $\Lambda = 32$ μm.

The acoustic wave was not perturbed by the glass film since the film thickness was much less than an acoustic wavelength. Bragg diffraction was preferred to normal incidence diffraction, since at the Bragg angle the light is scattered into only one diffracted beam of noticeable intensity. The scheme of the experiment is shown in Fig. 24.

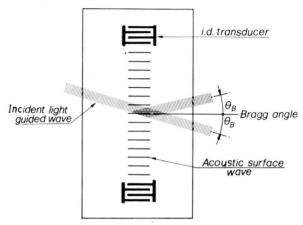

Fig. 24. – Schematic diagram for guided light diffraction by surface acoustic waves.

The Bragg angle between the light and the acoustic waves is given by

$$\theta_B = \arcsin \frac{\lambda_g}{2\Lambda},$$

where λ_g is the wavelength of the guided light wave and Λ the wavelength of the acoustic surface wave. When this angle is reached, a deflected beam appears in the film at twice the Bragg angle from the original beam. For a He-Ne laser, the result was $2\theta_B = 1.36°$.

In spite of poor transducer conversion efficiency (-11 dB electrical to acoustic), KUHN observed 66% deflection efficiency for an input electrical power of 2.5 W.

Promising applications of surface opto-acoustic interactions are the implementation of integrated deflection devices as switching elements in optical communication systems and the implementation of planar analogue optical-signal processors, similar to those developed in last decade using the interaction between laser light and bulk acoustic waves.

8. – Display systems based on surface acoustic waves.

Because of the planar nature of acoustic surface waves, it is quite natural that acoustic systems able to convert planar optical images to electrical signals have been largely investigated. So far, in spite of the considerable effort in this direction, these systems are still in a laboratory stage. We mention here some among several systems that have been proposed.

8.1. *Cameras*. – This system, investigated by KAUFMAN at the Arizona State University, consists of a matrix of output transducers, the rows of which are scanned in turn by a short acoustic pulse. The optical pattern is imaged over the matrix, and each transducer is connected to a photocell so that the spatial variation of optical intensity is converted into a time-varying electric signal. It was estimated that arrays with more than 250 elements per line and an overall length of 5 cm should be possible. The key question, however, is whether a practical integrated device of this type could ever be economical.

8˙2. *Image scanning by acousto-electro-optic interactions*. – Several display systems have been devised based on the properties of photoconductor films.

We mention here the system recently described by LUUKKALA in Finland. A piezoelectric substrate is coated with a thin (0.5 μm) photoconductive film, as CdSe. By illuminating a point of the film with a laser beam, charge carriers are released that couple to the piezoelectric field of a propagating acoustic surface wave, causing a change in acoustic impedance from which the surface wave will reflect. By scanning the photoconductive film along the path of the surface wave, accordingly variable delays of the reflected surface wave are obtained.

By illuminating the surface wave propagation path with the light intensity distribution obtained by focusing a light sheet on a transparent image (see Fig. 25), an acoustic impedance variation along the line arises that causes an accordingly variable amplitude variation of a reflected acoustic pip. These amplitude variations are used to z-modulate the oscilloscope intensity. Thus the light pattern is obtained as an intensity modulation on the oscilloscope face. Two-dimensional pictures have been obtained by mechanically scanning

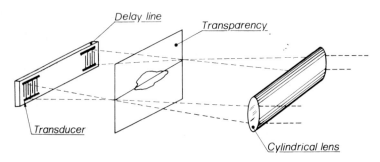

Fig. 25. – Schematic diagram of the display system based on acousto-electro-optic interaction.

the transparency. It was observed that the relation between the reflection factor and the light intensity is almost linear at low intensities. Under strong illumination the reflected pip will be severely attenuated, producing distortion of the image.

8'3. Touch-sensitive digitizers. – A commercial success has been obtained by a new s.a.w. device, known as the touch-sensitive digitizer. It consists of two wedge transducers (Fig. 26) extended in the X and Y directions, on a non-

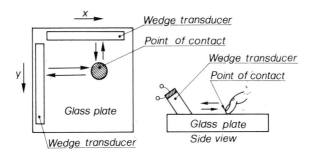

Fig. 26. – Schematic diagram of the touch-sensitive digitizer operation.

piezoelectric glass plate. The transducers launch pulses at 4 MHz frequency. When an object touches the surface, the pulses are reflected, and the delays along the X and Y directions (and hence the position of the reflector) are accurately detected and put in digital form. The system thus converts the co-ordinate of a selected point on a display into digital information, and so provides a direct interface between man and computer, or another digital system that uses visual display.

REFERENCES

[1] Special issue on microwave acoustics, in IEEE Transactions on Microwave Theory and Techniques, Vol. MTT-**17**, No. 11 (November, 1969).
[2] R. M. WHITE: Proc. IEEE, **58**, 1238 (1970).
[3] Special issue on microwave acoustic signal processing, in IEEE Transactions on Sonics and Ultrasonics, Vol. SU-**20**, No. 2 (April, 1973).
[4] D. P. MORGAN: Surface acoustic wave devices and applications, in Ultrasonics (May, 1973), p. 121.
[5] I. D. MAINES and E. G. S. PAIGE: IEE Rev., **120**, 1078 (1973).

[6] R. G. LEAN and A. N. BROERS: *Microwave acoustic delay lines*, in *Microwave Journal* (March, 1970), p. 97.

[7] C. ATZENI and L. MASOTTI: *IEEE Trans. Microwave Theory Techn.*, MTT-**8**, 505 (1973).

[8] E. G. H. LEAN and C. G. POWELL: *Proc. IEEE*, **58**, 939 (1970).

[9] L. KUHN, M. L. DAKSS, P. F. HEINDRICH and B. A. SCOTT: *Appl. Phys. Lett.*, **17**, 265 (1970).

[10] M. LUUKKALA and P. NERILAINEN: *Electronics Lett.*, **10**, 80 (1974).

[11] I. KAUFMAN and J. W. FOLTZ: *Proc. IEEE*, **57**, 2081 (1969).

Acoustics in Space.

T. G. Wang

Jet Propulsion Laboratory, California Institute of Technology - Pasadena, Cal.

1. – Introduction.

Many of the experiments to be carried out in the laboratory in space require the manipulation and control for weightless liquid systems. In these processes the liquid is to be positioned and formed within a container but without making contact with the container walls. Electromagnetic methods of positioning and forming are limited to materials which are electrically conducting.

An acoustical method is useful in the control of any liquid system including those that are electrically nonconducting. The method utilizes the static pressure produced by an acoustical standing wave excited within an enclosure. At the nodes of the wave the pressure is greater than at the antinodes. Consequently there is the tendency for liquids and particles introduced into the enclosure to be driven toward the antinodes where they collect, there to remain until the excitation ceases. The method also allows collected material to be rotated by slightly modifying the acoustic field.

This technique has recently been proven in our laboratory. A styrofoam ball placed within a resonator driven at its lowest mode was indeed levitated and rotated. Moreover we were able to levitate and rotate soap bubbles and water droplets within the resonator.

2. – Theory.

The laboratory model of the rectangular resonant chamber as shown in Fig. 1 consisted of a 4 inch $\times 4\frac{1}{2}$ inch $\times 5$ inch plexiglass rectangular box and three commercially made speaker drivers. Those drivers are mounted at the center of the orthogonal planes of the box. In order to maximize the efficiency of the system, three aluminum spacers were used to make the total distances from the diaphragms of the speakers to the inner walls of the chamber equal to the inner dimensions of the resonant chamber.

The resonant chamber and the speaker driver units are acoustically coupled through the twelve $\frac{1}{8}$ inch holes which are drilled radially symmetrically around the center of three orthogonal planes. When the chamber was driven at one of its resonance nodes by acoustical compressional drivers, the ambient pressure is maximum at the nodes of the wave and is minimum at the anti-nodes. Consequently there is a tendency for liquids and particles introduced

Fig. 1. – Triaxial acoustic positioning chamber.

into such enclosures to be driven toward the antinodes where the materials collect and remain until excitation ceases.

The average ambient pressure change ΔP in a sound wave is

$$(1) \qquad \Delta P = \frac{\overline{P^2}}{2\xi C^2} - \frac{1}{2}\xi\overline{U^2},$$

where P is the excess pressure, U is the particle velocity and ξ is the density of the medium.

The pressure profile in our system can be derived as follows.

The wave equation φ for our rectangular chamber can be expressed as

$$(2) \qquad \varphi = \varphi_x \cos k_x \chi \exp[i\omega_x t] + \varphi_y \cos k_y Y \exp[i\omega_y t] + \varphi_z \cos k_z Z \exp[i\omega_z t],$$

where $\varphi_{x,z,y}$ are the complex velocity potential amplitudes of standing waves of frequency $\omega_{x,z,y}$ and wavelength constant $k_{x,z,y}$ with velocity C.

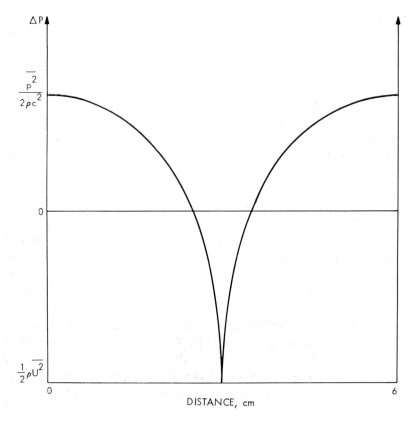

Fig. 2. – The one-dimensional pressure profile.

The particle velocity U by definition is

(3) $$U = \nabla\varphi .$$

The momentum conservation condition gives us

(4) $$P = -\xi\dot{\varphi} .$$

Thus, ΔP is minimum at the center of the chamber as shown in Fig. 2, when drivers are operated at the lowest resonance modes of the chamber.

3. – Experiments.

There are many possible experiments in the future NASA Space Flight Program which would be benefitted significantly by the unique capability of an acoustic chamber. However, here, we will be discussing those experiments that pertain to JPL's effort only.

3`1. *Superfluid helium.* – Superfluid helium has important applications throughout many fields of research and technology both because of its extensive use as a cryogenic coolant and because it is the only fluid which exhibits quantum-mechanical effects on a macroscopic scale. The investigations outlined in this proposal are directed toward a fundamental understanding of superfluidity utilizing experiments which are impossible to implement on Earth.

Heretofore, the hydrodynamics of superfluid helium have always been observed with the superfluid held in a container. In the present experiments, the hydrodynamics will be observed in a free drop. The only surface is then the free surface and vorticity cannot be introduced through relative motion of the superfluid helium and the container walls. Instead, vortices can only be introduced by relative motion of different portions of the superfluid helium itself, and its vapor. Consequently, this experiment—the first to measure critical velocities of superfluid helium in the absence of constraining walls—will investigate the role that the walls play in the generation of vortices.

When superfluid helium is moving at critical velocities, the superfluid state in some sense breaks down. Therefore, to understand why certain velocities are critical is to understand the stability of the superfluid state, and to understand this stability is to understand more about the nature of superfluidity itself. Reproducible experiments on critical velocities are notoriously difficult. At least one reason is that, at the walls, vorticity is easily introduced; because of image effects, the energy of a vortex near a wall has a lower energy than a vortex in the body of a fluid. Moreover, unavoidable imperfections in the wall

allow some sites to be favored over others for vortex formation, and this leads
to the lack of reproducibility of critical velocity experiments. The present experiment attempts to circumvent the latter and modify the former by using
a free drop.

While the word « drop » normally evokes the image of a spherical or sphe-
roidal drop, a rotating drop can assume other shapes which may not be simply
connected. These configurations should allow a superfluid helium current to
flow which, unless it exceeds a critical value, should be persistent. On Earth
such nonsimply connected flows can only be contained by constraining walls.
In the proposed experiment a toroidal drop will allow such a flow to take place
in the absence of walls.

3'2. *Drop dynamics.* – The problem of the equilibrium figures of rotating
self-gravitating liquid masses originated from the Newton-Cassini controversy
about the shape of Earth. In the centuries since the problem has become one
of the most famous problems of mathematical physics. The basic mathematical
properties of instabilities and bifurcating solutions have been discovered and
will be first investigated in detail in this experiment. The fission theories of the
formation of planetary satellites and double stars were deduced from the studies
of the bifurcating solutions for the equilibrium figures of rotating masses. As
earlier as in the 1840's it was recognized by PLATEAU that a rotating liquid drop
may offer a close experimental model for the theoretical problem. The experiments performed by PLATEAU indicated, indeed, some of the expected effects.
However, experiments with a rotating drop suspended in a fluid of the same
density are not possible without introducing differential rotation in the surrounding medium which qualitatively alters the experiment. Hence a weightless environment appears to be necessary to perform this important experiment.
The close dynamical similarity between surface tension and self-gravitation of a
liquid mass has been emphasized more recently by CHANDRASEKHAR. Even though
his detailed mathematical analysis is restricted to the axissymmetric case,
CHANDRASEKHAR comments that the remarkable similarity with the MacLaurin
sequency of rotating liquid masses will extend to the Jacobi sequence of triaxial
ellipsoids branching off the MacLaurin sequence and the sequence of pearshaped
configurations found by POINCARÉ which branches off the Jacobi sequence.

A reasonably complete theoretical discussion of the small-amplitude dynamics of freely suspended liquids under the sole influence of surface tension
forces has been developed. This work has, however, not had the benefit of detailed quantitative comparison with experiment and is incomplete in that there
is as yet no adequate theory for oscillations of large amplitude, criteria for rup-
ture or coalescence of liquid masses, or of the transition region between the high-
viscosity and low-viscosity regimes. An attempt to study the dynamics of
oscillations of finite amplitude was begun by BENEDIKT, but there is no experimental study to indicate whether the simple modes treated by him are

realizable in practice for any analysis of the limits to stability for such simple oscillations.

The energy dissipation mechanism for a droplet oscillating within another fluid of appreciable density has been shown by MILLER and SCRIVEN to be significantly different from the case where the second fluid has negligible density. Unfortunately, experiments with droplets in air or vacuum have generally been limited to drop sizes in the millimeter diameter range where it has been difficult to obtain accurate quantitative information for comparison with hydrodynamic theory. This limitation is also associated with the generally very short experimental times available.

3'3. *Space processing.* – Space processing is a fledgling technology. The philosophy underlying the space-processing concept is that of producing economically valuable products in space for use on Earth. Two criteria identify candidate products for space production: 1) the product must either be impossible to produce on Earth, or 2) it must be produced in space more inexpensively than on Earth.

Many of the processes to be carried out in space processing require the manipulation and control of weightless molten material. In these processes the melt is to be positioned and formed within a container but without making contact with the container walls. Electromagnetic methods of positioning and forming are limited to melts which are electrically conducting. This acoustic method is useful in the control of any molten material including material that is nonconducting.

The acoustical method described in this text can be used for levitating and positioning a melt within an enclosure, for zone melting, casting, crystal growing, casting of composites and casting materials with dispersed voids. For example, by using resonators of varying shapes, the antinodal region can be made to assume varying shapes thus forming the melt. For example, a cylindrical shape is particularly suited for zone refining. Moreover, by introducing sound waves at frequencies corresponding to the resonant modes of the melt, the melt can be made to oscillate and thus be stirred.

4. – Conclusion.

The application of acoustics in space technology has been proven to be successful in our laboratory. The acoustical positioning chamber we developed has the potential of becoming one of the most versatile tools for space research and application. At the present time, NASA is considering to incorporate the acoustic chamber into the first United States and European joint space shuttle flight to accommodate not only JPL's experiments, but also the possible experiments proposed by the scientific community of United States and Europe.

PROCEEDINGS OF THE INTERNATIONAL SCHOOL OF PHYSICS
« ENRICO FERMI »

Tipografia Compositori   Bologna - Italy